# Topics in Applied Physics
Volume 123

Available online at
SpringerLink.com

Topics in Applied Physics is part of the SpringerLink service. For all customers with standing orders for Topics in Applied Physics we offer the full text in electronic form via SpringerLink free of charge. Please contact your librarian who can receive a password for free access to the full articles by registration at:

springerlink.com → Orders

If you do not have a standing order you can nevertheless browse through the table of contents of the volumes and the abstracts of each article at:

springerlink.com → Browse Publications

# Topics in Applied Physics

Topics in Applied Physics is a well-established series of review books, each of which presents a comprehensive survey of a selected topic within the broad area of applied physics. Edited and written by leading research scientists in the field concerned, each volume contains review contributions covering the various aspects of the topic. Together these provide an overview of the state of the art in the respective field, extending from an introduction to the subject right up to the frontiers of contemporary research.

Topics in Applied Physics is addressed to all scientists at universities and in industry who wish to obtain an overview and to keep abreast of advances in applied physics. The series also provides easy but comprehensive access to the fields for newcomers starting research.

Contributions are specially commissioned. The Managing Editors are open to any suggestions for topics coming from the community of applied physicists no matter what the field and encourage prospective editors to approach them with ideas.

**Managing Editor**

Dr. Claus E. Ascheron

Springer-Verlag GmbH
Tiergartenstr. 17
69121 Heidelberg
Germany
Email: claus.ascheron@springer.com

**Assistant Editor**

Adelheid H. Duhm

Springer-Verlag GmbH
Tiergartenstr. 17
69121 Heidelberg
Germany
Email: adelheid.duhm@springer.com

For further volumes:
http://www.springer.com/series/560

Roberto Osellame • Giulio Cerullo
Roberta Ramponi
Editors

# Femtosecond Laser Micromachining

Photonic and Microfluidic Devices in Transparent Materials

With 305 Figures

*Editors*
Dr. Roberto Osellame
Consiglio Nazionale delle Ricerche
Istituto di Fotonica e Nanotecnologie
(IFN-CNR)
Piazza Leonardo da Vinci 32
20133 Milano
Italy
roberto.osellame@ifn.cnr.it

Dr. Giulio Cerullo
Dr. Roberta Ramponi
Politecnico di Milano
Department of Physics
Piazza Leonardo da Vinci 32
20133 Milano
Italy
giulio.cerullo@polimi.it
roberta.ramponi@polimi.it

ISSN 0303-4216
ISBN 978-3-642-23365-4         e-ISBN 978-3-642-23366-1
DOI 10.1007/978-3-642-23366-1
Springer Heidelberg Dordrecht London New York

Library of Congress Control Number: 2012932953

© Springer-Verlag Berlin Heidelberg 2012
This work is subject to copyright. All rights are reserved, whether the whole or part of the material is concerned, specifically the rights of translation, reprinting, reuse of illustrations, recitation, broadcasting, reproduction on microfilm or in any other way, and storage in data banks. Duplication of this publication or parts thereof is permitted only under the provisions of the German Copyright Law of September 9, 1965, in its current version, and permission for use must always be obtained from Springer. Violations are liable to prosecution under the German Copyright Law.
The use of general descriptive names, registered names, trademarks, etc. in this publication does not imply, even in the absence of a specific statement, that such names are exempt from the relevant protective laws and regulations and therefore free for general use.

Printed on acid-free paper

Springer is part of Springer Science+Business Media (www.springer.com)

# Preface

In 1996, a seminal paper by Hirao and coworkers demonstrated that tightly focused femtosecond laser pulses could modify in a permanent way the optical properties of a small volume inside the bulk of a transparent substrate. By a suitable choice of the irradiation conditions, one can induce a refractive index increase localized in the focal volume, thus enabling direct optical waveguide writing by simple translation of the substrate. This discovery has developed, in little over a decade, into a burgeoning research field with many groups working on the understanding and optimization of the process and even more researchers interested in exploiting its unique capabilities for microfabrication.

Femtosecond laser waveguide writing has some clear advantages over competing techniques, such as silica-on-silicon, ion exchange, or sol-gel:

(1) It is a direct, maskless fabrication technique, i.e., in a single step one can create optical waveguides or more complicated photonic devices (splitters, interferometers, etc.) by simply moving the sample with respect to the laser focus, avoiding complex clean room facilities; this is particularly important for rapid prototyping or on-demand production of a small number of devices, since it does not require the fabrication of dedicated photolithographic masks.
(2) In contrast with other techniques, which are optimized for specific substrates, it is highly flexible; with a suitable choice of irradiation parameters (wavelength, pulse energy, repetition rate, focusing conditions, translation speed, etc.) one can inscribe waveguides in almost any type of glass. In addition, the technique has been extended to crystalline materials (silicon, lithium niobate) and to polymers.
(3) It is a truly three-dimensional technique, since it allows one to define waveguides at arbitrary depths inside the substrate, according to geometries that would be impossible with standard fabrication techniques. This added degree of freedom is important because it allows developing innovative and unique device architectures.

Initial interest in femtosecond waveguide writing was mainly focused on the fabrication of devices for optical communications (splitters, interleavers, directional couplers, fiber and waveguide Bragg gratings, waveguide amplifiers and lasers, etc.). The field was considerably extended when, in another seminal paper in 2001, Marcinkevicius et al. demonstrated that femtosecond laser irradiation followed by etching in hydrofluoric acid solution enables the fabrication of directly buried microfluidic channels in fused silica. The etching rate in the laser irradiated regions is in fact enhanced by up to two orders of magnitude with respect to the pristine material, thus enabling the manufacturing of microchannels with high aspect ratio. This added capability broadens the scope of femtosecond micromachining to microfluidics, and microsystems in general. In particular, the possibility of combining optical waveguides with microchannels, both fabricated by femtosecond laser micromachining, opens exciting perspectives for optofluidics, a novel research field exploiting the synergy of optics and fluidics for the realization of completely new functionalities. With this addition, the arsenal of tools provided by femtosecond laser microstructuring becomes of interest not only for optics and photonics, but also for a wide range of disciplines, from biology to analytical chemistry.

The interest in femtosecond laser microfabrication has been boosted by the parallel developments in laser technology that have occurred in the first decade of the third millennium. Ultrafast lasers were initially very sophisticated devices that could only be operated by highly trained users in a research laboratory environment. Progress in laser media, optical pumping schemes, and mode-locking techniques has resulted in a new generation of compact, rugged, and reliable ultrafast laser systems, which guarantee turn-key operation. This makes the technique of femtosecond laser microfabrication amenable to real-world applications in an industrial environment.

This book, featuring contributions from the most renowned experts, has the ambition of providing a comprehensive introduction to the field of femtosecond laser micromachining of glass, starting from the basic concepts and progressing to the more advanced applications. The book will be of interest for experienced researchers working on femtosecond microstructuring, who will find advanced technical discussions and detailed descriptions of the various applications. It will also serve as an introductory textbook for graduate students or researchers approaching the field for the first time, since it also provide an entry-level coverage of the basic principles and techniques. Finally, and most importantly, it will be an invaluable reference for those researchers working in other disciplines and who will be the final users of the femtosecond laser microfabricated devices, making them aware of the new technological opportunities offered by the technique.

The book is structured in four parts. Part I deals with the fundamentals of femtosecond laser micromachining of transparent materials, describing the underlying physical mechanisms and the technical details of the fabrication process. Part II reviews the properties of waveguides and optical devices, both passive and active, fabricated in glass. Part III deals with refractive index modifications in materials different from glass, such as crystals and polymers. Finally, Part IV describes the more advanced applications to microsystems, including micro- and optofluidic devices.

The editors wish to acknowledge valuable help in the editing process by Shane M. Eaton and technical assistance by Sara Lo Turco.

Milano  
August 2011

*Roberto Osellame*  
*Giulio Cerullo*  
*Roberta Ramponi*

# Contents

**Part I Introductory Concepts, Characterization and Optimization Strategies**

1 **Fundamentals of Femtosecond Laser Modification of Bulk Dielectrics** ................................................. 3
  Shane M. Eaton, Giulio Cerullo, and Roberto Osellame
  1.1 Introduction .................................................. 3
  1.2 Femtosecond Laser–Material Interaction ........................ 4
      1.2.1 Free Electron Plasma Formation ......................... 4
      1.2.2 Relaxation and Modification ............................ 7
  1.3 Exposure Variables and Considerations ......................... 11
      1.3.1 Focusing ............................................... 11
      1.3.2 Writing Geometry ....................................... 13
      1.3.3 Influence of Exposure Variables Within Low- and High-Repetition Rate Regimes .............. 14
  1.4 Summary ....................................................... 16
  References ........................................................ 16

2 **Imaging of Plasma Dynamics for Controlled Micromachining** ........ 19
  Jan Siegel and Javier Solis
  2.1 Introduction .................................................. 19
  2.2 Assessment of the Interaction of Ultrashort Pulses with Dielectrics Using Optical Probes ................... 20
      2.2.1 Interaction Mechanisms and Characteristic Time Scales ............................................. 20
      2.2.2 Time-resolved Optical Techniques ....................... 22
      2.2.3 Exploiting Spatial Resolution .......................... 23
      2.2.4 Basic Models for Quantitative Analysis of Experimental Data ................................. 25

|  |  |  |  |
|---|---|---|---|
| | 2.3 | Ultrafast Imaging at the Surface of Dielectrics.................... | 27 |
| | | 2.3.1 Experimental Configurations and Constraints............................................. | 27 |
| | | 2.3.2 Transient Plasma Dynamics and Permanent Material Modifications .................................. | 29 |
| | 2.4 | Ultrafast Imaging in the Bulk of Dielectrics....................... | 32 |
| | | 2.4.1 Experimental Configurations and Constraints............................................. | 32 |
| | | 2.4.2 Transient Plasma Dynamics in Glasses Under Waveguide Writing Conditions: Role of Pulse Duration, Energy, Polarization, and Processing Depth ..................................... | 35 |
| | 2.5 | Outlook and Conclusions ........................................... | 38 |
| | References................................................................... | | 40 |
| **3** | **Spectroscopic Characterization of Waveguides** ......................... | | 43 |
| | Denise M. Krol | | |
| | 3.1 | Introduction ......................................................... | 43 |
| | 3.2 | Spectroscopic Analysis of Glass ................................... | 44 |
| | | 3.2.1 Fluorescence Spectroscopy............................... | 44 |
| | | 3.2.2 Raman Spectroscopy .................................... | 47 |
| | | 3.2.3 Confocal Imaging ........................................ | 49 |
| | 3.3 | Experimental Equipment and Procedures......................... | 49 |
| | | 3.3.1 Femtosecond Laser Systems and Micromachining Procedures ....................... | 50 |
| | | 3.3.2 Confocal Microscope System and Spectroscopy Procedures .......................... | 51 |
| | 3.4 | Spectroscopic Analysis of fs-laser Modification in Fused Silica .................................................... | 52 |
| | | 3.4.1 Fluorescence Spectroscopy and Imaging of Waveguides and Bragg Gratings ..................... | 52 |
| | | 3.4.2 Photobleaching of Defects............................... | 55 |
| | | 3.4.3 Raman Spectroscopy and Imaging ..................... | 55 |
| | 3.5 | Spectroscopic Analysis of Waveguides in Phosphate Glasses................................................ | 57 |
| | | 3.5.1 Fluorescence Spectroscopy and Imaging of IOG-1 ................................................... | 57 |
| | | 3.5.2 Comparison of Waveguides in Fused Silica and IOG-1 ................................................ | 59 |
| | | 3.5.3 Raman Spectroscopy and Imaging of Rare Earth-doped Phosphate Glass ........................... | 60 |
| | 3.6 | Summary ............................................................. | 63 |
| | References................................................................... | | 63 |

| 4 | **Optimizing Laser-Induced Refractive Index Changes in Optical Glasses via Spatial and Temporal Adaptive Beam Engineering** .................................................. | 67 |
|---|---|---|
| | Razvan Stoian | |
| | 4.1 Introduction ............................................................... | 67 |
| | 4.2 Mechanisms of Laser-Induced Refractive Index Changes ........ | 69 |
| | 4.3 Experimental Implementations for Pulse Engineering ............ | 73 |
| |     4.3.1 Spatio-Temporal Beam Shaping ....................... | 73 |
| |     4.3.2 Microscopy Based Adaptive Loops .................... | 75 |
| | 4.4 Material Interaction with Tailored Pulses ......................... | 76 |
| |     4.4.1 Refractive Index Engineering by Temporally Tailored Pulses ......................... | 76 |
| |     4.4.2 Energy Confinement and Size Corrections ............ | 78 |
| |     4.4.3 Adaptive Correction of Wavefront Distortions ........ | 80 |
| |     4.4.4 Dynamic Parallel Processing ........................... | 84 |
| | 4.5 Outlook and Conclusions ............................................. | 87 |
| | References ...................................................................... | 88 |
| 5 | **Controlling the Cross-section of Ultrafast Laser Inscribed Optical Waveguides** ........................................................ | 93 |
| | Robert R. Thomson, Nicholas D. Psaila, Henry T. Bookey, Derryck T. Reid, and Ajoy K. Kar | |
| | 5.1 Introduction ............................................................... | 93 |
| | 5.2 The Effect of the Waveguide Cross-section on the Properties of the Guided Modes ......................... | 94 |
| | 5.3 The Importance of Controlling the Waveguide Cross-section from a Device Engineering Perspective ............ | 97 |
| |     5.3.1 Effect of Mode Field Distribution on Waveguide Coupling Loss ......................... | 97 |
| |     5.3.2 Effect of Mode Field Distribution on Waveguide Propagation Loss ...................... | 98 |
| |     5.3.3 Effect of Mode Field Distribution on Evanescent Coupling ................................. | 99 |
| |     5.3.4 Effect of Waveguide Asymmetry on Polarisation Dependent Guiding Properties ............ | 100 |
| | 5.4 Experimental Techniques for Measuring the Refractive Index Profile of Ultrafast Laser Inscribed Waveguides ......................................... | 100 |
| |     5.4.1 Refracted Near-field (RNF) Method .................... | 100 |
| |     5.4.2 Micro-reflectivity ........................................ | 101 |
| |     5.4.3 Quantitative Phase Microscopy ........................ | 102 |
| |     5.4.4 Inverse Helmholtz Technique ........................... | 104 |

|  | 5.5 | Effect of Inscription Parameters on the Waveguide Cross-section | 105 |
|---|---|---|---|
|  | 5.6 | Experimental Techniques for Controlling the Waveguide Cross-section | 108 |
|  |  | 5.6.1 The Astigmatic Beam Shaping Technique | 108 |
|  |  | 5.6.2 The Slit Beam Shaping Technique | 111 |
|  |  | 5.6.3 Waveguide Shaping Using Active Optics | 114 |
|  |  | 5.6.4 Spatiotemporal Focussing | 118 |
|  |  | 5.6.5 The Multiscan Technique | 120 |
|  | 5.7 | Conclusions and Outlook | 121 |
|  | References | | 122 |
| **6** | **Quill and Nonreciprocal Ultrafast Laser Writing** | | 127 |
|  | Peter G. Kazansky and Martynas Beresna | | |
|  | 6.1 | Introduction | 128 |
|  | 6.2 | Quill Writing | 128 |
|  | 6.3 | Anisotropic Bubble Formation | 137 |
|  | 6.4 | Nonreciprocal Writing | 139 |
|  | 6.5 | Conclusion | 149 |
|  | References | | 150 |

**Part II  Waveguides and Optical Devices in Glass**

| **7** | **Passive Photonic Devices in Glass** | | 155 |
|---|---|---|---|
|  | Shane M. Eaton and Peter R. Herman | | |
|  | 7.1 | Introduction | 155 |
|  | 7.2 | Characterization of Femtosecond Laser-Written Waveguides | 157 |
|  |  | 7.2.1 Microscope Observation | 158 |
|  |  | 7.2.2 Insertion Loss | 158 |
|  |  | 7.2.3 Mode Profile and Coupling Loss | 159 |
|  |  | 7.2.4 Propagation Loss | 160 |
|  |  | 7.2.5 Refracted Near Field Method | 161 |
|  | 7.3 | Femtosecond Laser Microfabrication of Optical Waveguides | 161 |
|  |  | 7.3.1 Low-Repetition Rate Regime | 162 |
|  |  | 7.3.2 High-Repetition Rate Regime | 164 |
|  | 7.4 | Devices | 172 |
|  |  | 7.4.1 Y-Junctions | 173 |
|  |  | 7.4.2 Directional Couplers | 175 |
|  |  | 7.4.3 Mach–Zehnder Interferometers | 186 |
|  |  | 7.4.4 Other Devices | 186 |
|  | 7.5 | Summary and Future Outlook | 192 |
|  | References | | 193 |

## 8 Fibre Grating Inscription and Applications ... 197
Nemanja Jovanovic, Alex Fuerbach, Graham D. Marshall, Martin Ams, and Michael J. Withford

- 8.1 Introduction ... 197
- 8.2 Review of Gratings Types ... 199
  - 8.2.1 Long Period Gratings ... 199
  - 8.2.2 Fibre Bragg Gratings ... 200
- 8.3 Point-by-Point Inscribed Gratings ... 202
  - 8.3.1 Fabrication Methods ... 202
  - 8.3.2 Development of Femtosecond Laser Direct-Write LPGs ... 204
  - 8.3.3 Development of Femtosecond Laser Direct-Write FBGs ... 205
- 8.4 Phase Mask Inscribed Gratings ... 206
  - 8.4.1 Fabrication Method ... 206
  - 8.4.2 Development of Femtosecond Laser-Phase Mask Inscription ... 207
- 8.5 Properties of Femtosecond Laser Written Gratings ... 209
  - 8.5.1 Thermal Stability ... 209
  - 8.5.2 Stress and Birefringence ... 212
  - 8.5.3 Photoattenuation ... 214
- 8.6 Applications ... 216
  - 8.6.1 Fibre Lasers ... 216
  - 8.6.2 Sensors ... 218
  - 8.6.3 Other Applications ... 220
- 8.7 Novel Fibre Types and Challenges ... 220
  - 8.7.1 Microstructured Optical-Fibres (MOFs) ... 220
  - 8.7.2 Polymer and Non-linear Fibres ... 222
  - 8.7.3 Through Jacket Grating Writing ... 222
- 8.8 Summary ... 222
- References ... 223

## 9 3D Bragg Grating Waveguide Devices ... 227
Haibin Zhang and Peter R. Herman

- 9.1 Introduction ... 227
- 9.2 Bragg Grating Waveguide Fabrication ... 230
  - 9.2.1 BGW Fabrication Method 1: Single-pulse Writing ... 231
  - 9.2.2 BGW Fabrication Method 2: Burst Writing ... 238
  - 9.2.3 BGW Thermal Stability ... 244
- 9.3 BGW Devices ... 247
  - 9.3.1 Multi-wavelength BGWs ... 247
  - 9.3.2 Chirped BGWs ... 250
  - 9.3.3 3D BGW Sensor Network ... 253
- 9.4 Summary and Future Outlook ... 259
- References ... 261

## 10 Active Photonic Devices ... 265
Giuseppe Della Valle and Roberto Osellame

- 10.1 Introduction ... 265
- 10.2 Active Ions for Waveguide Devices ... 266
- 10.3 Gain Definitions and Measurement Technique ... 270
  - 10.3.1 Definition of the Main Figures of An Active Waveguide ... 270
  - 10.3.2 The On/Off Technique for Gain Measurement ... 274
- 10.4 Active Waveguides and Amplifiers ... 277
  - 10.4.1 Internal Gain in Nd-Doped Active Waveguides ... 277
  - 10.4.2 Waveguide Amplifier in Er:Yb-Doped Phosphate Glass ... 277
  - 10.4.3 Waveguide Amplifier in Er:Yb-Doped Oxyfluoride Silicate Glass ... 278
  - 10.4.4 Active Waveguides in New Glass Materials ... 281
- 10.5 Waveguide Lasers ... 281
  - 10.5.1 Waveguide Lasers in Er:Yb-Doped Phosphate Glass ... 281
  - 10.5.2 Waveguide Laser in Er:Yb-Doped Oxyfluoride Silicate Glass ... 283
- 10.6 Advanced Waveguide Lasers ... 283
  - 10.6.1 Single-Longitudinal-Mode Operation ... 283
  - 10.6.2 Mode-Locking Regime ... 288
- 10.7 Outlook and Conclusions ... 289
- References ... 290

## Part III Waveguides and Optical Devices in Other Transparent Materials

## 11 Waveguides in Crystalline Materials ... 295
Matthias Heinrich, Katja Rademaker, and Stefan Nolte

- 11.1 Origins of Refractive Index Changes ... 295
- 11.2 Waveguides Characteristics in Various Crystals ... 298
  - 11.2.1 Waveguide Fabrication in $LiNbO_3$ ... 299
  - 11.2.2 Other Crystals ... 300
  - 11.2.3 Actively Doped Crystals and Ceramics ... 302
- 11.3 Nonlinear Properties ... 304
  - 11.3.1 Lithium Niobate ... 304
- 11.4 Integrated Optical Devices ... 306
  - 11.4.1 Mach–Zehnder Interferometer ... 306
  - 11.4.2 Electrooptic Modulator ... 307
  - 11.4.3 Waveguide Lasers ... 308
- 11.5 Conclusion ... 310
- References ... 311

| | | | |
|---|---|---|---|
| **12** | **Refractive Index Structures in Polymers** | | 315 |
| | Patricia J. Scully, Alexandra Baum, Dun Liu, and Walter Perrie | | |
| | 12.1 | Introduction | 315 |
| | 12.2 | Motivation for Refractive Index Structures in Polymers | 316 |
| | 12.3 | Laser Photomodification of PMMA | 317 |
| | | 12.3.1 Continuous Wave UV Laser Sources | 317 |
| | | 12.3.2 Long Pulse (ns, ps) Laser Sources | 317 |
| | | 12.3.3 Ultrashort fs Laser Sources | 318 |
| | 12.4 | Waveguiding and Positive/Negative Refractive Index | 319 |
| | 12.5 | Direct Writing | 320 |
| | | 12.5.1 Simple Transmission Gratings (2D) | 321 |
| | | 12.5.2 Production of Waveguides (1D) | 322 |
| | 12.6 | Holographic Writing | 323 |
| | 12.7 | Comparisons of Commercial and Clinical Grade PMMA | 325 |
| | 12.8 | Pulse Duration, Wavelength, and Bandgap Dependence of Refractive Index Modification | 325 |
| | | 12.8.1 Pulse Duration Dependence of Refractive Index Modification | 326 |
| | | 12.8.2 Effect of Bandgap and Wavelength on Refractive Index Modification | 331 |
| | 12.9 | Relating Photochemistry to Writing Conditions | 332 |
| | | 12.9.1 Size Exclusion Chromatography | 334 |
| | | 12.9.2 Thermal Desorption Volatile Analysis | 334 |
| | | 12.9.3 Thermogravimetric Analysis | 334 |
| | | 12.9.4 Optical Spectroscopy | 335 |
| | | 12.9.5 Etching of Structures | 335 |
| | | 12.9.6 Summary of Photochemical Analysis | 336 |
| | 12.10 | Effect of Self-Focusing | 336 |
| | 12.11 | Effects of Depth | 339 |
| | 12.12 | Parallel Processing Using Spatial Light Modulator | 340 |
| | 12.13 | Applications of Refractive Index Structures in Polymers | 342 |
| | | 12.13.1 Polymer Optical Fibre Sensors and Devices | 343 |
| | 12.14 | Summary | 343 |
| | References | | 344 |

## Part IV  Microsystems and Applications

**13 Discrete Optics in Femtosecond Laser Written Waveguide Arrays** ... 351
Alexander Szameit, Felix Dreisow, and Stefan Nolte
- 13.1 Introduction to Waveguide Arrays ................................. 351
- 13.2 Fundamental Principles of Discrete Light Propagation .......... 353
- 13.3 Basic Experimental Techniques .................................... 355
  - 13.3.1 Evanescent Coupling ...................................... 355
  - 13.3.2 Waveguide Imaging Microscopy ....................... 356
  - 13.3.3 Multi-waveguide Excitation ............................ 357
- 13.4 Linear Propagation Effects ......................................... 358
  - 13.4.1 Straight Lattices ........................................... 358
  - 13.4.2 Curved Lattices ........................................... 365
  - 13.4.3 Quantum-Mechanical Analogies ....................... 370
- 13.5 Nonlinear Propagation Effects ..................................... 373
  - 13.5.1 Nonlinear Refractive Index ............................. 373
  - 13.5.2 One-Dimensional Solitons .............................. 374
  - 13.5.3 Two-Dimensional Solitons .............................. 381
- 13.6 Conclusions .......................................................... 385
- References ................................................................... 386

**14 Optofluidic Biochips** ..................................................... 389
Rebeca Martínez Vázquez, Giulio Cerullo, Roberta Ramponi,
and Roberto Osellame
- 14.1 Introduction .......................................................... 389
- 14.2 Femtosecond Laser Microfluidic Channel Fabrication ........... 392
  - 14.2.1 Fundamental Physical Mechanisms .................... 392
  - 14.2.2 Microchannel Properties ................................ 394
  - 14.2.3 Integration of Optical Waveguides and Microfluidic Channels .............................. 398
- 14.3 Integration of Photonic Sensors in LOCs ........................ 400
  - 14.3.1 Cell Sorting .............................................. 400
  - 14.3.2 Microchip Capillary Electrophoresis ................... 402
  - 14.3.3 Label-Free Sensing with Mach–Zehnder Interferometers ........................................... 405
- 14.4 Femtosecond Laser Fabrication of Optofluidic Devices .......... 409
  - 14.4.1 Flow Cytometry .......................................... 409
  - 14.4.2 Label-Free Sensing with Bragg Gratings ............... 411
  - 14.4.3 Cell Trapping and Stretching ........................... 413
- 14.5 Outlook and Conclusions .......................................... 416
- References ................................................................... 417

## 15  Microstructuring of Photosensitive Glass .............................. 421
Koji Sugioka
- 15.1  Introduction ...................................................... 421
- 15.2  Photosensitive Glass ............................................. 422
  - 15.2.1  Characteristics ........................................... 422
  - 15.2.2  Microstructuring Procedure ............................ 424
  - 15.2.3  Microstructuring Mechanism........................... 425
- 15.3  Fabrication of Microfluidic Structures ........................... 428
- 15.4  Fabrication of Micro-Optic Structures ........................... 430
  - 15.4.1  Micro-Optics ............................................. 430
  - 15.4.2  Optical Waveguides..................................... 431
  - 15.4.3  Integration of Optical Microcomponents ............... 432
- 15.5  Fabrication of Microchip Devices ................................ 433
  - 15.5.1  Microfluidic Dye Laser ................................. 433
  - 15.5.2  Optofluidics .............................................. 435
  - 15.5.3  Nano-Aquarium ......................................... 436
- 15.6  Summary ......................................................... 439
- References............................................................... 439

## 16  Microsystems and Sensors ............................................... 443
Yves Bellouard, Ali A. Said, Mark Dugan, and Philippe Bado
- 16.1  Introduction ...................................................... 443
- 16.2  Micro- and Nano-Systems......................................... 444
  - 16.2.1  A Brief Overview of Microsystems ..................... 444
  - 16.2.2  Issues on Microsystems Integration and Fabrication............................................ 444
- 16.3  Microsystems Fabricated Using Femtosecond Lasers: Review and State of the Art ...................................... 446
  - 16.3.1  Specificities of Femtosecond Laser–Matter Interaction from the View-Point of Microsystems Design ................................. 446
  - 16.3.2  Integrated Optics Devices .............................. 447
  - 16.3.3  Opto-Fluidics ............................................ 448
  - 16.3.4  Micromechanical Functionality ........................ 448
- 16.4  Multifunctional Monolithic System Integration .................. 448
  - 16.4.1  Concept.................................................. 448
  - 16.4.2  Taxonomy of Individual Elements Used in a Monolithic Design ................................. 449
  - 16.4.3  System Integration: Design Strategies and Interfacing ........................................ 456
  - 16.4.4  Illustration: Micro-Displacement Sensors and Micro-Force Sensors ............................. 456
- 16.5  Summary, Benefits, Future Prospects, and Challenges............ 464
- References............................................................... 464

## 17 Ultrashort Laser Welding and Joining ... 467
Wataru Watanabe, Takayuki Tamaki, and Kazuyoshi Itoh

 17.1 Introduction ... 467
 17.2 Laser Welding ... 468
 17.3 Ultrashort Laser Welding of Transparent Materials ... 469
  17.3.1 Ultrashort Laser Welding with Low-Repetition Rate ... 469
  17.3.2 Ultrashort Laser Welding with High-Repetition Rate ... 473
 17.4 Outlook and Conclusions ... 474
 References ... 476

**Index** ... 479

# Part I
# Introductory Concepts, Characterization and Optimization Strategies

# Chapter 1
# Fundamentals of Femtosecond Laser Modification of Bulk Dielectrics

Shane M. Eaton, Giulio Cerullo, and Roberto Osellame

**Abstract** Femtosecond laser pulses focused beneath the surface of a dielectric are absorbed through nonlinear photoionization mechanisms, giving rise to a permanent structural modification with dimensions on the order of a micrometer. At low pulse energies, the modification in many glasses is a smooth refractive index change, enabling photonic device fabrication. At higher pulse energies, the laser-induced modification may contain birefringent, periodic nanoplanes which align themselves orthogonally to the laser polarization. These nanogratings are not ideal for most waveguide devices but when the sample is exposed to hydrofluoric acid after writing, preferential chemical etching along the direction of the nanoplanes forms several millimeter-long buried microchannels which are useful for microfluidic applications. At even higher pulse energies, ultrahigh pressures within the focal volume lead to microexplosions causing empty voids which can be used for three-dimensional photonic bandgap devices and memories. In addition to pulse energy, other parameters have been shown to strongly influence the resulting morphology after femtosecond laser exposure including repetition rate, scan speed, focusing condition, polarization, pulse duration, depth, and direction.

## 1.1 Introduction

In 1996, Hirao's group showed that by focusing subpicosecond pulses in the bulk of transparent glass, the modification induced beneath the sample surface could be tailored to produce a permanent refractive index increase [1]. Because of the nonlinear nature of the interaction, absorption is confined to the focal volume inside

---

S.M. Eaton (✉) · G. Cerullo · R. Osellame
Istituto di Fotonica e Nanotecnologie - Consiglio Nazionale delle Ricerche (IFN-CNR), and Department of Physics - Politecnico di Milano, Piazza Leonardo da Vinci 32, 20133, Milan, Italy
e-mail: shane.eaton@ifn.cnr.it; giulio.cerullo@fisi.polimi.it; roberto.osellame@ifn.cnr.it

the bulk material. By scanning the sample relative to the laser focus with computer-controlled motion stages, a region of increased refractive index could be formed along an arbitrary three-dimensional path, unlike traditional photolithography, which is limited to fabricating devices in-plane. In this chapter, the current understanding of the femtosecond laser–material interaction physics in the bulk of dielectrics is discussed and the important exposure conditions influencing the resulting waveguide properties are reviewed.

## 1.2 Femtosecond Laser–Material Interaction

Peak intensities on the order of 10 TW/cm$^2$ can be readily produced by focused femtosecond laser pulses from today's commercial laser systems. Such intensities result in strong nonlinear absorption, allowing for localized energy deposition in the bulk of glasses. After several picoseconds, the laser-excited electrons transfer their energy to the lattice, leading to a permanent material modification. While a complete physical model of the laser–material interaction has thus far eluded researchers, the process can be simplified by subdivision into three main steps: the initial generation of a free electron plasma followed by energy relaxation and modification of the material.

### 1.2.1 Free Electron Plasma Formation

Focused femtosecond laser pulses, with wavelengths in the visible or near-infrared spectra, have insufficient photon energy to be linearly absorbed in glasses. Valence electrons are instead promoted to the conduction band through nonlinear photoionization, which proceeds by multiphoton ionization and/or tunneling photoionization pathways depending on the laser frequency and intensity [2, 3]. If nonlinear photoionization were the only absorption process, the threshold intensity for optical breakdown would vary greatly with bandgap due to the large variation in absorption probability with bandgap (multiphoton absorption order). However, avalanche photoionization is also present and since it depends only linearly on laser intensity, there is only a small variation in optical breakdown threshold intensity with material bandgap energy [4]. Because of this low dependence of the breakdown threshold on the bandgap energy, femtosecond laser microfabrication can be applied in a wide range of materials.

#### 1.2.1.1 Nonlinear Photoionization

Multiphoton absorption occurs due to the simultaneous absorption of multiple photons by an electron in the valence band (Fig. 1.1a). The number of photons $m$

**Fig. 1.1** Nonlinear photoionization processes underlying femtosecond laser machining. (**a**) Multiphoton ionization, (**b**) tunneling ionization, and (**c**) Avalanche ionization: free carrier absorption followed by impact ionization [5]

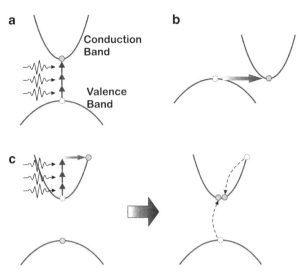

required to bridge the bandgap must satisfy $mh\nu > E_g$, where $E_g$ is the bandgap, and $\nu$ is the frequency of light. Multiphoton ionization is the dominant mechanism at low laser intensities and high frequencies (but below that which is needed for single photon absorption). At high laser intensity and low frequency, nonlinear ionization proceeds via tunneling, as shown in Fig. 1.1b. The strong field distorts the band structure and reduces the potential barrier between the valence and conduction bands. Direct band to band transitions may then proceed by quantum tunneling of the electron from the valence to conduction band. As shown by Keldysh [6], multiphoton and tunneling photoionization can be described in the same theoretical framework. The transition between the processes is described by the Keldysh parameter:

$$\gamma = \frac{\omega}{e}\sqrt{\frac{m_e c n \epsilon_0 E_g}{I}} \qquad (1.1)$$

where $\omega$ is the laser frequency, $I$ is the laser intensity at the focus, $m_e$ is the effective electron mass, $e$ is the fundamental electron charge, $c$ is the speed of light, $n$ is the linear refractive index and $\epsilon_0$ is the permittivity of free space. If $\gamma$ is much less (greater) than 1.5, tunneling (multiphoton) ionization dominates. For $\gamma \sim 1.5$, photoionization is a combination of tunneling and multiphoton ionization. For waveguide fabrication in dielectrics, typical laser, and material properties result in $\gamma \sim 1$, so that nonlinear ionization is a combination of both processes [3].

### 1.2.1.2 Avalanche Photoionization

Electrons present in the conduction band may also absorb laser light by free carrier absorption (Fig. 1.1c). After sequential linear absorption of several photons, a conduction band electron's energy may exceed the conduction band minimum by more than the band gap energy and the hot electron can then impact ionize a bound electron in the valence band, resulting in two excited electrons at the conduction band minimum. These two electrons can undergo free carrier absorption and impact ionization and the process can repeat itself as long as the laser field is present and strong enough, giving rise to an electron avalanche.

Avalanche ionization requires that sufficient seed electrons are initially present in the conduction band. These seed electrons may be provided by thermally excited impurity or defect states, or direct multiphoton or tunneling ionization. For subpicosecond laser pulses, absorption occurs on a faster time scale than energy transfer to the lattice, decoupling the absorption and lattice heating processes [3]. Seeded by nonlinear photoionization, the density of electrons in the conduction band grows through avalanche ionization until the plasma frequency approaches the laser frequency, at which point the plasma becomes strongly absorbing. For 1-$\mu$m wavelength laser radiation, the plasma frequency equals the laser frequency when the carrier density is on the order of $10^{21}$ cm$^{-3}$, which is known as the critical density of free electrons. At this high carrier density, only a few percent of the incident light is reflected by the plasma, so that most of the energy is transmitted into the plasma where it can be absorbed through free carrier absorption [3]. It is usually assumed that optical breakdown occurs when the number of carriers reaches this critical value. In glass, the corresponding intensity required to achieve optical breakdown is $10^{13}$ W/cm$^2$.

Since the lattice heating time is on the order of 10 ps, the absorbed laser energy is transferred to the lattice long after the laser pulse is gone. Because short pulses need less energy to achieve the intensity for breakdown and because the absorption is decoupled from the lattice heating, more precise machining is possible relative to longer pulses. Further, because nonlinear photoionization can seed electron avalanche with femtosecond laser pulses, this results in deterministic breakdown. This is in contrast to the stochastic breakdown with longer pulses which rely on the low concentration of impurities ($\sim$1 impurity electron in conduction band per focal volume) randomly distributed in the material to seed an electron avalanche [7]. Lenzner et al. found that for very short pulses ($<$ 10 fs and 100 fs in fused silica and borosilicate glasses, respectively), photoionization can dominate avalanche ionization and produce sufficient plasma density to cause damage by itself [8]. Schaffer et al. showed that the contribution from avalanche ionization is greater at longer pulse durations and for materials with greater band gap energies such as fused silica and sapphire [3].

## 1.2.2 Relaxation and Modification

It is well accepted that nonlinear photoionization and avalanche ionization from absorbed femtosecond laser pulses are responsible for the creation of a free electron plasma. However, once the electrons have transferred their energy to the lattice, the physical mechanisms for material modification are not fully understood. Of the hundreds of published works citing the pioneering work by Davis et al. [1], the observed morphological changes can be generally classified into three types of structural changes: a smooth refractive index change [9], a form birefringent refractive index modification [10–12], and microexplosions leading to empty voids [13]. The regime of modification and resulting morphological change depends not only on many exposure parameters (energy, pulse duration, repetition rate, wavelength, polarization, focal length, scan speed, and others) but also on material properties (bandgap, thermal conductivity, and others). However, in pure fused silica which is the most commonly processed material for waveguide writing, these three morphologies can be observed by simply changing the incident laser energy [14] as illustrated in Fig. 1.2. These morphologies are discussed below with emphasis on fused silica glass; further insight, also related to other glasses, will be provided in Chaps. 2, 3 and 4, while discussion on the main modification mechanisms in crystals and polymers can be found in Chaps. 11 and 12, respectively.

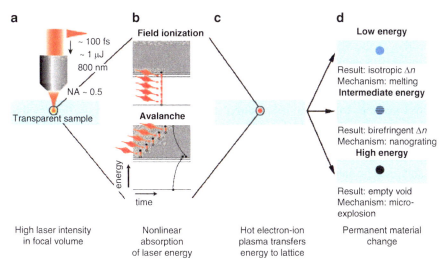

**Fig. 1.2** Illustration of the interaction physics of focused femtosecond laser pulses in bulk fused silica. (**a**) The laser is focused below the sample surface resulting in a high intensity in the focal volume. (**b**) The energy is nonlinearly absorbed and a free electron plasma is created by multiphoton/tunneling and avalanche photoionization. (**c**) The plasma transfers its energy to the lattice on a ~10 ps time scale resulting in one of three types of permanent modification (**d**): isotropic refractive index change at low pulse energy, sub-wavelength birefringent nanostructures at moderate energy and empty voids at high pulse energy [14]

### 1.2.2.1 Smooth Refractive Index Change

An isotropic regime of modification is useful for optical waveguides, where smooth and uniform refractive index modification is required for low propagation loss. At low pulse energies just above the modification threshold (~100 nJ for 0.6-NA focusing of 800-nm, 100-fs pulses), a smooth refractive index modification has been observed in fused silica, which is attributed to densification from rapid quenching of the melted glass in the focal volume [15]. In fused silica, the density (refractive index) increases when glass is rapidly cooled from a higher temperature [16]. Micro-Raman spectroscopy has confirmed an increase in the concentration of 3 and 4 member rings in the silica structure in the laser exposed region, indicating a densification of the glass [15]. Shock waves generated by focused femtosecond laser pulses giving rise to stress have been shown to play a role in causing densification under certain conditions [17].

It has been argued that laser-induced color centers may be responsible for the laser-induced refractive index change through a Kramers–Kronig mechanism (a change in absorption leads to a change in refractive index) [18]. Although induced color centers have been observed in glasses exposed to femtosecond laser radiation [19, 20], to date there has been no experimental evidence of a strong link between color center formation and the induced refractive index change. Waveguides formed in fused silica with an infrared femtosecond laser [21] were found to exhibit photo-induced absorption peaks at 213 nm and 260 nm corresponding to SiE' (positively charged oxygen vacancies) and non-bridging oxygen hole centers (NBOHC) defects, respectively. However, both color centers were completely erased after annealing at 400°C, although waveguide behavior was still observed up to 900°C. Therefore, it is unlikely that color centers played a significant role in the refractive index change [21]. Other research found that the thermal stability of color centers produced in borosilicate and fused silica glasses by femtosecond laser irradiation is not consistent with that of the induced refractive index change [20]. Recently, Withford's group has shown for Yb-doped phosphate glasses used in waveguide lasers, femtosecond laser-induced color centers contribute approximately 15% to the observed refractive index increase [22]. Using integrated waveguide Bragg gratings, the authors were able to accurately study the photobleaching and thermal annealing of the induced color centers. The color centers were stable for temperatures below 70°C, which is below the operating point during lasing. However, the green luminescence generated by the Yb ions results in a photobleaching of the color centers during laser operation, resulting in reduced lifetime which must be corrected by pre-aging techniques [22].

Although a complete understanding of the femtosecond laser material interaction in forming optical waveguides has presently eluded researchers, it is certain that densification and color centers play a role. However, their contributions will vary depending on the glass composition and the femtosecond laser exposure conditions, further adding to the complexities in modeling femtosecond laser waveguide writing. In glasses with structures that are more complex than fused silica, further contributions must be considered. For example, in multicomponent crown glass, the authors concluded that the ring-shaped refractive index profile during femtosecond laser irradiation was the result of ion exchange between network formers and network modifiers [23].

### 1.2.2.2 Birefringent Refractive Index Change

For higher pulse energies (∼150–500 nJ for 0.6-NA focusing of 800-nm, 100-fs pulses), birefringent refractive index changes have been observed in the bulk of fused silica glass, as first reported by Sudrie et al. [10]. Kazansaky et al. argued that the birefringence was due to periodic nanostructures that were caused by interference of the laser field and the induced electron plasma wave [11]. In similarly exposed fused silica samples, Taylor's group observed periodic layers of alternating refractive index with sub-wavelength period that were clearly visualized after etching the laser-written tracks with HF acid (Fig. 1.3a and b). The orientation of the nanogratings was perpendicular to the writing laser polarization in all cases. The period of the nanostructures was found to be approximately $\lambda/2n$, regardless of scan speed, which implies a self-replicating formation mechanism [24]. However, new research suggests a slight variation of the nanograting period with exposure parameters [25]. Taylor's group proposed that inhomogeneous dielectric breakdown results in the formation of a nanoplasma resulting in the growth and self-organization of nanoplanes [24]. The model was found to accurately predict the experimentally measured nanograting period, but further development is needed to explain why nanostructures have not been observed in borosilicate glasses, and why they only form in a small window of pulse energy and pulse duration in fused silica [24]. Preferential HF etching of laser-written tracks was observed when the nanogratings were parallel to the writing direction (polarization perpendicular to scan direction) allowing the HF acid to diffuse more easily in the track, as shown in Fig. 1.3c. This effect can be exploited to fabricate buried microchannels for microfluidic applications[24, 26–28], which are discussed in detail in Chaps. 14 and 16.

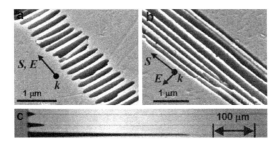

**Fig. 1.3** Scanned electron microscope image of nanogratings formed at 65-$\mu$m depth (sample cleaved and polished at writing depth) with polarization parallel (**a**) and perpendicular (**b**) to the scan direction. Overhead view (**c**) of etched microchannels demonstrating polarization selective etching with parallel (*top*), 45° (*middle*) and perpendicular (*bottom*) linear polarizations [24]

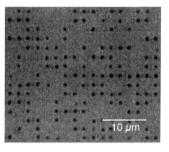

**Fig. 1.4** Binary bits (voids) recorded in fused silica with 100 fs laser [29]

### 1.2.2.3 Void Formation

At high pulse energies (>500 nJ for 0.6-NA focusing of 800-nm, 100-fs pulses) giving peak intensities greater than $\sim 10^{14}$ W/cm$^2$, pressures greater than Young's modulus are generated in the focal volume, creating a shockwave after the electrons have transferred their energy to the ions ($\sim 10$ ps). The shockwave leaves behind a less dense or hollow core (void), depending on the laser and material properties [13]. By conservation of mass, this core is surrounded by a shell of higher refractive index. Such voids may be exploited for 3D memory storage [29] (Fig. 1.4) or photonic bandgap materials [30], but are not suitable for optical waveguides.

### 1.2.2.4 Multiple Pulse Interaction

The above interpretations for the structural changes induced by focused femtosecond lasers were based on single pulse interactions, but can likely be extended to explain modification from multiple pulses within the same laser spot, assuming the repetition rate is low enough that thermal diffusion has carried the heat away from the focus before the next pulse arrives [14]. In this situation, the ensuing pulses may add to the overall modification, but still act independently of one another.

For high-repetition rates (>100 kHz), the time between laser pulses is less than the time for heat to diffuse away, resulting in an accumulation of heat in the focal volume. If the pulse energy is sufficient, the glass near the focus is melted and as more laser pulses are absorbed, this melted volume increases in size until the laser is removed, at which point and the melt rapidly cools into a structure with altered refractive index. For a scanned waveguide structure, the size of the melted volume can be controlled by the effective number of pulses in the laser spot size, $N = 2w_0 R/v$, where $2w_0$ is the spot size diameter ($1/e^2$), $R$ is the repetition rate and $v$ is the scan speed. For cumulative heating, the morphology of the structural change is dominated by the heating, melting, and cooling dynamics of the material in and around the focal volume [14]. Laser repetition rate, along with other exposure variables in femtosecond laser waveguide writing are discussed in more detail in the following section.

## 1.3 Exposure Variables and Considerations

Below we review the important exposure parameters in femtosecond laser processing, with an emphasis on glass substrates. The important experimental variables in processing crystals and polymers are discussed in Chaps. 11 and 12, respectively.

### 1.3.1 Focusing

Linear optical effects such as dispersion, diffraction, aberration, and nonlinear effects such as self focusing, plasma defocusing, and energy depletion influence the propagation of focused femtosecond laser pulses in dielectrics, thereby altering the energy distribution at the focus and the resulting refractive index modification.

#### 1.3.1.1 Linear Propagation

Incident femtosecond laser pulses are focused with an external lens to achieve a small micrometer-sized focal spot and drive nonlinear absorption. Neglecting spherical aberration and nonlinear effects, the spatial intensity profile of a femtosecond laser beam can be well represented by the paraxial wave equation and Gaussian optics. The diffraction-limited minimum waist radius $w_0$ (1/2 the spot size) for a collimated Gaussian beam focused in a dielectric is given by:

$$w_0 = \frac{M^2 \lambda}{\pi \text{NA}} \quad (1.2)$$

where $M^2$ is the Gaussian beam propagation factor (beam quality) [31], NA is the numerical aperture of the focusing objective and $\lambda$ is the free space wavelength. The Rayleigh range $z_0$ (1/2 the depth of focus) inside a transparent material of refractive index $n$ is given by:

$$z_0 = \frac{M^2 n \lambda}{\pi \text{NA}^2} \quad (1.3)$$

Chromatic and spherical aberration cause a deviation in the intensity distribution near the focus such that (1.2), and (1.3) may no longer be valid approximations. Chromatic aberration as the result of dispersion in the lens is corrected by employing chromatic aberration-corrected microscope objectives for the wavelength spectrum of interest. For lenses made with easily-formed spherical shapes, light rays which are parallel to the optic axis but at different distances from the optic axis fail to converge to the same point, resulting in spherical aberration. This issue can be addressed by using multiple lenses such as those found in microscope objectives, or employing an aspheric focusing lens. In waveguide writing where light is focused inside glass, the index mismatch at the air-glass interface introduces additional spherical aberration. As a result, there is a strong depth dependence for femtosecond-laser written buried structures [32–34]. This depth dependence is more pronounced for higher NA objectives [3], except in the case of oil-immersion lenses [3,35,36] or dry objectives with collars that enable spherical aberration correction at different focusing depths [33]. Other solutions to the problem of spherical aberration are described in Chaps. 4 and 5.

Dispersion from mirror reflection and transmission through materials can broaden the pulse width [37] which can reduce the peak intensity and alter the energy dissipation at the focus. For a typical Yb-based amplified femtosecond laser with 1-$\mu$m wavelength and 10-nm bandwidth, the dispersion in glass is -50 ps/km/nm and the pulse duration increase per length is 5 fs/cm. Since these sources have pulse durations >200 fs, the dispersion is negligible for most laser micromachining beam delivery systems which have less than 1 cm of transmission through glass. It is only for short pulse <40-fs oscillators with large bandwidths that dispersion becomes an issue. In this case, pre-compensation of the dispersion through the microscope objective is required to obtain the shortest pulse at the focus [37].

### 1.3.1.2 Nonlinear Propagation

When light propagates through a dielectric material, it induces microscopic displacement of bound charges, forming oscillating electric dipoles that add up to the macroscopic polarization. In amorphous glass which has an inversion center ($\chi^{(2)} = 0$), the polarization vector is given by:

$$\mathbf{P} = \epsilon_0 \left[ \chi^{(1)} + \frac{3}{4}\chi^{(3)}|\mathbf{E}|^2 \right] \mathbf{E} \tag{1.4}$$

where $\mathbf{E}$ is the electric field vector and $\chi^{(i)}$ is the $i$-th order susceptibility, with 4th and higher orders left out of (1.4) due to negligible contribution. The refractive index can be identified from (1.4) as:

$$n = \sqrt{1 + \chi^{(1)} + \frac{3}{4}\chi^{(3)}|\mathbf{E}|^2} = n_0 + n_2 I \tag{1.5}$$

where $n_0 = \sqrt{1 + \chi^{(1)}}$ is the linear refractive index, $n_2 = 3\chi^{(3)}/4\epsilon_0 c n_0^2$ is the nonlinear refractive index and $I = \frac{1}{2}\epsilon_0 n_0 c |\mathbf{E}|^2$ is the laser intensity.

The spatially varying intensity of a Gaussian laser beam can create a spatially varying refractive index in dielectrics. Because $n_2$ is positive in most materials, the refractive index is higher at the center of the beam compared to the wings. This variation in refractive index acts as a positive lens and focuses the beam inside a dielectric with a strength dependent on the peak power. If the peak power of the femtosecond laser pulses exceeds the critical power for self-focusing [3]:

$$P_c = \frac{3.77\lambda^2}{8\pi n_0 n_2} \quad (1.6)$$

the collapse of the pulse to a focal point is predicted. However, as the beam self focuses, the increased intensity is sufficient to nonlinearly ionize the material to produce a free electron plasma, which acts as a diverging lens that counters the Kerr lens self-focusing. A balance between self focusing and plasma defocusing leads to filamentary propagation, which results in axially elongated refractive index structures, which are undesirable for transversely written waveguide structures described in the next section. Self-focusing can be suppressed in waveguide fabrication by tightly focusing the laser beam with a microscope objective to reach the intensity for optical breakdown without exceeding the critical power for self focusing.

In fused silica, $n_0 = 1.45$ and $n_2 = 3.5 \times 10^{-20}$ m$^2$/W [38] so that for $\lambda = 800$ nm, the critical power is ~1.8 MW. From (1.6), the critical power is proportional to the square of the laser wavelength; therefore, lower critical powers result when working with the second and third harmonic frequencies of femtosecond lasers. Also, the critical power is inversely related to the nonlinear (and linear) refractive index, presenting a challenge in forming waveguides in nonlinear materials such as heavy metal oxide ($n_0 \sim 2$, $n_2 \sim 10^{-18}$ m$^2$/W [39]) and chalcogenide glasses ($n_0 \sim 2.5$, $n_2 \sim 10^{-17}$ m$^2$/W [40]), polymers ($n_0 \sim 1.5$, $n_2 \sim 10^{-18}$ m$^2$/W), lithium niobate ($n_0 \sim 2.3$, $n_2 \sim 10^{-19}$ m$^2$/W [41]) and silicon ($n_0 \sim 3.5$, $n_2 \sim 10^{-18}$ m$^2$/W [42]) crystals.

### 1.3.2 Writing Geometry

The standard configurations for laser-writing of optical waveguides are shown in Fig. 1.5. In longitudinal writing, the sample is scanned parallel, either toward or away from the incident laser. In this configuration, the resulting waveguide structures have cylindrical symmetry, owing to the transverse symmetry Gaussian intensity profile of the laser beam. The main disadvantage of the longitudinal writing geometry is that the waveguide length is limited by the working distance of the lens, which for a typical focusing objective with NA = 0.4, is approximately 5 mm. To overcome this issue, researchers have employed looser focusing lenses (NA ~ 0.2) [43], requiring higher laser power to reach the intensity required for optical breakdown. At such peak powers (~1 MW), the optical Kerr effect results in self focusing, producing filaments which yield refractive index change structures elongated in the axial direction by up to several hundred microns [43]. Despite the long length of the filaments, fabrication speeds are still relatively slow at ~1 $\mu$m/s to build up enough refractive index increase to efficiently guide light [43].

In the transverse writing scheme of Fig. 1.5, the sample is scanned orthogonally relative to the incoming laser. The working distance no longer restricts the

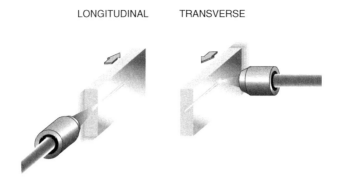

**Fig. 1.5** Longitudinal and transverse writing geometries for femtosecond laser waveguide fabrication in the bulk of transparent materials. In transverse (longitudinal) writing, the sample is scanned transverse (parallel) with respect to the incident femtosecond laser [4]

waveguide's length and structures may be formed over a depth range of several millimeters, which is sufficient flexibility for many applications to provide 3D optical circuits. The disadvantage of the transverse geometry is that the waveguide cross section is asymmetric due to the ratio between depth of focus and spot size $2z_0/2w_0 = n/\text{NA}$, where $n$ is the refractive index and NA is the numerical aperture. Since waveguides are formed in glasses with $n = 1.5$ with typical NA of 0.25 to 0.85, the focal volume asymmetry $n/\text{NA}$ varies from 6.0 to 1.8. This asymmetry results in elliptical waveguide cross sections with elliptical guided modes, which couple poorly to optical fibers. To overcome this focal volume asymmetry, special beam shaping methods are required as described in Chap. 5.

### 1.3.3 Influence of Exposure Variables Within Low- and High-Repetition Rate Regimes

The advent of high-repetition rate femtosecond lasers is opening new avenues for manipulating thermal relaxation effects [34–36, 44–46] that control the properties of optical waveguides formed when the ultrashort pulse lasers are focused inside dielectrics. As repetition rate increases, the time between laser pulses becomes shorter than the time for the absorbed laser radiation to diffuse out of the focal volume and heat builds up around the focal volume. This effect was first exploited in the surface micromachining of glass to form smooth, crack-free holes due to laser interaction with a thin sheath of ductile pre-heated glass [47]. Schaffer et al. reported a dramatic increase in the size of laser-modified structures formed in the bulk of glass under strong heat accumulation effects with a 25-MHz laser oscillator over structures formed by diffusion-only processes [44]. The combination of high-repetition rate and heat accumulation offers fast writing speeds and cylindrically symmetric waveguides together with benefits of annealing and decreased thermal cycling that are associated with low propagation and coupling loss to standard optical fiber [45]. Detailed discussion on the resulting waveguide characteristics within low (1–100 kHz) and high (>100 kHz) repetition rate regimes as a function of scan speed and pulse energy can be found in Chap. 7.

In addition to pulse energy, scan speed, and focusing, several other exposure parameters have been found to influence the resulting properties of femtosecond laser-written waveguides. These factors include pulse duration [48, 49], polarization [10, 50], direction [51], wavelength [52], spatiotemporal beam shaping, and waveguide cross section engineering (i.e., slit focusing, astigmatic focusing, and multiscan writing) [53, 54]. The effect of these parameters are briefly described below but in detail in the subsequent chapters.

Waveguides formed in fused silica with moderate fluence show no evidence of heat accumulation even as the repetition rate is raised from 1 kHz [50] to 1 MHz [34] and the waveguide properties were found to be strongly dependent on the incident writing polarization. In contrast, no detectable difference in insertion loss or mode size was found when waveguides were formed with different polarizations in borosilicate glass within the heat accumulation regime at MHz repetition rates [34]. In addition, the waveguide properties in borosilicate glass were invariant to pulse duration when varied from 300 to 700 fs, which is in contrast to results in fused silica, where pulse duration was observed to strongly affect waveguide mode size and loss [50]. The sensitivity to pulse duration and polarization in fused silica is associated with form birefringence arising from nanogratings formed within the laser-modified volume [12]. In borosilicate glass, nanogratings have not been observed [24], and such polarization and pulse duration dependence may possibly be erased by the strong thermal annealing [24] within the heat accumulation regime.

Due to energy depletion, self focusing and plasma defocusing, pulse duration strongly also influences the spatial distribution of the energy density in the focal volume [55, 56]. At 1-kHz repetition rate, where heat accumulation is not present, the dependence of waveguide properties on pulse duration in lithium niobate [56] and fused silica glass [57] was attributed to nonlinear pulse propagation. However, in the heat accumulation regime, nearly spherically symmetric thermal diffusion washes out the elliptical distribution of energy in the focal volume to yield waveguides with cross sections that are relatively circular. Therefore, one would expect pulse duration, despite its effect on the energy distribution at the focus, to play a lesser role on the properties of waveguides fabricated in the heat accumulation regime.

Kazansky et al. recently discovered the quill effect [51], in which laser material modification is influenced by the writing direction, even in amorphous glass with a symmetric laser intensity distribution. The researchers conclusively showed that the cause of the directional writing dependence is due to a pulse front tilt in the ultrafast laser beam [51]. Although any material should show a direction dependence due to a pulse front tilt, the effect was found to be almost negligible when processing borosilicate glass within the heat accumulation regime as evidenced by a directional coupler formed by arms written in opposite direction but showing a remarkably high peak coupling ratio of 99%, [58]. Further details of the quill effect are provided in Chap. 6.

As described in Chap. 7, wavelength is an important variable when processing high bandgap materials such as pure fused silica. Due to its large bandgap and low melting point, the increased fluence and lower order of multiphoton absorption provided by the second harmonic wavelength enabled stronger index contrast and lower loss waveguides in this material [52]. An infrared wavelength of 1.5 $\mu$m was applied to fused silica [21], revealing a very wide energy processing window of 1 to 23 $\mu$J in forming smooth waveguides compared to 0.5 to 2.0 $\mu$J at 800-nm wavelength. Such a large processing window is desirable, but the added complexity of using an optical parametric amplifier has dissuaded researchers from adopting this approach for waveguide device fabrication. By tuning the wavelength of a standard 800-nm Ti:Sapphire amplifier to the mid infrared (2.4 $\mu$m), linear absorption was avoided in silicon, making it possible to form buried waveguides in silicon via a 3-photon multiphoton absorption process [42].

To correct for asymmetric waveguide cross sections produced by low repetition rate fabrication as described above, several solutions have been proposed including astigmatic focusing [59] and slit focusing [60], deformable mirror beam reshaping [61] and multiscan writing [53]. More details on these techniques can be found in Chap. 5. The effects of varying the spatial and temporal properties of a laser beam for improving waveguide formation are discussed in Chap. 4.

## 1.4 Summary

It was shown that femtosecond laser pulses focused beneath the surface of a dielectric are initially absorbed through nonlinear and avalanche photoionization. After energy relaxation, the material is permanently modified within the small laser focal volume. If the laser pulse energy is just above the optical breakdown threshold, the modification can be tailored to be a smooth refractive index change, which is useful for optical waveguide devices. In addition to pulse energy, the waveguide properties depend on many other exposure variables, but principally on repetition rate, which determines whether the modification regime is due to individual pulses or cumulative pulse heating. The subsequent chapters will give more insight into the interaction of femtosecond laser pulses with glasses, crystalline, and polymer materials and highlight the many exciting devices created by the femtosecond laser micromachining method.

## References

1. K. Davis, K. Miura, N. Sugimoto, K. Hirao, Opt. Lett. **21**(21), 1729 (1996)
2. B.C. Stuart, M.D. Feit, S. Herman, A.M. Rubenchik, B.W. Shore, M.D. Perry, Phys. Rev. B **53**(4), 1749 (1996)
3. C. Schaffer, A. Brodeur, E. Mazur, Meas. Sci. Technol. **12**(11), 1784 (2001)

4. R.R. Gattass, E. Mazur, Nat. Photon. **2**(4), 219 (2008)
5. M. Ams, G.D. Marshall, P. Dekker, M. Dubov, V.K. Mezentsev, I. Bennion, M.J. Withford, IEEE J. Sel. Top. Quantum Electron. **14**(5), 1370 (2008)
6. L.V. Keldysh, Sov. Phys. JETP **20**(5), 1307 (1965)
7. D. Du, X. Liu, G. Korn, J. Squier, G. Mourou, Appl. Phys. Lett. **64**(23), 3071 (1994)
8. M. Lenzner, J. Krger, S. Sartania, Z. Cheng, C. Spielmann, G. Mourou, W. Kautek, F. Krausz, Phys. Rev. Lett. **80**(18), 4076 (1998)
9. K. Miura, J.R. Qiu, H. Inouye, T. Mitsuyu, K. Hirao, Appl. Phys. Lett. **71**(23), 3329 (1997)
10. L. Sudrie, M. Franco, B. Prade, A. Mysyrewicz, Opt. Commun. **171**(4-6), 279 (1999)
11. Y. Shimotsuma, P.G. Kazansky, J. Qiu, K. Hirao, Phys. Rev. Lett. **91**(24), 247405 (2003)
12. C. Hnatovsky, R.S. Taylor, P.P. Rajeev, E. Simova, V.R. Bhardwaj, D.M. Rayner, P.B. Corkum, Appl. Phys. Lett. **87**(1) (2005). 014104
13. S. Juodkazis, K. Nishimura, S. Tanaka, H. Misawa, E.G. Gamaly, B. Luther-Davies, L. Hallo, P. Nicolai, V.T. Tikhonchuk, Phys. Rev. Lett. **96**(16), 166101 (2006)
14. K. Itoh, W. Watanabe, S. Nolte, C.B. Schaffer, MRS Bull. **31**(8), 620 (2006)
15. J. Chan, T. Huser, S. Risbud, D. Krol, Opt. Lett. **26**(21), 1726 (2001)
16. R. Bruckner, J. Non-Cryst. Solids **5**(2), 123 (1970)
17. M. Sakakura, M. Shimizu, Y. Shimotsuma, K. Miura, K. Hirao, Appl. Phys. Lett. **93**(23), 3 (2008)
18. K. Hirao, K. Miura, J. Non-Cryst. Solids **239**(1-3), 91 (1998)
19. J. Chan, T. Huser, S. Risbud, D. Krol, Appl. Phys. A **A76**(3), 367 (2003)
20. A. Streltsov, N. Borrelli, J. Opt. Soc. Am. B **19**(10), 2496 (2002)
21. A. Saliminia, R. Vallee, S.L. Chin, Opt. Commun. **256**(4-6), 422 (2005)
22. P. Dekker, M. Ams, G.D. Marshall, D.J. Little, M.J. Withford, Opt. Exp. **18**(4), 3274 (2010)
23. S. Kanehira, K. Miura, K. Hirao, Appl. Phys. Lett. **93**(2), 3 (2008)
24. C. Hnatovsky, R.S. Taylor, E. Simova, P.P. Rajeev, D.M. Rayner, V.R. Bhardwaj, P.B. Corkum, Appl. Phys. A **84**(1-2), 47 (2006)
25. L.P.R. Ramirez, M. Heinrich, S. Richter, F. Dreisow, R. Keil, A.V. Korovin, U. Peschel, S. Nolte, A. Tunnermann, in *Frontiers in Ultrafast Optics: Biomedical, Scientific, and Industrial Applications X*, vol. 7589 (SPIE, San Francisco, California, USA, 2010), vol. 7589, pp. 758,919–8
26. A. Marcinkevicius, S. Juodkazis, M. Watanabe, M. Miwa, S. Matsuo, H. Misawa, J. Nishii, Opt. Lett. **26** (2001)
27. Y. Bellouard, A. Said, M. Dugan, P. Bado, Opt. Exp. **12**(10), 2120 (2004)
28. V. Maselli, R. Osellame, G. Cerullo, R. Ramponi, P. Laporta, L. Magagnin, P.L. Cavallotti, Appl. Phys. Lett. **88**(19), 191107 (2006)
29. E. Glezer, M. Milosavljevic, L. Huang, R. Finlay, T.H. Her, T. Callan, E. Mazur, Opt. Lett. **21**(24), 2023 (1996)
30. S. Juodkazis, S. Matsuo, H. Misawa, V. Mizeikis, A. Marcinkevicius, H.B. Sun, Y. Tokuda, M. Takahashi, T. Yoko, J. Nishii, Appl. Surf. Sci. **197-198**, 705 (2002)
31. T.F. Johnston, Appl. Opt. **37**(21), 4840 (1998)
32. A. Marcinkevicius, V. Mizeikis, S. Juodkazis, S. Matsuo, H. Misawa, Appl. Phys. A. **76**(2), 257 (2003)
33. C. Hnatovsky, R.S. Taylor, E. Simova, V.R. Bhardwaj, D.M. Rayner, P.B. Corkum, J. Appl. Phys. **98**(1), 013517 (2005). 013517
34. S.M. Eaton, H. Zhang, M.L. Ng, J. Li, W.J. Chen, S. Ho, P.R. Herman, Opt. Exp. **16**(13), 9443 (2008)
35. R. Osellame, N. Chiodo, G. Della Valle, G. Cerullo, R. Ramponi, P. Laporta, A. Killi, U. Morgner, O. Svelto, IEEE J. Sel. Top. Quantum Electron. **12**(2), 277 (2006)
36. K. Minoshima, A. Kowalevicz, I. Hartl, E. Ippen, J. Fujimoto, Opt. Lett. **26**(19), 1516 (2001)
37. R. Osellame, N. Chiodo, V. Maselli, A. Yin, M. Zavelani-Rossi, G. Cerullo, P. Laporta, L. Aiello, S. De Nicola, P. Ferraro, A. Finizio, G. Pierattini, Opt. Exp. **13**(2), 612 (2005)
38. L. Sudrie, A. Couairon, M. Franco, B. Lamouroux, B. Prade, S. Tzortzakis, A. Mysyrowicz, Phys. Rev. Lett. **89**(18) (2002). 186601

39. J. Siegel, J.M. Fernandez-Navarro, A. Garcia-Navarro, V. Diez-Blanco, O. Sanz, J. Solis, F. Vega, J. Armengol, Appl. Phys. Lett. **86**(12) (2005). 121109
40. V. Ta'eed, N.J. Baker, L. Fu, K. Finsterbusch, M.R.E. Lamont, D.J. Moss, H.C. Nguyen, B.J. Eggleton, D.Y. Choi, S. Madden, B. Luther-Davies, Opt. Exp. **15**(15), 9205 (2007)
41. J. Burghoff, C. Grebing, S. Nolte, A. Tuennermann, Appl. Phys. Lett. **89**(8) (2006). 081108
42. A.H. Nejadmalayeri, P.R. Herman, J. Burghoff, M. Will, S. Nolte, A. Tunnermann, Opt. Lett. **30**(9), 964 (2005)
43. K. Yamada, W. Watanabe, T. Toma, K. Itoh, J. Nishii, Opt. Lett. **26**(1), 19 (2001)
44. C. Schaffer, J. Garcia, E. Mazur, Appl. Phys. A **A76**(3), 351 (2003)
45. S.M. Eaton, H. Zhang, P.R. Herman, F. Yoshino, L. Shah, J. Bovatsek, A.Y. Arai, Opt. Exp. **13**(12), 4708 (2005)
46. R. Osellame, N. Chiodo, G. Valle, S. Taccheo, R. Ramponi, G. Cerullo, A. Killi, U. Morgner, M. Lederer, D. Kopf, Opt. Lett. **29**(16), 1900 (2004)
47. P. Herman, A. Oettl, K. Chen, R. Marjoribanks, in *Proceedings of the SPIE - The International Society for Optical Engineering*, vol. 3616 (SPIE, San Jose, CA, USA, 1999), vol. 3616, pp. 148–55
48. H. Zhang, S.M. Eaton, J. Li, A.H. Nejadmalayeri, P.R. Herman, Opt. Exp. **15**(7), 4182 (2007)
49. T. Fukuda, S. Ishikawa, T. Fujii, K. Sakuma, H. Hosoya, Proceedings of the SPIE - The International Society for Optical Engineering **5279** (2004)
50. D.J. Little, M. Ams, P. Dekker, G.D. Marshall, J.M. Dawes, M.J. Withford, Opt. Exp. **16**(24), 20029 (2008)
51. W. Yang, P.G. Kazansky, Y. Shimotsuma, M. Sakakura, K. Miura, K. Hirao, Appl. Phys. Lett. **93**(17), 171109 (2008)
52. L. Shah, A. Arai, S. Eaton, P. Herman, Opt. Exp. **13**(6), 1999 (2005)
53. Y. Nasu, M. Kohtoku, Y. Hibino, Opt. Lett. **30**(7), 723 (2005)
54. R.R. Thomson, H.T. Bookey, N.D. Psaila, A. Fender, S. Campbell, W.N. MacPherson, J.S. Barton, D.T. Reid, A.K. Kar, Opt. Exp. **15**(18), 11691 (2007)
55. D.M. Rayner, A. Naumov, P.B. Corkum, Opt. Exp. **13**(9), 3208 (2005)
56. J. Burghoff, H. Hartung, S. Nolte, A. Tuennermann, Appl. Phys. A **86**(2), 165 (2007)
57. H. Zhang, S.M. Eaton, P.R. Herman, Opt. Exp. **14**(11), 4826 (2006)
58. S.M. Eaton, W. Chen, H. Zhang, R. Iyer, M.L. Ng, S. Ho, J. Li, J.S. Aitchison, P.R. Herman, IEEE J. Lightwave Technol. **27**(9) (2009)
59. R. Osellame, S. Taccheo, M. Marangoni, R. Ramponi, P. Laporta, D. Polli, S. De Silvestri, G. Cerullo, J. Opt. Soc. Am. B **20**(7), 1559 (2003)
60. M. Ams, G.D. Marshall, D.J. Spence, M.J. Withford, Opt. Exp. **13**(15), 5676 (2005)
61. R.R. Thomson, A.S. Bockelt, E. Ramsay, S. Beecher, A.H. Greenaway, A.K. Kar, D.T. Reid, Opt. Exp. **16**(17), 12786 (2008)

# Chapter 2
# Imaging of Plasma Dynamics for Controlled Micromachining

Jan Siegel and Javier Solis

**Abstract** Femtosecond (fs) microscopy constitutes a powerful tool for imaging laser-induced plasmas, assessing their dynamics (formation and decay), and spatial distribution, as well as quantifying their density. This experimental technique can be applied in different modalities to fs laser processing, revealing a wealth of information on the interaction mechanisms. For the case of surface processing, fs-resolved microscopy has the capability of enabling a link between the plasma properties and the induced material modifications. Fs microscopy inside dielectric materials is similarly a powerful tool for optimizing the structures produced. The technique unveils a number of complex interaction mechanisms taking place, including multiple beam filamentation (MBF), and pre-focal energy depletion that act as important energy loss channels, which deteriorate the spatial distribution of the deposited laser energy. Detailed studies performed with this technique show how these undesirable effects can be minimized by adjusting the processing parameters: pulse duration, energy, polarization, and processing depth. As a consequence, the energy is deposited more efficiently and confined to the focal volume region, leading to the production of structures with optimized performance.

## 2.1 Introduction

Nonlinear processing of dielectrics with femtosecond (fs) laser pulses has opened new routes for the production of functional structures for applications in fields like microfluidics and integrated optics. Yet, nonlinear propagation and other undesirable effects can severely compromise the efficient and localized deposition of laser energy in the desired volume. An effective control of the spatio-temporal

J. Siegel (✉) · J. Solis (✉)
Laser Processing Group, Instituto de Optica, CSIC, Serrano 121, 28006 Madrid, Spain
e-mail: j.siegel@io.cfmac.csic.es; j.solis@io.cfmac.csic.es

properties of the fs laser-induced plasmas and their density is thus crucial for producing optimal functional structures.

The chapter describes the use of fs microscopy in different modalities for imaging laser-induced plasmas in dielectrics. After a brief introduction, the first section is devoted to a short overview of the interaction mechanisms and their relevant time scales, time-resolved optical measurements and how they can be combined with spatial resolution, and the basic models required for the quantitative analysis of experimental data. The following sections are devoted to describe suitable experimental configurations and to explain their advantages and shortcomings. These sections are grouped in studies of surface and bulk interactions in dielectrics, accompanied in each case by results that illustrate the enormous potential of this temporally and spatially resolved technique for optimizing the processing conditions of dielectrics with ultrashort laser pulses.

## 2.2 Assessment of the Interaction of Ultrashort Pulses with Dielectrics Using Optical Probes

### 2.2.1 Interaction Mechanisms and Characteristic Time Scales

In order to describe the assessment of laser-induced plasmas with time-resolved techniques it is useful to consider first the mechanisms of energy deposition and relaxation, and the characteristic time scales involved. With the exception of direct excitation of lattice vibrations, all other absorption processes in defect-free dielectrics involve interactions with electronic states in the valence and conduction bands [1]. Typical timescales and processes occurring during and after irradiation with an ultrashort laser pulse are illustrated in Fig. 2.1 (Left) [2, 3], which illustrates the differences in interaction and transformation mechanisms involved in metals, semiconductors, and dielectrics. It is worth emphasizing in this context that the transitions in terms of bandgap from semiconductors to semi-insulators and then to dielectrics have no sharp limits.

A comprehensive description of laser-induced processes in dielectrics, including non-linear propagation effects, can be found in [2, 4, 5] and references quoted therein. In these materials, energy coupling involves *multi-photon absorption* (MPA), *free carrier absorption* (FCA), and *avalanche ionization* (AI) [6]. MPA at visible and near IR wavelengths involves the simultaneous absorption of multiple photons, which is possible due to the intense electric fields provided by ultrashort laser pulses. Tunneling ionization requires even stronger fields to deform the atomic potential, which is possible for pulses typically shorter than 10 fs [7]. Once a free electron is produced by MPA, it can absorb further energy through single photon absorption processes. Finally, its kinetic energy can be sufficient to exceed the bandgap energy and it can generate another free electron by impact ionization, resulting in two quasi-free electrons at the bottom of the conduction band. These

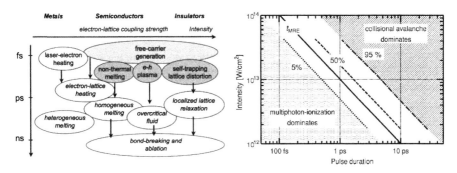

**Fig. 2.1** *Left* (taken from [2], adapted from [3]): typical timescales of processes occurring during and after irradiation of a solid with an ultrashort laser pulse of about 100 fs duration. *Right* (taken from [8]): percentiles of the fraction of impact-ionized electrons $n_{imp}/n_{total}$ as a function of laser intensity and pulse duration in a dielectric. The regions where either of the ionization processes is dominating are shaded. The transition time $t_{MRE}$ marks the transition between photoionization-dominated regime and avalanche-dominated regime

electrons can absorb again radiation by FCA leading to a free carrier avalanche. If a free carrier density $>10^{21}$ cm$^{-3}$ is reached already at the leading edge of the pulse, the material turns opaque to the incoming radiation and the remaining part of the incident pulse can be very efficiently absorbed.

The relative importance of the different electron excitation mechanisms (MPA and AI) obviously depends on the laser pulse duration and fluence and is still a subject of active discussion [8]. The controversy has in part a historical origin since in the mid-to-late nineties several works tried to explain the observed experimental deviation in the ablation threshold of fused silica from the expected $\tau^{1/2}$ scaling law for pulses shorter than 10 ps [6, 9, 10]. The temporal evolution of the free electron density is often described in terms of simple rate equations describing the total density of free electrons [6]. Recently, a model considering not only the density but also the energy of free electrons has been developed [11], showing that the contribution of impact ionization depends on the product of intensity and pulse duration, as illustrated in Fig. 2.1 (Right) [8].

After excitation, the laser-deposited energy is redistributed over different energy states of the system, mostly by carrier–carrier, carrier–phonon, and phonon–phonon interactions. The time required to reach a thermal energy distribution in a dielectric is typically in the order of $10^{-11}$ s, depending on the interaction strengths (i.e. electron–phonon coupling) and carrier density. It is worth noting though that in many dielectrics a number of metastable defects can be produced by laser irradiation [2]. In the particular case of fused silica, self-trapped excitons may be formed, with an electron trapping time as fast as 150 fs [12, 13]. Other defects leading to permanent coloration have been reported in borosilicate and alkali silicate glasses, which can be annihilated by thermal annealing [14]. These and other defects play an important role in the refractive index modification of the irradiated region and in diminishing the material damage threshold upon multiple laser pulse exposure.

## 2.2.2 Time-resolved Optical Techniques

The use of time-resolved optical measurements to investigate the structural evolution of solids under laser radiation is nearly as old as lasers are. The first transient reflectivity measurements during the irradiation of Si with a high power laser pulse date back to 1964 [15]. Afterward, the advent of ps and fs-pulsed laser sources paved the way for measurements of the material response to transient optical stimuli with unprecedented resolution [16]. Although *sensu stricto* time-resolved optical measurements cover a variety of techniques aimed at determining the evolution of the optical properties of the excited material, we will concentrate on so-called *pump–probe* techniques, given the time-scales involved in the case of non-linear processing of dielectrics. This kind of measurement has contributed in a crucial manner to our comprehension of the interaction of ultrashort laser pulses with matter. Still, most of the progress made in the development of sub-ns resolution optical techniques derives from the work performed in the early 80s in the field of laser-induced phase transitions in semiconductors [16–18].

In the most general situation, an energetic, main laser pulse excites the surface of a solid material while a weaker, secondary pulse of comparable duration probes an optical parameter of the excited region at a time delayed with respect to the arrival of the excitation pulse. The delay is normally controlled by an optical delay line that adjusts the time-of-flight difference of both pulses to reach the sample surface. This general scheme is illustrated in Fig. 2.2. The probe pulse can be, for instance, a laser pulse of the same or different wavelength [18], or a pulse of white light [19] while the detected signal can be the light reflected or transmitted by the sample, or the second harmonic signal generated at its surface [17]. For certain modalities it is even possible to determine the temporal evolution of the whole dielectric function, like in the case of two-angle reflectometry [19], although in the most common approaches, only information regarding the complex part of the refractive index ($k$) is accessible. In order to access the evolution of the real term ($n$), interferometric layouts have been successfully applied [4, 12, 13, 20]. In most configurations, the probe beam intensity is detected by means of a point detector (i.e. a photodiode) that time-integrates the signal over the duration of the probe pulse and spatially-integrates the material response over the region covered by the probe beam. We will see in next section that time-and-space resolved measurements provide a particularly powerful tool for the global assessment of the interaction.

Finally, it must be emphasized that only few special techniques, such as second harmonic generation and recently developed sub-ps X-rays and electron probes [21–23] yield direct information about the structural evolution of the material. In contrast, the vast majority of optical probes is essentially sensitive to the kinetics of electronic states, including phenomena that modulate the local electron density (or the dielectric function of the material) such as the excitation of coherent phonons [24, 25]. As we will show in Sects. 2.3 and 2.4, the sensitivity of optical probes to the presence of free electrons makes them particularly useful for analyzing the dynamics of excitation and relaxation of photo-generated carriers in dielectrics

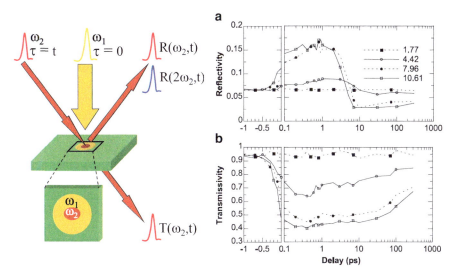

**Fig. 2.2** *Left*: scheme of an optical pump-probe experiment using a pump pulse at a frequency $\omega_1$ and a probe pulse at $\omega_2$, incident on the excited region at a delay $\tau = t$. The transient effect of the pump pulse can be detected, for instance, in the reflectivity $R(\omega_2)$ or the transmissivity $T(\omega_2)$ of the sample, or the second harmonic generation at its surface, denoted as $R(2\omega_2)$. The lower sketch illustrates the spatial overlap of the pump and probe pulses at the sample and the regions *excited* and *probed* by either pulse. *Right* (taken from [26]): time-resolved reflectivity and transmissivity of a probe laser pulse ($\lambda = 800$ nm) for different values of pump intensity (in $10^{13}$ W cm$^{-2}$) for fused silica with $\omega_1 = \omega_2$

[4, 12, 13]. An example of this is shown in Fig. 2.2, revealing a pronounced, ultrafast change in reflectivity and transmissivity in fused silica upon irradiation at fluences below and above the ablation threshold [26].

### 2.2.3 Exploiting Spatial Resolution

A source of error inherent to the pump-probe configuration is the relative size of pump and probe beams. In order to ensure true point-probing of a region exposed to a given peak fluence, the size of the probe spot has to be much smaller than the pump spot (typically <1:10). If this condition is not fulfilled, the measured transient optical change corresponds to a mean value that is averaged over a given pump intensity distribution. Since most fs lasers have a Gaussian-like rather than a top-hat intensity distribution the measured transient change is an average over regions exposed to different fluences. On the other hand, even if a spot size ratio of 1:10 is fulfilled, smallest misalignments of the probe beam unavoidably lead to a considerable error in the estimated fluence values.

An elegant solution to circumvent this experimental problem and add spatial resolution to the data recorded was first proposed by Downer and co-workers in

1985 [27]. They modified the basic pump-probe scheme shown in Fig. 2.2 by replacing the optical system that collects the reflected probe pulse with an optical microscope and the photodiode with a photographic film. The microscope is aligned such that the sample plane is imaged onto the film, recording an image of the irradiated region. In addition, the probe spot is now chosen to be much larger than the pump spot (typically >10:1) leading to a nearly homogeneous illumination of the region excited by the pump pulse. This configuration allows a parallel acquisition of transient optical changes for the entire pump spot intensity distribution. This technique, referred to in the following as fs microscopy, allows snapshots of the surface reflectivity to be recorded at different delay times after the pump pulse has reached the sample surface. The huge potential of this configuration, revealing a wealth of information hardly accessible with point-probing measurements, is already apparent from the data shown in their original paper showing results of plasma formation, melting, and ablation of silicon [27].

The work of von der Linde, Sokolowski-Tinten, and co-workers further refined the initial layout of this technique, not only replacing the recording medium (photographic film) by a modern charge coupled device (CCD). They also added new imaging modalities, which decisively contributed to our comprehension of ultrafast laser-induced ablation phenomena in semiconductors and metals [28]. Of particular interest was the discovery of Newton fringes in the images of the ablating surface (cf. Fig. 2.3, left) within a narrow fluence and delay time window. The fringes are caused by the interference of probe beam light reflected at the interfaces of a thin, semitransparent, and optically flat ablating liquid–gas layer with a high refractive index. Yet, the authors have applied fs microscopy only on a few occasions to the study of dielectric materials [29–31]. The characteristics of the ablation process in dielectrics were found to be essentially similar to the one observed in semiconductors and metals, except for two main differences (cf. Fig. 2.3, right). First, no Newton rings are observed, which can be understood by the much higher degree of ionization achieved in dielectrics, where MPA and impact ionization are the dominant free carrier generation mechanisms. Second, the sudden reflectivity

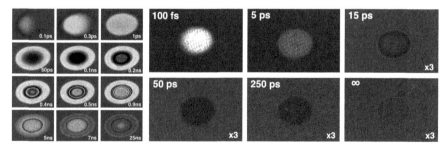

**Fig. 2.3** Taken from [31]. *Left set of frames*: surface of a Si wafer at different delay times after irradiation with a fs laser pulse. Frame size: 300 × 220 μm. *Right set of frames*: surface of BK7 glass after irradiation with a fs laser pulse. Frame size: 200 × 130 μm

change upon formation of free-electron plasma is much higher, due to the much lower initial reflectivity of dielectrics.

The recent work by our group using fs microscopy [32–34], including a variety of dielectric materials [35–37], has contributed to an improved comprehension of the phase transformation mechanisms and dynamics as well as to the discovery of new phenomena. Some results achieved within these studies and those reported by other groups are summarized and included in the following Sects. 2.3 and 2.4. An important tool used in these works was, along with time-resolved imaging, the use of numerical modeling to estimate the transient electron densities achieved at different stages of the interaction process. Basic models that can be used for quantifying the results obtained with time-resolved optical techniques are summarized in the next subsection.

## 2.2.4 Basic Models for Quantitative Analysis of Experimental Data

The Drude model is a classical approximation to describe the optical response of a free electron gas that can be used to estimate in a simple way the contribution of the photo-generated free electrons ($n_e$) to the dielectric function of a laser-excited dielectric. The complex Drude dielectric function of the excited material can be written as [38]:

$$\varepsilon = \varepsilon_m - \frac{n_e e^2 \tau}{\varepsilon_0 m \omega (i + \omega \tau)} \quad (2.1)$$

where $\varepsilon_m$ is the dielectric function of the material without free electrons (=square of its refractive index), $\varepsilon_0$ is the vacuum permittivity, $m$ the electron mass, $e$ the electron charge and $\omega$ the frequency of the incident light wave. The damping time $\tau$ represents the electron–electron scattering time that has been reported to scale inversely with $n_e$ [39] as $\tau = \tau_1 n_c / n_e$, yielding values <1 fs for dense plasmas [40]. The optical reflectivity of a material at normal incidence is then given by the Fresnel equation [41]:

$$R = \left| \frac{k_1 - k_2}{k_1 + k_2} \right|^2, \quad (2.2)$$

with the following expressions for the wave vectors $k_1$ and $k_2$:

$$k_1^2 = \omega^2 \mu \varepsilon = (n_1 \omega / c)^2, \quad k_2^2 = (\omega/c)^2 \left[ n_2^2 - \left( n_e e^2 / \omega^2 m \varepsilon_0 (1 + i/\omega \tau) \right) \right], \quad (2.3)$$

where $n_1$ is the refractive index of the medium of incidence, $n_2$ is the refractive index of the dielectric, $c$ the vacuum speed of light. At the so-called critical electron density

$$n_c = \varepsilon_0 \cdot m \cdot \omega^2 / e^2 \quad (2.4)$$

the plasma becomes absorbing and the reflectivity starts to increase. In the above equations, the value of $\omega$ to be considered depends on the effect to be analyzed. If one is considering, for instance, the effect of an increased reflectivity/absorption toward the trailing end of the pump pulse caused by MPA, FCA, and AI, $\omega$ would have the value corresponding to the pump laser beam [42]. Instead, it should refer to the probe light $\omega_{probe}$ if one is analyzing the transient reflectivity evolution of a plasma that has independently been produced using a pump laser pulse at $\omega_{pump}$. Throughout this chapter we will deal with the second case, i.e. $\omega = \omega_{probe}$. On the other hand, the transmittance at the plasma boundary can be found from the Fresnel formulas as [42,43] $T = 1 - R$. It is worth emphasizing that $T$ does not provide the overall transmittance of the laser-excited sample. Instead, $T$ provides the fraction of incident light that passes through the boundary of a plasma with a reflectivity $R$ and further propagates into the bulk. The skin depth of the plasma layer is given by

$$1/\alpha = 1/\text{Im}(2k_2), \tag{2.5}$$

with $\alpha(\omega)$ being the absorption coefficient of the Lambert–Beer law. The non-absorbed portion of pump-light will generate further free carriers that will be absorbing. For a full description of the transmission of the laser-excited sample, simple or multiple rate equations derived from kinetic theory [6, 8, 42] need to be then solved numerically. With a known surface electron density, the beam is then propagated into the bulk and the laser pulse intensity $I(t, z)$ recalculated at the new position, solving the rate equation again for the new, more attenuated intensity distribution [43]. Repeating this sequence iteratively, $n_e(z)$ as well as the total transmitted and reflected intensities are obtained. However, the exact expressions and terms to include in the rate equation and laser propagation are in many cases controversial and different approaches [8, 11, 38, 42] yield considerably different results.

Figure 2.4 displays the calculated results for reflectivity and skin depth at 400 nm and 800 nm wavelengths in fused silica, using an empirical factor $\tau_1 = 1$ fs to

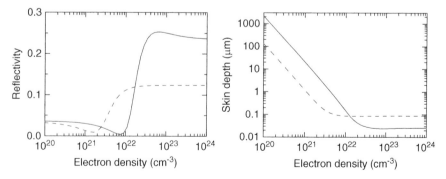

**Fig. 2.4** Reflectivity and skin depth at 400 nm (*solid*) and 800 nm (*dashed*) wavelengths as a function of the free-electron plasma density in fused silica, according to (2.1)–(2.5)

calculate the damping constant $\tau$. The reflectivity begins to increase (following an initial weak decrease) as the corresponding critical plasma frequency is approached and surpassed, accompanied by an enormous reduction of the skin depth over many orders of magnitude down to values of a few tens of nanometers. It emerges from the figure that, as expected from (2.4), infrared light is far more sensitive in detecting the presence of free-electron plasmas than short-wavelength light due to the corresponding shift of the critical electron density $n_c$ (cf. (2.4)). For the 800 nm fundamental radiation of commonly used Ti:Sa amplifiers $n_c = 1.7 \times 10^{21}$ cm$^{-3}$, whereas for a frequency-doubled radiation at 400 nm $n_c = 7.0 \times 10^{21}$ cm$^{-3}$. The influence of $\tau_1$ on the $n_e$-dependence of $R$ and skin depth is limited to the amplitude and does not shift the curves. As it will be shown in the next sections, a comparison of experimental data to results obtained using this simple model provides a means to estimate the maximum and transient electron density achieved under given irradiation conditions.

## 2.3 Ultrafast Imaging at the Surface of Dielectrics

### 2.3.1 Experimental Configurations and Constraints

As mentioned in Sect. 2.2.3, the experimental configurations for ultrafast time-resolved imaging of laser-induced plasmas at the surface are conceptually similar to those of standard, point-probing measurements. The main difference is that now the simple lens system collecting the probe beam and the photodiode are replaced by microscope optics in order to form an image of the surface, illuminated by the probe beam, on the chip of a CCD camera. The main requirements to fulfill are: (a) reasonably high numerical aperture of the objective lens to support good spatial resolution of small features; (b) optimum magnification of the region of interest onto the CCD chip; (c) sufficient density and number of pixels of the CCD chip to avoid reducing the spatial resolution attainable with the chosen objective lens. Further requirements for the CCD camera are a short exposure time (typically tens of microseconds) in order to discriminate against background light, as well as a high dynamic range and low read-out noise.

A typical configuration that can be rapidly switched between reflectivity and transmission modes is displayed in Fig. 2.5. It uses a loosely focused fs pump beam at 800 nm and 54° incidence combined with a 400 nm probe beam at normal incidence that has been obtained by frequency-doubling a fraction of the pump beam [36]. The combination of a polarizing beam splitter that reflects the probe light toward the sample and a quarter wave plate properly adjusted allows the beam reflected from the sample surface to be transmitted efficiently through the beamsplitter toward the CCD. Convenient discrimination against scattered pump light can be achieved by means of a narrow bandpass filter (Filter 3 in Fig. 2.5), which also strongly reduces the contribution from plasma emission [44] to the

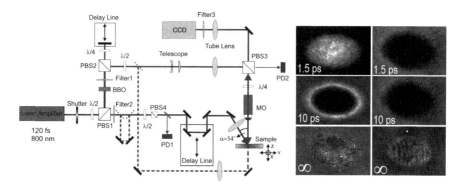

**Fig. 2.5** Taken from [37]. *Left*: femtosecond dual-mode pump-probe microscope. The reflection mode is drawn with solid lines, the transmission mode with dashed lines. A polarizing beamsplitter (PBS1) reflects a small fraction of the pump beam (800 nm, approximately 100 fs) to generate the probe beam, which is then frequency-doubled to 400 nm. *Right set of frames*: Time-resolved surface reflectivity (*left column*) and transmission (*right column*) images recorded for fused silica at different delay times (see labels) after exposure to a pump pulse with a peak fluence of 11.9 J cm$^{-2}$, above the ablation threshold

signal. While it is in principle possible to use the same wavelength for the pump and probe beams and discriminate their contributions using crossed polarizations, in practice this requires an amplified fs laser source free of any postpulse. Even very small postpulses (<0.1%) greatly disturb the image produced on the CCD camera because they probe the surface region while it is ablating and thus experience efficient scattering toward the collection optics. Scattering is accompanied by a loss of polarization, making the scattered postpulse to pass the barrier of the polarizing cube splitter and to contribute to the counts recorded by the CCD camera.

An additional peculiarity of the configuration shown in Fig. 2.5, along with the use of *two colors* for pumping and probing, is the existence of two optical delay lines, one in the pump arm that is used to provide a fine adjustment of the overall delay over 1 ns and a second one in the probe arm to extend the overall delay up to 20 ns. Such long delays are important for studying the dynamics of processes in the ablation regime and the propagation of shockwaves.

Dispersion of the probe and pump beams, as they pass through the multiple optical components of the set-up, compromises the optimal temporal resolution achievable. Dispersion leads to pulse broadening. This issue is particularly relevant for pulses shorter than 100 fs due to their large spectral bandwidth and for short-wavelength light, since dispersion in optical materials increases for decreasing wavelengths. If measurements with high temporal resolution are aimed at care should be taken to use thin optical elements with low dispersion. For instance, the dominant contribution that limits the temporal resolution in the setup of Fig. 2.5 comes from the probe arm, which is frequency doubled to 400 nm and passes through a total of approximately 40 mm glass. While this amount of glass has little effect on IR pulse (e.g. a transform-limited 100 fs pulse at 800 nm gets broadened to 110 fs), it considerably broadens a 400 nm pulse to 270 fs due to the higher material dispersion.

Representative time-resolved images recorded with the configuration shown in Fig. 2.5 are included on the right side of the figure, showing reflectivity and transmission at specific delay times upon laser-induced surface ablation in fused silica. The labels "∞" correspond to images recorded several seconds after the arrival of the pump laser pulse and therefore correspond to the final state of the sample surface. The images shown are obtained by dividing the corresponding single image recorded at a given delay by the image recorded, at the same sample position, before exposure to the pump laser pulse. Using this method, spatial inhomogeneities in the illumination field are efficiently reduced and the reflectivity and transmission values for each pixel are normalized to 1. Typically, between 20 and 40 images are recorded for each irradiation condition, scanning the delay time. It is advisable to use an approximately logarithmic spacing of the chosen delay times to account for the widely different time windows of the individual mechanisms monitored (cf. Fig. 2.1, left). Obviously, the exact selection of delay times should be adapted to the focus of interest of the study.

## 2.3.2 Transient Plasma Dynamics and Permanent Material Modifications

A quantitative analysis of the transient plasma dynamics can be performed in a straight-forward way from a full sequence of normalized time-resolved images, obtained as described in the previous section. Figure 2.6 shows the temporal evolution of reflectivity and transmission from a sequence of 28 images recorded in fused silica. In both cases, two different spatial positions of the images were analyzed. The center of the laser-irradiated region, corresponding to the peak laser fluence, exhibits an ultrafast rise in reflectivity and decay in transmission, consistent with the formation of a dense free-electron plasma at the sample surface. The maximum electron density $n_e$ of the plasma can be estimated using the Drude model described in Sect. 2.2.4 which, in this particular case, yields an approximate value $n_e \approx 3.3 \times 10^{22}$ cm$^{-3}$. Following the maximum, the reflectivity decays rapidly, reaching values below the one of the unexposed surface at $t = 10$ ps, which indicates the occurrence of surface ablation [31] lasting for a few ns before the reflectivity recovers. The corresponding recovery of transmission is very similar, as expected.

The second spatial position of the image sequence analyzed was chosen to lie outside the visible ablation crater determined from measurements at $t = \infty$ (dashed ellipse in the inset of Fig. 2.6). This position corresponds to a fluence of 6.4 J cm$^{-2}$. Remarkably, a plasma with an estimated density of $n_e \approx 1.6 \times 10^{22}$ cm$^{-3}$ is produced even below the visible ablation threshold. Using the Drude model, the minimum skin depth of the probe light at this fluence can be estimated to be ~60 nm, indicating the formation of a very dense and shallow plasma region. Since this particular spatial position lies outside the ablation crater, it is not affected by the ablation process and the corresponding plasma lifetime can be correctly measured. An approximate value

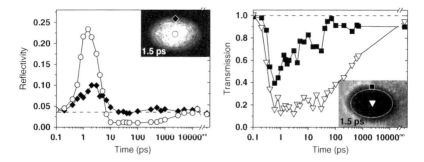

**Fig. 2.6** (Taken and modified from [37]): transient reflectivity and transmission at a probe wavelength of 400 nm as a function of delay time, extracted from specific positions of time-resolved images recorded during surface ablation of fused silica with single 800 nm, approximately 100 fs laser pulses. *Circles* and *triangles* correspond to the peak fluence in the center (11.9 J cm$^{-2}$), whereas diamonds and squares correspond to a spatial off-center position equivalent to local fluence of 6.4 J cm$^{-2}$. The spatial positions are shown in the *inset*, which includes also a *dashed ellipse* marking the extension of the visible ablation crater. The *horizontal dashed lines* indicate the initial reflectivity of fused silica at the probe wavelength and the normalized transmission

of 50 ps (extracted from the more sensitive transmission data) is obtained under these conditions of strong excitation.

The identification and characterization of this annular plasma regime, corresponding to a very narrow fluence window (6.4 ± 1.0 J cm$^{-2}$) just below the ablation threshold, is a good example of the enormous potential of fs-resolved microscopy. Using a point-probing configuration and measuring at discrete fluences this regime is likely to pass unnoticed, as can be seen in Fig. 2.2, right [26]. In contrast, the microscopy approach provides access to the continuous fluence range underneath the Gaussian intensity distribution of the pump pulse.

In addition to the detailed analysis of the plasma dynamics as a function of laser fluence, a correlation between transient plasma evolution and permanent material modifications can be established in a straightforward fashion, performing complementary imaging measurements of certain properties of the irradiated spot. Of particular interest are in this context topographical, morphological, structural, and optical properties. Suitable techniques for providing 2D maps of these properties include atomic force microscopy (AFM), white light interferometry (WLI), scanning electron microscopy (SEM), and white light microscopy (WLM). All of these techniques have comparable or superior spatial resolution than fs-resolved microscopy and the complementary data they provide enable a better interpretation of the results. As an example, Fig. 2.7 (left) shows a single time-resolved image of the surface plasma recorded in fused silica at $t = 10$ ps in comparison with steady-state images of different properties recorded after irradiation of the same spot, showing the same field of view.

One of them has been recorded with WLM using a standard optical microscope with a numerical aperture (NA = 0.85) higher than that used in the fs microscopy configuration (NA = 0.42), and thus providing superior spatial

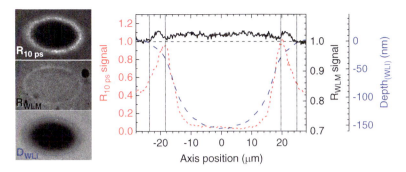

**Fig. 2.7** (Taken and modified from [36]). *Left*: transient reflectivity image at $t = 10$ ps ($R_{10ps}$), static white light reflection image ($R_{WLM}$) and surface topography from white light interferometry ($D_{WLI}$) of the same irradiated region in fused silica. *Right*: horizontal cross sections through the $R_{10ps}$ image (*dotted line*), $R_{WLM}$ image (*solid line*) and $D_{WLI}$ image (*dashed line*)

resolution. In addition, the standard microscope provides a better signal-to-noise ratio. The recorded image has been contrast-adjusted to reveal the presence of an annular feature of slightly increased reflectivity, which matches very well the spatial position of the transient plasma ring in $R_{10ps}$, surrounding the crater. The amount of relative increase ($\Delta R \approx 3\%$) is consistent with a refractive index increase of $\Delta n/n \approx 0.01$, as reported to occur upon fs laser irradiation inside bulk fused silica; an effect that can be exploited for the production of optical waveguides [45]. The left part of Fig. 2.7 also includes an image of the surface topography after irradiation, ($D_{WLI}$), showing a central dark region corresponding to the visible ablation crater, which matches the spatial extension of the dark ablating region seen in the time-resolved image $R_{10ps}$. In addition to the main crater seen in $D_{WLI}$, a surrounding light grey ring, consistent with a subtle surface depression, can be observed with a spatial extension comparable to the transient plasma in $R_{10ps}$ and to the increased reflectivity region in $R_{WLM}$. The good spatial overlap of the three annular features in the different images can be clearly seen in the right part of Fig. 2.7, showing the corresponding cross sections through the center of the images. The match between reflectivity increase and surface depression further supports the above-made assumption that a refractive index change is the underlying reason for the observed reflectivity increase, since a local material densification would lead to a depression of the surface. The results are thus consistent with a plasma-induced densification of the material below the visible ablation threshold.

The presence of an annular highly-reflective transient plasma region observed at a delay of typically 10 ps at peak fluences below the ablation threshold might be a general feature in dielectric materials. In particular we have found this phenomenon to occur in all the dielectrics we have studied so far, covering a broad range of optical band gaps $E_g$ and refractive indices $n$ (given for 800 nm); fused silica ($E_g = 7.2$ eV, $n = 1.45$), sapphire ($E_g = 9.9$ eV, $n = 1.76$), phosphate glass ($E_g = 3.6$ eV, $n = 1.56$), and SF57 ($E_g = 3.2$ eV, $n = 1.85$). However, the temporal and spatial evolution of the plasma produced upon irradiation differs noticeably and

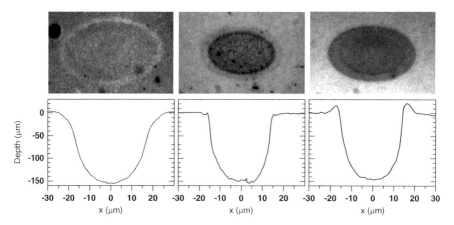

**Fig. 2.8** Static white light reflection images ($R_{WLM}$) and surface topography profiles from white light interferometry ($D_{WLI}$) of laser-irradiated regions in fused silica (*left*, 11.9 J cm$^{-2}$), sapphire (*middle*, 12.5 J cm$^{-2}$) and phosphate glass (*right*, 5.2 J cm$^{-2}$)

so does the type of material modification induced. Figure 2.8 shows, as an example, the comparison of steady-state reflection images ($R_{WLM}$) and topography profiles obtained upon irradiation of fused silica, sapphire, and phosphate glass with single laser pulses (800 nm, ∼100 fs).

The fluence of the laser pulse was chosen for each material such as to achieve comparable crater depths. Despite this common feature, the crater profile differs markedly, ranging from smooth walls for fused silica to steep walls for sapphire and phosphate glass, the latter displaying in addition an elevated crater rim. Complementary information is provided by the optical changes observed in the $R_{WLM}$ images, ranging from a featureless crater surrounded by a ring of increased reflectivity for fused silica, a crater with signs of fracture surrounded by a ring of slightly reduced reflectivity for sapphire and a dark featureless crater surrounded by a multiple ring structure for phosphate glass. This comparison emphasizes the strong influence of material parameters on the structural, topographical, and optical changes induced under the same irradiation conditions.

## 2.4 Ultrafast Imaging in the Bulk of Dielectrics

### 2.4.1 Experimental Configurations and Constraints

Femtosecond microscopy at the surface of dielectrics provides valuable information regarding plasma dynamics and its relation to the final state of the irradiated region. Yet, when fs-laser structuring is used for the production of functional structures for microfluidics [46, 47] or integrated optics [45, 48, 49] the interaction of the laser

## 2 Imaging of Plasma Dynamics for Controlled Micromachining

pulse with the material occurs in the bulk and is conditioned not only by non-linear absorption but also by other effects that greatly condition the efficiency of the energy deposition process and the shape/size of the transformed volume [50]. Spherical aberration [51–54], non-linear propagation [55–57], and prefocal depletion effects [58] will have a deep impact on the characteristics and spatial extension of the induced transformations. In this context, information obtained from time-resolved imaging of the accompanying transient plasma distribution provides a powerful means for optimizing the processing conditions. For such a purpose, the setup shown in Fig. 2.5 needs to be modified in order to enable the acquisition of time-resolved images of the focal region.

A suitable configuration can be seen Fig. 2.9, where the probe beam illuminates the interaction zone at normal incidence with respect to the pump beam. This configuration is often referred to as shadowgraphy and has the advantage of imaging the laser-induced plasma along its longitudinal dimension. As a consequence, it enables a determination of the actual shape and size of the focal volume, which can be compared to the one expected for focusing in air, thus visualizing the impact and relative contribution of the effects mentioned above, causing distortions of the focal volume. Moreover, it enables studying the temporal evolution of the photo-induced free-electron plasma and estimating the electron density from a comparison of

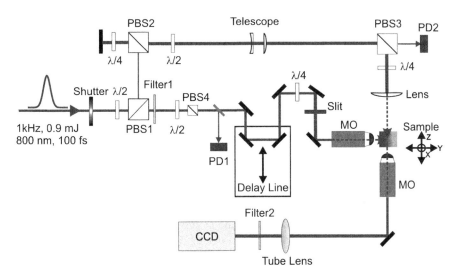

**Fig. 2.9** (Taken from [50]). Femtosecond pump–probe bulk microscopy in transmission mode. The focal region of the pump beam (800 nm, 100 fs) is illuminated by a 800 nm probe beam at an angle of incidence of 90°. The light transmitted by the sample is detected by a CCD camera after passing through a narrow band-pass filter (Filter 2) in order to remove the plasma emission contribution. If the probe beam is blocked and the band-pass filter removed, the signal corresponds to the time-integrated plasma emission image of the focal region. The variable slit in front of the focusing optics enables focal volume shaping

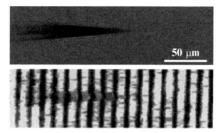

**Fig. 2.10** (Adapted from [59]). *Top image*: time-resolved transmission image of a fs laser-induced plasma in fused silica at a delay of 400 fs. The processing depth is several hundreds of microns. The laser pulse is incident from the left. *Bottom image*: interferometric image at the same delay showing the characteristic interference fringes and their shift in the focal region caused by the presence of a free-electron plasma

transmission measurements with calculations of the plasma skin penetration depth (see Sect. 2.2.4).

With a few additional minor changes the configuration shown in Fig. 2.9 can be used to probe the sample transmission at the second harmonic wavelength of the pump beam (see Fig. 2.5) [35], or to acquire time-integrated plasma emission images [50]. In addition, the extension to time-resolved interferometry measurements is feasible, providing in principle access to the evolution of the real part of the refractive index of the laser-excited region via phase shift measurements. This can be done, for instance, by using a double probe pulse and a spectrometer [4, 12]. With this technique it has been possible to analyze in great detail the dynamics of self-trapped excitons in fused silica irradiated with fs laser pulses, although one spatial dimension has to be traded in for acquiring the spectral dimension.

In a somewhat simpler approach, maintaining both spatial dimensions, interferometric transmission images can be acquired by inserting a Wollaston prism (with its optical axis at 45° with respect to the pump beam polarization axis) and an analyzer between the imaging lens and the CCD camera in the scheme of Fig. 2.9. Illustrative examples of images acquired in such a configuration are shown in Fig. 2.10. The shift observed in the interference fringes at the focal region can be correlated to the local plasma density and, making use of the Drude model, related to the electron–electron collision time ($\tau$) [59]. The so-obtained values for $\tau$ must be taken, though, with caution since $\tau$ itself is energy-dependent. For instance, in the case of fused silica, different works using different techniques and pulse peak powers have reported values ranging from less than one to tens of fs. Alternatively, phase contrast microscopy also provides valuable information regarding the evolution of the properties of the irradiated region, such as the time-resolved changes of the refractive index, as reported recently in fused silica [60].

It must be emphasized that ultrafast imaging in the bulk of dielectrics suffers from the same limitations as surface imaging in terms of dispersion experienced by the probe pulse in the optical components of the set-up. As above-indicated, these elements can cause substantial temporal spreading of the probe pulse, compromising the time resolution achievable. Still, due to the transverse incidence and bulk irradiation configuration, scattering of the pump light is reduced, which allows performing transmission measurements using the same IR wavelength for pump and probe beams. This substantially reduces the impact of dispersion when compared to measurements using a probe beam at the second harmonic of the pump frequency. Moreover, the use of infrared probes has the additional advantage of increasing the sensitivity for the detection of low-density plasmas compared to visible probes (cf. Fig. 2.4).

## 2.4.2 Transient Plasma Dynamics in Glasses Under Waveguide Writing Conditions: Role of Pulse Duration, Energy, Polarization, and Processing Depth

While the above-described methods and works have contributed strongly to improve the comprehension of fundamental interaction processes inside dielectrics, the focusing geometry used there is not optimal for subsurface laser processing. The reason is that in conventional focusing geometries the focal volume is elongated along the propagation direction of the pump beam. For low NA values of the focusing lens this is caused by the different dependences of the confocal parameter and transverse spot size on the NA. In contrast, for high NA values elongation is caused by spherical aberration. Different solutions to this problem, making use of beam shaping techniques, have been proposed and successfully implemented. They are based on the use of elliptical or astigmatic elliptical beams produced either with cylindrical telescopes [61] or slits [62] inserted into the beam path, as described in Chap. 5. In that way the focus is expanded transversely to the beam propagation axis, producing a disk-shaped focal volume, which allows the writing of transversal structures with circular cross-sections. This technique has been successfully used to produce waveguides with circular cross-sections in a variety of materials and has been shown to be robust against the influence of spherical aberration due to the compatibility with low NA optics, thus enabling the writing of circular cross-section structures at depths beyond 7 mm [54].

In order to reproduce the conditions used during waveguide writing in glasses using elliptical beams and low repetition rate fs laser pulses, the setup shown in Fig. 2.9 includes a slit to shape the focal volume. Typical parameters in this case are 1 kHz repetition rate for the fs amplifier and $\sim 100\,\mu\text{m s}^{-1}$ for the scanning speed of the sample along the $X$-axis, as reported in [54, 62, 63]. A sketch of the relevant writing beam intensity cross sections for the slit-shaped beam inside the material is shown in Fig. 2.11. For the indicated beam dimensions, focusing optics, and scanning speed the dose is 10 pulses per $\mu$m and thus the focal volume receives an average of 20 pulses during the writing procedure. In order to approximately match this condition for the acquisition of time-resolved images, the unexposed

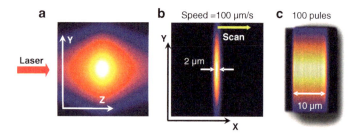

**Fig. 2.11** (**a** and **b**) Sketch of the calculated cross sections of the beam intensity distribution in the focal region for a beam shaped with a slit. The sample is scanned along the $X$-direction. The propagation axis is taken as $Z$ while the slit short axis is taken as $Y$. The beam (800 nm, 1 kHz rep. rate, 7.4 mm ($1/e^2$) diameter) passes through a 350 µm slit before being focused with a 0.26 NA lens. The material refractive index used in the calculation is $n = 1.56$ and the processing depth is $d = 1.56$ mm. Intensity cross sections are shown along the $Y - Z$ plane (**a**) and $Y - X$ (**b**) plane. During the writing procedure the sample scanning speed is 100 µm s$^{-1}$. (**c**) Shows schematically the $Y - X$ cross section of the volume affected by 100 pulses during the scan

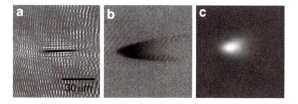

**Fig. 2.12** (Taken from [50]). Time-resolved transmission image of the spatial plasma distribution recorded in fused silica upon irradiation with a 100 fs laser pulse (incident from the left) focused with a 0.26 NA lens at a processing depth of 1.45 mm for a delay time of $t = 0.2$ ps without the slit (**a**) and with a 0.35 mm slit (**b**) and pulse energies of 0.25 and 6 µJ, respectively. (**c**) Static white light trans-illumination image of a femtosecond laser-written waveguide under similar conditions as in (**b**)

sample was pre-irradiated with 19 pulses before an image corresponding to the 20th pulse was recorded at a given time delay.

The effect of the slit on the shape of the laser-induced plasma inside fused silica can be clearly appreciated in Fig. 2.12, corresponding to transmission images acquired without (a) and with (b) slit for a delay time of 200 fs. In (a) the plasma distribution (dark region) shows the expected elongated cross section due to both the low NA of the focusing optics and a small contribution of spherical aberration at the corresponding processing depth [54]. The image corresponding to a slit width of 350 µm in (b) shows that the laser-induced plasma is now stretched perpendicularly to the propagation direction. When comparing this image to the calculated intensity distribution in Fig. 2.11(a), it is clear that self-focusing (SF) effects are clearly affecting the energy deposition profile. This effect is also evident in the corresponding trans-illumination image 2.12(c) of the waveguide written with the same parameters.

2 Imaging of Plasma Dynamics for Controlled Micromachining 37

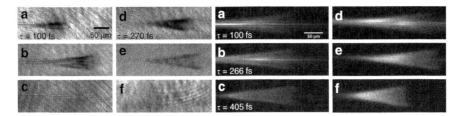

**Fig. 2.13** (Taken from [50]). *Left two columns*: time-resolved transmission images recorded in bulk Er:Yb codoped phosphate glass at 1.56 mm depth with circularly polarized transform limited, 100 fs pump laser pulses of 8 μJ energy [(**a**)–(**c**)] and stretched pulses (270 fs) [(**d**)–(**f**)] at several time delays $t = 0.65$ ps in (**a**), (**d**), 1.2 ps in (**b**), (**e**), and 50 ps in (**c**), (**f**). In all images the laser beam impinges from the left. *Right two columns*: time-integrated plasma emission images recorded for various pump pulse durations $\tau_{pump}$ under the same experimental conditions, i.e., pump energy, processing depth. In (**a**)–(**c**) plasma emission images for horizontal polarization are shown, whereas (**d**)–(**f**) show plasma emission images for circularly polarized laser light with the same values of $\tau_{pump}$ as in (**a**)–(**c**)

These images clearly demonstrate the sensitivity of optical transmittance to the modification of the focal volume shape and how a given spatial plasma distribution is translated into a corresponding region of modified refractive index. The asymmetry of the plasma region and the region of increased refractive index, resembling an arrow head rather than a disk, is caused by a variety of non-linear effects, including self-focusing and prefocal depletion, as discussed in the following paragraphs.

Figure 2.13 (left two columns) shows several illustrative images recorded in a different material, Er:Yb co-doped phosphate glass, irradiated with fs laser pulses at several delay times [50]. The free-electron plasma produced (dark region) shows the characteristic arrow head shape also observed in fused silica, indicating the presence of non-linear SF. The maximum electron density of these plasmas has been quantified approximately by estimating their skin depth and comparing it to the values of $(1/\alpha)$ calculated as a function of $n_e$ (see Sect. 2.2.4). An approximate value of the skin depth can be obtained from the transmittance values determined from time-resolved images and an estimation of the thickness of the plasma disk using the calculated (linear) beam intensity distribution in the focal region [54] (see Fig. 2.11 (b)). The obtained maximum values are in the order of $n_e \approx 3 \times 10^{20}$ cm$^{-3}$, even smaller than a previous estimate ($\sim 2 \times 10^{21}$ cm$^{-3}$, [50]). The plasma densities observed in Fig. 2.13 (Left) are thus below the critical plasma density at 800 nm ($1.7 \times 10^{21}$ cm$^{-3}$).

The transmission images in Fig. 2.13 (Left two columns) also evidence the appearance of multiple beam filamentation (MBF) in the prefocal region. The position and number of filaments induced by different pulses is quite reproducible as can be seen in the images corresponding to different temporal delays. They can be already observed at the earliest time delays and their effect on the transient transmission changes can persist up to 1 ns (not shown). MBF is a direct consequence of the input beam aspect ratio (quasi-elliptical) [64] in presence of non-linear self focusing and plasma self-defocusing. The material response upon irradiation with a longer (270 fs) pulse shows two important features: the plasma region is now more absorbing (darker) and spatially broader, and MBF is suppressed. Both effects, caused by the reduced peak power of the pulse, indicate improved energy coupling of the laser energy to the focal region. Prefocal depletion of the beam energy [58] (appearance of non-linear absorption before the beam reaches the focus) as well as formation of multiple filaments clearly act as important energy loss channels that limit the energy available for inducing structural modifications in the focal volume.

Since SF and nonlinear MBF are inherently bound to nonlinear propagation in a Kerr medium, a substantial change in the observed plasma behavior occurs when the pulse energy or the processing depth are reduced. Indeed, filamentation is strongly enhanced at larger processing depths. Similarly, laser polarization influences the MBF process by modifying the critical SF power via the polarization-dependent non-linear refractive index $n_2$. This can be seen in the plasma emission images for both polarization states of the pump pulse and increasing pump pulse durations in Fig. 2.13 (right two columns). Although plasma emission allows only accessing the time-integrated behavior of the plasma spatial distribution, the figure clearly shows that MBF effects are minimized by using circular polarization ($n_2$ decreases by a factor of 1.5) and nearly suppressed, for the specified pulse energy and processing depth, for laser pulses longer than $\sim$400 fs.

The importance of laser polarization in multi-photon ionization mechanisms in dielectrics has been recognized earlier [65]. Indeed, waveguides written with circularly polarized laser pulses show lower propagation losses than those produced with linearly polarized light [66]. This behavior was attributed to the suppression of periodic nanostructures (cracks – refer to Chap. 14) often observed in glasses upon irradiation with linearly polarized fs lasers. Our findings suggest that circular polarization not only avoids the generation of nanofractures but also minimizes prefocal depletion and MBF effects, and improves the spatial distribution of the transient plasma, which directly leads to a better quality of the laser-written structures. These findings are of fundamental relevance to many applications where fs laser pulses are employed for processing of dielectrics.

## 2.5 Outlook and Conclusions

An illustrative example of how fs-microscopy enables optimizing the performance of fs-laser written functional structures is shown in Fig. 2.14. The left pair of images corresponds to a trans-illumination image (a) and a guided mode (b) of a waveguide written in Er:Yd-doped phosphate glass in conventional conditions (100 fs, linear polarization), similar to the ones of Fig. 2.13 [right, (a)]. It can clearly be seen that the region where the material is finally transformed is essentially the same as the region corresponding to plasma emission, with a length ($\sim$200 $\mu$m) extending far beyond the confocal parameter of the beam for the corresponding focusing optics. The pulse energy is spread over a stretched region caused by non-linear self focusing and beam filamentation, leading to an important amount of non-linear absorption in the prefocal region (see the arrow-head black region in Fig. 2.14(a)).

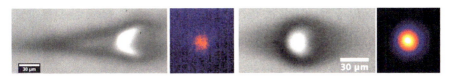

**Fig. 2.14** (Taken and adapted from [67]). Static white light trans-illumination images (**a**, **c**) of femtosecond laser-written waveguides in Er:Yb co-doped phosphate glass at a depth of 1.56 mm using a 1 kHz repetition rate and slit shaping (slit width = 350 µm). (**a**) Pulse duration = 100 fs, pulse energy = 6.4 µJ, linear polarization; (**c**) 250 fs, 4.0 µJ, circular polarization. Near field images of the corresponding guided modes at 1550 nm are shown in (**b**, **d**)

These effects are enhanced by the use of linearly polarized light and a relative large processing depth, even when spherical aberration effects are essentially negligible. The near field image of the corresponding guided mode at 1550 nm shows poor guiding performance with propagation losses of several dB cm$^{-1}$. From the study on time-resolved and time-integrated plasma images using fs- microscopy it was possible to optimize the processing parameters at the same processing depth, minimizing undesirable non-linear propagation effects using circular polarization, a longer pulse (250 fs) and a smaller pulse energy (4 µJ). Under these conditions, SF is strongly reduced and beam filamentation effects are totally suppressed. The lower pulse energy makes it necessary to write the waveduide using several scans to improve the refractive index change at the focal region. The result is shown Fig. 2.14(c and d) with the corresponding transillumination and guided mode images. The transformed zone has a well-defined circular symmetry with a refractive index increase in the bright region of $\Delta n = 1.5 \times 10^{-3}$ and losses well below 1 dB cm$^{-1}$ [67].

As a general conclusion it can be said that fs microscopy in dielectrics is capable of providing a detailed and quantitative description of spatio-temporal properties of laser-induced transient plasmas under real processing conditions. For the case of surface imaging, it has been possible to correlate the appearance of a transient plasma region below the ablation threshold of fused silica to a local increase of refractive index, comparable to that obtained upon subsurface writing of waveguides. Plasma observations in the bulk of dielectric allow identifying beam filamentation and prefocal depletion as important energy loss channels which deteriorate the spatial distribution of the laser-deposited energy. In the near future, controlled micromachining assisted by functional imaging of laser-induced plasmas will enable the in-situ, real-time optimization of functional structures produced by fs-laser writing.

**Acknowledgements** This chapter would have not been possible without the enthusiastic collaboration of current and former members of the "ultrafast team" of the Laser Processing Group (LPG) at the Instituto de Óptica-CSIC, namely (in alphabetical order) Guillaume Bachelier, Jörn Bonse, Victor Diez-Blanco, Hector Fernández, Andrés Ferrer, Marcial Galvan-Sosa, Wojciech Gawelda, Jose Gonzalo, Daniel Puerto, and Alejandro Ruiz, joining us in the quest for controlled micromachining assisted by functional imaging of laser-induced plasmas. We wish to acknowledge the fruitful collaboration with the other members of the LPG, in particular Carmen Nieves Afonso and Jose Maria Fernandez Navarro, and also with Fidel Vega and Jesus Armengol at Universitat

Politècnica de Catalunya. The work of the LPG was partially supported by the Spanish Ministry of Science and Innovation under TEC 2005–00074, and 2008–01183 projects and by the EU under TMR project "FLASH" (MRTNCT-2003–503641).

# References

1. M. von Allmen, A. Blatter, *Laser-Beam Interactions with Materials: Physical Principles and Applications, Springer Series in Materials Science*, vol 2 (Springer, Berlin, 1998)
2. H. Misawa, S. Juodkazis (eds.), *3D Laser Microfabrication. Principles and Applications*, ed. by R.F. Haglund, Jr., *Photophysics and Photochemistry of Ultrafast Laser Materials Processing* (Wiley, KGaA, Weinheim, 2006)
3. B. Rethfeld, K. Sokolowski-Tinten, D. von der Linde, S.I. Anisimov, Appl. Phys. A **79**, 767 (2004)
4. S.S. Mao, F. Quéré, S. Guizard, X. Mao, R.E. Russo, G. Petite, P. Martin, Appl. Phys. A **79**, 1695 (2004)
5. S.W. Winkler, I.M. Burakov, R. Stoian, N.M Bulgakova, A. Husakou, A. Mermillod-Blondin, A. Rosenfeld, D. Ashkenasi, I.V. Hertel, Appl. Phys. A **84**, 413 (2006)
6. B.C. Stuart, M.D. Feit, A.M. Rubenchik, B.W. Shore, M.D. Perry, Phys. Rev. Lett. **74**, 2248 (1995)
7. L.V. Keldysh, Sov. Phys. JETP **20**, 1307 (1965)
8. B. Rethfeld, Phys. Rev. B **73**, 035101 (2006)
9. D. Du, X. Liu, G. Korn, J. Squire, G. Mourou, Appl. Phys. Lett. **64**, 3071 (1994)
10. M. Lenzner, J. Krüger, S. Sartania, Z. Cheng, Ch. Spielmann, G. Mourou, W. Kautek, F. Krausz, Phys. Rev. Lett. **80**, 4076 (1998)
11. B. Rethfeld, Phys. Rev. Lett. **92**, 187401 (2004)
12. P. Martin, S. Guizard, Ph. Daguzan, G. Petite, P. D'Oliveira, P. Meynadier, M. Perdrix, Phys. Rev. B **55**, 5799 (1997)
13. G. Petite, P. Daguzan, S. Guizard, P. Martin, J. Phys. III **6**, 1647 (1996)
14. O.M. Efimov, K. Gabel, S.V. Garnow et al., J. Opt. Soc. Am. B **15**, 193 (1998)
15. W.R. Sooy, M. Geller, D.P. Bortfeld, Appl. Phys. Lett. **5**, 54 (1964)
16. C.V. Shank, Science **219**, 1027 (1983)
17. C.V. Shank, R. Yen, C. Hirlimann, Phys. Rev. Lett. **51**, 900 (1983)
18. C.V. Shank, R. Yen, C. Hirlimann, Phys. Rev. Lett. **50**, 454 (1983)
19. L. Huang, J.P. Callan, E.N. Glezer, E. Mazur, Phys. Rev. Lett. **80** 185 (1997)
20. V.V. Temnov, K. Sokolowski-Tinten, P. Zhou, D. von der Linde, J. Opt. Soc. Am. B **23**, 1954 (2006)
21. A. Rousse et al., Nature **410**, 65 (2001)
22. A.M. Lindenberg et al., Science **308**, 392 (2005)
23. M.S. Grinolds, V.A. Lobastov, J. Weissenrieder, A.H. Zewail, Proc. Natl. Acad. Sci. U.S.A. **103**, 18427 (2006)
24. H.J. Zeiger, J. Vidal, T.K. Cheng, E.P. Ippen, G. Dresselhaus, M.S. Dresselhaus, Phys. Rev. B **45**, 768 (1992)
25. D.M. Fritz et al., Science **315**, 633 (2007)
26. I.H. Chowdhury, A.Q. Wu, X. Xu, A.M. Weiner, Appl. Phys. A **81**, 1627 (2005)
27. M.C. Downer, R.L. Fork, C.V. Shank, J. Opt. Soc. Am. B **4**, 595 (1985)
28. K. Sokolowski-Tinten, J. Bialkowski, A. Cavalleri, D. von der Linde, A. Oparin, J. Meyer-ter-Vehn, S.I. Anisimov, Phys. Rev. Lett. **81**, 224 (1998)
29. D. von der Linde, H. Schüler, J. Opt. Soc. Am. B **13**(1), 2216 (1996)
30. D. von der Linde, K. Sokolowski-Tinten, J. Bialkowski, Appl. Surf. Sci. **109/110**, 1–10 (1997)
31. K. Sokolowski-Tinten, J. Bialkowski, SPIE Proc. **3343**, 46–57 (1998)
32. J. Bonse, G. Bachelier, J. Siegel, J. Solis, Phys. Rev. B **74**, 134106 (2006)
33. J. Bonse, G. Bachelier, J. Siegel, J. Solis, H. Sturm, J. Appl. Phys. **103**, 054910 (2008)

34. J. Siegel, W. Gawelda, D. Puerto, C. Dorronsoro, J. Solis, C.N. Afonso, J.C.G. de Sande, R. Bez, A. Pirovano, C. Wiemer, J. Appl. Phys. **103**, 023516 (2008)
35. J. Siegel, D. Puerto, W. Gawelda, G. Bachelier, J. Solis, L. Ehrentraut, J. Bonse, Appl. Phys. Lett. **91**, 082902 (2007)
36. D. Puerto, W. Gawelda, J. Siegel, J. Solis, J. Bonse, Appl. Phys. Lett. **92**, 219901 (2008)
37. D. Puerto, W. Gawelda, J. Siegel, J. Bonse, G. Bachelier, J. Solis, Appl. Phys. A: Mater. Sci. Process. **92**, 803 (2008)
38. A.Q. Wu, I.H. Chowdhury, X. Xu, Phys. Rev. B **72**, 085128 (2005)
39. C. Quoix, G. Hamoniaux, A. Antonetti, J.-C. Gauthier, J.-P. Geindre, P. Audebert, J. Quant. Spectr. Rad. Transf. **65**, 455 (2000)
40. N. Bloembergen, IEEE J. Quant. Electron. **10**, 375 (1974)
41. E. Hecht, *Optics*, 4th edn. (Addison-Wesley, Boston, 2002)
42. B.C. Stuart, M.D. Feit, S. Herman, A.M. Rubenchik, B.W. Shore, M.D. Perry, Phys. Rev. B **53**, 1749 (1996)
43. M.D. Feit, A.M. Komashko, A. M. Rubenchik, Appl. Phys. A **79**, 1657 (2004)
44. C.W. Carr, M.D. Feit, A.M. Rubenchik, P. De Mange, S.O. Kucheyev, M.D. Shirk, H.B. Radousky, S.G. Demos, Opt. Lett. **30**, 661 (2005)
45. K.M. Davis, K. Miura, N. Sugimoto, K. Hirao, Opt. Lett. **21**, 1729 (1996)
46. K. Sugioka, Y. Cheng, K. Midorikawa, Appl. Phys. A **81**, 1–10 (2005)
47. R.M. Vazquez, R. Osellame, D. Nolli, C. Dongre, H. van den Vlekkert, R. Ramponi, M. Pollnau, G. Cerullo Source, Lab Chip. **9**, 91–96 (2009)
48. K. Hirao, T. Mitsuyu, J. Si, J. Qiu, *Active Glass For Photonic Applications: Photoinduced Structures and Their Application* (Springer, Berlín, 2001)
49. H. Misawa, S. Juodkazis (ed.), in *3D Laser Microfabrication. Principles and Applications*, ed. by V. Mizeikis, S. Matsuo, S. Juodkazis, H. Misawa. Femtosecond Laser Microfabrication of Photonic Crystals (Wiley, KGaA, Weinheim, 2006)
50. W. Gawelda, D. Puerto, J. Siegel, A. Ferrer, A. Ruiz de la Cruz, H. Fernández, J. Solis, Appl. Phys. Lett. **93**, 121109 (2008)
51. Marcinkevicius, V. Mizeikis, S., Joudkazis, S. Matsuo, H. Misawa, Appl. Phys. A: Mater. Sci. Process. **76**, 257 (2003)
52. C. Hnatovsky, R.S. Taylor, E. Simova, V.R. Bhardwaj, D.M. Rayner, P.B. Corkum, J. Appl. Phys. **98**, 013517 (2005)
53. Q. Sun, H. Jiang, Y. Liu, Y. Zhou, H. Yang, Q. Gong, J. Opt. A, Pure Appl. Opt. **7**, 655 (2005)
54. V. Diez-Blanco, J. Siegel, A. Ferrer, A. Ruiz de la Cruz, J. Solis, Appl. Phys. Lett. **91**, 051104 (2007)
55. H.R. Lange, G. Grillon, J.F. Ripoche, M.A. Franco, B. Lamouroux, B.S. Prade, A. Mysyrowicz, E.T. Nibbering, A. Chiron, Opt. Lett. **23**, 120 (1998)
56. V. Kudriasov, E. Gaizauskas, V. Sirutkaitis, J. Opt. Soc. Am. B **22**, 2619 (2005)
57. J. Siegel, J.M. Fernandez-Navarro, A. Garcia-Navarro, V. Diez-Blanco, O. Sanz, J. Solis, F. Vega, J. Armengol, Appl. Phys. Lett. **86**, 121109 (2005)
58. D.M. Rayner, A. Naumov, P.B. Corkum, Opt. Express **13**, 3208 (2005)
59. Q. Sun, H. Jiang, Y. Liu, Z. Wu, H. Yang, Q. Gong, Opt. Lett. **30**, 320 (2005)
60. A. Mermillod-Blondin, C. Mauclair, J. Bonse, R. Stoian, E. Audouard, A. Rosenfeld, I.V. Hertel, Rev. Sci. Instrum. **82**, 033703 (2011)
61. R. Osellame, S. Taccheo, M. Marangoni, R. Ramponi, P. Laporta, D. Polli, S. De Silvestri, G. Cerullo, J. Opt. Soc. Am. B **20**, 1559 (2003)
62. M. Ams, G.D. Marshall, D.J. Spence, M.J. Withford, Opt. Express **13**, 5676 (2005)
63. A. Ferrer, V. Diez-Blanco, A. Ruiz, J. Siegel, J. Solis, Appl. Surf. Sci. **254**, 1121 (2007)
64. A. Dubietis, G. Tamošauskas, G. Fibich, B. Ilan, Opt. Lett. **29**, 1126 (2004)
65. V.V. Temnov, K. Sokolowski-Tinten, P. Zhou, A. El-Khamhawy, D. von der Linde, Phys. Rev. Lett. **97**, 237403 (2006)
66. M. Ams, G.D. Marshall, M.J. Withford, Opt. Express **14**, 13158 (2006)
67. A. Ferrer, A. Ruiz de la Cruz, D. Puerto, W. Gawelda, J.A. Vallés, M.A. Rebolledo, V. Berdejo, J. Siegel, J. Solis, In situ assessment and minimization of nonlinear propagation effects for femtosecond-laser waveguide writing in dielectrics. J. Opt. Soc. Am. B **27**(8), 1688–1692 (2010)

# Chapter 3
# Spectroscopic Characterization of Waveguides

Denise M. Krol

**Abstract** Since the first experiments on femtosecond laser waveguide writing the question on which mechanisms are responsible for the refractive index change immediately arose. Several efforts have been made in that direction but no conclusive answer has been achieved yet. In fact, it has been observed that several factors determine the actual mechanism dominating the refractive index change, such as the irradiation conditions and the material composition. Understanding the materials change at the microscopic level is however important in terms of optimization of both the fs-laser processing conditions and the material composition. It also can provide more detailed insight into the physical mechanisms involved in the fs-laser modification process to enhance its capabilities. Confocal fluorescence and Raman spectroscopy are powerful tools to investigate the material structure. This chapter will review the results obtained by using these techniques to characterize fs-laser induced structural changes in glass. The focus will be on structures related to waveguides and refractive index changes, since this has been the most active research area of fs-laser processing in glass to date.

## 3.1 Introduction

Since the initial reports on fs-laser micromachining *inside* glass in the mid 1990s there has been a rapidly growing activity in this area with applications in photonics (Chaps. 7–13) as well as optofluidics for lab-on-chip devices (Chaps. 14 and 15). In all these applications the use of fs-laser micromachining as a fabrication technique relies on the ability to engineer pre-designed 3D-structures in the glass with great

---

D. M. Krol (✉)
Department of Applied Science, University of California Davis, 1 Shields Ave, Davis, CA 95616, USA
e-mail: dmkrol@ucdavis.edu

spatial precision. And ultimately the presence of a modified structure in the glass, whether for the purpose of creating waveguides, diffraction gratings, or microfluidic channels, is synonymous with a change in materials properties. The modified glass can differ from the unmodified material in a wide variety of properties including refractive index, absorption coefficient, nonlinear optical susceptibility, crystal structure, morphology, etc. Understanding the materials change at the microscopic level is important in terms of optimization of the fs-laser processing conditions and materials. It also can provide more detailed insight into the physical mechanisms involved in the fs-laser modification process.

Confocal laser spectroscopy can be used to characterize the micron-scale structural changes that are associated with fs-laser micromachining. Fluorescence spectroscopy is especially sensitive to the presence of optically active defects, such as the non-bridging oxygen hole center (NBOHC), that are created as a result of fs-laser exposure. Raman spectroscopy, which basically measures the vibrational spectrum, is sensitive to overall changes in glass network structure. Interpretation of the fluorescence and Raman changes requires comparison of the spectra with spectroscopic studies that have been done in bulk glass. In addition to spectroscopy, confocal imaging makes it possible to image the fluorescence and Raman changes in the glass and by combining spectroscopic information with imaging it is possible to determine how the glass structure is changed with great spatial sensitivity. It turns out that both fluorescence and Raman spectroscopy can detect very small changes both in terms of concentration in defects as well as small changes in bonding. In the remainder of this chapter I will review our work on the use of confocal fluorescence and Raman spectroscopy to characterize fs-laser induced structural changes in glass. The focus will be on structures related to waveguides and refractive index changes, since this has been the most active research area of fs-laser processing in glass to date.

## 3.2 Spectroscopic Analysis of Glass

### *3.2.1 Fluorescence Spectroscopy*

Fluorescence of simple oxide glasses, such as fused silica, as well as soda lime silicate, borate, and phosphate glasses is typically caused by the presence of color centers or defects. One can distinguish extrinsic color centers, caused by the presence of transition metal or rare-earth impurities, from intrinsic color centers or defects, which are atomic-scale structures within the glass network involving dangling bonds associated with trapped holes and electrons. The formation of color center defects results in the creation of new electron trap states near, but below, the conduction band of the material that introduce new absorption and fluorescence bands. These defects can be intrinsically found in the glass or can be created in the material under applied external conditions, such as irradiation with γ-rays or neutrons. There is an extensive literature on the characterization of color center defects for many glass systems, where experimental techniques such as electron paramagnetic resonance (EPR), electron spin resonance (ESR), and of particular interest to us, fluorescence and absorption spectroscopy have been used in combination with computational modeling to identify and characterize defects in many types of glasses [1]. In the following some of the defect structures that are relevant within the context of fs-laser modification in oxide glasses will be discussed.

#### 3.2.1.1 Color Centers in Fused Silica

The common defects found in vitreous silica, either intrinsic or created through different modification techniques, can be classified as either oxygen-deficiency defects or oxygen-excess defects. An example of the former is the $E'$ center, which can be represented by the structure $\equiv$ Si• ($\equiv$are three covalent bonds attached to three oxygen atoms, • is a lone pair electron). Examples of the latter include the peroxy radical (POR) $\equiv$ Si–O–O• and the non-bridging oxygen hole center (NBOHC) $\equiv$ Si–O•. These and other known defects in fused silica have characteristic optical absorption/excitation and fluorescence/photoluminescence (PL) bands. Skuja provides a summary of the major bands of known defects in vitreous silica [2].

Of specific interest here is a well-known PL band centered at approximately 1.9 eV (∼650 nm), which has been observed in many spectroscopic studies of fused silica modified by neutrons, electrons, X-rays, and other radiation [3]. The band has also been detected in silica glass after fiber drawing and ion implantation. Historically, there has been controversy over the origin of this band. The exact peak position of the 1.9 eV band may vary from 1.8 to 2.0 eV depending on the exact excitation conditions. It has been shown by several studies [3–7] that the absorption bands responsible for the 1.9 eV emission are centered at 2 eV (FWHM 0.2 eV) and 4.8 eV (FWHM 1.05 eV). There is a general agreement that the 1.9 eV emission band is associated with oxygen excess related defects. Two models for the defects include the NBOHC defect and the interstitial ozone $O_3$ molecule [4, 8]. Recent studies confirm that the 1.9 eV band is, in fact, related to the NBOHC defect. Skuja et al. [3, 5, 7, 9] show that the contribution of ozone molecules to the 4.8 eV absorption band is negligible compared to NBOHC defects and that the 1.9 eV emission is independent of the concentration of ozone molecules. Figure 3.1 shows a typical PL spectrum of the NBOHC defect in fused silica.

Studies [10–13] show that a variety of defects are induced in vitreous silica exposed to femtosecond laser pulses. Strong PL bands at 280 nm (4.4 eV), 470 nm (2.6 eV), 560 nm (2.2 eV), and 650 nm (1.9 eV) have been reported [10] for an excitation wavelength of 250 nm. The 280 nm band is assigned to the oxygen vacancy $V_o$, $\equiv$ Si–Si $\equiv$ defects that are induced by the fs laser pulses. The generation of the defect is accomplished by the reaction Si–O–Si + $h\nu \rightarrow (V_o; O_i)$, where the oxygen vacancy is achieved by optical damage and the oxygen $O_i$ may link with nearby oxygen atoms. A similar reaction involving two Si–O–Si bonds,

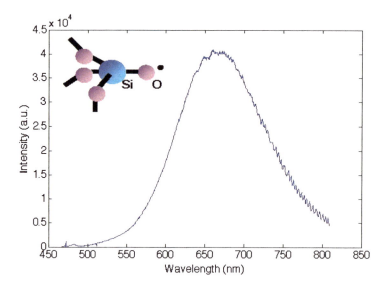

**Fig. 3.1** PL spectrum of NBOHC defects in fused silica. The *inset* shows an illustration of this defect

$2(Si-O-Si) + h\nu \rightarrow V_o + (V_o; (O_2)_i)$ results in the production of a vacancy defect and a vacancy paired with an $(O_2)_i$ created by reaction of two oxygen interstitials $O_i$. The bands at 470 nm and 558 nm have been attributed to the radiative recombination of electrons and holes that are separated on these defect pairs. Finally, the 650 nm PL band is assigned to the well-known NBOHC defect described above. It should be noted that in fs laser modified silica glass, the 650 nm band is only observed for silica irradiated with higher intensity fs laser pulses.

#### 3.2.1.2 Color Centers in Phosphate Glass

Radiation (excimer, femtosecond, X-rays, $\gamma$ rays) induced defects have previously been studied in phosphate and phosphorus-containing glasses. It has been determined [14–17] that several color center species are created upon exposure to irradiation, namely the $PO_3^{2-}$ (phosphoryl), $PO_4^{4-}$ (phosphoranyl), $PO_2^{2-}$ (phosphinyl), and $PO_4^{2-}$ (POHC, phosphorus-oxygen-hole-center) defects. These color centers are responsible for induced absorption bands in the various phosphate glasses. The $PO_3^{2-}$, $PO_4^{4-}$, and $PO_2^{2-}$ defects absorb in the 4.6–5.9 eV range. The room temperature stable POHC defect has broad overlapping absorption bands at 2.2, 2.5, and 5.3 eV. This defect is characterized by an unpaired electron shared between two orbitals of two non-bridging oxygens bound to a phosphorus atom (Fig. 3.2). Another form of the POHC defect is only stable at low temperatures and involves a hole metastably trapped on a lone NBO of a normal $PO_4$ group. This

## 3 Spectroscopic Characterization of Waveguides

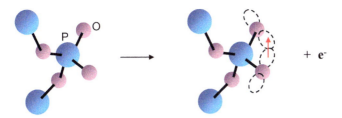

**Fig. 3.2** Proposed mechanism for the production of POHC defects. The precursor consists of two non-bridging oxygen atoms connected to a phosphorus atom. A hole gets trapped on two orbitals of the two oxygen atoms to form the POHC defect

defect absorbs at 3.1 eV (400 nm). No data on the fluorescence bands are given for these defects.

### 3.2.2 Raman Spectroscopy

Raman scattering is based on the phenomenon of inelastic scattering of light by vibrational modes of a material. Raman spectroscopy has become a very common laser technique used to probe the atomic-scale structure of materials. For glasses the spectra typically consist of a number of (often overlapping) bands that can be assigned to vibrations of different types of bonds in the glass network. The lack of long-range order in glass leads to a wide distribution of bond-lengths and -angles compared to crystals of the same composition, and as a consequence Raman bands in glass are much broader than those in crystals.

#### 3.2.2.1 Structure of Fused Silica

The atomic-scale structure of vitreous silica can be described as a continuous random network of cross-linked $SiO_4$ tetrahedra. Included in this cross-linked network are closed paths of repeated Si–O segments, resulting in the formation of $n$-membered ring structures, where n is the number of Si–O segments. Pasquarello and Car [18] have confirmed through computer simulation the existence of these ring structures and determined that there is a distribution of ring sizes in fused silica ranging from three- to nine-membered rings, with the five and sixfold structures being the most predominant. In experimental studies by Galeener and others [19–22] it has been proposed that the two Raman peaks at 495 and 605 $cm^{-1}$ correspond to the Raman active symmetric breathing modes of the oxygen atoms in the four- and three-membered ring structures, respectively. The modes involves the symmetric motion of the oxygen atoms, as depicted in Fig. 3.3. The decoupling of this mode from the rest of the network vibrational modes gives rise to the sharp profiles of the 490 $cm^{-1}$ and 605 $cm^{-1}$ peaks, which have been referred to as $D_1$ and $D_2$ defect

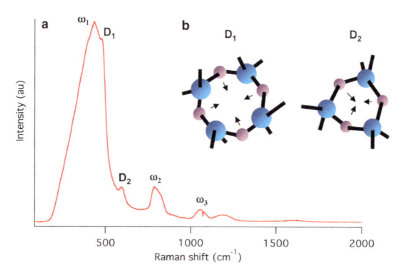

**Fig. 3.3** (a) Raman spectrum of fused silica. The $\omega_1$, $\omega_2$, and $\omega_3$ peaks are vibrational modes of the silica network, $D_1$ and $D_2$ are (b) Raman active symmetric breathing modes of four- and three-membered ring structures

peaks. Pasquarello and Car [18] have confirmed the assignment of these two peaks to the four- and three-membered ring structures in their molecular modeling work. The broad peaks observed at roughly 440, 790, 1060, and 1200 cm$^{-1}$ in the Raman spectrum of fused silica are due to vibrations of the cross-linked glass network. The broadness of these peaks is an indication of the wide distribution of Si–O–Si angles in the network.

### 3.2.2.2 Structure of Phosphate Glass

The structure of phosphate glasses can be described as a network of phosphate tetrahedra that are linked together through covalent bonding. Concentrations of network modifiers in the glass will break up, or depolymerize, the phosphate chains. The linked phosphate tetrahedra will have one, two, three, or four bridging oxygen atoms. These units can be classified using $Q^i$ terminology, where $i$ represents the number of bridging oxygens per tetrahedron [23]. The relative concentrations of the various $Q^i$ units depend on glass composition. Since the IOG-1 glass composition has a O/P ratio of roughly 3 it is classified as a metaphosphate glass [23].

The Raman spectrum of IOG-1, which is shown in Fig. 3.4, is very similar to that of other metaphosphate glasses [23–27]. The first broad Raman signal in the low wavenumber region between 200 cm$^{-1}$ and 600 cm$^{-1}$ is due to internal deformation bending modes of phosphate chains (in-chain PO$_2$ as well as OPO bending modes). The large band at 700 cm$^{-1}$ is caused by the symmetric stretching

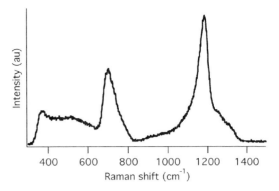

**Fig. 3.4** Raman spectrum of IOG-1 glass

mode of bridging oxygens between two phosphate tetrahedra, (POP)$_{sym}$. The large band at $1,180\,cm^{-1}$ comes from the symmetric stretching of the O–P–O non-bridging oxygens on $Q^2$ phosphate tetrahedra, (PO$_2$)$_{sym}$. The $1,260\,cm^{-1}$ peak, which sits on the shoulder of the $1,180\,cm^{-1}$ peak, is the asymmetric stretching vibration of non-bridging oxygens in O–P–O groups, (PO$_2$)$_{asym}$.

### 3.2.3 Confocal Imaging

Fs laser micromachining inside a bulk glass substrate induces modifications that have very small dimensions (on the order of microns). To obtain spectroscopic signals from only these small, modified regions embedded in a large, bulk unmodified glass is impossible with conventional spectroscopy since signals from the unmodified regions of the glass will dominate the signals from the modified regions. Spectroscopy using a confocal microscope, however, combines the ability to tightly focus the incident radiation into micron-size regions in a sample and spatially filter the detected spectroscopic signals to reject any background signals outside of the diffraction-limited focal volume of the microscope objective. This technique can enable the acquisition of signals from the small fs laser modified volumes embedded inside a bulk glass material. Several review papers [28, 29] have detailed the properties and advantages of confocal spectroscopy.

## 3.3 Experimental Equipment and Procedures

A combined femtosecond laser micromachining/confocal microscope setup is used to both perform laser processing of the samples and spectroscopic characterization without having to remove the sample. A schematic diagram of the set-up (not to scale) is shown in Fig. 3.5. The main components of the setup are the fs "processing"

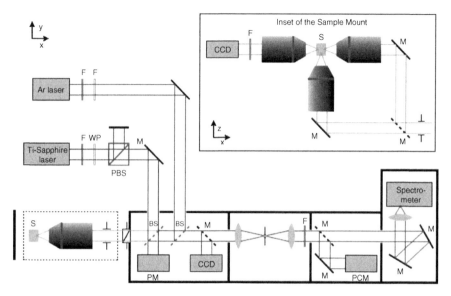

**Fig. 3.5** Schematic diagram of the combined femtosecond laser writing/confocal microscope setup. The dashed lines indicate flip mounts that allow for easy removal of the optical element. F = filter, M = mirror, WP = half-wave plate, PBS = polarization beam splitter, BS = beam splitter, S = sample on stepper motor stages, PM = power meter, PCM = photon counting module

laser, a cw laser used for spectroscopic characterization, microscope focusing and collection objectives, a sample mount attached to translation stages, and a grating spectrometer.

### 3.3.1 Femtosecond Laser Systems and Micromachining Procedures

For most experiments described in this chapter a commercially available laser system (Spectra-Physics), which can deliver up to $\sim 0.5$ mJ of 130 fs laser pulses at 800 nm with a repetition rate of 1 kHz, was used. The system consists of a Tsunami 3941-Mis Ti-sapphire oscillator pumped by a Millennia V solid-state laser, in combination with a Spitfire LCX amplifier pumped by a Merlin laser. In some experiments fs-laser processing was performed with a different laser system, as indicated in the text.

For the micromachining/waveguide writing experiments the fs laser beam is directed through a microscope objective and focused into polished glass cubes of roughly $1 \times 1 \times 1$ cm$^3$. The following glass compositions were investigated: (1) fused silica (Corning 7940), (2) a phosphate glass with a nominal composition (mole %) of 60% $P_2O_5$ 24% $Na_2O$ 13% $Al_2O_3$ 3% $La_2O_3$ (Schott IOG-1), and

(3) a phosphate glass co-doped with 1.8 wt% $Er_2O_3$ and 4.9 wt% $Yb_2O_3$ (Kigre MM-2a60). There are three objective mounts for looking at the sample from different directions. Mounts (a) and (b) are used for transverse and longitudinal waveguide writing respectively. Mounts (b) and (c) are used together to couple a cw laser into the resulting waveguides. Various objectives (10 × /0.25 NA, 20 × /0.4 NA, 50 × /0.55 NA, or 100 × /0.70 NA) were used for different experiments. Adjustment of the fs laser power is made using the half wave plate and polarizing cube beamsplitter. Typical pulse energies at the sample lie in the range of 0.05–5 µJ. The glass samples are mounted on a sample holder attached to a high precision alignment stage (Newport ULTRAlign series) that permits manual translation along all three axes as well as tilt around two of the axes. This stage is placed on computer controlled motorized translation stages to allow scanning along two axes. With this set-up it is possible to employ both the transverse and longitudinal writing geometries. In addition a wide range of scan speeds from 0.01 to 5 mm s$^{-1}$, is available. The structural modification can be monitored in situ by using white light to image onto a CCD camera.

## 3.3.2 Confocal Microscope System and Spectroscopy Procedures

The confocal fluorescence/Raman microscope system uses an excitation laser at 488 nm (Ar+ laser from Uniphase) or 473 nm (cw diode laser, LRS-473 from Laserglow, Ltd). The excitation beam is directed through one of the microscope objective using a 50/50 broadband dielectric or dichroic beam splitter and focused into the glass sample. Backscatter (PL and Raman) signals produced by the laser excitation are collected by the same objective and directed through the beam splitter. A pinhole conjugate to the focal point is used to ensure signals are only collected from the focal volume of the objective. A holographic notch filter in combination with a high quality long pass filter is used to filter out any excitation light that passes through the beamsplitter. The signal is then directed into an Oriel MS257 spectrometer equipped with a liquid nitrogen cooled CCD array (Roper Spec-10:100B) to obtain the spectral profiles. A grating with 300 grooves per mm is used for fluorescence spectroscopy and a grating with 1200 grooves per mm for Raman spectroscopy.

Fluorescence images of the modified glass are obtained using an avalanche photodiode detector (Perkin Elmer SPCD-AQR-141), which detects the total fluorescence signal from a localized region in the material. The detector is also used to perform temporal studies on the fluorescence signal. The imaging is achieved by using the avalanche photodiode detector in tandem with the computer controlled precision xyz translation stage in the confocal microscope setup. The glass is scanned in the plane perpendicular to the tightly focused 488 nm beam (100× objective) and the spectral signals are obtained using the APD detector to map out the fluorescence profile. In the case of Raman imaging, a full spectrum is collected for each pixel in the image.

## 3.4 Spectroscopic Analysis of fs-laser Modification in Fused Silica

### 3.4.1 Fluorescence Spectroscopy and Imaging of Waveguides and Bragg Gratings

Figure 3.6 shows white-light microscopic images, captured by the CCD camera in the confocal microscopy setup, of single lines written in fused silica at a scan rate of 40 μm s$^{-1}$ by delivering fs pulses through a 50×, 0.55 NA objective. The sample was scanned in the direction of the laser beam. The glass morphology is strongly dependent on the amount of energy deposited into the sample.

For pulse energies below 0.25 μJ the lines appear to be smooth and continuous; these smooth lines are indicative of good waveguides with modest changes in the optical properties of the glass. For higher pulse energies, more extensive/severe damage occurs to the glass, with the lines being rough and discontinuous. The damage threshold between the smooth lines and the damaged tracks occurs at roughly 8 J cm$^{-2}$. Subsequent experiments [30, 31] showed that when light from a He–Ne laser was launched into these types of structures good waveguiding behavior was only observed for fluences between 3 and 8 J cm$^{-2}$, i.e., for smooth lines.

Laser spectroscopy of the lines (Fig. 3.7a) shows the presence of new PL bands that are not present in the unmodified glass. The overall fluorescence intensity increases with fs laser pulse energy. Further inspection of the spectral shape shows that for low fs pulse energies the spectrum is dominated by a band centered at 530 nm, whereas for higher pulse energies a band centered at 630 nm dominates. This is more clearly seen when the spectra are all normalized to the 530 nm band (Fig. 3.7b). The fluorescence peaks that we observe after modification originate from defects that are formed in the glass as a result of the femtosecond-laser exposure. The fluorescence peak centered at 630 nm matches the characteristics of NBOHC defects. The peak centered at 530 nm matches the characteristics of self-trapped exciton defects from very small silicon nanoclusters that several groups have

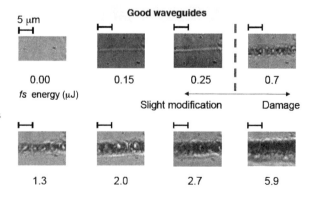

**Fig. 3.6** White-light microscopy images showing the morphology of waveguides and damage lines written longitudinally inside fused silica at 40 μm s$^{-1}$ using different laser pulse energies through a 50× objective

# 3 Spectroscopic Characterization of Waveguides

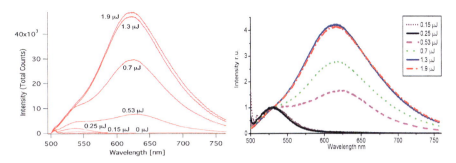

**Fig. 3.7** (a) PL spectra of fs-laser modified fused silica; 488 nm light from an Ar+ laser was used as the excitation source; the different curves in (a) represent different fs pulse energies used for modification. (b) Same spectra as in (a) normalized to the 530 nm band

reported seeing in γ-irradiated fused silica [32, 33]. Comparison of the waveguide morphology with the fluorescence spectra shows that only the 530 nm PL band is present for the structures that have a smooth morphology and for which good waveguiding of a He–Ne beam can be observed. These results suggest that there are two regimes in which different modification processes occur, depending on the intensity of the femtosecond laser. We have also observed a distinct difference in the light that is emitted during modification [34].

A similar trend in fluorescence spectroscopy is observed when measuring fiber Bragg gratings that had been fabricated using fs-laser pulses [34, 35]. Relatively low pulse energies corresponding to peak intensities of approximately $4.5 \times 10^{13}$ W cm$^{-2}$ produced smooth grating lines, as shown in Fig. 3.8a. These gratings exhibit a PL spectrum that is again dominated by a band at 530 nm, as shown in Fig. 3.8c. Greater pulse energies corresponding to peak intensities of $12-16 \times 10^{13}$ W cm$^{-2}$ produced the lines shown in Fig. 3.8b. These lines have a fluorescence peak at 630 nm (Fig. 3.8d).

Fluorescence imaging is performed to determine the spatial location of the color center defects that are responsible for the fluorescence. Figure 3.9a shows the fluorescence image for the end face of two waveguides written in fused silica using ~1 μJ pulse energy with a 10x objective.

By comparing the fluorescence image with the white light image, it is clear that the defects are created in the waveguide core regions. This indicates that the waveguide is created through direct exposure to the fs pulses. Exposure of the glass to fs laser pulses leads to the formation of a plasma, i.e., many broken bonds. After the laser pulse is gone, the plasma quickly dissipates its energy and the material in the exposed region cools down, but in doing so not all broken bonds are reformed and as a result defects are present in the exposed region. Higher fs pulse energies induce a larger number of these defects, resulting in higher fluorescence intensity.

In the case of fs-laser fabricated Bragg gratings, fluorescence imaging reveals the grating pattern and periodicity (cf. Fig. 3.9b and c) indicating that the fluorescence is indeed associated with the formation of the grating structures.

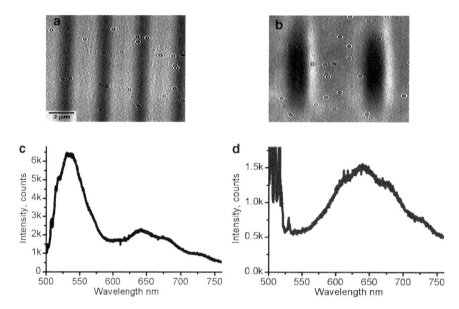

**Fig. 3.8** White-light transmission images (**a**) and (**b**) and PL spectra (**c**) and (**d**) for smooth (*left panels*) and rough (*right panels*) femtosecond-laser-written Bragg gratings

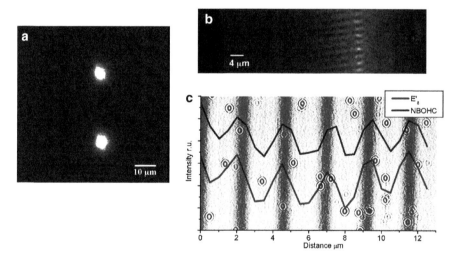

**Fig. 3.9** (**a**) PL image for the end face of two waveguides written in fused silica using $\sim 1\,\mu J$ pulse energy with a 10× objective. (**b**) PL image of the fiber Bragg grating shown in Fig. 3.8a. (**c**) PL intensity variation as a function of spatial position, (top trace for $E'_\delta$, bottom trace for NBOHC defects) overlaid by a white light image of the grating

3 Spectroscopic Characterization of Waveguides

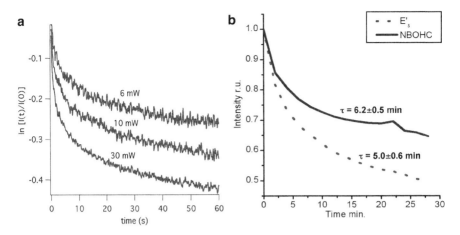

**Fig. 3.10** (a) Decay of NBOHC fluorescence from modified fused silica continuously exposed to different powers of 488 nm light. (b) Decay of $E$ and NBOHC fluorescence in fiber Bragg gratings

## 3.4.2 Photobleaching of Defects

In addition to the fluorescence that is observed when the modified glasses are illuminated with 488 nm light, we also observe that the fluorescence decays with prolonged exposure to the excitation light (photobleaching). To accurately determine the decay behavior, an avalanche photodiode detector in the microscope setup is used to measure the total intensity of the fluorescence signal as a function of total continuous exposure time to the 488 nm excitation light. Figure 3.10a shows the fluorescence intensity as a function of time for different 488 nm powers used to probe a silica glass sample modified with 2 µJ of fs laser pulses.

It is clear that the defects are bleached by the visible light and that the decay process is dependent on the intensity of the probe beam. Apparently the broken bonds associated with the presence of defects are repaired under influence of the modest excitation energy from the cw 488 nm light.

In the case of the Bragg gratings (Fig. 3.10b) the photobleaching was measured over a longer time period for the two defect PL bands separately. Exponential curve fits were used to calculate bleaching rates of $5.0 \pm 0.6$ and $6.2 \pm 0.5$ min for the 530 and 630 nm peaks, respectively.

## 3.4.3 Raman Spectroscopy and Imaging

In addition to the broad fluorescence band in Fig. 3.7, there are much lower intensity signals near the excitation line, which are due to Raman scattering. These signals sit on the edge of the broad fluorescence band. For analyzing the Raman signals the

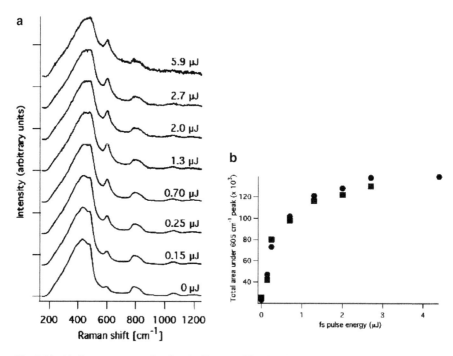

**Fig. 3.11** (a) Raman spectra for fused silica modified by fs pulses of different energies (cf. Fig. 3.5) and (b) the total area under the 605-cm 1 Raman peak as a function of fs pulse energy. The two symbols denote data from two experimental runs

fluorescence background is subtracted and the resulting spectra are normalized using the 445-cm$^{-1}$ peak to obtain the final corrected Raman spectra shown in Fig. 3.11a.

The most noticeable changes in the Raman spectra are increases in the intensities of the two peaks at 490 and 605 cm$^{-1}$, which are due to the three- and four-membered rings discussed in Sect. 3.2.2, with increasing fs pulse energy. To quantify the results, Fig. 3.11b shows the area under the 605 cm$^{-1}$ peak as a function of fs pulse energy. Raman changes in the intensities of the 490 and 605 cm$^{-1}$ peaks have also been observed for fused silica subjected to other forms of treatment such as neutron radiation [36], shock wave compression [37] and UV irradiation [38], with the intensity of the two Raman peaks increasing with increasing radiation and shock pressures. It has also been shown that both the density and the refractive index of the glass increase as the 490 and 605 cm$^{-1}$ peaks in the Raman spectra increase [39]. The fused silica structural network typically has predominantly large five and sixfold ring structures. An increase in the number of smaller three and fourfold ring structures leads to a decrease in the overall bond angle and a densification of the glass. The refractive index change is directly related to this densification of the glass. In summary, there is a direct relationship between the Raman changes in the spectra (changes in the glass structural network) and the changes in the physical and optical properties of the glass.

By comparing our Raman results on waveguides with those obtained from bulk samples [31], we can conclude that for low fs pulse fluences (8–10 J cm$^{-2}$), the changes in the Raman spectra correspond to modest changes in the structural network of the glass and small changes in the refractive index of the material ($<10^{-4}$). This magnitude of the Raman and index changes is also observed for bulk glass that has undergone fast quenching, resulting in a glass with a 1,500 °C fictive temperature. For higher fs laser fluences (40 J cm$^{-2}$), the larger increase in the Raman peak corresponds to a greater perturbation in the glass network and much larger changes in the refractive index ($10^{-2}$). These changes are comparable to those from the shock wave and neutron radiation experiment.

As with the PL signals, we can get information about the regions in the sample where Raman changes occur by collecting data while scanning the sample [40]. Finally, it should be noted that although the intensity of the fluorescence band at 630 nm decreases with exposure time to the 488-nm light, the Raman changes do not. This indicates that the NBOHC defects can be photobleached by the 488-nm beam, but the overall changes in the network structure, which we probe with Raman scattering are stable under 488 nm exposure.

## 3.5 Spectroscopic Analysis of Waveguides in Phosphate Glasses

### 3.5.1 Fluorescence Spectroscopy and Imaging of IOG-1

Figure 3.12a shows white light microscope images of the end faces of lines written in IOG-1 glass using 3.5 µJ fs energy and scanning the glass along the $z$-axis, parallel to the laser beam axis at a rate of 20 µm s$^{-1}$. The images show a central region where the fs laser beam is focused that is dark and circular with a diameter of roughly 8 µm surrounded by bright regions that appear to guide the white light used to illuminate the sample. If the laser beam axis is only slightly misaligned from the scan axis, the shape of the laser-induced damage is significantly different.

Figure 3.12b shows the end face image of a line written when the laser beam axis is slightly misaligned with the $z$ scan axis in the $xz$ plane. The central, dark region becomes elongated horizontally in the direction of the skewed beam axis. Two distinct regions above and below the central, elongated damage region guide the white light illuminating the sample. Experimental tests have determined that there is no correlation between the induced damage profile and the polarization of the fs laser beam. Rather, the elongation in the central region can be explained by the geometry of the laser interaction volume [41].

Figure 3.12e shows the PL spectrum for modified IOG-1 phosphate glass along with the initial spectrum of unmodified IOG-1. The initial spectrum for unmodified IOG-1 glass (with 488 nm excitation light) shows only Raman signals in the 500–525 nm range due to the phosphate glass network and no fluorescence bands. After modifying the glass with fs pulses, a fluorescence band centered at roughly

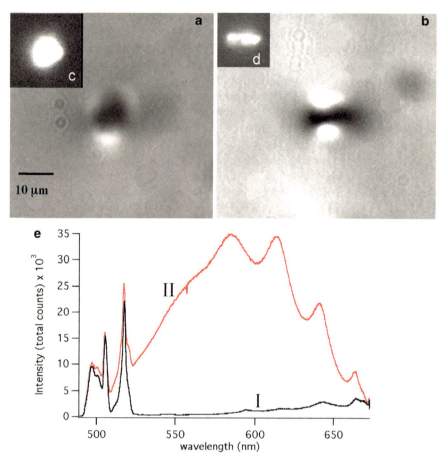

**Fig. 3.12** (**a** and **b**) White light microscopy images of the end face of laser written waveguides in IOG-1. (**c** and **d**) Fluorescence images of color centers created by the fs laser. (**e**) PL spectra of IOG-1 (*I*) before and (*II*) after laser modification

600 nm is observed along with the Raman signals. The peaks at roughly 590, 615, and 640 nm are artifacts of the transmission profile of a 488 nm dichroic filter used in the confocal microscope setup. It should be pointed out that the same fluorescence is observed for fs laser modified lanthanum phosphate glasses (80% $P_2O_5$–20% $La_2O_3$) that we have synthesized ourselves.

The fact that we observe similar fluorescence behavior in both modified IOG-1 phosphate glass and our synthesized lanthanum phosphate glasses leads us to believe that the defect centers that are created are, indeed, phosphorus related. We propose that the tightly focused femtosecond pulses create POHC color center defects that absorb the 488 nm (2.5 eV) excitation light and fluoresce at a longer wavelength of ∼600 nm (2.1 eV). Our assignment is supported by a comparison of the similarities in the atomic arrangements between the POHC defect and the equivalent NBHOCs

3 Spectroscopic Characterization of Waveguides

in fused silica. Both involve a trapped hole on oxygen atom(s) covalently bonded to a phosphorus/silicon atom. Aside from the similarity in their atomic structure, the two defects have spectroscopic similarities. Both have similar absorption bands (POHC – 2.2, 2.5, 5.3 eV, NBOHC – 2.0, 4.8 eV) and fluorescence bands (NBOHC – 1.9 eV emission, POHC – 2.1 eV from our study). The defect fluorescence in IOG-1 glass also exhibits a decay upon continuous exposure to 488 nm light [42], similar to what is observed in fused silica.

Fluorescence images (Fig. 3.12c and d) are nearly identical in both shape and dimension to the central, dark regions in the white light images. It is clear that, based on the comparison between the fluorescence and white light images, the color center defects responsible for the fluorescence are located primarily within the central, non-waveguiding regions. No color center defects are formed in the waveguide regions surrounding the central region. Raman experiments also confirm that structural changes take place in the ventral, dark region, while no noticeable Raman changes are observed when the waveguide (non-fluorescent) regions are probed [43].

### 3.5.2 Comparison of Waveguides in Fused Silica and IOG-1

Our results show that the waveguides that we have fabricated using fs laser pulses, behave differently in fused silica and IOG-1 glass. Figure 3.13 (top images) shows typical transmission white light images of end faces of 5 mm long lines written in fused silica and IOG-1.

In the case of fused silica waveguiding takes place in the central region where the writing beam was focused, but in IOG-1 the central (exposed) region is dark

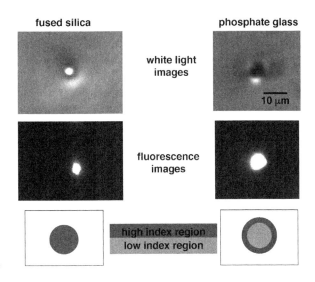

**Fig. 3.13** White light (*top*) and fluorescence (*center*) images for fs-laser modified fused silica and IOG-1 glass together with an illustration of the induced index changes

and surrounded by bright regions that appear to guide the white light. Fluorescence imaging shows that in both glasses defects (NBOHC) are formed in the regions exposed to the fs-laser. The fact that the exposed region in fused silica guides light indicates that fs-laser exposure leads to an increase in refractive index, whereas in IOG-1 the results show that the exposed region has a lower index surrounded by a stress-induced higher-index region, which can guide light. This difference in behavior can be understood by considering the processes that occur in the glass after the fs laser energy is deposited into the material. In both glasses the fs laser energy is absorbed in the focal region and results in the formation of a localized hot plasma. Because the focal region is very small it will cool down/quench very fast before the arrival of the next pulse (1 ms later since the laser rep rate is 1 kHz). Therefore, the exposed region and resulting modified spot consists of glass with a much higher quenching rate and has a structure that is different from the original glass. This is confirmed by the fact that the modified spot shows fluorescence from NBOHC defects (resulting from bond-breaking) and a Raman spectrum that resembles a glass with high quench rates. Although the exposed regions in both fused silica and IOG-1 have localized defects, the resulting change in refractive index is positive for fused silica, but negative for IOG-1. This is consistent with the dependence of the index on cooling rate. In fused silica the index increases with cooling rate [39] whereas in IOG-1 it decreases [41]. It should be pointed out that the lower index in the exposed region of IOG-1 is associated with an expansion of the structure and the waveguides that are observed around the exposed (central) region are the result of stress created by the expansion of the central region. The fact that color centers are not found in the waveguide regions of IOG-1, but only in the central region, lends supporting evidence that the waveguides are not created via direct exposure to the fs laser beam. Other studies [44–46] have shown that many other glass compositions also show an index decrease in the focal region of the fs laser, similar to what is observed for IOG-1.

Finally we note that, although our results can be understood in terms of the dependence of the refractive index on cooling rate, we do not claim that the processes occurring in the exposed region are merely identical to those occurring during fast thermal quenching. It is well possible that other phenomena, such as shockwave propagation, are involved, especially at high fs pulse energies. For the relatively low fs pulse energies necessary to fabricate good waveguides the cooling rate model might be adequate and we are currently investigating this further.

### 3.5.3 *Raman Spectroscopy and Imaging of Rare Earth-doped Phosphate Glass*

Rare-earth doped phosphate glasses are well suited for the fabrication of active waveguide devices [47]. As shown in Sect. 3.5.1 fs-laser fabrication of waveguides in phosphate glass using a low rep rate (1 kHz) laser results in waveguides with

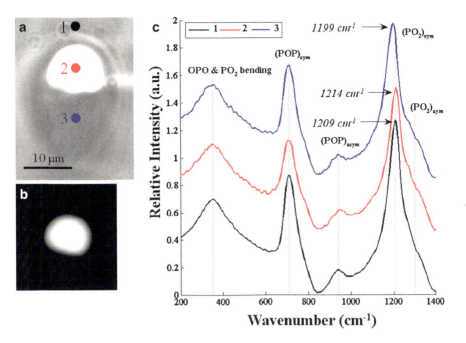

**Fig. 3.14** (**a**) White-light transmissions image of modification induced under fs laser condition of 885 kHz rep rate, 320 nJ pulse energy, and 50 μm s$^{-1}$ scan speed. *1*. Unmodified bulk glass; *2*. modified high index guiding region; *3*. modified non-guiding low index region; (**b**) He–Ne transmission near field image. (**c**) Normalized Raman spectra of Er-Yb doped phosphate glass for: *1*. unmodified bulk glass; *2*. modified high index region; *3*. modified low index region

less than ideal profiles. However, single-mode waveguides can be fabricated using astigmatic beam shaping techniques or by the use of oil-immersion focusing objectives [48–50]. High quality, low loss waveguides have also been reported in Er–Yb doped phosphate glass using fs lasers with higher repetition rates leading to conditions where cumulative thermal effects take place [48]. High rep rate micromachining also offers the advantage of faster overall processing speeds. We have studied the structural changes of Er-Yb doped phosphate glass (Kigre MM-2a60) after fs-laser irradiation at both the macroscopic scales and atomic scales using in situ confocal white light microscopy and Raman spectroscopy [51,52].

As is discussed in detail elsewhere [48, 51] relatively low loss waveguides can be fabricated when using pulse energies lower than 240 nJ, repetition rates lower than 885 kHz, and scan speeds faster than 10 mm/s. Figure 3.14a shows a white light image of such a waveguide and Fig. 3.14b shows the near-field mode profile for a wavelength of 633 nm. It should be noted that at such high repetition rate heat accumulation takes place and the modified region becomes much larger than the focal spot size of the laser.

The Raman spectrum of Er–Yb doped phosphate glass (Fig. 3.14c) consists of a number of bands that are caused by specific vibrations associated with the

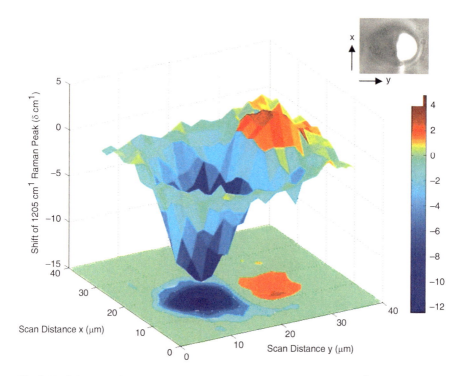

**Fig. 3.15** Color map of shifts in the relative spectral position of the 1,209 cm$^{-1}$ Raman peak as a function of the spatial position within the modified area (see *inset*)

phosphate glass matrix. The general spectral characteristics are very similar to those described in Sect. 3.2.2, suggesting that the glass composition is very close to a metaphosphate composition. In order to analyze changes in the Raman spectra for the laser-modified glass, all plots in Fig. 3.14c are normalized to the 360 cm$^{-1}$ peak for a relative comparison. Raman spectra collected form the modified regions (regions 2 and 3 in Fig. 3.14a) demonstrate a noticeable shift relative to the spectrum of the bulk material (region 1 in Fig. 3.14a) for the 1,209 cm$^{-1}$ and 710 cm$^{-1}$ peaks [51,52]. Here we will only discuss the behavior of the 1,209 cm$^{-1}$ peak. This peak shows a shift of $-10$ cm$^{-1}$ toward lower wavenumbers inside the non-guiding region, followed by a shift of $+5$ cm$^{-1}$ toward higher wavenumbers in the guiding region. To better understand the overall nature of the 1,209 cm$^{-1}$ shift, a series of Raman spectra were collected over a waveguide cross-section of 40 μm × 40 μm. The spectral position of the normalized 1,209 cm$^{-1}$ peak was measured for every individual spectrum collected (2 μm × 2 μm region per spectrum) and compared to the normalized 1,209 cm$^{-1}$ Raman peak position of the unmodified bulk material. The measured difference between the spectral positions of the two Raman peaks is represented in a 3D color map illustrated in Fig. 3.15.

An overall spectral shift toward lower wavenumbers is observed spatially in the non-guiding region of the waveguide, and an overall spectral shift toward higher wavenumbers is observed in the guiding region of the waveguide. The measured spectral shift of the 1,209 cm$^{-1}$ Raman peak correlates with the relative changes in refractive index of the modification. Our results show that a shift to lower wavenumbers is associated with a negative change in index and vice versa. A lowering of the index is equivalent to a lower density, i.e., overall expansion of the network, resulting in an overall lengthening of the P–O bonds, leading to a lower frequency for the (PO$_2$)$_{sym}$ vibrational mode.

## 3.6 Summary

In order to fully exploit fs-laser processing it is necessary to understand how the glass is structurally modified after exposure to the fs laser pulses. Confocal laser spectroscopy is a powerful tool to locally probe the fs-laser processed regions in the glass with micron size precision. Thus we can obtain crucial information about the nature and spatial extent of the structural changes in the glass. By measuring the fluorescence spectra of the waveguide regions we can determine whether optically active defects are formed as a result of fs laser modification. Raman spectroscopy of the waveguides is used to characterize changes in the glass network resulting from fs laser modification. In the imaging mode the fluorescence and Raman signals can be used to provide a spatial map of the laser-induced modification. This information helps us to further unravel the mechanism by which fs laser pulses structurally alter the glass.

**Acknowledgements** The author would like to thank James Chan, Luke Fletcher, Wilbur Reichman, and Jon Witcher for their contributions to this chapter. The author acknowledges financial support from the National Science Foundation under Grant No. DMR-0801786.

## References

1. G. Pacchioni, L. Skuja, D.L. Griscom (eds.), *Defects in SiO$_2$ and Related Dielectrics: Science and Technology, NATO Science Series*, Kluwer, Dordrecht, 2000)
2. L. Skuja, in *Defects in SiO$_2$ and Related Dielectrics: Science and Technology, NATO Science Series II*, vol. 2, ed. by G. Pacchioni, L. Skuja, D.L. Griscom. Optical Properties of Defects in Silica (Kluwer, Dordrecht, 2000), pp. 73–116
3. L. Skuja, T. Suzuki, K. Tanimura, Site-selective laser-spectroscopy studies of the intrinsic 1.9-eV luminescence center in glassy SiO$_2$. Phys. Rev. B: Condens. Matter **52**(21), 15208–15216 (1995)
4. L. Skuja, The origin of the intrinsic 1.9 eV luminescence band in glassy SiO$_2$. J. Non-Cryst. Solids **179**, 51–69 (1994)
5. L. Skuja, K. Tanimura, N. Itoh, Correlation between the radiation-induced intrinsic 4.8 eV optical absorption and 1.9 eV photoluminescence bands in glassy SiO$_2$. J. Appl. Phys. **80**(6), 3518–3525 (1996)
6. M. Cannas, M. Leone, Photoluminescence at 1.9 eV in synthetic wet silica. J. Non-Cryst. Solids **280**(1–3), 183–187 (2001)
7. L. Skuja, M. Mizuguchi, H. Hosono, H. Kawazoe, The nature of the 4.8 eV optical absorption band induced by vacuum-ultraviolet irradiation of glassy SiO$_2$. Nucl. Instrum. Meth. Phys. Res., Sect. B **166**, 711–715 (2000)

8. K. Awazu, H. Kawazoe, $O_2$ molecules dissolved in synthetic silica glasses and their photochemical reactions induced by arf excimer laser radiation. J. Appl. Phys. **68**(7), 3584–3591 (1990)
9. L. Skuja, M. Hirano, H. Hosono, Oxygen-related intrinsic defects in glassy $SiO_2$: interstitial ozone molecules. Phys. Rev. Lett. **84**(2), 302–305 (2000)
10. M. Watanabe, S. Juodkazis, H.B. Sun, S. Matsuo, H. Misawa, Luminescence and defect formation by visible and near-infrared irradiation of vitreous silica. Phys. Rev. B **60**(14), 9959–9964 (1999)
11. S. Juodkazis, M. Watanabe, H.B. Sun, S. Matsuo, J. Nishii, H. Misawa, Optically induced defects in vitreous silica. Appl. Surf. Sci. **154**, 696–700 (2000)
12. H.B. Sun, S. Juodkazis, M. Watanabe, S. Matsuo, H. Misawa, J. Nishii, Generation and recombination of defects in vitreous silica induced by irradiation with a near-infrared femtosecond laser. J. Phys. Chem. B **104**(15), 3450–3455 (2000)
13. M. Watanabe, S. Juodkazis, H.B. Sun, S. Matsuo, H. Misawa, M. Miwa, R. Kaneko, Transmission and photoluminescence images of three-dimensional memory in vitreous silica. Appl. Phys. Lett. **74**(26), 3957–3959 (1999)
14. D.L. Griscom, E.J. Friebele, K.J. Long, J.W. Fleming, Fundamental defect centers in glass – electron-spin resonance and optical absorption studies of irradiated phosphorus-doped silica glass and optical fibers. J. Appl. Phys. **54**(7), 3743–3762 (1983)
15. D. Ehrt, P. Ebeling, U. Natura, UV transmission and radiation-induced defects in phosphate and fluoride-phosphate glasses. J. Non-Cryst. Solids **263**(1–4), 240–250 (2000)
16. U. Natura, D. Ehrt, Modeling of excimer laser radiation induced defect generation in fluoride phosphate glasses. Nucl. Instrum. Meth. Phys. Res., Sect. B **174**(1–2), 151–158 (2001)
17. U. Natura, D. Ehrt, Generation and healing behavior of radiation-induced optical absorption in fluoride phosphate glasses: the dependence on UV radiation sources and temperature. Nucl. Instrum. Meth. Phys. Res., Sect. B **174**(1–2), 143–150 (2001)
18. A. Pasquarello, R. Car, Identification of Raman defect lines as signatures of ring structures in vitreous silica. Phys. Rev. Lett. **80**(23), 5145–5147 (1998)
19. A.E. Geissberger, F.L. Galeener, Raman studies of vitreous $SiO_2$ versus fictive temperature. Phys. Rev. B **28**(6), 3266–3271 (1983)
20. J.C. Mikkelsen Jr., F.L. Galeener, Thermal equilibrium of Raman active defects in vitreous silica. J. Non-Cryst. Solids **37**(1), 71–84 (1980)
21. F.L. Galeener, Raman and ESR studies of the thermal history of amorphous $SiO_2$. J. Non-Cryst. Solids **71**(1–3), 373–386 (1985)
22. F.L. Galeener, Planar rings in vitreous silica. J. Non-Cryst. Solids **49**(1–3), 53–62 (1982)
23. R.K. Brow, Review: the structure of simple phosphate glasses. J. Non-Cryst. Solids **263 & 264**, 1–28 (2000)
24. S.H. Morgan, R.H. Magruder III, E. Silberman, Raman spectra of rare-earth phosphate glasses. J. Am. Ceram. Soc. **70**, 378–380 (1987)
25. D. Ilieva, B. Jivov, G. Bogachev, C. Petkov, I. Penkov, Y. Dimitriev, Infrared and Raman spectra of $Ga_2O_3$–$P_2O_5$ glasses. J. Non-Cryst. Solids **283**, 195–202 (2001)
26. J.J. Hudgens, R.K. Brow, D.R. Tallant, S.W. Martin, Raman spectroscopy study of the structure of lithium and sodium ultraphosphate glasses. J. Non-Cryst. Solids **223**, 21–31 (1998)
27. R. Lebullenger, L.A.O. Nunes, A.C. Hernandes, Properties of glasses from fluoride to phosphate composition. J. Non-Cryst. Solids **284**, 55–60 (2001)
28. R.H. Webb, Confocal optical microscopy. Rep. Progr. Phys. **59**(3), 427–471 (1996)
29. D.R. Sandison, W.W. Webb, Background rejection and signal-to-noise optimization in confocal and alternative fluorescence microscopes. Appl. Opt. **33**(4), 603–615 (1994)
30. J.W. Chan, T.R. Huser, S.H. Risbud, D.M. Krol, Structural changes in fused silica after exposure to focused femtosecond laser pulses. Opt. Lett. **26**(21), 1726–1728 (2001)
31. J.W. Chan, T.R. Huser, S.H. Risbud, D.M. Krol, Modification of the fused silica glass network associated with waveguide fabrication using femtosecond laser pulses. Appl. Phys. A **76**, 367–372 (2003)

32. H. Nishikawa, E. Watanabe, D. Ito, Y. Sakurai, K. Nagasawa, Y. Ohki, Visible photoluminescence from Si clusters in irradiated amorphous $SiO_2$. J. Appl. Phys. **80**, 3513–3519 (1996)
33. S. Demos, M. Staggs, K. Minoshima, J. Fujimoto, Characterization of laser induced damage sites in optical components. Opt. Express **10**, 1444–1450 (2002)
34. W.J. Reichman, J.W. Chan, C.W. Smelser, S.J. Mihailov, D.M. Krol, Spectroscopic characterization of different femtosecond laser modification regimes in fused silica. J. Opt. Soc. Am. B **24**, 1627 (2007)
35. W.J. Reichman, D.M. Krol, C.W. Smelser, S.J. Mihailov, Fluorescence spectroscopy of fiber gratings written with an ultrafast infrared laser and a phase mask 2005 conference on lasers and electro-optics (CLEO). IEEE **2**, 1106 (2005)
36. J.B. Bates, R.W. Hendricks, L.B. Shaffer, J. Chem. Phys. **61**, 4163 (1974)
37. M. Okuno, B. Reynard, Y. Shimada, Y. Syono, C. Willaine, Phys. Chem. Minerals **26**, 304 (1999)
38. S.G. Demos, L. Sheehan, M.R. Kozlowski, Proc. SPIE **3933**, 316 (2000)
39. R. Bruckner, J. Non-Cryst. Solids **5**, 123 (1970)
40. W.J. Reichman, D.M. Krol, L. Shah, F. Yoshino, A. Arai, S.M. Eaton, P.R. Herman, A spectroscopic comparison of femtosecond-laser-modified fused silica using kilohertz and megahertz laser systems. J. Appl. Phys. **99**, 123112 (2006)
41. J.W. Chan, T.R. Huser, S.H. Risbud, J.S. Hayden, D.M. Krol, Waveguide fabrication in phosphate glasses using femtosecond laser pulses. Appl. Phys. Lett. **82**, 2371 (2003)
42. J.W. Chan, T. Huser, J.S. Hayden, S.H. Risbud, D.M. Krol, Fluorescence spectroscopy of color centers generated in phosphate glasses after exposure to femtosecond laser pulses. J. Am. Ceram. Soc. **85**(5), 1037–1040 (2002)
43. D.M. Krol, J.W. Chan, T.R. Huser, S.H. Risbud, J.S. Hayden, *Fs-Laser Fabrication of Photonic Structures in Glass: the Role of Glass Composition. Fifth International Symposium on Laser Precision Microfabrication*, vol. 5662, ed. by I. Miyamoto, H. Helvajian, K. Itoh, K.F. Kobayashi, A. Ostendorf, K. Sugioka. Proceedings of SPIE (2004), p. 30
44. V.R. Bhardwaj, E. Simova, P.B. Corkum, D.M. Rayner, C. Hnatovsky, R.S. Taylor, B. Schreder, M. Kluge, J. Zimmer, Femtosecond laser-induced refractive index modification in multicomponent glasses. J. Appl. Phys. **97**, 083102-1 – 083102-9 (2005)
45. W. Reichman, C.A. Click, D.M. Krol, Femtosecond laser writing of waveguide structures in sodium calcium silicate glasses. Proc. SPIE **5714**, 238 (2005)
46. M. Ams, G.D. Marshall, P. Dekker, M. Dubov, V.K. Mezentsev, I. Bennion, M.J. Withford, Investigation of ultrafast laser–photonic material interactions: challenges for directly written glass photonics. IEEE J. Sel. Top. Quant. Electron. **14**, 1370 (2008)
47. S. Taccheo, G. Della Valle, R. Osellame, G. Cerullo, N. Chiodo, P. Laporta, O. Suelto, A. Killi, U. Morgner, M. Lederer, D.l. Kopf, Er:Yb-doped waveguide laser fabricated by femtosecond laser pulses. Opt. Lett. **29**, 2626–2628 (2004)
48. R. Osellame, N. Chiodo, G. Della Valle, G. Cerillo, R. Ramponi, P. Laporta, A. Killi, U. Morgner, O. Suelto, Waveguide lasers in the C-band fabricated by laser inscription with a compact femtosecond oscillator. J. Sel. Top. Quant. Electron. **12**, 277–285 (2006)
49. R. Osellame, N. Chiodo, G. Della Valle, S. Taccheo, R. Ramponi, G. Cerullo, A. Killi, U. Morgner, M. Lederer, D.l. Kopf, Optical waveguide writing with a diode-pumped femtosecond oscillator. Opt. Lett. **29**, 1900–1902 (2004)
50. M. Ams, G.D. Marshall, D. Spence, M.J. Withford, Slit beam shaping method for femtosecond laser direct-write fabrication of symmetric waveguides in bulk glasses. Opt. Express **13**, 5676–5681 (2005)
51. L.B. Fletcher, J.J. Witcher, W.J. Reichman, J. Bovatsek, A. Arai, D.M. Krol, Structural modifications in Er–Yb doped phosphate glass induced by femtosecond laser waveguide writing. Proc. of SPIE **6881**, 688111-1 (2008)
52. L.B. Fletcher, J.J. Witcher, W.B. Reichman, A. Arai, J. Bovatsek, D.M. Krol, Changes to the network structure of Er–Yb doped phosphate glass induced by femtosecond laser pulses, submitted for publication

# Chapter 4
# Optimizing Laser-Induced Refractive Index Changes in Optical Glasses via Spatial and Temporal Adaptive Beam Engineering

**Razvan Stoian**

**Abstract** The result of ultrafast laser processing of embedded refractive index changes in optical materials depends on the material relaxation paths, as well as on the spatio-temporal characteristics of the writing beam. Recently, new beam manipulation concepts were developed which allow a modulation of the energy feedthrough enabling a synergetic interaction between light and matter and, therefore, improved results. We discuss here the possibility of managing laser-induced physical phenomena employing automated temporal pulse shaping. In reviewing some of the control factors we indicate the potential of regulated energy input in triggering thermo-mechanical pathways that may establish desired refractive index distributions. The adaptive techniques indicate as well an engineering aspect related to efficient processing of structural modifications in three-dimensional arrangements. Here, a feasible solution is represented by dynamic spatial shaping techniques. The approach includes corrections for beam propagation errors and spatial intensity design in desired patterns. Insights into parallel writing techniques for complex structures using wavefront engineering will be given, with the purpose of achieving performant optical functions in a time-effective way.

## 4.1 Introduction

Photoinscription techniques employing ultrashort laser pulses have already demonstrated increased potential for three-dimensional (3D) optical functionalization of bulk transparent materials. Some of the basic and applicative aspects and their importance for optical and analytic technologies are reviewed throughout the

---

R. Stoian (✉)
Laboratoire Hubert Curien, UMR 5516 CNRS, Université Jean Monnet, 42000 Saint Etienne, Université de Lyon, 42023 Saint Etienne, France
e-mail: razvan.stoian@univ-st-etienne.fr

chapters of this book. The fundament is represented by localized refractive index changes which constitute the building blocks of more complex photonic systems embedded in optical materials. Photonic functions were demonstrated in various glasses of optical relevance and the optimal performance requires a precise adjustment of the refractive index. The laser-induced structural modifications associated with refractive index changes and the degree of processing precision can be severely influenced by the irradiation conditions. Here we essentially mean, among other factors, the rate of energy delivery. The main question relates to how fast and in which quantity the energy is accumulated and released inside the material as a function of the laser delivery rate. The second aspect, determinant for applications, is how the irradiation result can be improved considering various criteria for quality processing. From the perspective of influencing the material reaction by excitation control, this chapter focuses on material structuring approaches based on adaptive spatio-temporal pulse manipulation for optimizing ultrafast laser-induced processes inside transparent materials. Consequently, laser radiation profiles designed in spatial and temporal domains can assist light-induced photo-transformations and control the material response toward user-determined directions. This indicates the potential of improving processing results according to the chosen application. Two main strategies for gaining impact on material transformations under light exposure will be presented, adaptive temporal tailoring techniques and spatial beam forming, together with their possible integration in feedback-driven approaches.

The temporal beam tailoring techniques and the implicit controlled energy dynamics are natural candidates to influence the physical behavior of the interaction and to dominantly determine the material modification. Kinetic processes related to excitation and energy relaxation can be guided on molecular and mesoscopic scales in a controllable manner leading to an upgrade in the processing flexibility. We will review the possibility to design structural modifications with predictable properties by taking advantage of adaptive irradiation systems and self-optimization loops. The original idea was suggested by previous successful efforts in other science and technology fields, particularly in femtochemistry, to apply a new approach, i.e., the concept of optimal control involving the intelligent assistance of light, in determining specific reaction evolutions [1–3]. Widely used for control of molecular dissociation, selective bond breaking, or charge transfer in biochemical complexes, tailored pulses were integrated in a broader attempt of mastering complex systems using laser radiation. The technique can provide benefits for material processing, having in mind that similar molecular processes occur on larger dimensional scales. At the same time, the employment of temporally tailored pulses shows potential for regulating intensity-dependent phenomena as light interacts with nonlinear media. The pertinent questions are to which extent are laser-induced transitions controllable and how can a desired final state be achieved? The answer is not straightforward. However, the above observations indicate flexibility in manipulating propagation, ionization, and energy gain events generated by ultrashort laser pulses in nonlinear environments using judicious intensity adjustments. This creates material transformations that are not easily accessible and generates additional possibilities for structuring materials otherwise difficult to process in standard irradiation conditions.

These previous efforts and the development of efficient light modulators allows us to believe that a processing technique may be developed that is flexible and adaptable, responding "intelligently" to material reactions and taking advantage of the inherent changes in material properties during irradiation.

Additionally, from an engineering perspective, spatial pulse forming is equally able to promote improvements in the structuring process. It can rectify propagation errors, compensate for optical aberrations, or design the interaction region, mastering both the nonlinearity and the geometry of interaction. Time-effective concepts including the possibility of parallel processing will be consequently discussed. Dynamic control of the spatial phase can efficiently respond to expectations and become an effective way to fulfill corrective functions or to create multispot irradiation and arbitrary excitation designs. This offers multiple possibilities for nonlinearity control and proposes in turn viable concepts for upgrading laser interactions. Adaptive schemes in spatio-temporal domains can thus have an extended range of action, enabling applications that can largely benefit from light-matter synergies.

The chapter is organized as follows. A first Sect. 4.2 will review fundamental aspects related to the primary physical factors prone to play a fundamental role in controlling energy coupling and the time evolution of the excited matter. These aspects determine the resulting path for the refractive index change. Practical concepts of pulse manipulation using programable light-modulation devices will be presented in Sect. 4.3, with a focus on Fourier synthesis of light components. An overview on the achieved spatio-temporal flexibility in irradiation, the detection method, and the subsequent adaptive loops will be given. Section 4.4 will present selected aspects of application of these techniques in processing transparent materials. Insights will be given on how the dynamic light regulation creates the premises to upgrade the degree of process control concerning the relevant refractive index change, reversal of structural transitions, or energy confinement to smallest scales. Section 4.5 will consider practical implementation of spatial beam shaping for structuring bulk dielectrics, including correction of propagation errors or parallel multispot processing. The potential and the perspectives derived from these techniques will be discussed in the conclusion section.

## 4.2 Mechanisms of Laser-Induced Refractive Index Changes

In bulk transparent materials, the ability to locally design the dielectric function is based on the potential balance of electronic and structural transformations associated with the refractive index change (see e.g., [4] and the references therein). These aspects will be largely discussed in this book and we only make here a succinct presentation which delivers a basis for further argumentation. Upon focusing, ultrashort near-infrared laser pulses create physico-chemical material changes, building up on deviations from charge, thermal, or mechanical equilibrium. The high absorption nonlinearity in the normally transparent material initiates the development of bulk electron-hole plasmas by complex photoionization events. The excitation scenario leading to the creation of energetic free carriers involves laser-induced electronic transitions, interband multiphoton or field ionization, collisional heating, and avalanche multiplication [5–7]. The corresponding photoionization cross-sections or collisional rates are explicitly dependent on the value and the direction of the assisting electric field, leading to a transient nature of the optical properties [8]. This controls the efficiency of excitation and the level of the emerging electronic population up to the point of catastrophic damage. It is then the balance between photo and collisional ionization that mediates the localized formation of a hot electron population and determines how much energy is coupled in the system on ultrafast scales. Equally, bulk excitation involves light transport in a nonlinear environment defined by the dielectric material. Especially at peak intensities associated to ultrashort laser pulses, the laser pulse propagation is strongly affected by Kerr self-focusing and by defocusing or diffraction on the laser generated carrier plasma, being intrinsically related to the efficiency of the primary excitation processes [9–11]. This has consequences for the amount of energy stored in local perturbations of the dielectric matrix, as well as on the heating and cooling rates, the geometry of the heat source, and the conversion in mechanical forms via compression and rarefaction pressure waves. The strong energy deposition is then able to induce unique structural material phases due to extreme pressure and temperature regimes generated inside the material [12], controllable by the irradiation conditions.

The local permanent modification of the dielectric function can thus be viewed as interplay of several factors, including generation of defects, modifying the local atomic structure, or accumulating stress. These transitions occur on characteristic timescales and a certain control of their competition may be established, if not directly, then by manipulating the original energy source. For example, charge trapping in self-induced matrix deformations can be seen almost as fast as the laser excitation, being then followed by slower structural readjustments while the transient defect states relax [13–15]. Energy conversion in vibration modes may take several picoseconds, increasing the local temperature. Pressure waves can be generated due to swift heating in less than one nanosecond, inducing compaction and rarefaction processes [12, 16, 17], while subsequent mechanical relaxation may last up to several microseconds. These complex behaviors deliver specific changes on the real and the imaginary part of the dielectric function. The relative importance of these individual transformation channels can then be of assistance in engineering particular index changes in the situation where the transformation sequence is jointly determined by the material response and the spatio-temporal character of excitation [11]. The emerging idea is that suitable light delivery can enable the control of the aforementioned processes with consequences for the final modification. Modulated light can act on different timescales, either corresponding to electronic excitation and exploiting fast changes of the dielectric function via electron population levels, temperature and density effects on collisional rates and defect generation, or taking advantage of structural phase transitions where the relevant parameter becomes the timescale of electron-vibration coupling and lattice deformations. The photoinscription techniques may therefore be conveniently tuned to determine a suitable material behavior, i.e., positive refractive index changes in materials otherwise difficult to process, with strong fundamental and technological consequences.

Waveguiding structures were obtained in different optical materials by translating a focused laser beam longitudinally or transversally with respect to the propagation axis. Current photoinscription techniques (for a brief review see [4, 18–23] and the references therein) aim at producing positive refractive index changes in optical glasses, e.g., for waveguiding applications. Waveguiding generation was reported in many materials, dominantly in fused silica where the material quenches to a higher density state, however, the processing window is critically dependent on the material properties. As a function of the irradiation regime, different waveguide writing regimes may be identified based on the corresponding structural morphologies. They are namely related to the amount of energy deposition and can be divided in two general categories. First, gentle interactions, capable of triggering soft electronic alterations by an "erosion" process produced by underdense plasmas and weak index changes, can be observed at low irradiation doses. Strong thermo-mechanical effects associated with a higher index contrast via repeated expansion or compaction cycles [13,24] may appear in energetic irradiation conditions. An example of the irradiation result in phase-contrast microscopy is given in Fig. 4.1 showing different types of modifications achieved in two optical glasses; fused silica (a-SiO$_2$) and borosilicate crown (BK7). One can note that, as a function of the energy accumulation dose, the index change can be conveniently tuned from positive to negative variations (represented as different colors in the figure as detailed in Sect. 4.3.2).

The irradiation result in terms of refractive index changes can be seen as the essential construction element of more complex photonic devices. Relying on this processing principle, several groups were able to achieve rather complex embedded optical components, particularly in fused silica [4, 25, 26] but also in

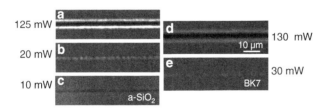

**Fig. 4.1** The effect of the irradiation dose on the structure morphology during longitudinal photoinscription of refractive index modification traces using focused ultrafast laser radiation. The structural changes are observed in phase-contrast microscopy and the black color denotes a positive index change. The left parts (**a–c**) describe various trace types in fused silica at different input powers and a scan velocity of 50 m/s. The photoinscription character varies from an incubative type at low intensities, relying on electronically-triggered structural effects which determine positive index changes, to the onset of thermo-mechanical effects at moderate and higher incident powers resulting in visible dense and rarefied regions. The right parts (**d,e**) indicate morphological changes in borosilicate BK7, where most observable are rarefaction and compaction effects. The incident pulse duration is 150 fs and the pulse repetition rate is 100 kHz. The focusing numerical aperture for the incident 800 nm laser radiation is NA = 0.42

other media of interest such as doped materials, metallic oxides or chalcogenide materials [22, 27–29], fabricating gain [30] or frequency-doubling waveguides [31], and femtosecond-written waveguide lasers [32], all requiring in general positive index changes. Among the main characteristics of interest are the amplitude of the refractive index increase [33], propagation losses [34], and symmetry [35, 36]. Early attempts of implementation of specific pulse forms approached this problem by using pulse sequences with variable separations and amplitudes. Minimization of optical losses [37] or augmentation of the positive refractive index value [33] were observed in fused silica and interpreted as preconditioning of the excited region, matrix softening, and possible involvement in different proportions of transient effects induced by optical centers or structural rearrangements. Novel approaches with extended nonlinearity control build up on a coupled spatio-temporal regulation of the laser pulse via spatial chirping, spatio-temporal focusing, non-diffractive propagation, or front tilt design [39–42]. Further control of the pulse duration was recently implemented [35] in photoinscription techniques complementary to spatial beam modulation resulting in uniform irradiation regions, suitable for waveguide writing in, for example, phosphate glasses. Improvements can be expected likewise in materials with high nonlinear refractive index such as heavy metal oxide glasses [38]. Nevertheless, as a function of the amount of energy accumulation and the relaxation properties of different materials, the result may not always be a positive index change. Irradiation generally determines a complex interplay of electronic and structural alterations associated with either increasing or decreasing the refractive index under light exposure. The latter is usually detrimental for waveguiding. In many glassy materials, especially multicomponent glasses where the laser action is more difficult to decipher, the standard ultrafast radiation induces merely a decrease of the refractive index at higher irradiation levels. This is particularly valid for "thermal" glasses, characterized by slow relaxation, low transition points, and, eventually, large expansion coefficients. Speculatively, this is related to a strong volumetric expansion and subsequent rarefaction. Guiding regions may thus be restricted to stressed regions around the excitation area. One relevant question is how to improve such a situation and realize suitable index changes directly in the central irradiation region. From this perspective, the possibility to reverse the natural tendency to rarefaction toward compaction carries then fundamental and technological significance.

The follow-up idea is to design irradiation patterns that are able to generate material phases with desired optical properties. A potential solution is the control of the excitation dynamics while triggering transformation factors that induce densification. This can be achieved in self-optimization, programable feedback loops acting on the energy delivery rate. New functional degrees of freedom are in principle possible by adjusting the energy flow to the material properties, easing the achievement of desired levels of deposition and conversion of energy. The irradiation design exercise an implicit control on the excitation and relaxation channels. This can presumably overturn an unsatisfactory standard material response, with the purpose, for example, of producing positive refractive index changes in materials where the regular response is rarefaction. It will be shown below that the standard response of materials can be tailored, indicating how plasticity can be used for materials with high volume expansion by enforcing key control factors; size, geometry, and the temperature level of the heat source. We will describe how management of nonlinearities and wavefront engineering can impose limitations on filamentation and energy deposition spatial scales. Furthermore, concepts will be presented to boost the photoinscription techniques using parallelized approaches.

## 4.3 Experimental Implementations for Pulse Engineering

### 4.3.1 Spatio-Temporal Beam Shaping

Dealing with complex interaction processes, self-improving material structuring approaches were developed to achieve the stipulated objectives. These are based on adaptive pulse spatio-temporal manipulation for optimizing ultrafast laser-induced processes. They require real-time pulse control and fast detection means, and, consequently, challenges were taken in precisely evaluating the laser action for the establishment of self-learning feedback loops. The pulse engineering technique, which exploits the laser pulse coherence, facilitates energy transfer and control on structural and thermodynamic phase transitions, attempting to manipulate matter properties. As the characteristics are defined by the type and the scale of the modification, it is worth noting that the technique offers beam manipulation capabilities in temporal and spatial domains at the same time, putting forward two perspectives for material modification. First to be noted is the phenomenological aspect related to controlling laser-induced physical phenomena by temporal synchronization between radiation and the material reaction. This may accompany a material response away from the standard relaxation, toward a constrained behavior that responds to specific user requirements. Second, the engineering aspect is related to the potential of designing irradiation geometries by spatial tailoring of excitation in arbitrary patterns or corrective procedures for distortions inherent to optical beam propagation.

In practice, the creation of complex and flexible shaped laser pulses with respect to phase and amplitude, both in the temporal and spatial domains relies dominantly on programable pulse shaping techniques [43–47], able to generate desired optical transfer functions. Several approaches exist with various degrees of appropriateness as a function of the chosen application. For the sake of simplicity, we concentrate on the case relying on Fourier synthesis of light pulse components. Without entering into details which can be found in various text books [43, 47, 48], we mention that the operation principle is based on optical Fourier transformations of the electric field $E(x, y, z, t)$, from time and space to frequency and spatial frequency domains, respectively. In Fig. 4.2a standard design of a conceptual Fourier synthesis pulse shaper is sketched. The electric field can be conveniently manipulated via spectral

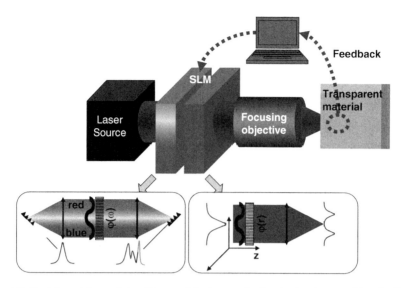

**Fig. 4.2** Basic layout for spatial and temporal femtosecond pulse shaping integrated in a feedback loop. The left inset sketches the concept of temporal pulse tailoring via Fourier synthesis of frequency components. The right inset indicates the Fourier transformer for spatial beam shaping

or spatial masks. A recent review of pulse shaping techniques with relevance for material processing applications is given in [49].

A similar irradiation system is used in most of the after-coming discussion. It includes a programable liquid-crystal pulse-shaping apparatus, which realizes temporal pulse tailoring using spectral phase filtering [45]. This standard device has the role to manipulate the spatially dispersed spectral frequency components of the pulse, allowing for spectral phase modulation and subsequent pulse temporal design. Additionally, the laser system incorporates, within the pulse control unit, a programable, optically addressed two-dimensional beam-forming system which performs spatial pulse tailoring using spatial phase control and modulation of the propagation vector for the incident laser pulse [50]. We will focus in the following on manipulating the pulse phase only in order to conserve the energy on the laser beam.

The advantage of programable modulators either based on liquid-crystal arrays [45, 50, 51], acousto-optics devices [52–56], or deformable mirrors [57, 58] is that they can react as a function of an evaluation of the irradiation result. The subsequent optimization problem is related to finding an optimal solution in a complex interaction search space, which is directly connected to an improvement criterion. This can be done by educated guesses using simple pulse forms derived from previous information concerning the interaction and designed to gain a positive influence on the laser fabrication process. Alternatively, it can be realized by performing closed-loop approaches, driven by global optimization nature-mimetic strategies. In the latter case the solution is achieved without prior knowledge of the

## 4.3.2 Microscopy Based Adaptive Loops

The efficiency of the optimization procedure is relying on the existence of an evaluation tool for the laser-affected region. An appropriate way to evaluate optical phase objects such as local changes of the material dielectric function is represented by phase-contrast microscopy. This is a convenient technique to directly monitor relative refractive index changes with respect to the dielectric environment, which transforms optical retardation introduced by the phase object into amplitude modulations. In the following examples a positive phase contrast microscope (PCM) was employed to image the interaction region in a side-view geometry. In this arrangement a two-dimensional (2D) map of the refractive index is given, where the relative positive index changes are appearing dark on a gray background, while white zones indicate negative index variations or scattering centers. It is to be noted that other measurable processes that relate to the final result of irradiation may as well be used for optimizing the interaction (e.g., plasma emission, spectroscopy probing, harmonic generation, etc). In the present case a feedback loop connects the microscopy detection and the pulse control unit (Fig. 4.3), being guided by adaptive optimization algorithms of genetic character (see e.g., [59] and references therein). The pulse tailoring unit performs the variation of the incoming intensity either in the temporal or in the spatial domain and the detection of the refractive

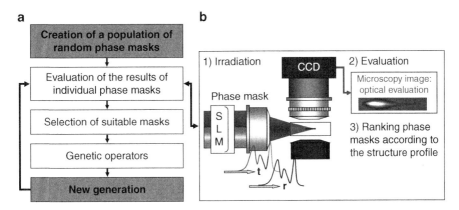

**Fig. 4.3** (**a**) Sketch of an adaptive optimization process with the main steps of the self-improvement approach; evaluation of the phase masks and generation of new solutions. The strategy involves applying and comparing each phase mask according to its ability to address a specific irradiation problem. This has the purpose to select the most fitted patterns for a new generation of an improved phase pattern family. (**b**) Example of an adaptive optimization with the aforementioned steps where the feedback is derived from microscopy observations

index delivers the quantitative evaluation of the laser action. Retrieving the 2D map of the laser generated structure, an objective functional is defined by analyzing the axial morphology of the photo-inscribed phase object and comparing it to a user-designed profile. Based on the resemblance a note (fitness) is assigned. The laser pulse envelope is then iteratively changed to increase the success of particular laser pulses in maximizing the functional fitness. For each optimization attempt described in the following, particular functional objectives will be discussed. The optimal outcome is an intensity shape that produces index patterns close to desired profiles.

## 4.4 Material Interaction with Tailored Pulses

### 4.4.1 Refractive Index Engineering by Temporally Tailored Pulses

For demonstration purposes the case of Schott BK7 borosilicate crown glass will be discussed due to the fact that this material shows a narrow laser processing window for positive refractive index changes upon the laser scanning [17, 60]. This requires fine tuning of the energy density which enhances the demands for precise processing and, therefore, for determining optimal irradiation conditions. This particular glass, used here as a model material with high expansion coefficient and low softening point, usually shows a decrease of the refractive index under standard intense tightly focused ultrafast laser excitation above the critical self-focusing power. This behavior is associated with the formation of a hot region, where, due to rapid thermal expansion, the material is quenched to a low-density phase, rich in oxygen centers. The result is a lower refractive index region in the focal spot, being an outcome that requires further improvements.

The temporal shaping approach indicated the possibility to flip the refractive index in borosilicate crown BK7 [17] from the standard negative change to a significant region of index increase observed in static structures (Fig. 4.4a,b). In the optimization loop a corresponding functional target was chosen that augments the positive index change according to a user profile defined by a Gaussian mask. The optimal pulse (OP) is a radiation sequence extending on several picoseconds. The control mechanism is related to the design of the resultant heat source which influences the subsequent stress-induced plasticity driving axial compaction. As compared to the standard short pulse (SP) irradiation, the ps sequence allows higher energy concentration and the achievement of an elevated temperature due to a less efficient plasma generation and light defocusing on free carriers. This leads to plastic deformation accompanied by partial healing of the lateral stress due to preferential heat flow. As a result, a transition from a radial expansion regime characteristic to ultrashort pulse excitation to directional compaction was observed. The matter momentum relaxation leads to axial densification and to a

4 Optimizing laser-induced Refractive Index Changes in Optical Glasses    77

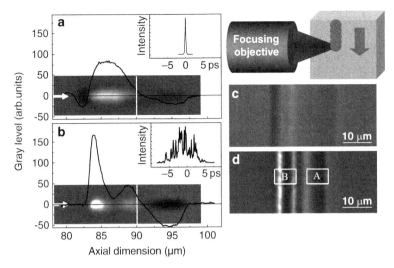

**Fig. 4.4** Refractive index flip in borosilicate crown BK7 under multipulse (**a**) short (150 fs) and (**b**) optimally tailored irradiation (4.5 ps). Pulse energy 0.17 μJ, irradiation dose $10^5$ pulses at 100 kHz, with the 800 nm pulses focused from the left via an NA of 0.42. Phase contrast microscopy images of refractive index changes (white and black colors represent negative and positive refractive index changes, respectively) with corresponding pulse envelopes given in the insets. The axial index profiles are indicated by the solid lines. The onset of a significant compression zone is visible under optimal conditions (intense black color). Transverse trace written by (**c**) short pulses and (**d**) optimal laser pulses at a scan velocity of 50 μm/s and pulse energy of 1.1 μJ. Two regions of positive refractive index changes are indicated by the labels A and B. Data were extracted from [17]

positive refractive index change. This scenario was confirmed by simulating the nonlinear pulse propagation and the subsequent elasto-plastic material behavior [17]. It appears that the adaptive technique was able to determine an excitation sequence which induces a thermo-mechanical path leading to compaction. This is particularly interesting for laser repetition rates on the timescale of mechanical relaxation (hundreds of kHz) and shows the importance of the heating and relaxation rates for defining proper processing windows. The result of static irradiation was then replicated by transverse translation of the laser beam where short pulse and optimal pulse irradiation were compared (Fig. 4.4c,d). Compressed regions are observable in the optimal transverse trace [17]. The influence one can exercise on the refractive index distributions indicates the possibility to create waveguide structures in materials that do not easily allow it in standard ultrafast irradiation conditions.

One important observation in this thermo-mechanical regime of interaction is that the energy should be efficiently restricted to the interaction zone, realizing a maximum effect with minimal energy costs. However, two mechanisms control the spread of energy: beam filamentation and wavefront distortion.

## 4.4.2 Energy Confinement and Size Corrections

Based on the idea that the nonlinearity of interaction can be influenced, adaptive control of pulse temporal forms was recently used to regulate filamentary propagation in nonlinear environments [61, 62]. The location and the spectral properties of the ionization region were shown to be modulative. The key factor is the intensity feedthrough which determines the competition between self-focusing and ionization. Breakdown probability was equally observed to be controllable via temporal envelopes [63] which determines the time development of excitation. The nonlinear aspect of interaction may also suggest the temporal form as a control knob to regulate the propagation effects while restricting the energy deposition volume. The nonlinear control has consequently proven its capability to induce energy confinement even in the presence of wavefront distortions [64] and we have noticed before the importance of limiting energy dispersion and confining it to minimal scales. High energy concentrations judiciously deposited can generate positive refractive index changes based on material compaction, thus enabling guiding using this regime of photoinscription [59]. However, 3D structuring of bulk dielectrics requires focusing through air-dielectric interfaces. Consequently, depth-dependent spherical aberration appears due to distinct ray paths upon refraction converging in different spatial locations. This determines an elongation of the energy deposition area and restricts the structuring accuracy.

With a focus on refractive index modification processed in optical materials we analyze below some aspects generating spatial energy confinement and structural modification in dielectric media by suitable temporal pulse forming when the illuminated zone is elongated due to spatial wavefront distortions [64]. As in the preceding case a programable temporal shaping apparatus is used. It will be shown that temporal forming of ultrafast laser pulses can restrict the energy spread, leading to a higher confinement. We indicate the role of reduced nonlinearity in plasma formation as a control factor coupling the spatial and temporal response of the material. The size corrections enable higher processing accuracy.

Let us analyze below the single pulse irradiation effects observable in Fig. 4.5a. The permanent traces indicate a refractive index modulation which is generated by an axial variation of excitation. This modulation derives from a dynamic filamentation balance between dispersion, nonlinear focusing, and ionization, resulting in a varying refractive index structure along the propagation axis. This shows a central positive (dark) index change bordered by white regions of decreased density and increased inhomogeneity. The white domains were previously identified with regions of maximum energy deposition [11], where material expands thermally. The left-side white dot is presumably a self-focusing effect [11, 65, 66]. The black, positive refractive index region involves a complex mixture of thermo-mechanical phenomena and defect generation. If the spatial index modulation accentuates with the depth, the modulation domain appears to be defined by the longitudinal aberration [67]. When aberrations occur, the length of the laser-induced structure augments (Fig. 4.5a), being detrimental to the photoinscription precision. The focal

4  Optimizing laser-induced Refractive Index Changes in Optical Glasses         79

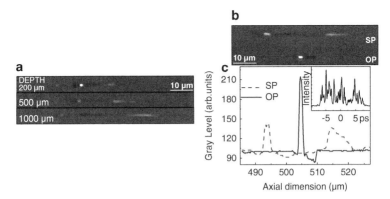

**Fig. 4.5** (**a**) PCM images of static permanent structures induced by a single short pulse (SP) in bulk irradiated $a$-SiO$_2$ for various working depths (NA=0.45). The elongation of the focal region becomes visible, increasing with the processing depth. Irradiation parameters: 160 fs and 1 µJ for 800 nm radiation wavelength. (**b**) Short pulse (SP) and optimal pulse (OP) induced structures at a depth of 500 µm and 1 µJ input energy indicating the spatial confinement of the OP structure even in the presence of aberrations. (**b**) The corresponding axial cross-sections of the resultant index patterns. The inset shows the optimal pulse shape [64]

elongation influences the nonlinear energy deposition and stronger modulation of the refractive index appears in the exposed region. Relying on intensity to control the energy spread during propagation, it was indicated that the energy can be confined using adaptive temporal pulse shaping, delivering in addition desired changes of the refractive index [64]. The result is presented in Fig. 4.5b. The decreased nonlinearity and the lower ionization efficiency of the optimal pulse assist the energy confinement, regulating the structuring precision in the presence of wavefront distortions [64]. These aspects are detailed below.

Preserving positive index changes, an attempt to reduce the spatial extent of the white regions was made using the temporal shaping strategy. The objective functional was defined accordingly as the ratio between the number of black and white pixels, respectively, detected in the PCM image within a narrow axial region. The automated temporal feedback loop was applied with the purpose of maximizing the index ratio. The optimization result is an intensity shape that has produced refractive index patterns with maximum contrast. The result of the optimal pulse compared to the 160 fs short pulse irradiation is shown in Fig. 4.5b for a depth of 500 µm, where wavefront distortions due to spherical aberrations are present. The irradiation conditions are given in the figure caption. The optimal sequence takes the form of a structured intensity envelope extending on several picoseconds scale, suggesting that the control parameter lies in the extended envelope. This particular sequence provides a more than two times increase for the objective functional associated with a drastic decrease in the size of the structure, while keeping the negative index region to a minimum. A smoother ps sequence is equally effective, with a minor role attributed to the multipeaks. The consequences of a ps envelope are twofold [11]. First, the ps envelope induces a retarded, low density, spatially modulated plasma. This creates a smaller negative shift for the incoming energy and less defocusing, thus helping to concentrate the energy in the region of best focus. Second, the nonlinearity of the excitation diminishes, allowing efficient absorption only in a restricted region around the geometric focal point, where the modification threshold is surpassed.

The nonlinear propagation coupled with the material response defines the topology of the solution space. In order to gain insights into the propagation and modification factors, simplifying the control landscape by using less complex pulse shapes appears to be a suitable approach. This brings interesting clues for explaining the partial filamentation control which is related to the decreased nonlinearity and the lower ionization efficiency. Simpler pulse designs, multiples pulses or pulse envelopes resembling the optimal sequences were used, confirming the presented factors [64]. The conclusion is that the pulse tailoring can assist the energy confinement while reducing the structure ellipticity and offers potential benefits for 3D optical processing [64].

### *4.4.3 Adaptive Correction of Wavefront Distortions*

In the above section the intensity factor was indicated for controlling nonlinear propagation in the presence of aberrations. We discuss below a natural strategy for counteracting wavefront distortion effects which occur during ultrafast laser-induced changes of refractive index as result of deep focusing. The proposed approach is based on programable spatial phase modulation and has the objective of concentrating the laser energy on minimal spatial scales. Using adaptive spatial tailoring of ultrashort laser pulses, spherical aberrations can be dynamically corrected in a direct manner in synchronization with the writing procedure. This facilitates for example optimal writing of homogeneous longitudinal waveguides over significant lengths over significant lengths and at various positions.

This is also the place to review current approaches to the problematic of wavefront distortions. The effect of spherical aberration in ultrafast laser material processing was previously analyzed via depth-dependent material modification thresholds and aspect-ratio measurements, underlining the importance of preserving excitation conditions [68–72]. Microscope objective collars delivering adjustable compensation of spherical aberration were employed for controlling the modification size in bulk optical materials [12]. Corrective functions using adaptive optics were applied to minimize aberrations [72, 73] for data storage, microscopy, and imaging applications, indicating spatial phase corrections as straightforward solutions to amend propagation errors. Consequences for quality waveguide writing were already indicated [68–72, 74]. With emphasis on applications in light-guiding technologies, we first review here a technique able to dynamically correct spherical aberration and to optimize the process of photowriting longitudinal waveguides in a dielectric environment using automated spatial phase filtering integrated in adaptive loops.

The problem of spherical aberration can be analytically addressed in an effective manner using Zernike or other forms of polynomial decomposition. A perspective on the issue is given in [75]. However, the approach requires calibration of the phase retardation induced by the shaping system and an accurate correspondence between the phase-manipulation plane and the pupils of the optical focusing system. Using a simple problem as a proof-of-principle demonstration, we intend to show that global search algorithms [76] represent an effective calibration-free technique, able to significantly improve the structuring process even when the nature of wavefront deformation is not accurately known or additional nonlinear effects are present. No prior assumptions have to be made on the processes of wavefront deformation upon focusing and advantages for material structuring applications may be derived when the feedback response is obtained directly from a laser processing result.

We have previously noticed the reduction of material density in BK7 static traces in standard ultrafast irradiation conditions and this specific topology can presumably be connected to a strong dilatation of the irradiated volume in glasses characterized by high thermal expansion. After the initial laser heating, the material expands while cooling, which inhibits the backward relaxation and freezes the material in a low density phase. However, high energy densities generate compressive pressure waves [12, 16, 17] and determine the formation of a strongly compacted region around the low density core. This indicates the achievement of a high temperature in the interaction region, followed by the onset of the surrounding mechanical densification. These observations provide the prerequisites for a significant positive change in the refractive index at the structure tip. An example of this topology is given in the left panel of Fig. 4.6, being especially visible at low processing depths, where the influence of aberrations is negligible and the confinement is maximal. Upon photoinscription, if the energy concentration stays sufficient, the high density positive index phase that surrounds the core can be replicated during the scan, leading to the formation of a waveguide. In order to observe this consequence for a dynamic regime of photoinscription and to create guiding elements, the focal point was rastered along the propagation axis.

Let us analyze the irradiation results in a regime where thermo-mechanical effects dominate. The longitudinal translation at high repetition rates (100 kHz) leads to specific traces upon scanning in the direction of the laser pulse [59], accounting for the effects of a moving heat source. For the high power regime, the photoinscribed longitudinal line delivers a contrasted region of positive index change. This is denoted by the intense black color, visible in the top inset of Fig. 4.6a and bordered by narrow lines of decreased index. This structure shows a high and homogeneous index contrast which may be connected as indicated to the presence of the surrounding high index, compressed region observable in the static trace. If the energy density diminishes (e.g., by focal spot elongation), this irradiation regime will determine a dominant negative refractive index change (represented by the perceptible white color in the bottom inset of Fig. 4.6a), which, in normal conditions, inhibits guiding. It appears then that a uniform region of

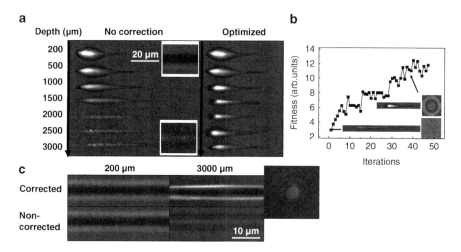

**Fig. 4.6** (**a**) Non-corrected (*left*) and spatially corrected (*right*) static PCM structures induced in BK7 at different depths. The structures are generated by $10^5$ pulses of 150 fs duration at 100 kHz and 125 mW average power. The input radiation wavelength is 800 nm and the effective focusing NA is 0.36. (**b**) Evolution of the trace fitness during the optimization run at the depth of 2500 $\mu$m. Example of traces and corresponding gray-level phase masks at different moments are given as well. (**c**) Longitudinal structures (PCM) at different working depths in corrected (*top*) and non-corrected (*bottom*) cases. The corrections enable a positive refractive index change over a distance of 3 mm. Scanning speed is 1 $\mu$m/s at 125 mW average power. Right, far-field pattern of the guided mode at 633 nm for the corrected guide [59]

positive refractive index change generated by mechanical compaction occurs during translation only when a critical density of energy was transported at the interaction place. For lower energy densities, moderate thermal expansion and rarefaction upon cooling determine to a large extent the material response, inducing a dominant low density low index phase. Inevitably upon irradiation at different depths, the energy density decreases due to spherical aberration and limits the possibility to trigger the positive index regime far from the air-glass interface.

Since this property degrades with the depth, we will focus below on the possibility to annihilate wavefront distortions and to restrict the energy spread by spatial phase adjustments. In order to reach the compressive regime at arbitrary depths, it is imperative that the energy delivery remains concentrated to the narrowest region. This was achieved by adaptively determining the corrective spatial phase masks in the microscopy-based feedback loop presented before [59], where the functional objective was associated with achieving a structure of minimal length and high contrast. Figure 4.6a shows the results of the optimization procedure for photoinscription as compared to the effect of the uncorrected pulse for different depths into the material. If the structures induced by the uncorrected pulse show a threefold increase in length down to a depth of 3000 $\mu$m (Fig. 4.6a left), the correction procedure has stabilized the structure length at almost the initial size (Fig. 4.6a right) irrespective of the processing depth. An example of the iterative improvement during the optimization run is also given in Fig. 4.6b. The effective minimization of the damaged region, sometimes better than the confinement achieved by the theoretically determined correction phase masks [59] demonstrates the efficiency of the loop, the apparently low nonlinear contribution for this specific material, and the fact that the adaptive strategy is suitable for more complex aberration correction problems, including wavefront distortions due to nonlinear effects. The success of the operation is verified by effectively writing longitudinal guiding structures. The spatial phase correction masks corresponding to correction solutions at different depths were gradually applied in synchronization with the advance of the structure inside the glass material. A complementary technique for further increasing the energy density deposited within the material, relying on pulse linear temporal chirping, was concomitantly applied to delay the plasma formation and limit light defocusing effects. The photoinscription outcome is depicted in the top part of Fig. 4.6c which synthesizes the result of the corrective procedure involving both spatial and temporal modulation. Both ends of the longitudinal structure are shown. As a result, a uniform dark structure is becoming visible, indicating a positive index change all along the trace length. This indicates that the dynamic correction enables the generation of a uniform cylindrical waveguide with a high positive index contrast for a length superior to standard irradiation and which allows symmetric guiding on structures as long as 3 mm in rather tight focusing conditions. For comparison, the uncorrected trace is shown as well in the bottom part of Fig. 4.6c. In this case, the guiding region is restricted to 1 mm, the rest of the trace showing a negative index change.

Apart from wavefront corrections, other techniques and beam shaping schemes emerged and were used to preserve a certain symmetry of writing [20, 21, 23, 35, 36, 77]. These approaches involve the design of the excitation region with respect to the elongation in directions parallel or perpendicular to the propagation axis and will be treated in detail in Chap. 5.

A further perspective, perhaps not fully exploited, lies in the use of focal spot engineering or apodization functions applied to the laser beam with the purpose to localize nonlinear excitation at and below the diffraction limit [78–80]. Successfully applied on surfaces, this may show benefits for bulk processing, with the further prospect of manipulating matter movement via field gradients [81]. Promising concepts for 3D designs [82] and z-invariant and non-diffractive [83, 84] beams can potentially be applied to volume processing. Powerful algorithms ranging from direct or iterative Fourier inversions to more complex non-deterministic approaches are then requested to steer these actions, a complex topic not approached here due to space limitations. Concerning the structuring aspect. Concerning the structuring aspect, research in laser shaping techniques proposed non-diffractive beams that can be directly employed to process materials [85, 87, 88]. Bessel beams were used to generate low-loss and birefringence-free waveguides in fused silica [89]. The potential of non-diffractive beams to control filamentation and optical damage on long distance due to the characteristic self-regenerative capabilities was recognized and applied in [90] to generate arrays of equidistant spots. The possibility to fabricate linear nanosized void arrangements was also demonstrated in the presence of spherically aberrated and truncated beams generating predictable axial modulation of propagation [86, 91, 92] beams generating predictable axial modulation of propagation [91, 92]. Recently, unusual filamentation, plasma generation, and propagation on bent trajectories were indicated in the presence of nonsymmetric laser beams [93, 94].

### 4.4.4 Dynamic Parallel Processing

As femtosecond laser radiation can generate localized increase of the refractive index, the simple translation of the laser spot can achieve complex light-guiding structures in the three dimensions. However, if a capability of spatial shaping is present, one of the most important aspects is the potential of performing several tasks in parallel by beam division and multispot processing. If extensively applied for surface processing, we will focus in the following on bulk applications. Whereas the 3D photoinscription access constitutes a clear advantage over multilayer lithographic techniques, the single laser focal spot must undergo potentially complicated movements with respect to the sample in order to draw a whole photonic structure. This single spot operation may involve long processing times when the writing of several complex structures is envisaged. The beam shaping properties allow easy access to the 3D space with variable patterns. A procedure of dynamic ultrafast laser beam tailoring for parallel photoinscription of photonic devices in fused silica using double spot operation [95, 96] will be analyzed in the following. A first concept will be presented where the wavefront of the beam is modulated with a simple periodical binary ($0$-$\pi$) phase mask to achieve multi spot operation. A schematic description is given in Fig. 4.7 and further details to the dynamic phase masking can be found in [95, 96]. The technique was applied for parallel fabrication of light-guiding structures in the bulk at moderate and high laser repetition rates [95, 96]. Light dividers in the three dimensions and wavelength-division demultiplexing devices relying on evanescent wave coupling were fabricated.

Dynamic multispot operation based on laser wavefront modulation constitutes a step forward as it enables the simultaneous processing of several embedded structures, thus lowering the process time demand and reducing the complexity of the mechanical support design. It becomes an obvious application for replicating structures of similar character where a good example is represented by regular arrays. Embedded waveguide arrays have been recently investigated, pointing out novel aspects of light propagation in discrete spaces [97] in the presence of evanescent coupling [98, 99]. 2D and 3D waveguide arrays sequentially written by femtosecond laser radiation were recently achieved showing interesting light evolution properties [38, 100, 101]. In particular, linear effects such as Bloch oscillations [102], light localization, and quasi-incoherent propagation [103] were theoretically predicted and experimentally verified. Nonlinear behaviors in the form of spatial solitons were indicated as well in such structures [104] opening a perspective for a new types of nonlinear components written by femtosecond lasers.

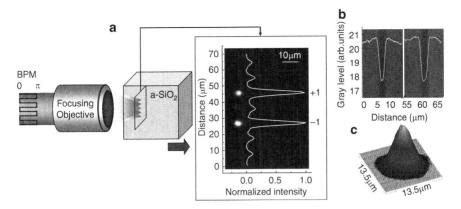

**Fig. 4.7** (**a**) Figure of merit of double spot operation. BPM: Binary phase mask. Inset: Double spot intensity profile in the sample at the focus of the objective generated by a step phase grating. The arrow shows the motion of the sample for longitudinal photoinscription. Right: the result of double spot operation through an effective NA of 0.3. (**b**) PCM picture of two simultaneously photowritten waveguides through double spot operation at 800 nm, 5 $\mu$m/s, 10 kHz, 24 mW, superimposed with horizontal cross section of the index increase. (**c**) near-field profile of one of the waveguides pictured in (**b**) at 633 nm. Results are taken from [95]

Considering multispot operation as a natural solution to achieve these types of structures, it will be demonstrated that complex 2D and 3D arrays of light-guiding structures can be efficiently written, with corresponding optical performances [95, 96]. In the following, examples of 2D photonic devices relying on evanescent coupling fabricated by double spot operation are presented, performing various optical functions. In the longitudinal configuration for waveguide writing, the resulting guides keep the rotational symmetry of the irradiation beam. In order to underline the 3D capabilities and the high flexibility of the double spot technique, various types of optical components are presented hereafter (Figs. 4.8 and 4.9). A first example (Fig. 4.8a–d) refers to a structure where the two arms were simultaneously written using continuous translation and a variation of the phase step cycling frequency to achieve variable spot separation via the wavefront control. Upon injection with two different radiation wavelengths, the device performed as a wavelength demultiplexer relying on the different coupling efficiencies between the two arms at various spectral contents as indicated in Fig. 4.8c. The spectral discrimination is demonstrated by the spatial separation of the two mode fields corresponding to each injection wavelength (Fig. 4.8d). Second, a twisted X-coupler is depicted in Fig. 4.8e–g. This device was photowritten in a single scan of the sample by translation and gradual rotation of the modulator binary phase mask. The structure dimensions (separation, overlapping length) were designed in order to regulate the transmission of the 633 nm light injected in the bottom arm to the top one. Figure 4.8g depicts the experimentally obtained near-field mode at the output of the device. We note that in all these operations the achieved index change is isotropic and has a low relative contrast ($10^{-4}$–$10^{-3}$) required to determine large mode fields.

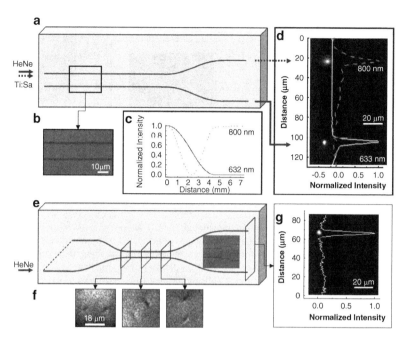

**Fig. 4.8** (a) Bulk photowritten wavelength demultiplexing device achieved through double spot operation in a single scan. Schematic view of the structure. Irradiation conditions are similar to Fig. 4.7. (b) Near-field profile under 633 nm (*solid*) and 800 nm (*dashed*) simultaneous injection in the top arm. (c) PCM side-image of the overlapping region. (d) Theoretical prediction of the 633 nm (*solid*) and 800 nm (*dashed*) intensity variations in the excited waveguide according to [105, 106] taking into account the wavelength dependance of the coupling coefficient. The structure length is 7.4 mm. (e) Schematic view of a bulk photowritten twisted X coupler achieved through double spot operation in a single scan. (f) Optical transmission microscopy pictures of the central region, showing 90° rotation. (g) Near-field profile at the output of the top arm under 633 nm injection in the bottom waveguide

A second type structure of interest achieved by multiple spot operation is represented by 2D and 3D light dividers. A first example indicates an uneven waveguide array fabricated by applying sequential translations of irradiation spots generated by binary phase patterns with different cycling frequencies. The light division function is depicted in Fig. 4.9a–c. The extension to 3D employing additionally a rotation of the phase mask consists of a one-to-seven hexagonal light divider for 633 nm (Fig. 4.9d). An additional central guide was photowritten without any wavefront modulation at lower power. The distance between two adjacent guides equals the separation between the central guide and the six others (being equal to 15 $\mu$m), the whole structure forming a regular hexagon. We note that recently diffractive approaches on regular grids were used to control multibeam break-up and the generation of filaments of remarkable spatial stability [107]. These examples show that complex 3D parallel fabrication techniques can be implemented to speed up the effectiveness as well as the degree of complexity for fabricated structures. Recently,

4 Optimizing laser-induced Refractive Index Changes in Optical Glasses

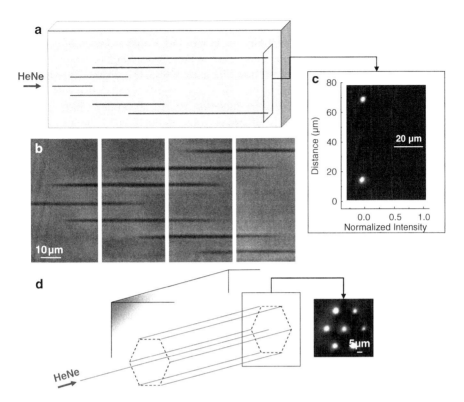

**Fig. 4.9** Bulk dividers based on evanescent wave coupling in partial arrays achieved through double spot operation. The irradiation conditions are similar to Fig. 4.7. (**a**) Schematic view of the structure. (**b**) Assemblage of PCM side-pictures of the device, its total length is 8.2 mm and the lateral separation between tracks is 9 μm. The overlapping distances were determined to induce efficient evanescent coupling. (**c**) Near-field profile under 633 nm light injection. (**d**) Schematic view of a bulk photowritten 1 to 7 light divider. The length of the structure were chosen to enable evanescent coupling from the central to the external guides. Inset: Near-field profile under 633 nm injection in the central waveguide

highly flexible 3D parallel techniques were proposed [108, 109] that may moreover combine beam demultiplexing approaches with aberration corrective techniques [110], particularly effective for structuring bulk materials.

## 4.5 Outlook and Conclusions

The present review has illustrated that control of laser-induced effects in bulk materials is possible by acting on the temporal and spatial development of the original excitation source in a regime where visible thermo-mechanical effects occur. The control could in principle be exercised on various mechanisms, including the efficiency of defect generation, heat transport, and stress accumulation. In all cases the energy coupling can be optimized accordingly. The outcome is a laser structuring approach showing additional flexibility, accuracy, and a higher degree of process control. This has relevance for upgrading current laser processing technologies and offers a better understanding of the laser-induced physical processes, the nature of material modification, including the ability to identify competitive relaxation processes. The regulation factors were identified in the absorption phase, in the degree of heating, or nonlinearity of propagation, as well as in the succession of phase transitions. Consequently, using temporal pulse forming, control may be achieved on the structural changes of the irradiated material and on the energy confinement on smallest spatial scales. The potential spectrum of applications is related to quality structuring for increased optical functionality in glasses which are otherwise more difficult to process.

We have also shown that adaptive optics in the spatial domain in connection to feedback loops are effective means to fulfill corrective functions during ultrafast laser photoinscription of waveguiding structures in optical glasses. A procedure of automated wavefront correction, namely spherical aberration, was discussed which can be correlated with the photoinscription process, requires no calibration, and allows extended versatility. The technique of spatial light modulation was extended to multifocus, parallel processing with 3D capacity. 2D and 3D light couplers and dividers relying on evanescent coupling were achieved with this method. All these manipulation capabilities constitute therefore strong arguments for developing smart processing techniques with high effectiveness. Together with potential control over the global, vectorial character of the laser pulse, access to sub-wavelength scales is possible and unprecedented flexibility and reactivity of machining techniques can thus be obtained.

**Acknowledgements** The author is deeply indebted to C. Mauclair, A. Mermillod-Blondin, and G. Cheng for their enthusiastic involvement in many of the presented studies. The participation of I. M. Burakov, N. M. Bulgakova, Yu. P. Meshcheryakov, A. Rosenfeld, I. V. Hertel, N. Huot, and E. Audouard is equally acknowledged.

# References

1. R.S. Judson, H. Rabitz, Phys. Rev. Lett. **68**, 1500 (1992)
2. A. Assion, T. Baumert, M. Bergt, T. Brixner, B. Kiefer, V. Seyfried, M. Strehle, G. Gerber, Science **282**, 919 (1998)
3. C. Daniel, J. Full, L. Gonzáles, C. Lupulescu, J. Manz, A. Merli, S. Vajda, L. Wöste, Science **299**, 536 (2003)
4. K. Itoh, W. Watanabe, S. Nolte, C. Schaffer, MRS Bull. **31**, 620 (2006)
5. L.V. Keldysh, Sov. Phys. JETP **20**, 1307 (1965)
6. B.C. Stuart, M.D. Feit, A.M. Rubenchik, B.W. Shore, M.D. Perry, Phys. Rev. Lett. **74**, 2248 (1995)
7. B. Rethfeld, Phys. Rev. Lett. **92**, 187401 (2004)
8. I.H. Chowdhury, X. Xu, A.M. Weiner, Appl. Phys. Lett. **86**, 151110 (2005)
9. A. Couairon, L. Sudrie, M. Franco, B. Prade, A. Mysyrowicz, Phys. Rev. B **71**, 125435 (2005)
10. D.M. Rayner, A. Naumov, , P.B. Corkum, Opt. Express **13**, 3208 (2005)

11. I.M. Burakov, N.M. Bulgakova, R. Stoian, A. Mermillod-Blondin, E. Audouard, A. Rosenfeld, A. Husakou, I.V. Hertel, J. Appl. Phys. **101**, 043506 (2007)
12. S. Juodkazis, K. Nishimura, S. Tanaka, H. Misawa, E. Gamaly, B. Luther-Davies, L. Hallo, P. Nicolai, V. Tikhonchuk, Phys. Rev. Lett. **96**, 166101 (2006)
13. A.M. Streltsov, N.F. Borrelli, J. Opt. Soc. Am. B **19**, 2496 (2002)
14. S.S. Mao, F. Quéré, S. Guizard, X. Mao, R.E. Russo, G. Petite, P. Martin, Appl. Phys. A Mat. Sci. Process. **79**, 1695 (2004)
15. C.W. Ponader, F.J. Schroeder, A. Streltsov, J. Appl. Phys. **103**, 063516 (2008)
16. M. Sakakura, M. Terazima, Y. Shimotsuma, K. Miura, K. Hirao, Opt. Express **15**, 5674 (2007)
17. A. Mermillod-Blondin, I.M. Burakov, Y.P. Meshcheryakov, N.M. Bulgakova, E. Audouard, A. Rosenfeld, A. Husakou, I.V. Hertel, Phys. Rev. B **77**, 104205 (2008)
18. H. Misawa, S. Juodkazis, (eds.), *3D Laser Microfabrication: Principles And Applications* (Wiley-VCH, Weinheim, 2006)
19. R.R. Gattass, E. Mazur, Nat. Photonics **2**, 219 (2008)
20. G. Cerullo, R. Osellame, S. Taccheo, M. Marangoni, D. Polli, R. Ramponi, P. Laporta, S. De Silvestri, Opt. Lett. **27**, 1938 (2002)
21. Y. Cheng, K. Sugioka, K. Midorikawa, M. Masuda, K. Toyoda, K. Kawachi, K. Shihoyama, Opt. Lett. **28**, 55 (2003)
22. S. Nolte, M. Will, J. Burghoff, A. Tünnermann, Appl. Phys. A Mat. Sci. Process. **77**, 109 (2003)
23. R.R. Thomson, A.S. Bockelt, E. Ramsay, S. Beecher, A.H. Greenaway, A.K. Kar, D.T. Reid, Opt. Express **16**, 12786 (2008)
24. J.W. Chan, T.R. Huser, S.H. Risbud, D.M. Krol, Opt. Lett. **26**, 1726 (2001)
25. K.M. Davis, K. Miura, N. Sugimoto, K. Hirao, Opt. Lett. **21**, 1729 (1996)
26. E.N. Glezer, M. Milosavljevic, H. L., R.J. Finlay, T.H. Her, J.P. Callan, E. Mazur, Opt. Lett. **21**, 2023 (1996)
27. A. Apostolopoulos, L. Laversenne, T. Colomb, C. Depeursinge, R.P. Salathé, M. Pollnau, R. Osellame, G. Cerullo, P. Laporta, Appl. Phys. Lett. **85**, 1122 (2004)
28. J. Siegel, J.M. Fernández, A. García-Navarro, V. Diez-Blanco, O. Sanz, J. Solis, Appl. Phys. Lett. **86**, 121109 (2005)
29. A. Zoubir, M. Richardson, C. Rivero, A. Schulte, C. Lopez, K. Richardson, Opt. Lett. **29**, 748 (2004)
30. Y. Sikorski, A.A. Said, P. Bado, R. Maynard, C. Florea, K.A. Winick, Electron. Lett. **36**, 226 (2000)
31. S. Campbell, R.R. Thomson, D.P. Hand, A.K. Kar, D.T. Reid, C. Canalias, V. Pasiskevicius, F. Laurell, Opt. Express **15**, 17146 (2007)
32. G. Della Valle, S. Taccheo, R. Osellame, A. Festa, G. Cerullo, P. Laporta, Opt. Express **15**, 3190 (2007)
33. D. Wortmann, M. Ramme, J. Gottmann, Opt. Express **15**, 10149 (2007)
34. H. Zhang, S.M. Eaton, P.R. Herman, Opt. Express **14**, 4826 (2006)
35. W. Gawelda, D. Puerto, J. Siegel, A. Ferrer, A. Ruiz de la Cruz, H. Fernandez, J. Solis, Appl. Phys. Lett. **93**, 121109 (2008)
36. M. Ams, G.D. Marshall, D.J. Spence, M.J. Withford, Opt. Express **13**, 5676 (2005)
37. T. Nagata, M. Kamata, M. Obara, Appl. Phys. Lett. **86**, 251103 (2005)
38. V. Diez-Blanco, J. Siegel, J. Solis, Appl. Surf. Sci. **252**, 4523 (2006)
39. G. Zhu, J. van Howe, M. Durst, W. Zipfel, C. Xu, Opt. Express **13**, 2153 (2005)
40. D.N. Vitek, E. Block, Y. Bellouard, D.E. Adams, S. Backus, D. Kleinfeld, C.G. Durfee, J.A. Squier, Opt. Express **18**, 24673 (2010)
41. F. He, H. Xu, Y. Cheng, J. Ni, H. Xiong, Z. Xu, K. Sugioka, K. Midorikawa, Opt. Lett. **35**, 1106 (2010)
42. W. Yang, P.G. Kazansy, Y.P. Svirko, Nat. Photon **2**, 99 (2008)
43. M. Wollenhaupt, A. Assion, T. Baumert, in *Springer Handbook of Lasers and Optics*, ed. by F. Träger (Springer Science + Business Media, New York, 2007)
44. A.M. Weiner, Prog. Quantum Electron. **19**, 161 (1995)

45. A.M. Weiner, Rev. Sci. Instrum. **71**, 1929 (2000)
46. L.A. Romero, F.M. Dickey, in *Laser beam shaping: Theory and techniques*, ed. by Dickey, F. M. and Holswade, S. C. (Marcel Dekker, Inc, New York, Basel, 2000), p. 21
47. W. Goodman, *Introduction to Fourier Optics*, 2nd edn. (McGraw-Hill, Singapore, 1996)
48. J.C. Diels, W. Rudolph, *Ultrashort Laser Pulse Phenomenon: Fundamentals,Techniques, and Applications on a Femtosecond Time Scale*, 2nd edn. (Academic Press, London, 2006)
49. R. Stoian, M. Wollenhaupt, T. Baumert, I.V. Hertel, in Laser Precision Microfabrication, ed. by K. Sugioka, M. Meunier, A. Pique (Springer Verlag, Heidelberg, 2010) vol. 135, p. 121
50. N. Sanner, N. Huot, E. Audouard, C. Larat, J.P. Huignard, B. Loiseaux, Opt. Lett. **30**, 1479 (2005)
51. M.M. Wefers, K.A. Nelson, Opt. Lett. **93**, 2032 (1993)
52. S.H. Shim, D.B. Strasfeld, E.C. Fulmer, M.T. Zanni, Opt. Lett. **31**, 838 (2006)
53. J.X. Tull, M.A. Dugan, W.S. Warren, in *Advances in Magnetic and Optical Resonance*, ed. by W.S. Warren (Academic Press, London, 1997)
54. D. Goswami, Phys. Rep. **374**, 385 (2003)
55. P. Tournois, Opt. Commun. **140**, 245 (1997)
56. F. Verluise, V. Laude, Z. Cheng, C. Spielmann, P. Tournois, Opt. Lett. **25**, 575 (2000)
57. E. Zeek, K. Maginnis, S. Backus, U. Russek, M.M. Murnane, G. Mourou, H.C. Kapteyn, Opt. Lett. **24**, 493 (1999)
58. F. Druon, G. Chériaux, J. Faure, J. Nees, M. Nantel, A. Maksimchuk, G. Mourou, J.C. Chanteloup, G. Vdovin, Opt. Lett. **23**, 1043 (1998)
59. C. Mauclair, A. Mermillod-Blondin, N. Huot, E. Audouard, R. Stoian, Opt. Express **16**, 541 (2008)
60. V.R. Bhardwaj, E. Simova, P.B. Corkum, D.M. Rayner, C. Hnatovsky, R.S. Taylor, B. Schreder, M. Kluge, J. Zimmer, J. Appl. Phys. **97**, 083102 (2005)
61. G. Heck, J. Sloss, R.J. Levis, Opt. Commun. **259**, 216 (2006)
62. R. Ackermann, E. Salmon, N. Lascoux, J. Kasparian, P. Rohwetter, K. Stelmaszczyk, S. Li, A. Lindinger, L. Wöste, P. Béjot, L. Bonacina, Appl. Phys. Lett. **89**, 171117 (2006)
63. M.Y. Shverdin, S.N. Goda, G.Y. Yin, S.E. Harris, Opt. Lett. **31**, 1331 (2006)
64. A. Mermillod-Blondin, C. Mauclair, A. Rosenfeld, J. Bonse, I.V. Hertel, E. Audouard, R. Stoian, Appl. Phys. Lett. **93**, 021921 (2008)
65. L. Sudrie, M. Franco, B. Prade, A. Mysyrowicz, Opt. Commun. **191**, 333 (2001)
66. C.B. Schaffer, A.O. Jamison, E. Mazur, Appl. Phys. Lett. **84**, 1441 (2004)
67. N. Huot, R. Stoian, A. Mermillod-Blondin, C. Mauclair, E. Audouard, Opt. Express **15**, 12395 (2007)
68. A. Marcinkevicius, V. Mizeikis, S. Juodkasis, S. Matsuo, H. Misawa, Appl. Phys. A: Mat. Sci. Process. **76**, 257 (2003)
69. Q. Sun, H.Y. Jiang, Y. Liu, Y. Zhou, H. Yang, Q. Gong, J. Opt. A: Pure and Appl. Opt. **7**, 655 (2005)
70. D. Liu, Y. Li, R. An, Y. Dou, H. Yang, Q. Gong, Appl. Phys. A: Mater Sci. Process. **84**, 257 (2006)
71. M. Ams, G.D. Marshall, P. Dekker, M. Dubov, V.K. Mezentsev, I. Bennion, M.J. Withford, IEEE J. Sel. Top. Quant. Elec. **14**, 1370 (2008)
72. M.J. Booth, M. Schwertner, T. Wilson, M. Nakano, Y. Kawata, M. Nakabayashi, S. Miyata, Appl. Phys. Lett. **88**, 031109 (2006)
73. M.A.A. Neil, R. Juškaitis, T. Wilson, Z.J. Laczik, V. Sarafis, Opt. Lett. **25**, 245 (2000)
74. C. Hnatovsky, R.S. Taylor, E. Simova, B.V. R., D.M. Rayner, P.B. Corkum, J. Appl. Phys. **98**, 013517 (2005)
75. M.J. Booth, M.A.A. Neil, T. Wilson, J. Micros **192**, 90 (1998)
76. J. Hahn, K. Kim, K. Choi, B. Lee, Appl. Opt. **45**, 915 (2006)
77. V. Diez-Blanco, J. Siegel, A. Ferrer, A. Ruiz de la Cruz, J. Solis, Appl. Phys. Lett. **91**, 051104 (2007)
78. A.P. Joglekar, H. Liu, G.J. Spooner, E. Meyhöfer, G. Mourou, A.J. Hunt, Appl. Phys. B: Laser Opt. **77**, 25 (2003)

79. N. Sanner, N. Huot, E. Audouard, C. Larat, P. Laporte, J.P. Huignard, Appl. Phys. B Laser Opt. **80**, 27 (2005)
80. S. Hasegawa, Y. Hayasaki, Opt. Lett. **34**, 22 (2009)
81. W. Watanabe, T. Toma, K. Yamada, J. Nishii, K. Hayashi, K. Itoh, Opt. Lett. **25**, 1669 (2000)
82. G. Sinclair, J. Leach, P. Jordan, G. Gibson, E. Yao, Z.J. Laczik, M.J. Padgett, J. Courtial, Opt. Express **12**, 1665 (2004)
83. N. Huot, N. Sanner, E. Audouard, J. Opt. Soc. Am. B **24**, 2814 (2007)
84. T. Cižmár, K. Dholakia, Opt. Express 17, 15558 (2009)
85. M.K. Bhuyan, F. Courvoisier, P.A. Lacourt, M. Jacquot, R. Salut, L. Furfaro, J.M. Dudley, Appl. Phys. Lett. 97, 081102 (2010)
86. C. Mauclair, A. Mermillod-Blondin, S. Landon, N. Huot, A. Rosenfeld, I.V. Hertel, E. Audouard, I. Myiamoto, R. Stoian, Opt. Lett. 36, 325 (2011)
87. P. Polesana, F. Faccio, P. Di Trapani, A. Dubietis, A. Piskarskas, A. Couairon, M.A. Porras, Opt. Express **13**, 6160 (2005)
88. R. Grunwald, M. Bock, V. Kebbel, S. Huferath, U. Neumann, G. Steinmeyer, G. Stibenz, J.L. Néron, M. Piché, Opt. Express **16**, 1077 (2008)
89. V. Zambon, N. McCarthy, M. Piché, in *Photonics North 2008 (Proceedings Volume)*, *Proceedings of SPIE*, vol. 7099, ed. by Vallée, R. and Piché, M. and Mascher, P. and Cheben, P. and Côté, D. and LaRochelle, S. and Schriemer, H. P. and Albert, J. and Ozaki, T. (SPIE, 2008), p.
90. E. Gaižauskas, E. Vanagas, V. Jarutis, S. Juodkazis, V. Mizeikis, H. Misawa, Opt. Lett. **31**, 80 (2006)
91. S. Kanehira, J. Si, J. Qiu, K. Fujita, K. Hirao, Nano Lett. **5**, 1591 (2005)
92. J. Song, X. Wang, X. Hu, Y. Dai, J. Qiu, Y. Cheng, Z. Xu, Appl. Phys. Lett. **92**, 092904 (2008)
93. P. Polynkin, M. Kolesik, J.V. Moloney, G.A. Siviloglou, D.N. Christodoulides, Science **324**, 229 (2009)
94. G.A. Siviloglou, J. Broky, A. Dogariu, D.N. Christodoulides, Phys. Rev. Lett. **99**, 213901 (2007)
95. C. Mauclair, G. Cheng, N. Huot, E. Audouard, A. Rosenfeld, I.V. Hertel, R. Stoian, Opt. Express **17**, 3531 (2009)
96. M. Pospiech, M. Emons, A. Steinmann, G. Palmer, R. Osellame, N. Bellini, G. Cerullo, U. Morgner, Opt. Express **17**, 3555 (2009)
97. D. Christodoulides, N. Demetrios, F. Lederer, Y. Silberberg, Nature (London) **424**, 817 (2003)
98. A. Yariv, *Optical Electronics*, 4th edn. (Saunders College Publ., Philadelphia, 1991)
99. A. Szameit, F. Dreisow, T. Pertch, S. Nolte, A. Tünnermann, Opt. Express **15**, 1579 (2007)
100. A. Szameit, D. Blömer, J. Burghoff, T. Pertsch, S. Nolte, A. Tünnermann, F. Lederer, Appl. Phys. B Laser Opt. **82**, 507 (2006)
101. A. Szameit, I.V. Kartashov, V.A. Vysloukh, M. Heinrich, F. Dreisow, T. Pertsch, S. Nolte, A. Tünnermann, F. Lederer, L. Torner, Opt. Lett. **33**, 1542 (2008)
102. T. Pertsch, P. Dannberg, W. Elflein, A. Bräuer, F. Lederer, Phys. Rev. Lett. **83**, 4752 (1999)
103. A. Szameit, F. Dreisow, H. Hartung, S. Nolte, A. Tünnermann, F. Lederer, Appl. Phys. Lett. **90**, 241113 (2007)
104. A. Szameit, D. Blömer, J. Burghoff, T. Schreiber, S. Nolte, A. Tünnermann, F. Lederer, Opt. Express **13**, 10552 (2005)
105. A.W. Snyder, J. Opt. Soc. Am. **62**, 1267 (1972)
106. P.D. McIntyre, A.W. Snyder, J. Opt. Soc. Am. **64**, 286 (1974)
107. O.G. Kosareva, T. Nguyen, N.A. Panov, W. Liu, A. Saliminia, V.P. Kandidov, N. Akozbek, M. Scalora, R. Vallee, S.L. Chin, Opt. Commun. **267**, 511 (2006)
108. H. Takahashi, Y. Hayasaki, Appl. Opt. 46, 5917 (2007)
109. M. Sakakura, T. Sawano, Y. Shimotsuma, K. Miura, K. Hirao, Opt. Lett. 36, 1065 (2011)
110. A. Jesacher, M. Booth, Opt. Express 18, 21090 (2010)

# Chapter 5
# Controlling the Cross-section of Ultrafast Laser Inscribed Optical Waveguides

Robert R. Thomson, Nicholas D. Psaila, Henry T. Bookey, Derryck T. Reid, and Ajoy K. Kar

**Abstract** The refractive index profile, or cross-section, of an optical waveguide is its most defining property. It directly determines the number of transverse modes supported by the waveguide and the properties of these modes. Proper control of the waveguide cross-section is therefore essential if the performance of the waveguide, or waveguide device, is to be optimised. This chapter describes how the waveguide cross-section affects the properties of the guided modes, why it is important to control the waveguide cross-section from a device engineering point of view and the various experimental techniques that have been developed to control the cross-section of ultrafast laser inscribed waveguides.

## 5.1 Introduction

The refractive index profile, or cross-section, of an optical waveguide is its most defining property. It directly determines the number of transverse modes supported by the waveguide, and the properties of these modes. These properties then directly affect the performance of the waveguide. For example, waveguides with too small a core and core-cladding refractive index contrast poorly confine light, resulting in high radiation losses and high coupling losses to single mode fibre. On the other hand, waveguides with asymmetric core shapes exhibit polarisation dependent guiding properties and increased coupling losses to single mode fibres. In order to create waveguides with optimum performance using ultrafast laser inscription, steps must be taken to appropriately control the spatial distribution of the refractive index modification induced by the inscription process.

R.R. Thomson (✉) · N.D. Psaila · H.T. Bookey · D.T. Reid · A.K. Kar
Scottish Universities Physics Alliance (SUPA), School of Engineering and Physical Sciences, Physics Department, Heriot Watt University, Riccarton Campus, Edinburgh, Scotland, EH14 4AS, UK
e-mail: R.R.Thomson@hw.ac.uk

In Sect. 5.2 of this chapter, the effect of core size, core shape and core-cladding refractive index on the properties of the guided modes is discussed. In Sect. 5.3, the importance of controlling the waveguide cross-section is discussed from a device engineering point of view. In Sect. 5.4, a review of the experimental techniques that can be used to measure the refractive index profile of ultrafast laser inscribed waveguides is given. In Sect. 5.5, the effect of the relevant inscription parameters on the waveguide cross-section is discussed. In Sect. 5.6, a review of the various experimental techniques that have been demonstrated to control the cross-section of ultrafast laser inscribed waveguides is presented. In Sect. 5.7, conclusions are drawn and an outlook for the field is presented.

## 5.2 The Effect of the Waveguide Cross-section on the Properties of the Guided Modes

In the simplest case, a channel optical waveguide consists of a high refractive index core material, surrounded on all sides by a low refractive index cladding material, as shown in Fig. 5.1. The waveguide supports a finite number of guided optical modes, each of which exhibits a distinct spatial distribution of optical energy. These modes are analogous to the vibration modes of a stretched membrane, but unlike the membrane, an optical waveguide supports only a finite number of modes.

The allowed modes of propagation, both guided and radiation, for any given waveguide structure are found by solving Maxwell's equations, subject to the appropriate boundary conditions which are the continuity of the tangential components of the electric and magnetic fields at the core-cladding interface. Following the approach given in [2], any mode must be a solution of (5.1) which is known as Maxwell's wave equation

$$\nabla^2 \vec{E}(r) - \left(\frac{n(r)^2}{c^2}\right) \frac{\partial^2 \vec{E}(r)}{\partial t^2} = 0, \tag{5.1}$$

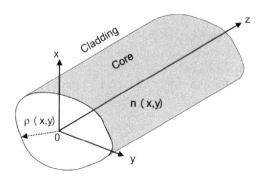

**Fig. 5.1** Diagram of a channel optical waveguide of arbitrary cross-section. Nomenclature taken from [1]

where $\vec{E}(r)$ is the electric field (E-field) vector at some point $r$, $n(r)$ is the refractive index of the medium at that point and $c$ is the speed of light in a vacuum. Equation (5.1) has solutions of the form shown in (5.2).

$$\vec{E}(r,t) = \vec{E}(r) \exp\{i\left[\omega t - \varphi(r)\right]\}, \quad (5.2)$$

where $\omega$ is the radian frequency of the light and $\varphi(r)$ is a phase function. Substituting (5.2) into (5.1), we arrive at (5.3)

$$\nabla^2 \vec{E}(r) + k^2 n^2(r) \vec{E}(r) = 0, \quad (5.3)$$

where $k \equiv \omega/c$.

For weakly guiding waveguides, where the maximum variation between the core and cladding is small (typically less than ~1%), the mode can be thought of as a plane wave travelling along the z-axis, i.e. $\varphi(r) = \beta z$, where $\beta$ is a propagation constant. Also assuming there is no $z$ dependence of $E(r)$, one arrives at the scalar wave equation or Helmholtz equation given by (5.4).

$$\nabla^2 E(x, y) + \left[k^2 n^2(x, y) - \beta^2\right] E(x, y) = 0, \quad (5.4)$$

where $E(x, y)$ is the magnitude of one of the Cartesian components of $\vec{E}(x, y)$, $n(x, y)$ is the refractive index profile transversely across the waveguide and $\beta$ is a propagation constant given by $\beta = n_{\text{eff}} k$, where $n_{\text{eff}}$ is the effective index for the mode.

The guided waveguide modes are solutions to (5.4), where the field distribution outside the core decays with increasing distance from the core, but varies in an oscillatory manner within the core region. Analytical solutions are not known for all but the simplest waveguide structures, and numerical methods must be used to find the guided modes for a given refractive index profile.

Formal descriptions of the mathematics behind the formation of waveguide modes are given in numerous books such as [1] in particular, and therefore are not discussed here. It is necessary here to acknowledge that from (5.4) there is a direct relationship between the field distribution of the guided modes $E(x, y)$ and the cross-sectional refractive index profile of the waveguide $n(x, y)$. Although this relationship is complex, it is important to understand at least qualitatively, how the waveguide cross-section affects the properties of the guided modes.

To demonstrate the effect of core-size and core-cladding refractive index contrast, Fig. 5.2 plots the $1/e^2$ mode field diameter (MFD) of the fundamental guided mode as a function of core diameter for step-index circular-core waveguides with different core-cladding refractive index contrast. The modes were modelled using a finite element modelling software package (Femlab). In the model, the refractive index of the cladding was set to 1.444 and the refractive index contrast was varied between 0.002 and 0.008. In all cases, the free space wavelength of the light in the model was 1.55 μm.

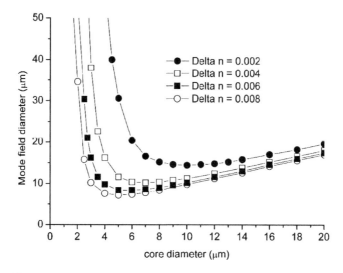

**Fig. 5.2** $1/e^2$ mode field diameter (MFD) of the fundamental mode as a function of the core diameter for circular-core step-index waveguides with different refractive index contrasts. The free-space wavelength of the guided light in the model was set to 1.55 μm

As can be seen in Fig. 5.2, the MFD of the fundamental mode decreases as the core size is reduced until it reaches a minimum. After this point, the MFD again increases as the core diameter is reduced further. Although the position of the minimum is different for each refractive index contrast, the trend is the same. The origin of this minimum can be understood intuitively as follows [1]. Since the waveguide mode is not infinite in space, it cannot be a perfect plane wave. It can therefore, according to Fourier theory, be decomposed into a series of plane-wave components. Each plane-wave component propagates at a different angle, and the mode can be thought of as being composed of a series of rays within a cone of angles ($\theta_d$). According to diffraction, this cone of angles increases as the size of the mode is reduced, but must always lie within the critical angle of the core-cladding interface ($\theta_c$) of the waveguide. As the waveguide size is reduced, the MFD of the fundamental mode decreases until a further reduction in the MFD would require that the mode be composed of additional plane waves propagating at angles larger than $\theta_c$. If the waveguide size is reduced further, the mode is pushed increasingly into the cladding region since it cannot be reduced in size any further. Since the field decays at a slower rate in the cladding region than in the core, the mode then expands rapidly as the core size is reduced further. Because the critical angle at the core-cladding interface increases with increasing refractive index contrast, waveguides with higher refractive index contrasts are able to confine the light more tightly, thus explaining the effect of the refractive index contrast apparent in Fig. 5.2.

As would be expected intuitively from the form of (5.4), any asymmetry in the waveguide cross-section results in guided modes with asymmetric field distributions. This relationship is not at all straightforward, however, since the confinement of the mode in a given axis is not only dependent on the refractive index profile of the waveguide along that axis. Asymmetry in the waveguide cross-section also induces form birefringence [3], particularly when the asymmetry is high and the mode confinement is low. In single mode waveguides, form birefringence breaks the degeneracy of the orthogonally polarised doublet of fundamental modes. Consequently, each mode exhibits a different mode field distribution and different propagation constant. Waveguide asymmetry must therefore be avoided if polarisation dependent guiding properties are undesirable.

The waveguide cross-section not only affects the shape of the guided modes, it also directly determines the number of guided modes. For example, the normalised frequency of a step-index waveguide with a circular core is given by (5.5). Single-mode behaviour, where only one mode is guided for each orthogonal polarisation, is only obtained when the normalised frequency is $<2.405$ [4].

$$V = \frac{2\pi a}{\lambda} \sqrt{n_1^2 - n_2^2}, \tag{5.5}$$

where $V$ is the dimensionless normalised frequency of the waveguide, $\lambda$ is the free space wavelength of the light, $n_1$ and $n_2$ are the refractive indices of the core and cladding materials respectively, and $a$ is the radius of the core.

An inspection of (5.5) reveals that the normalised frequency of a given waveguide increases if the wavelength of the guided light is decreased, the size of the waveguide core is increased or the refractive index contrast of the waveguide is increased.

## 5.3 The Importance of Controlling the Waveguide Cross-section from a Device Engineering Perspective

From the discussion given in Sect. 5.2, it is clear that the waveguide cross-section is central to the function of an optical waveguide. It directly determines how many transverse modes are supported by the waveguide and the spatial distribution of each guided mode. It also determines the polarisation dependent guiding properties of the waveguide through form birefringence. The following discusses why these properties are important from a device engineering point of view.

### 5.3.1 Effect of Mode Field Distribution on Waveguide Coupling Loss

The insertion loss of a waveguide is defined as the loss of signal induced by placing the waveguide in the beam path. The insertion loss is due to three factors, as shown by (5.6). These are the input and output coupling losses to and from the waveguide and the propagation losses along the waveguide. Further discussion of the characterisation techniques for measuring the insertion, propagation and coupling losses for optical waveguides may be found in Chap. 7.

$$\text{Insertion loss (dB)} = \text{input coupling loss (dB)} + \text{propagation losses (dB)} \\ + \text{output coupling loss (dB)}. \tag{5.6}$$

Many of the waveguide devices that have been created using ultrafast laser inscription have been intended for use in optical fibre transmission lines. In this case, light must be coupled from an optical fibre into the waveguide device. After the light has propagated through the device, the light must then be coupled into a second optical fibre. To reduce the device size and remove the need for free space optics, these fibres are commonly butt-coupled directly to the waveguide facet. Neglecting alignment inaccuracies and Fresnel losses, the coupling loss between a single mode waveguide and a single mode fibre is given by the overlap integral shown in (5.7) [5].

$$\eta(\text{dB}) = -10 \log_{10} \left[ \frac{\left( \int E_g E_f^* \, dx \, dy \right)^2}{\int E_g E_g^* \, dx \, dy \int E_f E_f^* \, dx \, dy} \right], \tag{5.7}$$

where $E_g$ is the E-field of the waveguide and $E_f$ is the E-field of the fibre.

In practice, the use of (5.7) may be time consuming, and a simplified expression (5.8) based on the Gaussian field approximation may also be used for the coupling loss between a single mode fibre with a circular mode and a single mode waveguide [6].

$$C(\text{dB}) \approx 10 \log_{10} \left( \frac{4a^2 xy}{(a^2 + x^2)(a^2 + y^2)} \right), \tag{5.8}$$

where $a$ is the fibre mode field radius and $x$ and $y$ are the mode field radii of the waveguide mode.

It is clear from both (5.7) and (5.8) that any mismatch in the mode fields of the waveguide and coupling fibre will result in an increased coupling loss to and from the waveguide. This will in turn increase the device insertion loss as dictated by (5.6). As shown in Fig. 5.2, the waveguide cross-section directly affects the spatial distribution of the fundamental mode supported by the waveguide. Interestingly, it is also evident from Fig. 5.2 that waveguides with different refractive index contrasts can support modes with the same MFD, if the core size is chosen appropriately. This flexibility allows the mode fields of the fibre and the waveguide to be matched, over a limited range, even though their refractive index contrasts may differ.

### 5.3.2 Effect of Mode Field Distribution on Waveguide Propagation Loss

As light travels down an optical waveguide, energy is lost from the mode through scattering, absorption and radiation. The loss coefficient of an optical waveguide is therefore dependent on the magnitude of the contribution from each of these processes as shown by (5.9).

$$\alpha_{\text{waveguide}} = \alpha_{\text{absorption}} + \alpha_{\text{scattering}} + \alpha_{\text{radiation}}, \tag{5.9}$$

where $\alpha$ is the loss coefficient in units of cm$^{-1}$.

The spatial properties of the guided modes may affect the magnitude of all three processes. For example, the numerical aperture (NA) and MFD of a single mode waveguide are inversely proportional. The smaller the MFD, the larger the waveguide NA and the larger the maximum exit angle of the waveguide. Tighter waveguide bends can therefore be created using higher NA waveguides without increasing radiation losses. It is also reasonable to expect that the mode field distribution will affect the degree of scattering and absorption in ultrafast laser inscribed waveguides. It has recently been shown for example that Rayleigh type scattering plays an important role in ultrafast laser inscribed waveguides [7]. Since the scattering centres are created by the inscription process itself, the spatial overlap of the mode field with the modified region will largely determine the magnitude of the scattering loss coefficient. It has also been shown that the inscription process can create colour centres [8], affect the elemental distribution in the modified region [9] and modify the valence state of the active ions in the modified region [10]. The spatial overlap of the guided mode with the modified region will therefore impact the magnitude of the absorption loss coefficient considerably.

### 5.3.3 *Effect of Mode Field Distribution on Evanescent Coupling*

The fabrication of coupled waveguide systems using ultrafast laser inscription has attracted considerable research attention in recent years. Practical devices realised so far include telecoms type devices such as wavelength division multiplexers (WDMs) [7], and broadband power splitters [11, 12], these are discussed further in Chap. 7. The unique fabrication capabilities of ultrafast laser inscription have enabled the fabrication of more exotic structures for studying fundamental optical phenomena such as one- and two-dimensional spatial solitons [13, 14], optical Bloch oscillations [15] and optical Bloch–Zener oscillations [16].

Regardless of the application, the physical operation of coupled waveguide systems is based on the coupling of light between waveguides through optical tunnelling, the nature of which is discussed extensively in Chap. 13. For the purposes of this chapter, it is important to note that the strength of the coupling between two single mode waveguides is strongly dependent on the difference in the propagation constants ($\beta$ in (5.4)) and the spatial overlap of the modes supported by each waveguide. Since the modal properties of any waveguide are directly determined by its cross-section, proper control of the waveguide cross-section is essential if the coupling is to be controlled effectively.

### 5.3.4 Effect of Waveguide Asymmetry on Polarisation Dependent Guiding Properties

As outlined briefly in Sect. 5.2, any asymmetry in the waveguide cross-section results in waveguide modes with asymmetric shapes. A single mode waveguide with an asymmetric waveguide cross-section also exhibits form birefringence which breaks the degeneracy of the two orthogonally polarised modes. Each mode therefore exhibits a different mode field distribution and propagation constant. According to the discussions presented in Sects. 5.3.1 to 5.3.3, it can be expected that asymmetry in the waveguide cross-section will result in increased coupling losses to standard single mode fibres, polarisation dependent losses and polarisation dependent evanescent coupling behaviour.

## 5.4 Experimental Techniques for Measuring the Refractive Index Profile of Ultrafast Laser Inscribed Waveguides

For ultrafast laser inscribed waveguides, the form of the refractive index modification greatly depends upon the material and inscription parameters. Obtaining accurate refractive index profiles therefore adds insight into the guiding properties and waveguide formation mechanisms. Many techniques exist to measure the refractive index profile of optical waveguides. Due to the often complex nature of the refractive index profiles of ultrafast laser inscribed structures, any technique used must be capable of measuring arbitrary refractive index profiles that do not rely on convenient core geometries, such as the cylindrical symmetry found in optical fibres. The following sections outline several refractive index profiling techniques applicable to ultrafast laser inscribed structures.

### 5.4.1 Refracted Near-field (RNF) Method

The refracted near-field (RNF) technique is one of the most widely used of all the refractive index profiling techniques. It operates by focussing a laser beam onto the polished facet of the waveguide under test, as shown in Fig. 5.3. The waveguide sample is mounted on computer controlled translation stages that allow the sample surface to be scanned across the laser beam focus. A detector with a mask measures the amount of light refracted through the sample with an exit angle greater than $\theta_{min}$. From Fig. 5.3, it can be seen that the minimum incident ray angle that hits the detector $\phi_{min}$, and hence the amount of light reaching the detector $I(x, y)$, is dependent on the refractive index of the material placed at the focus $n(x, y)$. Thus, by plotting $I(x, y)$, the refractive index profile of the waveguide can be recovered.

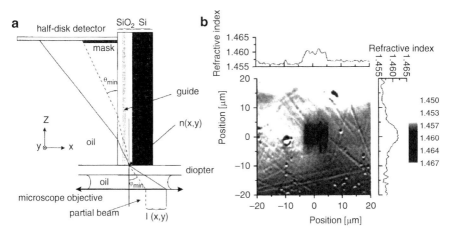

**Fig. 5.3** (**a**) Schematic diagram of a typical refracted near-field experimental setup. (Taken from [17].) (**b**) Refractive index profile of an ultrafast laser inscribed waveguide whose cross-section was controlled using the multiscan technique described in Sect. 5.6.5. (Taken from [18].)

The RNF technique is capable of high spatial resolution, typically the wavelength of light used to probe the structure. Refractive index resolutions of $\Delta n = 10^{-4}$ are also typically achievable. Several commercial devices exist that utilise the RNF technique for the purposes of measuring refractive index profiles of waveguides, fibres and fibre preforms. One particular drawback with the RNF technique is that it requires the sample to be immersed in a bath of potentially toxic refractive index oil, and depending on the material this could cause a chemical modification of the substrate.

The RNF technique has been used extensively by many groups to measure the refractive index profile of ultrafast laser inscribed waveguides [18–20]. As an example, Fig. 5.3(b) shows the refractive index profile of an ultrafast laser inscribed waveguide whose cross-section was controlled using the multiscan technique described later in Sect. 5.6.5.

## 5.4.2 Micro-reflectivity

The micro-reflectivity technique operates by focussing a laser beam onto the polished facet of the waveguide under test and measuring the change in the Fresnel reflection as the sample facet is scanned across the focus. Figure 5.4 shows a micro-reflectivity refractive index profiling experimental setup based on a confocal reflectance microscope [21]. As shown in Fig. 5.4, light is focussed onto the surface of the sample. The Fresnel reflection is then retro-reflected back along the beam and detected by detector D2. The power incident on the sample is monitored by detector D1. By monitoring the ratio of the incident to reflected power, the Fresnel reflection coefficient, and hence the refractive index profile of the waveguide can be constructed by scanning the sample across the focus in the $x - y$ plane.

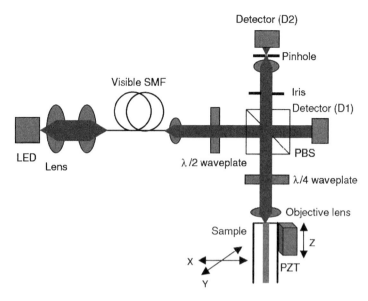

**Fig. 5.4** Micro-reflectivity refractive index profiling experimental setup based on a confocal reflectance microscope. (Taken from [21].)

Several different experimental configurations can be employed to measure the micro-reflectivity profile. Examples of such configurations include all fibre configurations such as shown by Park et al. [22], or using conventional bulk optics [23]. The micro-reflectivity technique does exhibit some limitations, however, such as the need for very high-quality surfaces and highly accurate $z$-axis positioning. Particular care must also be taken in order to avoid contributions from light that is coupled into the waveguide and reflected back from the opposite end of the waveguide [24]. This can be avoided by immersing the opposite end of the waveguide in carefully selected index matching fluid, or by angling the opposite waveguide facet appropriately.

The micro-reflectivity technique is able to provide high spatial resolutions, limited only by the focussed spot size, and refractive index resolutions of the order of $\Delta n = 10^{-5}$ [24]. Given the relative simplicity of the micro-reflectivity technique and its ability to provide high resolution measurements, it is surprising that very few have applied it to ultrafast laser inscribed waveguides. As an example, Fig. 5.5 shows the refractive index profile of an ultrafast laser inscribed structure in BK7 glass obtained using the micro-reflectivity technique.

### 5.4.3 Quantitative Phase Microscopy

When an optical wave passes through a transparent object, the phase of the wavefront is modulated by differences in the optical path length across the object. If

**Fig. 5.5** Refractive index profile of an ultrafast laser inscribed structure in BK7 glass measured using the micro-reflectivity technique. The laser modified region, represented by the *dark area*, exhibits a decrease in refractive index of ∼0.0035. (Taken from [25].)

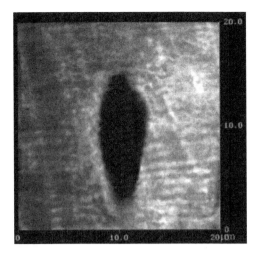

the object is physically flat, these differences are due to refractive index variations across the object. Quantitative phase microscopy (QPM) is the name given to a group of techniques that can use this phase change to recover the refractive index properties of the sample. Numerous different implementations of QPM can be used for analysing the refractive index profile of ultrafast laser inscribed optical waveguides.

Conventional phase contrast microscopy is not immediately applicable to quantitatively measuring refractive index changes. This is due to the ambiguity between amplitude and phase modulations in the images obtained. It is possible with considerable image processing to extract a pure phase map as shown in [26]. This has only been applied to optical fibres possessing cylindrical symmetry and when viewing the sample perpendicular to the fibre axis. The use of processed phase contrast microscopy to calculate the phase profile of ultrafast laser inscribed structures would be considerably more difficult unless the structure possessed the same rotational symmetry as optical fibres.

An alternate phase-contrast technique capable of achieving a phase-only image is that of Barty et al. [27] who, using a perpendicular viewing geometry, was able to obtain the refractive index profiles of optical fibres. Here, three images are captured with a conventional microscope operating in transmission mode – one in-focus, one slightly positively defocused and one slightly negatively defocused. Since these images give information about the wavefront evolution between the image planes, and hence the distortion of the wavefront by the object, together they can be post-processed to obtain a phase profile. This phase profile can then be converted into a refractive index contrast image. Again, the profiling of structures which are not symmetrical around the waveguide axis would be difficult given the ambiguity of the phase build up.

Interferometers can also be employed to measure refractive index profiles of waveguides. In this case, a thin cross-section of the waveguide (∼50 µm thick) is placed in one arm of the interferometer with the waveguide axis orientated along the beam path. Such thin samples are difficult to prepare, but are needed to avoid the phase ambiguity induced by optical phase differences greater than $2\pi$. It should be noted that multi-wavelength interferometry can also be employed to enable the use of thicker samples, but this introduces chromatic aberrations and circular fringes around structures which need to be corrected. The digital holography technique employed by Ferraro et al. is capable of correcting for such aberrations in the reconstruction of the phase profile [28] and has been successfully applied to the refractive index profiling of ultrafast laser inscribed waveguides as shown in Fig. 5.6.

### 5.4.4 Inverse Helmholtz Technique

In Sect. 5.2, it was shown that the guided modes for a weakly guiding waveguide may be found by solving the Helmholtz equation (5.4), which demonstrates the relationship between the refractive index profile of an optical waveguide and the field distribution of the guided modes. If the optical waveguide of interest is single mode, the Helmholtz equation can also be used to determine its refractive index profile by measuring the near-field profile of the guided mode.

Rearranging the Helmholtz equation for $n(x, y)$, one arrives at (5.10):

$$n^2(x, y) = (n_b + \Delta n(x, y))^2 = n_{\text{eff}}^2 - \frac{\lambda^2}{4\pi^2} \frac{\nabla^2 E(x, y)}{E(x, y)}, \qquad (5.10)$$

where $n_b$ is the refractive index of the bulk material.

The normalised E-field distribution $E(x, y)$ across the mode can be inferred by measuring the near-field intensity of the mode $I(x, y)$ and using the relationship $E(x, y) = \sqrt{I(x, y)}$. Although $n_{\text{eff}}$ is an unknown constant, the refractive index difference profile $\Delta n(x, y)$ is unaffected by the magnitude of $n_{\text{eff}}$. The

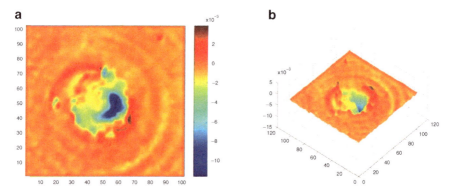

**Fig. 5.6** Refractive index profile of an ultrafast laser inscribed waveguide in borosilicate glass obtained using QPM with digital holography. (Taken from [28].)

5 Controlling the Cross-section of Ultrafast Laser Inscribed Optical Waveguides 105

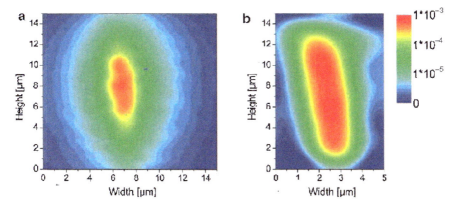

**Fig. 5.7** (**a**) Measured near-field mode profile for 800 nm light confined by an ultrafast laser inscribed waveguide in fused silica and (**b**) corresponding refractive index profile recovered using the inverse Helmholtz technique. (Taken from [14].)

refractive index difference profile can therefore be recovered by solving (5.10) using numerical methods.

Due to the second order differential in (5.10), the refractive index profile index profile obtained using the inverse Helmholtz technique is highly sensitive to noise on the measured data. Even so, the technique has been successfully applied to ultrafast laser inscribed waveguides [14] and provides results that compare well with profiles measured using techniques such as the RNF technique described in Sect. 5.4.1. As an example, Fig. 5.7 shows the measured near-field profile of an ultrafast laser inscribed waveguide and the refractive index profile recovered using the inverse Helmholtz technique.

## 5.5 Effect of Inscription Parameters on the Waveguide Cross-section

When an ultrashort laser pulse is focussed inside a transparent dielectric material, the pulse travels through the material until the intensity of the E-field is high enough to induce a plasma of electrons through a combination of multi-photon absorption, tunnelling ionisation and avalanche ionisation [29]. This plasma then transfers its energy to the material lattice through electron–phonon coupling. The transfer of energy heats the material and induces the observed structural modification. This structural modification manifests itself through the refractive index modification that is used to inscribe optical waveguides.

As discussed in Chap. 1 of this book, it is known that at least nine process parameters (laser polarisation, pulse energy, pulse duration, pulse shape, pulse repetition rate, focussing optics, translation speed, translation direction and wavefront tilt) play a role in the inscription process. Of these parameters, it is the focussing optics, translation direction and the pulse repetition rate that play the most important roles in defining the form of the waveguide cross-section. In the single pulse regime, using lasers operating at repetition rates lower than ∼100 kHz, the material in the focal region cools before the next pulse arrives. In the single pulse regime, the spatial distribution of the modified region is almost completely determined by the spatial distribution of the plasma induced by each pulse, and diffusion of thermal energy outside the focal region plays a minimal role. In the intermediate and cumulative regimes, using lasers operating at repetition rates between 100 kHz and 5 MHz and greater than 5 MHz, respectively, heat accumulation and diffusion play important roles in defining the spatial distribution of the modified region [19, 30].

Waveguides with excellent propagation characteristics have been fabricated using lasers operating in both the single pulse [18] and intermediate [30] regimes. Even so, this chapter will describe only the experimental techniques that have been developed to control the cross-section of waveguides inscribed in the single pulse regime. The reason for this is simple. Due to the role of heat accumulation and diffusion, the refractive index modification created by lasers operating in the intermediate and cumulative regimes can be extremely complicated and unpredictable. As a consequence, control of the modal properties of structures inscribed in these regimes is based mostly on experimental trial and error. In the single-pulse regime, however, the distribution of the modified region is almost solely determined by the distribution of the plasma, which is in turn directly determined by the E-field distribution in and around the focus. Assuming that the modified region then exhibits a positive refractive index change, as has been shown to be true in a number of cases [31, 32], the shape of the waveguide cross-section can be predicted and controlled via the shape of the focus. It is also important to note that in the single pulse regime, the magnitude of the photo-induced refractive index change can be readily controlled via the pulse energy and translation speed.

The substrate translation direction also plays an important role in determining the waveguide cross-section. As shown in Fig. 5.8, the sample can either be translated longitudinally along the direction of propagation of the inscription laser beam – the

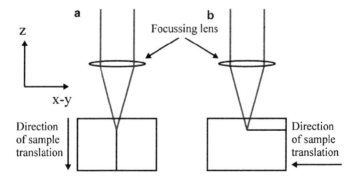

**Fig. 5.8** Schematic diagrams of (**a**) the longitudinal and (**b**) transverse writing geometries

longitudinal writing geometry, or transverse to the direction of propagation of the inscription laser beam – the transverse writing geometry.

When inscribing a waveguide in the single pulse regime using a circular Gaussian beam and a longitudinal writing geometry, the cross-sectional size of the modified region (the waveguide core) is directly related to the beam waist diameter as given by (5.11). For waveguides written using the same system but a transverse writing geometry, the size of the waveguide core in either the $x$- or $y$-axis is again directly related to the focussed beam waist diameter, but now the size of the waveguide core along the $z$-axis is directly related to the confocal parameter as given by (5.12).

$$\text{Beam waist diameter} = 2w_0 = \frac{2f\lambda}{\pi w(f)} \approx \frac{2\lambda}{\text{NA}\pi}, \tag{5.11}$$

$$b = 2R = \frac{2\pi n w_0^2}{\lambda} \tag{5.12}$$

where $w_0$ is the beam waist radius, $w(f)$ is the beam radius at the lens, $f$ is the focal length of the lens used to focus the beam, $\lambda$ is the free space wavelength of the light, $n$ is the refractive index of the substrate material, NA is the numerical aperture of the focussing, $b$ is the confocal parameter and $R$ is the Rayleigh range of the focus.

As would be expected intuitively, waveguides written using the longitudinal writing geometry are inherently circular due to the rotational symmetry of the laser beam, but the length of the waveguide is limited to the working distance of the focussing lens. Furthermore, the degree of spherical aberration imparted on the laser beam by the substrate changes as the depth of the focus inside the sample is varied. This in turn changes the E-field distribution in and around the focus, thus changing the form of the modification along the waveguide. It should be noted that techniques employing adaptive optics can be used to correct for this [33,34], but this considerably complicates the experimental setup. For these reasons, the transverse writing geometry is more versatile and has attracted more attention.

An examination of (5.11) and (5.12) reveals that waveguides inscribed in the single pulse regime using the transverse writing geometry exhibit a high degree of asymmetry unless the beam waist radius is close to $\lambda/n\pi$. Such a small spot size can only be created using a focussing NA of close to $n$. Since the maximum refractive index contrast of ultrafast laser inscribed waveguides is limited to $\sim$0.5% in most materials, creating waveguides this small results in poor confinement of the guided mode, particularly when the guided light is at telecommunication wavelengths (1.3 and 1.55 µm). It is clear therefore that for many applications, the width of the waveguide must be increased to properly support the mode. As discussed previously, Fig. 5.2 plots the MFD of the fundamental mode at 1.55 µm for step-index circular-core waveguides with different refractive index contrasts and core sizes. From this graph, it is logical to assume that a waveguide core in the region of at least $\sim$6.0 µm in diameter would be required to properly confine

1.55 µm light if the refractive index contrast is limited to ∼0.5%. According to (5.12), a beam waist diameter of 6.0 µm would result in a confocal parameter of ∼100 µm when using 800 nm light from a Ti:sapphire laser system to inscribe the waveguide inside a glass with a refractive index of $n = 1.5$, thus resulting in a considerable asymmetry in the waveguide cross-section. In the following section, the experimental techniques that have been developed to correct this asymmetry are presented and discussed.

## 5.6 Experimental Techniques for Controlling the Waveguide Cross-section

In this section, the various techniques that have been developed to control the cross-section of ultrafast laser inscribed waveguides are described and discussed. The first four sections (5.6.1, 5.6.2, 5.6.3 and 5.6.4) deal with the astigmatic beam shaping, slit-beam shaping, active optics and spatiotemporal focussing techniques, respectively. Each of these techniques involve shaping the inscription laser beam in some manner to control the cross-section of waveguide. Section 5.6.5 deals with the multiscan technique where the waveguide cross-section is constructed by scanning the substrate through the focus of the laser beam many times and building the desired cross-section from the individual modified regions created by each scan.

### 5.6.1 The Astigmatic Beam Shaping Technique

The first technique that was developed to correct the asymmetry of waveguides inscribed in the single pulse regime using a transverse writing geometry was the astigmatic beam shaping technique, first demonstrated by Cerullo et al. in 2002 [35]. As the name suggests, the technique relies on shaping the inscription laser beam to control the shape of the modified region. The basic idea behind the astigmatic beam shaping technique is that if a waveguide is inscribed using the transverse writing geometry by translating the sample along the $x$-axis shown in Fig. 5.9, the width of the waveguide is independent of the beam waist diameter in the $x$-axis. As detailed by Cerullo et al. [35] and Osellame et al. [36], this fact can be exploited by focussing tightly in this axis to reduce the Rayleigh range of the focus in the $x - z$ plane. The rapid divergence of the beam in this plane reduces the depth of focus of the beam, thus reducing the $z$-axis width of the waveguide. The size of the focus in the $y$-axis can then be used independently to control the width of the waveguide along the $y$-axis.

Although this technique would work well in theory, in practice a beam waist ratio ($w_{0y}/w_{0x}$) of ∼10 would be required to achieve a waveguide with a symmetric

**Fig. 5.9** Modelled beam profiles in the $y/z$ plane (*solid curves*) and the $x/z$ plane (*dashed curves*) for focussed astigmatic Gaussian beams with $w_{0x} = 1.0\,\mu m$, $w_{0y} = 3.0\,\mu m$ and (**a**) $z_0 = 0$, (**b**) $z_0 = 100\,\mu m$. The *shaded area* corresponds to waveguide shape that would be expected according to the modelled distribution of the electron plasma induced by the pulse. (Taken from [36].)

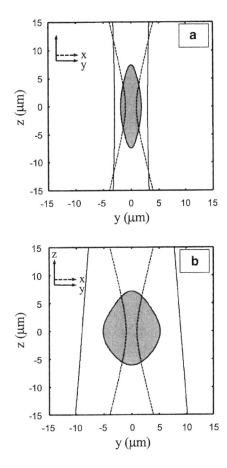

cross-section [35, 36]. Such a high ratio would require a high demagnification cylindrical telescope to adjust the beam waist ratio. This would be difficult to align, would introduce significant aberrations and would be difficult to optimise since changing the waveguide symmetry would require changing the magnification of the telescope. To solve this problem, Cerullo et al. suggested that by adjusting the astigmatic difference ($z_0$), the offset in the focal planes for $w_{0x}$ and $w_{0y}$, a symmetric waveguide could be inscribed using beam waist ratios of only 2 to 3. The basis of this suggestion is that for a focussed astigmatic beam, the region of highest intensity is normally positioned in the plane corresponding to the smallest spot, $w_{0x}$ in this case. As shown in Fig. 5.9b, the width of the modified region, and hence the waveguide core, can therefore simply be adjusted by varying the astigmatic difference without increasing the beam waist ratio. The astigmatic beam shaping technique allows one to vary not only the symmetry of the waveguide but also its size by adjusting the astigmatic difference and the size of $w_{0x}$. This can in theory be controlled via a spherical telescope placed before the cylindrical telescope.

**Fig. 5.10** Astigmatic beam shaping technique experimental setup. (Taken from [36].)

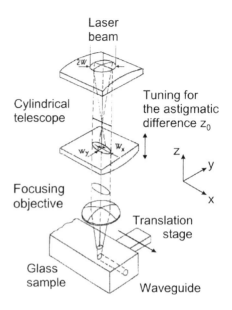

Figure 5.10 shows the astigmatic beam shaping experimental setup. As shown in the figure, a cylindrical telescope is used to create the required beam waist ratio. A microscope objective is used to focus the shaped beam inside the sample which is mounted on a translation stage. The position of the second cylindrical lens in the telescope is used to vary the astigmatic difference $z_0$.

The astigmatic beam shaping technique has been successfully used by Cerullo et al. [35] and Osellame et al. [36] to inscribe waveguides with symmetric cross-sections in ErYb-doped phosphate glass. The inscription laser used in these studies was a regeneratively amplified Ti:sapphire system emitting 130 fs pulses of 790 nm radiation at a repetition rate of 1 kHz. This inscription setup is therefore firmly in the single pulse regime. Figure 5.11 shows microscope images of waveguides inscribed in an ErYb-doped phosphate glass sample using the astigmatic beam shaping technique. Also shown in the figure are the electron density profiles simulated by the authors using the modelled E-field distribution. As shown in Fig. 5.11, the asymmetry of the waveguide can be readily controlled by varying the astigmatic difference, and a symmetric waveguide could be fabricated using an astigmatic difference of 180 μm.

Although the astigmatic beam shaping technique has been successfully applied to the fabrication of waveguides in a number of materials, including ErYb-doped phosphate glass [35, 36] and fused silica [37], its use in the community has been somewhat limited. This is most probably due to the subsequent development of simpler techniques. One should not underestimate the impact that the original astigmatic beam shaping papers [35, 36] made on the field. These papers clearly demonstrated that in the single pulse regime, the waveguide cross-section is directly related to the spatial distribution of the electron plasma induced by the pulse.

**Fig. 5.11** *Left column*: optical microscope images of the waveguide cross-section written using an astigmatic beam with $w_{0x} = 1.2\,\mu$m and $w_{0y} = 3.6\,\mu$m. *Right column*: corresponding simulated electron density profiles. (Taken from [36].)

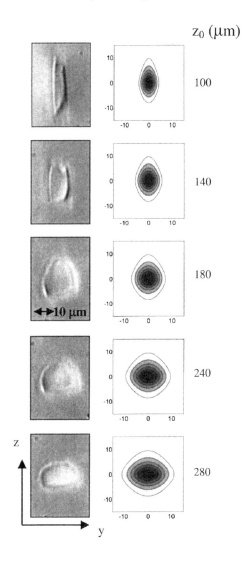

They therefore showed, for the first time, that the waveguide cross-section could be controlled in a predictable manner by shaping the inscription beam.

## 5.6.2 The Slit Beam Shaping Technique

As discussed in the previous section, it should be possible in theory to inscribe a waveguide with a symmetric cross-section simply by varying the beam waist ratio. It was also stated, however, that a beam waist ratio of $\sim$10 would be required to achieve a symmetric waveguide cross-section if no astigmatic difference was induced [35, 36], and that this would require a high demagnification cylindrical telescope which could pose significant practical issues. If, however, there was a simple way to control the beam waist ratio in an analogue manner, without the use of additional optics, then the alignment would be trivial. To this end, the slit beam shaping technique was developed, first applied to fabrication of embedded microchannels with symmetric cross-sections [38] and then soon after to the inscription of optical waveguides [39, 40].

The basic idea behind the slit beam shaping technique is that the focussed beam waist in each axis is directly dependent on the NA of the focussing in that axis, as defined by (5.11). The NA is in turn dependent on both the focal length of the focussing lens and the diameter of the beam at the lens. The slit beam shaping technique operates by placing a slit directly in front of the focussing lens. The slit is orientated along the sample translation direction to reduce the NA of the focussing in the plane perpendicular to the waveguide axis. The slit therefore controls the beam waist ratio, and hence the waveguide symmetry. Unlike the astigmatic beam shaping technique, the slit beam shaping technique allows the beam waist ratio to be controlled easily and in an analogue manner using a micrometer controlled adjustable slit. It should be noted that although this is the easiest and most intuitive way to understand the function of the slit, it has been shown that a more accurate, but considerably more complicated way to describe the function of the slit is to consider it as a diffractive optical element that redistributes optical energy around the focus [41]. Nevertheless, the idea in each case is the same.

If the effect of diffraction is neglected, the propagation of the laser beam can be modelled using simplified Gaussian beam propagation theory. It can be shown [39] that under this assumption, the condition for a focal region with a symmetric energy distribution, when viewed along the sample translation direction, is given by (5.13).

$$\frac{W_y}{W_x} = \frac{NA}{n}\sqrt{\frac{\ln 2}{3}}, \quad \text{if} \quad W_x > 3W_y, \tag{5.13}$$

where $W_y$ and $W_x$ are the width of the slit and the non-apertured laser beam, respectively, and the NA $\approx f/W_x$ where $f$ is the focal length of the focussing lens and $n$ is the refractive index of the substrate.

Figure 5.12 shows the modelled beam evolution and energy distribution for two focussed beams, each propagating through a glass with a refractive index of 1.54. The simulations shown in figure (a) and (b) are for a Gaussian beam with a $1/e^2$ width ($W_x/2$) of $\sim 2.2$ mm focussed with a microscope objective of NA $= 0.46$. The simulations shown in figure (c) and (d) are for the same Gaussian beam and lens but with a 500 μm slit, aligned along the $x$-axis, placed in front of the objective lens. The 500 μm width of the slit is close to the 630 μm value required, according to (5.13), to create a symmetric energy distribution. As is evident in the figure, the use of the slit creates a considerably more symmetric energy distribution when viewing the focal region along the $x$-axis.

5 Controlling the Cross-section of Ultrafast Laser Inscribed Optical Waveguides 113

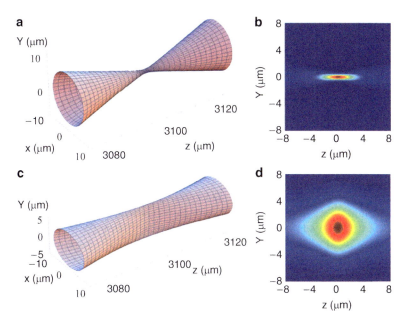

**Fig. 5.12** (**a**) Beam evolution near focus not using slit, (**b**) energy distribution in YZ plane not using slit, (**c**) beam evolution near focus using slit and (**d**) energy distribution in YZ plane using a 500 μm slit orientated along the x-axis. (Taken from [39].)

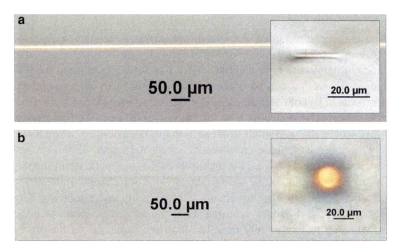

**Fig. 5.13** Differential interference contrast images of two structures. One inscribed (**a**) without the use of a slit, and the other inscribed (**b**) with the use of a 500 μm slit placed before the focussing lens. (Taken from [39].)

Ams et al. successfully applied the slit beam shaping technique to the inscription of symmetric waveguides in phosphate glass. Figure 5.13 shows differential interference contrast microscope images of two structures inscribed in the glass sample using a regeneratively amplified Ti:sapphire laser emitting 120 fs pulses of 800 nm light at a repetition rate of 1 kHz. Again, the inscription process lies firmly in the single pulse regime. The structure shown in figure (a) was inscribed using no slit before the lens. The structure shown in figure (b) was inscribed using a 500 μm slit placed before the lens. As is clearly evident in Fig. 5.13, the slit beam shaping technique has enabled the inscription of a structure with an almost perfectly symmetric circular cross-section. Ams et al. reported high quality guiding of both 633 nm and 1550 nm light for the structure inscribed using the slit beam shaping technique but no guiding for the structure inscribed without the use of the slit. This simple observation confirms the necessity for controlling the cross-section of the inscribed structure.

Due to the experimental simplicity and intuitive nature of the slit beam shaping technique, it has been utilised by several groups. It has been successfully applied to many different materials including poly(methyl methacrylate) (PMMA) [42], soda-lime silica glass [43] and bismuth-borate glass [44]. It has also very recently been applied to the creation of monolithic ErYb-doped [45] and Yb-doped [46] phosphate glass waveguide lasers. These pioneering devices were created using the slit beam shaping technique to shape the waveguide, and a Bragg-waveguide structure was created by modulating the laser power during the inscription process.

### 5.6.3 Waveguide Shaping Using Active Optics

The astigmatic beam shaping and slit beam shaping techniques described in Sects. 5.6.1 and 5.6.2 have been successfully applied to the inscription of symmetric waveguides in many different materials. They have clearly demonstrated that the waveguide cross-section can be controlled via the shape of the inscription laser beam. Unfortunately, both of these techniques in their current forms suffer from two drawbacks that may become important for future applications. The first drawback is inflexibility – once the waveguide cross-section has been defined, it is not possible to readily vary the waveguide cross-section along its length, or in fact between waveguides. The second drawback is that for both the astigmatic beam shaping and slit beam shaping techniques, the waveguide cross-section changes if the waveguide performs a bend. The origin of this change can be readily understood from Fig. 5.9 and Fig. 5.12 where it can be seen that both the astigmatic beam shaping and slit beam shaping technique rely on controlling the symmetry of the waveguide cross-section via the beam waist ratio. A symmetric cross-section is therefore only obtained when scanning the sample along one axis and the waveguide cross-section again becomes highly asymmetric if the sample translation direction is changed.

Recently, new waveguide shaping techniques based on the use of active optics have been proposed and demonstrated. The basic idea behind these techniques is that instead of using either a cylindrical telescope or slit, a computer controlled active optic, such as a deformable mirror or spatial light modulator (SLM), can be used to shape the inscription laser beam prior to entering the microscope objective.

# 5 Controlling the Cross-section of Ultrafast Laser Inscribed Optical Waveguides

Since the properties of the active optic can be readily controlled using a computer, it is envisaged that waveguide shaping techniques based on active optics would not suffer from the same drawbacks as the astigmatic beam shaping and slit beam shaping techniques described above.

The first use of active optics to control the cross-section of ultrafast laser inscribed waveguides was by Thomson et al. [47], who used a deformable mirror to control the laser beam shape. A schematic diagram of the deformable mirror technique experimental setup used by Thomson et al. is shown in Fig. 5.14. The inscription laser was a regeneratively amplified Ti:sapphire laser emitting 130 fs pulses of 800 nm light at a repetition rate of 5 kHz. The inscription setup therefore operated in the single pulse regime. The laser light was projected onto a two-dimensional "free-standing membrane" type deformable mirror where a very thin reflective membrane is suspended over an array of electrode actuators. Applying a voltage potential between the membrane and the actuators deforms the mirror by pulling the mirror towards the actuator.

In this proof-of-concept study, the deformable mirror was used to create a line focus directly in front of the microscope objective. In this way, Thomson et al. intended to use the deformable mirror to replicate the effect of the slit in the slit beam shaping technique discussed in Sect. 5.6.2.

Figure 5.15(a) presents transmission mode optical micrographs of the end facets of two structures, one inscribed using the deformable mirror set to flat, and the other inscribed using the deformable mirror shaped to create a line focus ∼700 μm wide directly in front of the microscope objective. Figure 5.15(b) shows beam profiles of the inscription laser beam using the flat and shaped mirror settings. As shown

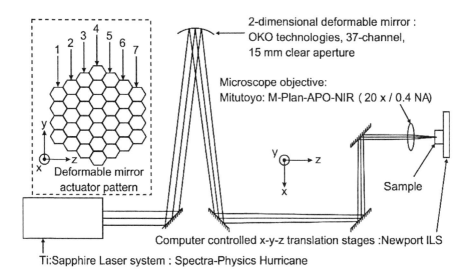

**Fig. 5.14** Schematic diagram of the experimental setup used to shape waveguides using the deformable mirror technique. (Taken from [47].)

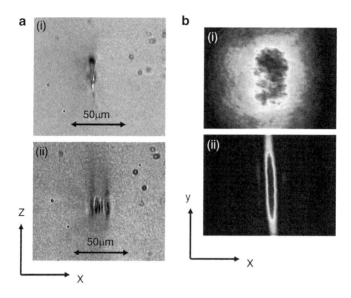

**Fig. 5.15** (**a**) Transmission mode optical micrographs of structures inscribed using (*i*) a flat deformable mirror and (*ii*) the deformable mirror set to create a line focus in front of the microscope objective. (**b**) Beam profile images of the inscription laser directly before entering the microscope objective using (*i*) a flat deformable mirror and (*ii*) the deformable mirror set to create a line focus. The field of view of each image is ∼6.5 × 4.8 mm in the *x*- and *y*-axes, respectively. (Taken from [47].)

in figure (a), the cross-section of the structure inscribed using the flat deformable mirror setting is highly asymmetric, as would be expected from the discussion given in Sect. 5.5. Thomson et al. reported that no guiding of light could be observed for any structure inscribed with the flat mirror setting. As also shown in figure (a), the cross-section of the structure inscribed using the shaped deformable mirror is considerably more symmetric. Thomson et al. reported that high confinement guiding of 1550 nm light was observed for the structure inscribed with the shaped mirror setting, thus confirming the requirement for waveguide shaping.

The application of active optics to waveguide shaping has been further explored by Ruiz de la Cruz et al. [48]. In this case, a computer controlled SLM, rather than a deformable mirror, was used to control the phase profile across the inscription laser beam. A schematic diagram of the experimental setup used in [48] is shown in Fig. 5.16. The inscription laser was a regeneratively amplified Ti:sapphire laser emitting 100 fs pulses of 800 nm light at a repetition rate of 1 kHz. A controlled amount of astigmatism was induced on the laser beam phase profile using a liquid crystal SLM. After the SLM, a fraction of laser light was coupled into a wavefront sensor which enabled the authors to monitor the phase profile across the laser beam and compensate for unwanted aberrations.

The combined effect of the SLM induced astigmatism and propagation from the SLM to the focussing objective on the laser beam can be seen in Fig. 5.17(a). The

5 Controlling the Cross-section of Ultrafast Laser Inscribed Optical Waveguides

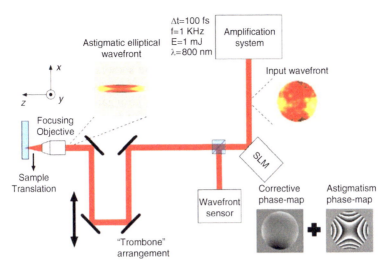

**Fig. 5.16** Schematic diagram of the experimental setup used to shape waveguides using the spatial light modulator technique. (Taken from [48].)

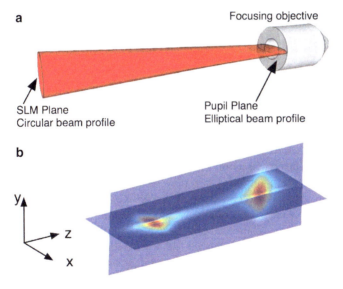

**Fig. 5.17** Schematic drawing of the astigmatism and beam propagation effects on the beam profile at the entrance pupil of the focussing objective (**a**) and on the irradiance distribution in the focal region, showing two foci lying in orthogonal planes (**b**). (Taken from [48].)

astigmatism results in an elliptical beam at the entrance to the microscope objective. From Fig. 5.17(a), it can be seen that the ellipticity of the beam in front of the objective can be controlled by varying the propagation distance from the SLM to the objective which can be varied via the "Trombone arrangement" shown in

Fig. 5.16. As shown in Fig. 5.17(b), the SLM induced astigmatism results in two spatially separated foci in each orthogonal plane. This splitting widens each focus in one axis and facilitates the control of the waveguide cross-section. In order to avoid the inscription of structures at each focus, the authors of [48] demonstrated that the beam ellipticity in front of the objective may be used to control the NA of the focussing for each foci. Using the trombone arrangement, the authors were successfully able to set the peak irradiance at the first focus to below the modification threshold, while still providing the E-field intensity required to modify the material at the second one. Using the SLM technique, the authors of [48] were able to successfully inscribe a waveguide in a phosphate glass that exhibited a symmetric cross-section.

The use of active optics for waveguide shaping is still in its infancy, having only first been demonstrated in late 2008. Unlike either the astigmatic beam shaping or slit beam shaping techniques, the use of active optics should enable waveguides with symmetric cross-sections to be inscribed, regardless of the sample translation direction, simply by synchronising the sample translation with changes in the phase and amplitude profile imprinted on the laser beam. Furthermore, the use of active optics should also enable the waveguide cross-section to be altered during the inscription process, thus enabling the fabrication of more advanced waveguide structures.

## 5.6.4 Spatiotemporal Focussing

All of the techniques described thus far have each been applied with success, but they are not without drawbacks. For example, the astigmatic and slit-beam shaping techniques are not well suited to the fabrication of waveguides with arbitrary paths since the waveguide cross-section would change as the sample translation direction changed. A possible solution to this may be to use active optics, as discussed in Sect. 5.6.3, but these techniques are expensive and complex to implement. A particularly promising and relatively simple technique which may overcome these drawbacks is the spatiotemporal focusing technique (STF) technique [49]. In contrast to the previously discussed techniques, the STF technique enables the control of the focal Rayleigh range, independent of the focussed beam waists. Consequently, the STF technique may enable the inscription of optical waveguides exhibiting a symmetric cross-section, regardless of the sample translation direction.

In the STF technique, first applied to ultrafast laser inscription by He et al. [49] using the experimental setup shown in Fig. 5.18, a spectral chirp is induced across the beam prior to entering the objective lens. The effect of this chirp is to reduce the spectral bandwidth of the pulse at each point across the beam. According to Fourier theory, this reduction stretches the pulse temporally, and its duration can only be fully restored in the focal region where the spectral components of the pulse are spatially reunited. In effect, the pulse is therefore focussed both spatially and temporally by the objective lens.

5 Controlling the Cross-section of Ultrafast Laser Inscribed Optical Waveguides

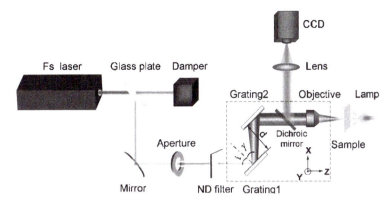

**Fig. 5.18** Schematic diagram of the experimental setup used by He et al. for spatiotemporal focussing. (Taken from [49].)

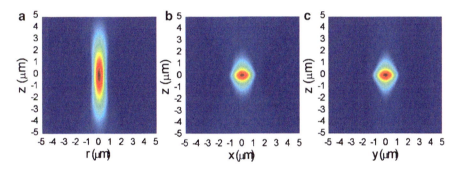

**Fig. 5.19** Numerically calculated laser intensity distributions at the focus produced by an objective lens (**a**) without and (**b** and **c**) with temporal focussing technique in *XZ* and *YZ* planes, respectively. (Taken from [49].)

Since the pulse duration directly determines the peak power of the pulse, the gradient of the temporal focussing can be used to control the Rayleigh range of the peak E-field distribution of the focus. The STF technique therefore uniquely enables the creation of a peak E-field distribution that is close to spherical in structure, as shown in Fig. 5.19. Under these circumstances, the cross-section of the inscribed structure would be symmetric, regardless of the sample translation direction.

It is important to acknowledge the fact that the STF technique is yet to be used to inscribe optical waveguides, although it has been successfully used to fabricate three-dimensional micro-channels by He et al. In this case, it was shown that through careful control of the STF parameters, micro-channels could be fabricated that exhibited a symmetric cross-section, regardless of the sample translation direction.

## 5.6.5 The Multiscan Technique

The multiscan technique is an alternative to the waveguide shaping techniques described in the preceding sections. In the multiscan technique, a rotationally-symmetric laser beam can be used to inscribe the waveguide. The waveguide asymmetry is then corrected by scanning the sample through the focus multiple times, normally on the order of 20 for a single mode waveguide. After each scan, the position of the sample is moved slightly in the axis perpendicular to both the sample translation axis and beam propagation axis. Over the course of the inscription process, the desired waveguide cross-section is built up by combining the numerous lines of modified material induced by each scan, as shown in Fig. 5.20.

Unlike the beam shaping techniques, the multiscan technique enables great flexibility in defining the waveguide cross-section. It allows, for example, the inscription of waveguides with an almost square and step-index refractive index profile as shown in Fig. 5.3(b). The flexibility of the multiscan technique has also enabled the fabrication of advanced waveguide structures such as multimode interference waveguides [50].

Due to its simplicity, the multiscan technique has been widely adopted and successfully applied to both active materials, such as ErYb-doped silicate glass [51], Er-doped bismuthate glass [52], Bi-doped silicate glass [20], and passive materials, such as fused silica [18]. In fact, the lowest propagation loss ever reported for an ultrafast laser inscribed optical waveguide ($0.12\,\text{dB}\,\text{cm}^{-1}$) was reported by Nasu et al. in 2005 for a multiscan fabricated waveguide in fused silica [18]. One reason for this low-loss may be that since the multiscan technique builds up the waveguide cross-section over many scans, the peak intensity can be kept to a minimum. This may reduce the number of scattering and absorbing defects induced in the modified material by the inscription process. The low fluence nature of the multiscan technique may also be responsible for its successful application

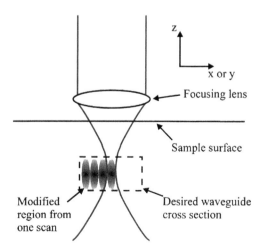

**Fig. 5.20** Schematic diagram of how the multiscan technique can be used to construct the desired waveguide cross-section

to waveguide inscription in crystalline materials such as periodically poled lithium niobate (PPLN) [53].

At this point it should be highlighted that although all the techniques described here are applicable to lasers operating in the single pulse regime, the multiscan technique has also been applied very successfully when using lasers operating in the intermediate regime [20, 51, 53, 54]. It should also be noted that in these cases, low pulse energies were used which minimised the effect of thermal diffusion and accumulation. It was seen for example in [20] that when using a 600 kHz laser and low energy pulses, the cross-section of the modified region induced by a single scan was highly asymmetric, as would be expected when using a laser operating strictly in the single pulse regime. When the pulse energy was increased, the role of thermal diffusion and accumulation increased, and the waveguide cross-section became considerably more symmetric.

## 5.7 Conclusions and Outlook

In this chapter, the importance of controlling the modal properties of optical waveguides was discussed. Through mainly qualitative arguments, it was shown that the modal properties are controlled via the waveguide cross-section. The problem of cross-section asymmetry for waveguides fabricated in the single pulse regime using a transverse writing geometry was described and experimental techniques were presented for controlling the cross-section of such waveguides. These included a number of techniques based on beam shaping, one that used spatiotemporal focusing, and the multiscan technique which operates by scanning the sample multiple times through the focus to construct the desired cross-section from the individual modified regions created by each scan.

Although the astigmatic beam shaping and slit beam shaping techniques described in Sects. 5.6.1 and 5.6.2, respectively, have enabled the inscription of waveguides with symmetric cross-sections, it was shown that they suffer from inflexibility. It may be possible in the future that the inflexibility of the astigmatic beam shaping technique may be overcome by synchronising the astigmatic difference and orientation of the cylindrical telescope with the sample translation. It has also recently been demonstrated that the inflexibility of the slit beam shaping technique can be overcome to some degree by synchronising the orientation of the slit with the sample translation [55]. Such techniques may in general be hard to implement in practice and could introduce mechanical instabilities. One way forward may be to use the active optic techniques discussed in Sect. 5.6.3. These techniques still operate using beam shaping principles, but rely on the use of a computer controlled active optics to shape the beam instead of bulk optics or apertures. Since the phase profile of the active optics can be readily controlled via a computer and synchronised with the sample translation, such techniques should not suffer the same inflexibility issues as the other beam shaping techniques.

Of all the techniques described in Sect. 5.6, it is the multiscan technique that provides the most flexibility in controlling the waveguide cross-section, but this flexibility comes at a penalty. By nature, the multiscan technique requires that the sample to be scanned multiple times through the focus. Since the number of scans is usually in the region of 20 for a single mode waveguide, the multiscan technique increases the fabrication time considerably. This time penalty is of little significance for research studies, but could be a considerable drawback for commercial applications where fabrication time is a primary concern. To overcome this issue, it is foreseeable that techniques utilising SLMs may be developed that produce a reconfigurable array of many focussed spots on the sample at one time. Such techniques would in effect reproduce the basic idea of the multiscan technique but using beam shaping to enable lower fabrication times. Techniques based on the use of SLMs have recently been demonstrated that enable multiple waveguides to be inscribed in a single scan [56]. Clearly, the number of spots that can be produced is dependent on the pulse energy available from the laser, and it may be unpractical to fabricate very wide structures, such as the multimode interference waveguide described in [50]. To overcome this problem, it is likely that future waveguide shaping approaches will utilise a combination of beam shaping and multiscan approaches.

# References

1. A.W. Snyder, J.D. Love, *Optical Waveguide Theory* (Chapman and Hall, New York, 1983)
2. R.G. Hunsperger, *Integrated Optics: Theory and Technology* (Springer, Berlin, 1982)
3. A.M. Zheltikov, D.T. Reid, Weak-guidance-theory review of dispersion and birefringence management by laser inscription. Laser Phys. Lett. **5**, 11–20 (2008)
4. J.M. Senior, *Optical Fiber Communications: Principles and Practice* (Prentice Hall Europe, Hertfordshire, 1992)
5. R. Osellame, N. Chiodo, G. Della Valle, S. Taccheo, R. Ramponi, G. Cerullo, A. Killi, U. Morgner, M. Lederer, D. Kopf, Optical waveguide writing with a diode-pumped femtosecond oscillator. Opt. Lett. **29**, 1900–1902 (2004)
6. X. Orignac, D. Barbier, X.M. Du, R.M. Almeida, O. McCarthy, E. Yeatman, Sol–gel silica/titania-on-silicon Er/Yb-doped waveguides for optical amplification at $1.5\,\mu m$. Opt. Mater. **12**, 1–18 (1999)
7. S.M. Eaton, W.J. Chen, H. Zhang, R. Iyer, J. Li, M.L. Ng, S. Ho, J.S. Aitchison, P.R. Herman, Spectral loss characterization of femtosecond laser written waveguides in glass with application to demultiplexing of 1300 and 1550 nm wavelengths. J. Lightwave Technol. **27**, 1079–1085 (2009)
8. W.J. Reichman, J.W. Chan, C.W. Smelser, S.J. Mihailov, D.M. Krol, Spectroscopic characterization of different femtosecond laser modification regimes in fused silica. J. Opt. Soc. Am. B **24**, 1627–1632 (2007)
9. Y. Liu, M. Shimizu, B. Zhu, Y. Dai, B. Qian, J. Qiu, Y. Shimotsuma, K. Miura, K. Hirao, Micromodification of element distribution in glass using femtosecond laser irradiation. Opt. Lett. **34**, 136–138 (2009)
10. J. Qiu, C. Zhu, T. Nakaya, J. Si, K. Kojima, F. Ogura, K. Hirao, Space-selective valence state manipulation of transition metal ions inside glasses by a femtosecond laser. Appl. Phys. Lett. **79**, 3567–3569 (2001)

11. W.J. Chen, S.M. Eaton, H. Zhang, P.R. Herman, Broadband directional couplers fabricated in bulk glass with high repetition rate femtosecond laser pulses. Opt. Express **16**, 11470–11480 (2008)
12. S.M. Eaton, W. Chen, L. Zhang, H. Zhang, R. Iyer, J.S. Aitchison, P.R. Herman, Telecom-band directional coupler written with femtosecond fiber laser. IEEE Photon. Technol. Lett. **18**, 2174–2176 (2006)
13. A. Szameit, D. Blomer, J. Burghoff, T. Schreiber, T. Pertsch, S. Nolte, A. Tunnermann, F. Lederer, Discrete nonlinear localization in femtosecond laser written waveguides in fused silica. Opt. Express **13**, 10552–10557 (2005)
14. A. Szameit, J. Burghoff, T. Pertsch, S. Nolte, A. Tuennermann, F. Lederer, Two-dimensional soliton in cubic fs laser written waveguide arrays in fused silica. Opt. Express **14**, 6055–6062 (2006)
15. N. Chiodo, G. Della Valle, R. Osellame, S. Longhi, G. Cerullo, R. Ramponi, P. Laporta, U. Morgner, Imaging of Bloch oscillations in erbium-doped curved waveguide arrays. Opt. Lett. **31**, 1651–1653 (2006)
16. F. Dreisow, A. Szameit, M. Heinrich, T. Pertsch, S. Nolte, A. Tunnermann, S. Longhi, Bloch–Zener oscillations in binary superlattices. Phys. Rev. Lett. **102**, 076802 (2009)
17. P. Oberson, B. Gisin, B. Huttner, N. Gisin, Refracted near-field measurements of refractive index and geometry of silica-on-silicon integrated optical waveguides. Appl. Opt. **37**, 7268–7272 (1998)
18. Y. Nasu, M. Kohtoku, Y. Hibino, Low-loss waveguides written with a femtosecond laser for flexible interconnection in a planar light-wave circuit. Opt. Lett. **30**, 723–725 (2005)
19. S.M. Eaton, H. Zhang, M.L. Ng, J. Li, W. Chen, S. Ho, P.R. Herman, Transition from thermal diffusion to heat accumulation in high repetition rate femtosecond laser writing of buried optical waveguides. Opt. Express **16**, 9443–9458 (2008)
20. N.D. Psaila, R.R. Thomson, H.T. Bookey, A.K. Kar, N. Chiodo, R. Osellame, G. Cerullo, G. Brown, A. Jha, S. Shen, Femtosecond laser inscription of optical waveguides in bismuth ion doped glass. Opt. Express **14**, 10452–10459 (2006)
21. Y. Youk, D.Y. Kim, A simple reflection-type two-dimensional refractive index profile measurement technique for optical waveguides. Opt. Commun. **262**, 206–210 (2006)
22. Y. Park, N.H. Seong, Y. Youk, D.Y. Kim, Simple scanning fibre-optic confocal microscopy for the refractive index profile measurement of an optical fibre. Meas. Sci. Technol. **13**, 695–699 (2002)
23. Y. Youk, D.Y. Kim, Tightly focused epimicroscope technique for submicrometer-resolved highly sensitive refractive index measurement of an optical waveguide. Appl. Opt. **46**, 2949–2953 (2007)
24. Y. Youk, D.Y. Kim, Reflection-type confocal refractive index profile measurement method for optical waveguides: effects of a broadband light source and multireflected lights. Opt. Commun. **277**, 74–79 (2007)
25. V.R. Bhardwaj, E. Simova, P.B. Corkum, D.M. Rayner, C. Hnatovsky, R.S. Taylor, B. Schreder, M. Kluge, J. Zimmer, Femtosecond laser-induced refractive index modification in multicomponent glasses. J. Appl. Phys. **97**, 083102 (2005)
26. B. Kouskousis, D.J. Kitcher, S. Collins, A. Roberts, G.W. Baxter, Quantitative phase and refractive index analysis of optical fibers using differential interference contrast microscopy. Appl. Opt. **47**, 5182–5189 (2008)
27. A. Barty, K.A. Nugent, D. Paganin, A. Roberts, Quantitative optical phase microscopy. Opt. Lett. **23**, 817–819 (1998)
28. P. Ferraro, L. Miccio, S. Grilli, M. Paturzo, S. De Nicola, A. Finizio, R. Osellame, P. Laporta, Quantitative phase microscopy of microstructures with extended measurement range and correction of chromatic aberrations by multiwavelength digital holography. Opt. Express **15**, 14591–14600 (2007)
29. C.B. Schaffer, A. Brodeur, E. Mazur, Laser-induced breakdown and damage in bulk transparent materials induced by tightly focused femtosecond laser pulses. Meas. Sci. Technol. **12**, 1784–1794 (2001)

30. S.M. Eaton, H. Zhang, P.R. Herman, Heat accumulation effects in femtosecond laser-written waveguides with variable repetition rate. Opt. Express 13, 4708–4716 (2005)
31. S. Nolte, M. Will, J. Burghoff, A. Tuennermann, Femtosecond waveguide writing: a new avenue to three-dimensional integrated optics. Appl. Phys. A 77, 109–111 (2003)
32. J. Liu, Z. Zhang, C. Flueraru, X. Liu, S. Chang, C.P. Grover, Waveguide shaping and writing in fused silica using a femtosecond laser. IEEE J. Sel. Top. Quant. Electron. 10, 169–173 (2004)
33. C. Mauclair, A. Mermillod-Blondin, N. Huot, E. Audouard, R. Stoian, Ultrafast laser writing of homogeneous longitudinal waveguides in glasses using dynamic wavefront correction. Opt. Express 16, 5481–5492 (2008)
34. A. Mermillod-Blondin, C. Mauclair, A. Rosenfeld, J. Bonse, I.V. Hertel, E. Audouard, R. Stoian, Size correction in ultrafast laser processing of fused silica by temporal pulse shaping. Appl. Phys. Lett. 93, 021921 (2008)
35. G. Cerullo, R. Osellame, S. Taccheo, M. Marangoni, D. Polli, R. Ramponi, P. Laporta, S. De Silvestri, Femtosecond micromachining of symmetric waveguides at 1.5 $\mu$m by astigmatic beam focusing. Opt. Lett. 27, 1938–1940 (2002)
36. R. Osellame, S. Taccheo, M. Marangoni, R. Ramponi, P. Laporta, D. Polli, S. De Silvestri, G. Cerullo, Femtosecond writing of active optical waveguides with astigmatically shaped beams. J. Opt. Soc. Am. B 20, 1559–1567 (2003)
37. R. Osellame, V. Maselli, R.M. Vazquez, R. Ramponi, G. Cerullo, Integration of optical waveguides and microfluidic channels both fabricated by femtosecond laser irradiation. Appl. Phys. Lett. 90, 231118 (2007)
38. Y. Cheng, K. Sugioka, K. Midorikawa, M. Masuda, K. Toyoda, M. Kawachi, K. Shihoyama, Control of the cross-sectional shape of a hollow microchannel embedded in photostructurable glass by use of a femtosecond laser. Opt. Lett. 28, 55–57 (2003)
39. M. Ams, G.D. Marshall, D.J. Spence, M.J. Withford, Slit beam shaping method for femtosecond laser direct-write fabrication of symmetric waveguides in bulk glasses. Opt. Express 13, 5676–5681 (2005)
40. S. Ho, Y. Cheng, P.R. Herman, K. Sugioka, and K. Midorikawa, Direct ultrafast laser writing of buried waveguides in Foturan glass, in *Conference on Lasers and Electro-Optics/International Quantum Electronics Conference and Photonic Applications Systems Technologies, Technical Digest (CD)* (Optical Society of America, Oxford, 2004), paper CThD6
41. K.J. Moh, Y.Y. Tan, X.C. Yuan, D.K.Y. Low, Z.L. Li, Influence of diffraction by a rectangular aperture on the aspect ratio of femtosecond direct-write waveguides. Opt. Express 13, 7288–7297 (2005)
42. S. Sowa, W. Watanabe, T. Tamaki, J. Nishii, K. Itoh, Symmetric waveguides in poly(methyl methacrylate) fabricated by femtosecond laser pulses. Opt. Express 14, 291–297 (2005)
43. R.R. Thomson, H.T. Bookey, N.D. Psaila, A. Fender, S. Campbell, W.N. MacPherson, J.S. Barton, D.T. Reid, A.K. Kar, Ultrafast-laser inscription of a three dimensional fan-out device for multicore fiber coupling applications. Opt. Express 15, 11691–11697 (2007)
44. W. Yang, C. Corbari, P.G. Kazansky, K. Sakaguchi, I.C.S. Carvalho, Low loss photonic components in high index bismuth borate glass by femtosecond laser direct writing. Opt. Express 16, 16215–16226 (2008)
45. G.D. Marshall, P. Dekker, M. Ams, J.A. Piper, M.J. Withford, Directly written monolithic waveguide laser incorporating a distributed feedback waveguide-Bragg grating. Opt. Lett. 33, 956–958 (2008)
46. M. Ams, P. Dekker, G.D. Marshall, M.J. Withford, Monolithic 100 mw Yb waveguide laser fabricated using the femtosecond-laser direct-write technique. Opt. Lett. 34, 247–249 (2009)
47. R.R. Thomson, A.S. Bockelt, E. Ramsay, S. Beecher, A.H. Greenaway, A.K. Kar, D.T. Reid, Shaping ultrafast laser inscribed optical waveguides using a deformable mirror. Opt. Express 16, 12786–12793 (2008)
48. A. Ruiz de la Cruz, A. Ferrer, W. Gawelda, D. Puerto, M.G. Sosa, J. Siegel, J. Solis, Independent control of beam astigmatism and ellipticity using a SLM for fs-laser waveguide writing. Opt. Express 17, 20853–20859 (2009)

49. F. He, H. Xu, Y. Cheng, J. Ni, H. Xiong, Z. Xu, K. Sugioka, K. Midorikawa, Fabrication of microfluidic channels with a circular cross section using spatiotemporally focused femtosecond laser pulses. Opt. Lett. **35**, 1106–1108
50. W. Watanabe, Y. Note, K. Itoh, Fabrication of multimode interference waveguides in glass by use of a femtosecond laser. Opt. Lett. **30**, 2888–2890 (2005)
51. N.D. Psaila, R.R. Thomson, H.T. Bookey, A.K. Kar, N. Chiodo, R. Osellame, G. Cerullo, A. Jha, S. Shen, Er:Yb-doped oxyfluoride silicate glass waveguide amplifier fabricated using femtosecond laser inscription. Appl. Phys. Lett. **90**, 131102 (2007)
52. R.R. Thomson, N.D. Psaila, S.J. Beecher, A.K. Kar, Ultrafast laser inscription of a high-gain Er-doped bismuthate glass waveguide amplifier. Opt. Express **18**, 13212–13219 (2010)
53. R. Osellame, M. Lobino, N. Chiodo, M. Marangoni, G. Cerullo, R. Ramponi, H.T. Bookey, R.R. Thomson, N.D. Psaila, A.K. Kar, Femtosecond laser writing of waveguides in periodically poled lithium niobate preserving the nonlinear coefficient. Appl. Phys. Lett. **90**, 241107 (2007)
54. N.D. Psaila, R.R. Thomson, H.T. Bookey, N. Chiodo, S. Shen, R. Osellame, G. Cerullo, A. Jha, A.K. Kar, Er:Yb-doped oxyfluoride silicate glass waveguide laser fabricated using ultrafast laser inscription. IEEE Photon. Technol. Lett. **20**, 126–128 (2008)
55. Y. Zhang, G. Cheng, G. Huo, Y. Wang, W. Zhao, C. Mauclair, R. Stoian, R. Hui, The fabrication of circular cross-section waveguide in two dimensions with a dynamical slit. Laser Phys. **19**, 2236–2241 (2009)
56. M. Pospiech, M. Emons, A. Steinmann, G. Palmer, R. Osellame, N. Bellini, G. Cerullo, U. Morgner, Double waveguide couplers produced by simultaneous femtosecond writing. Opt. Express **17**, 3555–3563 (2009)

# Chapter 6
# Quill and Nonreciprocal Ultrafast Laser Writing

Peter G. Kazansky and Martynas Beresna

**Abstract** Since the discovery of lasers, it was believed that a Gaussian mode of a laser beam interacting with an isotropic medium can produce only centrosymmetric material modifications. However, recent experiments provide the evidence that it is not always true. A remarkable phenomenon in ultrafast laser processing of transparent materials has been reported manifesting itself as a change in material modification by reversing the writing direction. The phenomenon has been interpreted in terms of plasma anisotropic trapping and heating by a tilted front of the ultrashort laser pulse. It has been experimentally demonstrated that indeed the pulse front tilt can be used to control material modifications and in particular as a new tool for laser processing and optical manipulation. Additionally, a new type of light-induced modification in a solid, namely an anisotropic cavitation, was observed in the vicinity of the focus at high fluences. The bubbles, formed in the bulk of the glass, can be trapped and manipulated in the plane perpendicular to the light propagation direction by controlling the laser writing direction relative to the tilt of the pulse front.

Another common belief was that in a homogeneous medium, the photosensitivity and corresponding light-induced material modifications do not change on the reversal of light propagation direction. Recently, it was demonstrated that when the direction of the femtosecond laser beam is reversed from $+Z$ to $-Z$ directions, the structures written in a lithium niobate crystal are mirror images when translating the beam along the $+Y$ and $-Y$ directions. In contrast to glass, the directional dependence of writing in lithium niobate depends on the orientation of the crystal with respect to the direction of the beam movement and the light propagation direction. A theoretical model was created to demonstrate how interplay of the crystal anisotropy and light-induced heat flow can lead to a new nonreciprocal nonlinear optical phenomenon, nonreciprocal photosensitivity. In the lithium niobate,

P.G. Kazansky (✉) · M. Beresna
Optoelectronics Research Centre, University of Southampton, SO17 1BJ, UK
e-mail: pgk@orc.soton.ac.uk; mxb@orc.soton.ac.uk

the nonreciprocal photosensitivity manifests itself as a changing the sign of the light-induced current when the light propagation direction is reversed. Therefore, in a non-centrosymmetric medium, modification of the material can be different when light propagates in opposite directions. Nonreciprocity is produced by magnetic field (Faraday effect) and movement of the medium with respect to the direction of light propagation: parallel (Sagnac effect) or perpendicular (KaYaSo effect).

## 6.1 Introduction

Writing is the main method for information storage, which has been known from the dawn of civilization. These days direct write technologies, including plasma spray, micropen, ink jet, e-beam, focused ion beam, and laser beam are of increasing importance in material processing [1]. More recently, modification of transparent materials with ultrafast lasers has attracted considerable interest due to a wide range of applications including laser surgery [2], integrated optics [3], optical data storage [4], and three-dimensional micro- [5] or nanostructuring [6]. Demonstration of femtosecond laser nano-surgery and optical tweezers made the combination of ultrafast laser processing and optical manipulation very attractive for bio-photonic applications [7, 8]. Moreover, modifications of materials by light span from photosynthesis and photography to material processing and laser writing, and there are only few independent parameters of irradiation which control material transformations, in particular wavelength, intensity, exposure time, and pulse duration. *For instance, a* key advantage of using femtosecond pulses, as opposed to longer pulses, for direct writing is that such pulses can rapidly and precisely deposit energy in solids [9]. The process, initiated by multiphoton ionization, exhibits a highly nonlinear dependence on the intensity of the light beam. The light is absorbed by photoelectrons and the optical excitation ends before the surrounding lattice is perturbed, resulting in highly localized breakdown without collateral material damage [10]. However, it is well recognized that reversing the writing direction should not affect material processing and associated modifications.

## 6.2 Quill Writing

Recently it was noticed that the regions of glass irradiated by femtosecond laser pulses with intensities from a certain range exhibit anisotropic scattering [11], reflection, and negative birefringence [12, 13]. It was shown that 3D self-assembled sub-wavelength planar structures, aligned perpendicular to the polarization direction of the writing laser, are responsible for these peculiar optical properties. Such birefringence has been identified as form birefringence with the optical axis aligned along the direction of the writing laser's polarization. Recently interesting applications of these self-assembled nanostructures for diffractive elements [14], integrated optics [15], nano-fluidics, and rewritable 3D optical memories have been demonstrated [16–18]. Apart from the anisotropic phenomena mentioned above, the anisotropy of laser machining parallel and perpendicular to the direction of the laser polarization is also known, which is explained by an optical anisotropy at the interface of the irradiated region [19]. Here we report the observation of a remarkable phenomenon in direct writing and ultrafast laser processing in transparent materials, in particular silica glass, manifested as a change in material modification by reversing the writing direction. The effect resembles writing with a quill pen and is interpreted in terms of anisotropic trapping of the electron plasma by a tilted front of the ultrashort laser pulse.

Experimental conditions were as follows. The laser radiation, in a Gaussian mode, produced by a regeneratively amplified mode-locked Ti: Sapphire laser system (Coherent RegA 9000) generating 150 fs pulses with a maximum energy of 2.4 µJ at a wavelength of 800 nm and a repetition rate of 250 kHz, was focused via a 50× (NA = 0.55) objective into the bulk of the sample. A series of lines were directly written by scanning in alternating directions at a depth of 0.5 mm below the front surface. The writing speed was 200 µm/s and each line was written with only one pass, in one direction, of the laser, with the polarization vector directed perpendicular to the line and pulse energy of 0.9 µJ. After writing, the structures were side-polished and imaged with a scanning electron microscope (SEM). The SEM images exposed tracks elongated in the direction of light propagation due to beam's confocal parameter, and enhanced by self-focusing effects, with a periodic structure in the direction of light polarization (Fig. 6.1). On closer inspection,

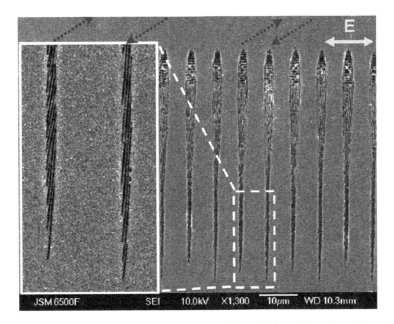

**Fig. 6.1** SEM images of cross sections of the structures formed 0.5 mm below the surface of fused silica with a 0.55 NA lens. The distance between lines is 7 µm

we were surprised to observe a difference in the structures written in opposite directions. This difference is revealed in small variations of the depth of the tracks and a tilt of the periodic structures written in the forward and reverse directions. The periodic planar nanostructures are aligned along the direction of the writing laser polarization and are responsible for form birefringence of the irradiated region.

In another experiment, we wrote a series of lines using an IMRA-FCPA μJewel D-400 amplified ytterbium fiber laser system, operating at 1045 nm, with pulse duration <500 fs and repetition rates ranging from 100 kHz to 1 MHz. The high stability of the FCPA laser system is crucial for systematic studies. The polarization of the laser was aligned perpendicular to the writing direction. The lines were written in alternating directions from forward to reverse at different pulse energies (from 0.2 to 1.8 μJ), and the product of repetition rate and scan speed was kept constant to maintain the same total fluence at different repetition rates. After writing, laser exposed areas were captured using both crossed-polarized (CP) and Nomarski-differential interference contract (DIC) illumination (both back-illumination). Composite images (Fig. 6.2) were created to show the same portion of each feature using the two illumination techniques. With these composite images, the amount of birefringence visible with the CP illumination can be compared with the texture of the feature using the DIC imaging technique. At low energies, lines written in both directions were the same (Fig. 6.2, 0.4–0.6 μJ at 500 kHz repetition rate). However, with an increase in energy, the appearance of a directional dependence (which was strongest at about 0.8–0.9 μJ) in the written lines was observed. The directional dependence is more clearly seen in the birefringence of the lines. This dependence can also be observed in the morphology (texture) of lines written in opposite directions, with a line written in one direction being rougher than a line written in the reversed direction (Fig. 6.2a). However, with further increase of

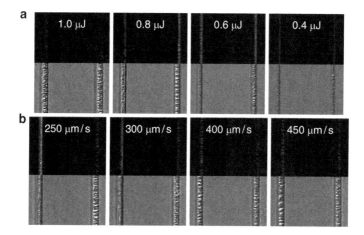

**Fig. 6.2** Images in crossed polarizers (*dark part*) and Nomarski-DIC (*light part*) of the lines written in glass in opposite directions at repetition rate of 500 kHz with (**a**) a writing speed of 500 μm/s and different energies and (**b**) a pulse energy of 0.9 μJ and different writing speeds

energy above a certain threshold value, both lines become uneven with indications of collateral damage, and the birefringence of the lines disappears as well as the directional dependence. We believe that the latter phenomenon can be explained by a cumulative thermal effect [20]. This is supported by the presence of modifications with rough features, much bigger than the spot of the beam, at high repetition rates (500 kHz to 1 MHz) and the absence of such features with collateral damage at low repetition rates (below 300 kHz). This agrees with the silica glass heat diffusion time (about 1 μs). We also tested the dependence of the observed effect on writing speed near the energy threshold for the disappearance of directional phenomenon and observed that the directional dependence becomes stronger at lower writing speeds (Fig. 6.2b).

When the beam was turned by 90°, using a two-mirror periscope while other writing parameters were maintained the same including the polarization of the laser beam perpendicular to the stage movement, we observed only a small difference in the structures written in opposite directions. This indicates that some asymmetry in the structure of the beam can be responsible for the observed phenomenon.

The other intriguing result is the observation of *different* textures in the processed material for laser polarizations perpendicular and parallel to the movement of the sample *in one direction* and the *same* textures for two polarizations when writing in the *opposite* direction (Fig. 6.3a). The SEM images of the cross sections of the lines written with 500 kHz repetition rate, 250 μm/s writing speed, and 0.9 μJ pulse energy revealed a different texture in the lines written in opposite

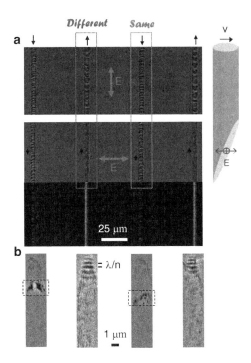

**Fig. 6.3** (a) CP and DIC images of the lines written with orthogonal polarizations with 500 kHz repetition rate, writing speed 250 μm/s, and pulse energy 0.9 μJ. The difference in texture for two polarizations is observed only for one writing direction. The tilted front of the pulse along writing direction is shown. (b) SEM images of cross sections of lines written with polarization perpendicular to writing direction are also shown. The regions of collateral damage are marked with *dashed lines*

directions (Fig. 6.3b). Remarkably, the nanograting of about 300 nm period, which is responsible for the form birefringence of irradiated regions, can be seen only in the initial part of cross sections of lines written in one of the two directions. This small area is followed by one with a collateral damage due to thermal effect, which correlates with a weak birefringence of these lines. It is also observed that in almost entire cross sections of the lines, written in opposite direction, there is the nanograting along the direction of light polarization with the period of about 250 nm together with the additional periodicity, along the direction of light propagation, of about 720 nm, which is of the wavelength of light ($\lambda/n$, $\lambda = 1045$ nm, $n = 1.45$) (Fig. 6.3b). These lines demonstrate no evidence of the collateral thermal damage and much stronger birefringence (Fig. 6.3a).

Lines written at a repetition rate of 100 kHz (Fig. 6.4) also clearly show different textures in opposite directions, without any evidence of collateral damage due to thermal effect. The SEM images reveal the presence of the nanograting in the direction of light polarization almost in the entire cross section for one writing direction and the nanograting, again with additional periodicity along the propagation direction of about the wavelength of light, for opposite writing direction (Fig. 6.4b).

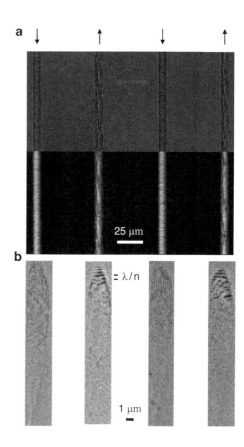

**Fig. 6.4** Optical microscope CP and DIC images (**a**) and corresponding SEM (**b**) images of cross sections of the lines written in glass in opposite directions with repetition rate 100 kHz, writing speed 100 μm/s and pulse energy 2 μJ

The writing anisotropy is observed only at a certain range of pulse energies, which excludes the stage movement as the cause. Inspection of the intensity distribution of the laser beam did not reveal any peculiarities in the shape of the beam, which was close to circular (Gaussian shape). The only possibility left to explain the puzzle of the writing direction anisotropy is related to the anisotropy of the frequency distribution (frequency chirp) in the beam. A spatial frequency chirp and related pulse front tilt are quite common in femtosecond laser systems [21]. Even a small delay across the beam that corresponds to ~10% of the pulse duration results in pulse tilt as strong as tens of degrees in the vicinity of the focal plane. The pulse front tilt is enhanced in dispersive media, as in the case of electron plasma close to plasma frequency, which is formed in the focus of the beam due to multiphoton ionization of glass. The pulse front tilt is a tilt in the intensity distribution in the front of the pulse. It is known that in the presence of intensity gradients, the charges (e.g., electrons) experience the ponderomotive force (light pressure), which expels the electrons from the region of high intensity [22]. Indeed, free electrons are affected by a variation in the laser intensity as they quiver in the electric field of the laser pulse. In the non-relativistic case, this can be expressed with the fluid equation of motion in an electromagnetic field by

$$\frac{\partial \mathbf{v}}{\partial t} + (\mathbf{v} \cdot \nabla)\mathbf{v} = -\frac{e}{m_e}(\mathbf{v} \times \mathbf{B} + \mathbf{E}), \tag{6.1}$$

where $\mathbf{v}$ is the electron velocity vector, $m_e$ is the electron mass, and $e$ is the electron charge. The ponderomotive force, $\mathbf{F_p}$, follows from this equation by time averaging the electric field as:

$$\mathbf{F_p} = -\frac{e^2}{4m_e\omega^2}\nabla E^2 = -\frac{e^2}{2c\varepsilon_0 m_e \omega^2}\nabla I, \tag{6.2}$$

where $c$ is the speed of light in vacuum, $\varepsilon_0$ is the permittivity of free space, $\omega$ is the frequency of light, and $I$ is the light intensity. It is possible to derive a ponderomotive potential, $\mathbf{U_p}$, as

$$\mathbf{U_p} = \frac{e^2 I}{2c\varepsilon_0 m_e \omega^2}. \tag{6.3}$$

For very short and intense laser pulses, the ponderomotive force can become very important, and the resulting acceleration will tend to push electrons in front of the laser pulse, as a kind of "snow-plough" effect [23]. Estimated intensity in the focus of a laser beam in our experiments is of about $3 \times 10^{14}$ W/cm$^2$, which will produce ponderomotive potential of 60 meV. This potential is higher than the electron energy at room temperature, which is of about 40 meV. Electron plasma in our experiments will still experience this kind of force in front of the pulse. Due to the tilt of the intensity distribution, the force will act on the electron plasma along the direction of the intensity gradient. By moving the beam, the ponderomotive force in the front of the pulse will trap and displace the electrons along the direction of movement of the beam and only in one direction corresponding to the tilt in the intensity distribution (we refer to this phenomenon as the "quill effect"). The electron plasma waves, excited in electron plasma, are responsible for the formation of the nanograting [13] and self-assembled form birefringence [17]. The trapping and displacement of the electrons with the movement of the beam affect the interference of plasma waves, and related form birefringence. The periodic structure, with the period of the wavelength of light, along the direction of light propagation is created as a result of the interference between plasma waves and plasma oscillation. Trapping of the electron plasma *damps plasma oscillation* and related interference producing longitudinal periodic structure with the wavelength of light. The observed difference in the onset of the collateral thermal damage for two writing directions is also the consequence of the anisotropic trapping effect (Fig. 6.3b). Further support of the proposed mechanism is the evidence of different textures of modified material for writing with light polarizations parallel and perpendicular to the movement in one of the writing directions (Fig. 6.3a). This observation is explained by the difference in boundary conditions for two orthogonal polarizations at the interface of the *tilted pulse front along the writing direction*.

The spatial chirp and the related pulse front tilt can change significantly along the beam propagation direction. For the interpretation of the experimental results, measurements of the pulse front tilt in the focus of the beam are necessary [24]. However, currently such measurements cannot be realized due to the small spot size and high light intensity in the focus of the beam. However, further experiments confirmed that indeed the pulse front tilt can be used to control material modifications and in particular as a new tool for laser processing and optical manipulation, e.g., for achieving calligraphic style of laser writing, when the appearance of a "stroke" varies in relation to its direction.

The experiments were carried out with an amplified, mode-locked Ti:sapphire laser operating at *800* nm wavelength with *70* fs pulse duration and a 250 kHz repetition rate. The linearly polarized laser beam was focused via a $50\times$ (NA $= 0.8$) objective at a depth of 60 μm beneath the surface of the fused silica sample. Line structures were written inside the bulk material by translating the sample perpendicularly to the light propagation direction, using a linear motorized stage (Aerotech ALS-130). After irradiation, the sample was inspected using an optical microscope. The first group of lines was written by scanning in alternating directions inside the sample with pulse energy of 2.6 μJ and scan speed of 50 μm/s (Fig. 6.5a). The temporal characteristic of the pulses, in particular the pulse front tilt before the focusing objective, was characterized using a GRENOUILLE device (8.50 Model) [21]. The measured pulse front tilt for the first group of lines was $4\times 10^{-2}$ fs/mm. As shown in Fig. 6.5a, the directional dependence can be clearly observed in the morphology of the lines written in opposite directions, with a line written in one direction being rougher than a line written in the reversed direction. The directional dependence can also be revealed by imaging the lines between crossed polarizers, in which only the smooth lines written in one direction show birefringence. Next the pulse front tilt was reversed by tuning the pulse

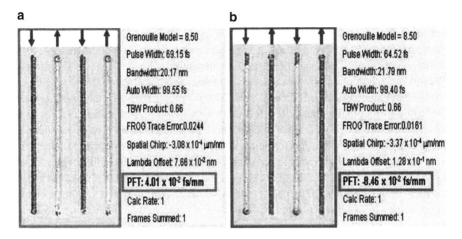

Fig. 6.5 Microscope bright field image of the line structures written using femtosecond pulses with positive pulse front tilt (**a**) or negative pulse front tilt (**b**). The distance between the lines is 25 μm. The writing direction is shown by the *arrow*. The respective screen shots containing measured laser pulse parameters by GRENOUILLE device are shown

compressor to the value of $-8.64 \times 10^{-2}$ fs/mm, and the second group of lines was imprinted by scanning in alternating directions (Fig. 6.5b). Comparing the structures written with the opposite sign of the pulse front tilt, the mirror change in the induced modifications becomes evident (Fig. 6.5). This experiment unambiguously demonstrates that the directional dependence of the writing process and the induced modification is determined by the pulse front tilt of the femtosecond laser pulses.

Smooth line structures written in one direction (Fig. 6.5) correspond to type 2 modifications with the evidence of form birefringence, while the rough structures produced by writing in the opposite direction belong to type 3 modifications, with the evidence of a void formation. At pulse energies below the threshold value of 2 μJ, both lines written in opposite directions reveal type 2 modifications. Moreover, when the pulse energy was increased to 3 μJ and above, both lines revealed type 3 modifications and the birefringence in the induced structures disappeared. The latter result indicates that the threshold energy for creating type 3 modification inside fused silica depends not only on the pulse energy but also on the writing direction. For example, the line structures, written toward the top (Fig. 6.5a), reveal higher threshold energy for the type 3 modifications than the structures written in the opposite direction. This threshold dependence reverses when the sign of the pulse front tilt is changed (Fig. 6.5b). Furthermore, no directional dependence could be observed when the pulse front tilt was minimized by tuning the pulse compressor. It should be highlighted that the quill writing effect *depends strongly* on the focusing depth of the laser irradiation under the sample surface. Changing the focusing depth by only 10%, which is from 60 to 55 or 65 μm, completely eliminated the quill writing effect and produced type 2 or type 3 structures in both directions.

An alternative experiment was also carried out to provide additional evidence that the pulse front tilt is responsible for the quill writing effect. A laser source, producing pulses of 150 fs duration at 250 kHz repetition rate and 800 nm, was used in this experiment. The laser beam profile, measured using a BeamScan system (Photon Inc.), had a well-defined Gaussian shape. A 50 × (NA = 0.55) objective was used to focus laser beam at a depth of 120 μm beneath the surface of the sample. The optimum depth for the quill effect in this experiment was different from the previous one due to the difference in laser beam parameters in the two experiments. The group of four lines with alternating writing directions was imprinted with the pulse energy of 1.4 μJ and scan speed of 50 μm/s (Fig. 6.6a). One additional mirror was added in the setup to reverse the direction of the pulse front tilt before writing the next group of lines (Fig. 6.6b). In this writing, configuration of the second group of four lines was imprinted with all other writing parameters identical to the previous experiment. It was observed that the structural modifications in the lines of the second group were mirrored when compared with the lines of the first group.

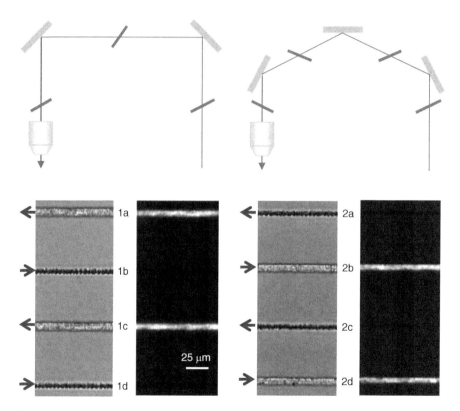

**Fig. 6.6** "Quill" effect – lines written in opposite directions have different morphology. In crossed polarizers, one can clearly see that birefringent modification is induced only in one direction. The effect can be reversed by reversing tilt direction by reflecting beam from a mirror

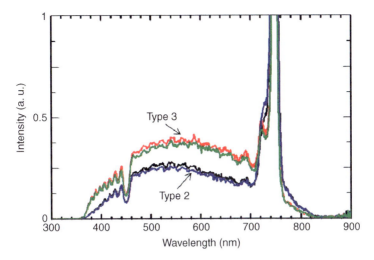

**Fig. 6.7** Spectra of the white emission collected in reflection during inscription of type 2 and type 3 structures. Two spectra corresponding to different writing experiments for each type of modification are shown

This result provides further evidence that the pulse front tilt is responsible for the directional dependence in the ultrafast laser writing.

We also observed strong white light emission (400–700 nm) corresponding to type 3 structural modifications and a weaker emission corresponding to the nanogratings formation (Fig. 6.7). This difference was reproducible in different writing experiments. Supercontinuum generation can be excluded as the explanation of this emission because the length of light propagation in the focus of high NA objective is too small to produce a significant spectral broadening. The observed white emission might be related to thermal radiation, e.g., bremsstrahlung radiation, which is emitted by electron plasma heated to thousands of degrees at high light intensities ($10^{14}$ W/cm$^2$) [25].

## 6.3 Anisotropic Bubble Formation

Another intriguing result is the observation of anisotropic directional cavitation in the irradiated region when the pulse energy was increased to the level of 2.4 µJ (Fig. 6.8). Bubbles with diameters of about 1–2 µm were formed on the sides of the irradiated line structures and about 10 µm upstream from the focus of the beam toward the laser (Fig. 6.8a). When the writing direction was reversed, the bubbles were shifted toward one side from the center of the beam (Fig. 6.8b). The bubbles have a lower refractive index in the center and higher refractive index in the surrounding, which indicates that these bubbles are cavities or voids surrounded by a densified material. Moreover, the bubbles show different color under transmission

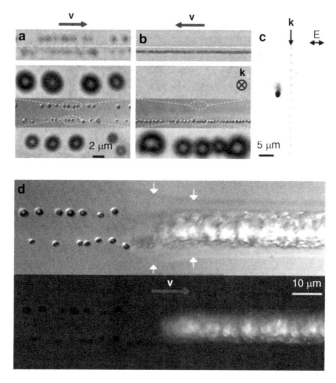

**Fig. 6.8** [a and b] Bright field images with different magnifications of the line structures fabricated in opposite directions (a) and (b) with 2.4 μJ pulse energy and scan speed of 50 μm/s. (c) Image of a cross section of the line structure shown in (b). (d) Images of the transition region without and with crossed polarizers. **V** is the writing velocity, **k** is the wave vector of the writing laser, and **E** is the electric field of the writing laser

microscopy, which could be explained by the interference between the light reflected at the front and back side of a bubble. The formation of multiple bubbles in the plane perpendicular to writing direction is difficult to explain by the micro-explosion mechanism [4, 26]. One should expect a micro-explosion to take place only in the region of highest intensity, which is in the center of the Gaussian beam. It is likely that the bubbles are created as the result of glass melting followed by cavitation. In femtosecond optical breakdown, laser pulse duration and the thermalization time of the energy of the free electrons are much shorter than the acoustic transit time from the center of the focus to its periphery. Therefore, no acoustic relaxation is possible during the thermalization time, and the thermo-elastic stresses caused by the temperature rise stay confined in the focal volume, leading to a maximum pressure rise. The tensile stress causes the formation of a cavitation bubble when the rupture strength of the molten glass is exceeded [27]. The observation of the shift of the bubble formation from the center of the beam can be explained by the fact that a particle with a refractive index lower than the surrounding medium, such as a void or bubble, is repulsed from the region of high intensity in the focus of a Gaussian beam. The component of the tilt, which is orthogonal to the directions of writing and the light propagation, will trap and shift the bubbles toward one side of the imprinted line structure, in a kind of a "snow plough" effect. It should be noted that this effect provides the first evidence of a possibility of manipulation of microscopic objects using pulses with a tilted intensity front.

Surprisingly, the bubble modifications can directly transfer into the birefringent features during the line writing process (Fig. 6.8). This transition can even be observed during the writing. At a certain distance, typically about 500 μm from the start of writing, the formation of bubbles abruptly terminated. Instead, evidence of self-assembled form birefringence appeared and continued until the end of line writing process. The transformation in the type of modification was correlated with the change in intensity of white light emission during the inscription process, which was similar to the emission described above. Strong white light emission during the bubble formation dropped at the modification transition, and only weak emission could be observed during the formation of birefringent structures. It should be noted that all writing parameters, including the pulse energy and scanning speed, were kept constant during the inscription process, and the phenomenon could be repeated in different areas of the sample. However, there is one parameter which is not constant, the temperature of the whole glass sample, which could increase as a result of light absorption during the writing process. It should be noted that self-assembled form birefringence structures and related nanogratings disappear at temperatures above the glass transition temperature [28]. This indicates that local temperature in the irradiated region should drop at modification transition from the glass melting temperature (1715°C), corresponding to the bubble formation, to the temperature below the glass transition temperature (1175°C), corresponding to the nanograting formation. This local temperature decrease can be explained by an increase in the heat capacity caused by the heating of the whole glass sample [29]. Indeed, given that $\Delta T = \Delta Q / m c_p$, where $\Delta T$ is the temperature difference, $\Delta Q$ is the heat energy, $m$ is the mass of the substance, and $c_p$ is the heat capacity, the temperature drops when heat capacity increases. The nanograting starts to form once the local temperature has dropped below the glass transition temperature. The energy is consumed by a nanograting formation process, which locks temperature below the glass transition temperature and explains why formation of nanogratings does not stop until the end of writing process.

## 6.4 Nonreciprocal Writing

It has also been a common belief that in a homogeneous medium, the photosensitivity and corresponding light-induced material modifications do not change on the reversal of light propagation direction [5]. However, we have demonstrated that when the propagation direction of the femtosecond laser beam is reversed from $+Z$ to $-Z$ directions, the structures written in a lithium niobate crystal ($LiNbO_3$) are

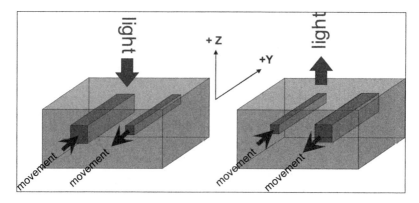

**Fig. 6.9** In a non-centrosymmetric medium, modification of the material can be different when femtosecond laser beam propagates in opposite directions. *Red arrow* indicates light propagation direction; *blue arrows* show direction of sample's movement

mirror images when translating the beam along the $+Y$ and $-Y$ directions (Fig. 6.9). Thus, the phenomenon of the laser writing in lithium niobate is nonreciprocal, i.e., a single light beam interacts with the crystal differently for opposite directions of light propagation. This experimental finding implies that forward and backward propagating intense light beams interact with the crystal differently resembling conventional Faraday effect. Therefore, laser writing in lithium niobate can be seen as a new nonreciprocal nonlinear optical phenomenon.

The experiments were performed using a mode-locked, regeneratively amplified Ti:Sapphire laser system (Coherent RegA 9000), operating at 800 nm with 150 fs pulse duration and 250 kHz repetition rate. The linearly polarized laser beam was focused via a 50× objective (NA = 0.55) 150 μm below the surface of a 1-mm-thick LiNbO$_3$ sample. The spot size in the focus of the beam was estimated to be ∼2 μm. The laser polarization was controlled using a half-wave plate, and the pulse energy was varied using a neutral density filter. The sample was mounted on a computer-controlled linear motorized stage (Aerotech ALS-130). Straight lines of modifications were written by translating the sample perpendicular to the propagation direction of the laser beam. The writing direction is defined as the scan direction of the laser spot with respect to the LiNbO$_3$ sample.

First, groups of lines with various pulse energies ranging from 1 to 2.4 μJ were written below the $-Z$ face of the LiNbO$_3$ sample. The scan speed was 200 μm/s and the laser beam was linearly polarized along Y axis. For each group, two lines were produced with opposite direction, in other words with writing direction along the $-Y$ and $+Y$ axis of LiNbO$_3$, respectively (Fig. 6.10).

After irradiation, the structural modifications were inspected by quantitative phase microscopy (QPM) [30], using an Olympus BX51 optical microscope equipped with a motorized focusing stage (Physik Instrumente). A quantitative phase map of the structures was produced using the QPM software (Iatia Vision Sciences). The refractive index change of the structure was estimated from

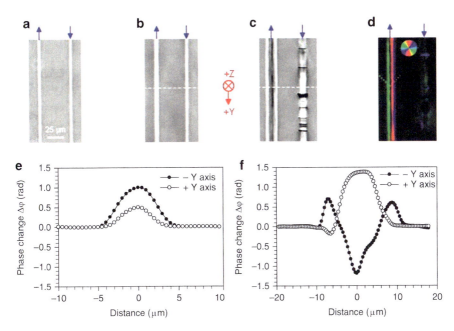

**Fig. 6.10** Line structures written along the Y axis of the LiNbO$_3$ sample. (**a–c**) QPM phase images of the lines written along the −Y axis and the +Y axis of LiNbO$_3$ with pulse energies of: 1.2 μJ (**a**); 1.8 μJ (**b**); 2.4 μJ (**c**). (**d**) Quantitative birefringence image of the line structures. The brightness represents the retardance magnitude while the color represents the slow axis of the birefringence region. The color of the circular legend shows the direction of slow axis. The *dashed lines* show the slow axis of the birefringence region. Quantitative phase change profiles of the line structures written along the −Y and +Y axes at the *dashed lines* in **b** (**e**) and **c** (**f**), respectively. Writing directions of the structures in (**a–d**) are shown by *arrows*

$\Delta n = \Delta \varphi \lambda / (2\pi d)$, where $\Delta \varphi$ is the phase shift of the light at wavelength of 550 nm and $d$ is the thickness of the structure along the light propagation direction. The stress region was analyzed using quantitative birefringence imaging system (CRi Abrio Imaging System).

We observed different material modifications depending on the pulse energy and the writing direction. For the lines written in both directions with pulses energies below 1.4 μJ, a positive phase change and thus a positive index change is created in the exposed region (Fig. 6.10a). The lines have the width of 2.5 μm, which is close to the focal spot size of the laser beam. No difference was observed for the lines written in the opposite directions. The lines imprinted with pulses energies between 1.4 and 2 μJ still have a positive index change in both directions (Fig. 6.10b). However, we observed the appearance of directional dependence in the structures written along the −Y and the +Y axis of LiNbO$_3$. The difference is first revealed in the morphology of the line structures. The lines inscribed along the −Y axis of LiNbO$_3$ have a larger width than those along the +Y axis (Fig. 6.10b). In addition, the directional difference can be revealed from the phase change in the structures.

The lines written along the $-Y$ axes of LiNbO$_3$ produce a larger positive phase change than those along the $+Y$ axis (Fig. 6.10e). With the further increase in pulse energy above $2\,\mu$J, this dependence could even be distinguished from the textures of the lines written in opposite directions. Some optical damage regions appeared in the lines written along the $+Y$ axis of LiNbO$_3$ (Fig. 6.10c). In contrast, no damage could be observed in the whole line written along the $-Y$ axis. The phase change ($\Delta\phi$) profile of the lines is shown in Fig. 6.1f. As a result, the line inscribed along the $+Y$ axis still has a positive phase change, similar to those created with lower pulse energies, and exhibits a maximum $\Delta\phi$ of 1.4 rad. By assuming the length of the structure along the Z axis of about $30\,\mu$m, the maximum $\Delta n$ is $4 \times 10^{-3}$. However, the line written along the $-Y$ axis exhibits a more complex phase change profile. A central dip region shows a negative phase change, while the surrounding regions have a positive phase change. This observation indicates that femtosecond laser modification results in a negative refractive index change of the central exposed region and a positive index change of the surrounding region. Stressed regions created as a result of the expansion of the material at the focus of the writing laser beam can account for the positive index change of the surrounding region [31–33]. This was confirmed by the quantitative birefringence imaging of the line structures (Fig. 6.10d). The line created along the $-Y$ axis of the sample shows two birefringent regions surrounding the central irradiated region, and the slow axis of the birefringent region is inclined about 45° toward the writing direction. The amorphization of the focal volume and subsequent stress-induced birefringence could account for this type of modification. In contrast, no birefringence features could be observed in the structures written along the $+Y$ axis. It is also noted that the birefringence becomes weaker when the pulse energy is decreasing, and no birefringence could be observed when the pulse energy was below $2\,\mu$J. Moreover, the directional dependence does not depend on the laser polarization. Compared to the structures imprinted when the crystal was translated along the Y axis, no directional dependence has been observed when the structures were written by translating the crystal along the X axis (Fig. 6.11). This indicates that the directional dependence of the light-induced modifications in LiNbO$_3$ is associated with the crystal symmetry.

To verify that the directional dependence of the modifications along the Y axis of LiNbO$_3$ is defined by the crystal axes, four groups of lines were fabricated in the sample. Two groups of the lines with opposite writing directions, along the $-Y$ and $+Y$ axis respectively, were written inside the sample with the pulse energies of $2.4\,\mu$J and $2\,\mu$J. (Fig. 6.12a). The sample was then rotated by 180° around the Z axis, and another two groups of structures, with the same writing parameters including the direction of the stage movement, was then imprinted in the sample (Fig. 6.12b). By comparing Fig. 6.12a, b, one can observe that the modification of the crystal structure is determined by the orientation of the writing direction with respect to the Y axis of the crystal.

Second, groups of structures were written by the laser beam propagating along the $+Z$ (Fig. 6.13a) and the $-Z$ (Fig. 6.13b) axes of the crystal. As expected, the created structures were different for writing directions along the $+Y$ and $-Y$ axes of

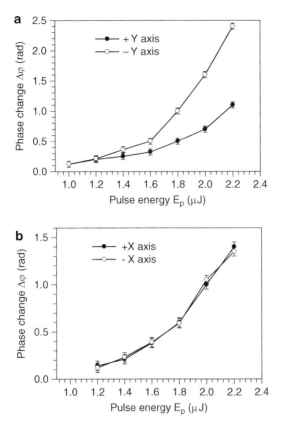

**Fig. 6.11** Comparison of the line structures imprinted along the Y axis and along the X axis. (**a, b**) Measured phase change of the line structures versus pulse energies for the lines written along the Y axis of the LiNbO$_3$ sample (**a**) and the X axis of the LiNbO$_3$ sample (**b**). The error bar is due to the variation in phase value for the unexposed region of the LiNbO$_3$ sample

the sample. However, we were surprised to observe a mirror change in the structural modifications in this case, when the propagation direction of the writing beam is reversed. Indeed, in contrast to the lines, written along the +Y axis with 2.4 μJ pulse energy by the beam propagating along +Z axis, which showed the optical damage features (Fig. 6.13a), no damage could be observed in the lines imprinted with the same fabrication parameters by the beam propagating along −Z axis (Fig. 6.13b). The change in the structural modifications between these lines is only produced by reversing the propagation direction of the focused laser beam with respect to the Z axis of the crystal. Moreover, the modification features, which appeared previously in the line written along the +Y axis by focusing below the −Z face of the crystal, could only be observed in the line written along the −Y axis when focusing below the +Z face. Therefore, a mirror change of the modifications in two similar structures written along the Y axis (+Y and −Y directions) with the

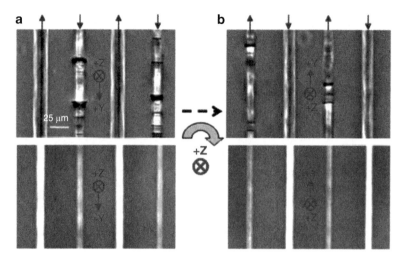

**Fig. 6.12** Phase images of line structures in the rotation experiment. (**a, b**) QPM phase images of lines imprinted along the Y axis with 2.4 µJ (*top*) and 2 µJ (*bottom*) pulse energies. The lines were written before (**a**) and after rotating by 180° around the Z axis (**b**) of the crystal, respectively

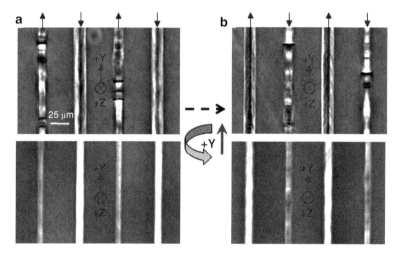

**Fig. 6.13** Phase images of line structures in the flip experiment. (**a, b**) QPM phase images of lines written along the Y axis with 2.4 µJ (*top*) and 2 µJ (*bottom*) pulse energies. The lines were written by propagating the laser beam along the +Z axes (**a**) and the −Z axes (**b**) of the crystal respectively

reversed beam propagating direction along the Z axis is discovered. Similar mirror change of phase profiles was also observed for the line structures without strong damage features written with the pulse energy of 2 µJ.

We also used different writing speeds and observed that the line width increases with reducing the scan speed. This indicates that a heat accumulation effect takes place during the writing process.

Groups of line structures were also written in Z-cut LiNbO$_3$ sample using the picosecond laser system (Lumera Super Rapid), operating at 1064 nm with 9 ps pulse duration, 250 kHz repetition rate, and with pulse energies comparable to those used with the femtosecond laser system. Modifications with crack features were observed at the same fluencies and at intensities of about one order of magnitude lower compared to the femtosecond laser system and with no evidence of the nonreciprocal writing phenomenon. This indicates that the phenomenon is accompanying only ultrashort pulses, possibly due to higher intensities which could be achieved with femtosecond pulses without strong damage of the sample.

It should be noted that the directional asymmetry of ultrashort-pulse-light-induced modifications in glass (femtosecond laser quill writing effect) does not depend on the orientation of a medium or the direction of light propagation and can be controlled by the tilt of a pulse front. In contrast to glass, the directional dependence of writing in lithium niobate depends on the orientation of a crystal with respect to direction of the beam movement and the light propagation direction. Specifically, the created structures possess a strong anisotropy only when the focused beam is translated along the Y axis, while no anisotropy is observed when the beam is translated along the X axis. The mirror symmetry of the structures created by the light beam propagating along the +Z and −Z axes of the crystal (see Fig. 6.13) indicates that the phenomenon observed in LiNbO$_3$ is not related to the tilt of the pulse front, as is the femtosecond laser quill writing effect in glass. We show in the following that, in a crystalline medium, even the pressure produced by a non-tilted pulse front can result in the dependence of the created structures on both the direction of writing and the propagation direction of the laser beam.

A heat current $\mathbf{J}$, which is carried by the electrons of plasma created by the femtosecond laser pulse, and generated by the ponderomotive force and the photon drag effect, can be phenomenologically presented in the following form:

$$J_i = \eta_{ijklmn} E_j E_k^* \nabla_n \left( E_l E_m^* \right) + i\zeta_{ijklmn} E_j E_k E_l^* E_m^* k_n \tag{6.4}$$

where subscripts label Cartesian indices, $\mathbf{E}$ is the complex amplitude of the light electric field, i.e., $E_k E_l^*$ is proportional to the light intensity and is responsible for heating via the plasma absorption. The first and second terms in the right-hand side of (6.4) describe pressure created by the front of the pulse and photon drag effect [34], respectively, $\eta_{ijklmn}$ and $\zeta_{ijklmn} = \zeta_{ikjlmn} = \zeta_{ilmjkn}$ are sixth rank tensors, and $\mathbf{k}$ is the wave vector. When the light beam propagates along the Z axis of the lithium niobate crystal, the current along the X axis is identically zero by symmetry, while that along the Y axis has symmetry allowed components of tensors $\eta_{ijklmn}$ and $\zeta_{ijklmn}$. For example, for the Y-polarized light beam, the thermal current along the Y axis is given by the following equation:

$$J_y = \left( \eta_{yyyyyz} \frac{\partial |E_y|^2}{\partial z} + ik \zeta_{yyyyyz} |E_y|^2 \right) |E_y|^2 \tag{6.5}$$

We will refer this phenomenon as the photothermal effect in noncentosymmetric media or the bulk photothermal effect to highlight that the light-induced heat current can be excited even under homogeneous illumination and in a homogeneous noncentosymmetric medium [35].

However, the developed phenomenological description of the observed effect does not explain why the heat current produced by laser beam can manifest itself in the modification of a moving sample. The major difficulty here is that the timescales of these processes are very different. Specifically, the laser field, which drives the heat current, described by (6.4), is around for only about 150 fs. The created electrons in the laser-produced plasma thermally equalize with lattice in few picoseconds and finally recombine with nearby ions in few nanoseconds or even faster. On the other hand, at the pulse repetition rate of 250 kHz, the time interval between laser pulses is at least 1000 times longer. What is there that "remembers" what direction the laser beam is being translated in 4 µs later, when the next laser pulse arrives?

The short answer is that anisotropic energy distribution created by the short lasting current described by (6.4) is finally imprinted in the anisotropy of lattice temperature across the irradiated area. This can been seen as more efficient heating of the material when subsequent pulses arrive in a region where the electric field-driven heat current has already pushed thermal energy from another part of the beam due to the crystal anisotropy. To clarify this, we would like to recall that our experiments were performed in the so-called thermal accumulation regime [36]. In this regime, the material temperature is determined by the cumulative effect of many laser pulses. This is because in our experimental conditions (beam diameter $d = 2\,\mu m$, thermal diffusivity $D = \kappa/\rho c_p \approx 10^{-2}\,cm^2/s$, thermal conductivity $\kappa \approx 2\,W/mK$, specific heat capacity $c_p \approx 714\,J/kg/K$, volume mass density $\rho \approx 4640\,kg/m^3$, pulse repetition rate $f = 250\,kHz$, sample velocity $v = 200\,\mu m/s$), the effective cooling time $\tau_c = d^2/D \approx 4\,\mu s$ coincides with $1/f$ and the thermal diffusion length $L_D = (4D/f)^{1/2} \approx 4\,\mu m$ is comparable with the beam diameter. The distance traveled by the sample between two consecutive pulses $h = v/f = 0.8\,nm$, i.e., the number of laser pulses overlapping within the beam diameter, is $N = d/h = 2{,}500$. Such a small beam shift between two pulses implies that the beam movement can be considered as continuous.

The absorption of the laser radiation results in the average heat production within the focus area with the rate

$$\overline{\left(\frac{\partial Q}{\partial t}\right)}_{hom} = a(\tau_p \times f)I \tag{6.6}$$

where $I$ is the laser intensity, $\tau_p$ is the pulse duration, and $a$ is determined by the absorption coefficient, volume of the irradiated area, etc. This heating homogeneously increases the temperature within the whole focal area. However, in the LiNbO$_3$ focal area there exists an average heat current along the Y axis (6.4) of

$$\overline{J_y} = b(\tau_p \times f)I^2 \qquad (6.7)$$

where $b$ is determined by relevant nonzero components of the material tensors introduced in (6.4). When the light beam is propagated along Z axis of the crystal, this flow will push the heat along Y axis at the transfer rate

$$\overline{\left(\frac{\partial Q}{\partial t}\right)}_{\text{anisotropic}} = A\overline{J_y} \qquad (6.8)$$

where $A$ is the cross section of the irradiated area in the XZ-plane. This heat transfer will produce the temperature difference between opposite sides of the beam:

$$\Delta T = \frac{d}{\kappa A} \overline{\left(\frac{\partial Q}{\partial t}\right)}_{\text{anisotropic}} = \frac{d}{\kappa} \overline{J_y} \qquad (6.9)$$

The sign of $\Delta T$ depends on the parameter $b$, i.e., on the relevant component of the material tensor responsible for the observed effect. In particular, if the laser beam is Y-polarized and $Im\{\zeta_{yyyyyz}\} > 0$, then $J_y < 0$. Therefore, if the Y axis is horizontal with negative direction on the right, the temperature of the right-hand side of the irradiated area will be higher than that of the left-hand side (Fig. 6.13). In such a case, *the direction of the beam movement along the Y axis does matter.* Specifically, the temperature of the crystal when the beam is translated along Y axis in the negative direction will be always higher than that when the beam is translated along Y axis in the positive direction.

The mechanism of this direction-dependent writing can be visualized using the simplified model presented in Fig. 6.14. In this model, for sake of simplicity we assume that the beam movement is discontinuous, i.e., that the beam movement along Y axis consists of jumps at length equal to its diameter. The conventional absorption described by (6.6) produces homogeneous heating of the irradiated area at temperature $T_0$. If no other heating mechanisms are present, the temperature is the same for any position of the beam and does not depend in which direction we move the beam. However, if the anisotropic heating mechanism described by (6.7) is involved, the temperature at the right side of the beam is higher than that of the left-hand side by $\Delta T = dJ_y/\kappa$. When the beam jumps right (to the negative direction of the Y axis) to its next placement, the temperature of the *left side* of the beam increases to $T + \Delta T$, while the temperature of the right side of the beam is $T$. The anisotropic heating mechanism is switched on and increases the temperature of the *right side* so that it will be by $\Delta T$ higher than that of the left side of the beam. As a result, the temperature of the *right side* of the beam will be equal to $T + 2\Delta T$. The same temperature increase will happen after the next jump. After $m$ jumps in the direction that coincides with the direction of the anisotropic heat flow, the temperature of the very right irradiated area will be equal to

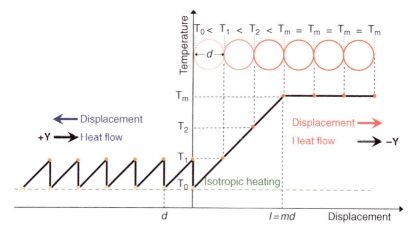

**Fig. 6.14** Illustration of the differential heating of a crystal as a result of the bulk photothermal effect. Heat flows in the −Y direction of the crystal (*black arrow*). Temperature of the crystal increases till saturation when the beam is displaced in the direction of heat flow (*red arrow*) and oscillates near the level defined by isotropic heating (*green line*) when displacement is opposite to the heat flow (*blue arrow*). *Big circles* illustrate the laser beam and *darker color* corresponds to higher temperature of the sample in the position of the beam

$$T^{\max}_{\text{parallel}} = T_0 + \frac{md}{\kappa} \times \overline{J_y} \qquad (6.10)$$

When the light beam moves in the positive direction along Y axis, the temperature of the left-side of the beam is the same after each jump. Therefore, although the anisotropic heating mechanism results in the increase in the temperature of the right side of the beam, the temperature of the crystal does not increase (Fig. 6.14):

$$T^{\max}_{\text{opposite}} = T_0 + \frac{d}{\kappa} \times \overline{J_y} \qquad (6.11)$$

Therefore, the anisotropic heating may result in a drastically different scenario of the laser writing, which we refer as *differential heating* of the sample. This increase in the temperature described by (6.10) will eventually slow down due to thermal diffusion and melting of the sample. Moreover, uneven heating of the sample when the beam moves opposite to the heat flow, which is described by (6.11) and illustrated in Fig. 6.14, may result in a shock-induced damage, which is evident for one of the writing directions (Figs. 6.12 and 6.13). Apparently in the real experimental conditions, the situation is more complicated because we need to account for the interplay of the thermal diffusion and accumulation; however, the simplified model above well illustrates how anisotropic heating can manifest itself in the crystal modification by a moving laser beam.

The restrictions imposed on the light-induced current by crystalline symmetry explain the observed directional dependence of writing. Specifically, since the light beam propagated along the Z axis does not produce thermal current along the X axis, the crystal modification is not sensitive to the translation of the light beam along the X axis. In contrast, when the beam is translated along the Y axis, the in-plane heat flow is either parallel or antiparallel to the beam velocity. Since the heating of crystal is stronger when the direction of the heat flow coincides with direction of the beam movement, the modification of the non-centrosymmetric crystal shows a pronounced directional dependence independently of the pulse front tilt. It should be also noted that we observed similar crystal modifications for the X- and Y-polarized beams. Such an experimental finding is also supported by crystal symmetry because in the $LiNbO_3$ crystal, the thermal current along Y axis is not forbidden for both X- and Y-polarized beams; however, the current is described by different independent components of the material tensor. The similarity of the crystal modifications just indicates that these components are of the same order in magnitude.

One can observe from (6.4) and (6.5) that interplay of the crystal anisotropy and light-induced heat flow gives rise to a new nonreciprocal nonlinear optical phenomenon, the nonreciprocal photosensitivity. In lithium niobate, the nonreciprocal photosensitivity manifests itself as changing the sign of the light-induced heat current when light propagation direction is reversed. This phenomenon is visualized when the modification of the crystal is performed by a moving light beam; in such a case the created pattern is mirrored when light propagation direction is reversed.

## 6.5 Conclusion

In conclusion, it is remarkable that a laser beam, one of the most modern writing tools, could be used for calligraphic inscription similar to writing with a quill pen, which is based on the anisotropy of a quill's tip shape. Moreover, modifications of materials by light span from photosynthesis and photography to material processing and laser writing, and there are only few parameters of the light beam which control material transformations, in particular wavelength, intensity, polarization, exposure time, and pulse duration. Our results add one more parameter to this list – direction of beam movement or pulse front tilt.

Also the first realization of nonreciprocal ultrafast laser writing in Z-cut lithium niobate crystal was demonstrated using a tightly focused ultrafast laser beam. We discovered that when the direction of the laser beam is *reversed* from +Z to −Z directions, the structures written in the crystal when translating the beam along the +Y and −Y directions are *mirrored*. Therefore, a single light beam interacts with the crystal differently for opposite directions of light propagation. This new nonreciprocal nonlinear optical phenomenon is interpreted in terms of light pressure at the front of ultrashort pulse, photon drag effect, and associated light-induced thermal current in crystalline media. We would like to point out that nonreciprocal phenomena are very rare in optics and are usually associated with breaking of the time-reversal symmetry due to the presence of the magnetic field. One may recall conventional Faraday effect and magneto-chiral dichroism [37], when interaction of a single light beam with a homogeneous material is different for two opposite directions similar to the nonreciprocal photosensitivity reported here.

Note: While this book was in preparation, it was found that pulse front tilt can also manifest a an anisotropic photosensitivity in isotropic material [38].

# References

1. D.B. Chrisey, "Materials processing - The power of direct writing," Science **289**, 879 (2000)
2. R. Birngruber, C.A. Puliafito, A. Gawande, W.Z. Lin, R.W. Schoenlein, J.G. Fujimoto, "Femtosecond laser tissue interactions - retinal injury studies," IEEE J. Quant. Electron. **23**, 1836–1844, (1987)
3. K.M. Davis, K. Miura, N. Sugimoto, K. Hirao, "Writing waveguides in glass with a femtosecond laser," Opt. Lett. **21**, 1729–1731 (1996)
4. E.N. Glezer, M. Milosavljevic, L. Huang, R.J. Finlay, T.H. Her, J.P. Callan, E. Mazur, "Three-dimensional optical storage inside transparent materials," Opt. Lett. **21**, 2023–2025 (1996)
5. W.J. Yang, P.G. Kazansky, Y.P. Svirko, "Non-reciprocal ultrafast laser writing," Nature Photon. **2**, 99–104 (2008)
6. Y. Shimotsuma, P.G. Kazansky, J.R. Qiu, K. Hirao, "Self-organized nanogratings in glass irradiated by ultrashort light pulses," Phys. Rev. Lett. **91**, 247405 (2003)
7. U.K. Tirlapur, K. Konig, "Cell biology - Targeted transfection by femtosecond laser," Nature **418**, 290–291 (2002)
8. Y.Q. Jiang, Y. Matsumoto, Y. Hosokawa, H. Masuhara, I. Oh, "Trapping and manipulation of a single micro-object in solution with femtosecond laser-induced mechanical force," Appl. Phys. Lett. **90**, 3 (2007)
9. B.C. Stuart, M.D. Feit, A.M. Rubenchik, B.W. Shore, M.D. Perry, "Laser-induced damage in dielectrics with nanosecond to subpicosecond pulses," Phys. Rev. Lett. **74**, 2248–2251 (1995)
10. D. Du, X. Liu, G. Korn, J. Squier, G. Mourou, "Laser-induced breakdown by impact ionization in SiO2 with pulse widths from 7 ns to 150 fs," Appl. Phys. Lett. **64**, 3071–3073 (1994)
11. P.G. Kazansky, H. Inouye, T. Mitsuyu, K. Miura, J. Qiu, K. Hirao, F. Starrost, "Anomalous anisotropic light scattering in Ge-doped silica glass," Phys. Rev. Lett. **82**, 2199–2202 (1999)
12. L. Sudrie, M. Franco, B. Prade, A. Mysyrowicz, "Writing of permanent birefringent microlayers in bulk fused silica with femtosecond laser pulses," Opt. Commun. **171**, 279–284 (1999)
13. E. Bricchi, B.G. Klappauf, P.G. Kazansky, "Form birefringence and negative index change created by femtosecond direct writing in transparent materials," Opt. Lett. **29**, 119–121 (2004)
14. E. Bricchi, J.D. Mills, P.G. Kazansky, B.G. Klappauf, J.J. Baumberg, "Birefringent Fresnel zone plates in silica fabricated by femtosecond laser machining," Opt. Lett. **27**, 2200–2202 (2002)
15. J.D. Mills, P.G. Kazansky, E. Bricchi, J.J. Baumberg, "Embedded anisotropic microreflectors by femtosecond-laser nanomachining," Appl. Phys. Lett. **81**, 196–198 (2002)
16. C. Hnatovsky, R.S. Taylor, E. Simova, V.R. Bhardwaj, D.M. Rayner, P.B. Corkum, "Polarization-selective etching in femtosecond laser-assisted microfluidic channel fabrication in fused silica," Opt. Lett. **30**, 1867–1869 (2005)
17. C. Hnatovsky, R.S. Taylor, P.P. Rajeev, E. Simova, V.R. Bhardwaj, D.M. Rayner, P.B. Corkum, "Pulse duration dependence of femtosecond-laser-fabricated nanogratings in fused silica," Appl. Phys. Lett. **87**, 3 (2005)
18. V.R. Bhardwaj, E. Simova, P.P. Rajeev, C. Hnatovsky, R.S. Taylor, D.M. Rayner, P.B. Corkum, "Optically produced arrays of planar nanostructures inside fused silica," Phys. Rev. Lett. **96**, 057404, (2006)
19. W.M. Steen, *Laser material processing*, (Springer, London; New York, 1998)

20. S.M. Eaton, H.B. Zhang, P.R. Herman, "Heat accumulation effects in femtosecond laser-written waveguides with variable repetition rate," Opt. Exp. **13**, 4708–4716 (2005)
21. S. Akturk, M. Kimmel, P. O'Shea, R. Trebino, "Measuring pulse-front tilt in ultrashort pulses using GRENOUILLE," Opt. Exp. **11**, 491–501 (2003)
22. W. L. Kruer, *The physics of laser plasma interactions*, (Westview Press, Boulder, Colorado, 2001)
23. M. Ashourabdalla, J.N. Leboeuf, T. Tajima, J.M. Dawson, C.F. Kennel, "Ultra-relativistic electromagnetic pulses in plasmas," Phys. Rev. A **23**, 1906–1914 (1981)
24. S. Akturk, M. Kimmel, P. O'Shea, R. Trebino, "Measuring spatial chirp in ultrashort pulses using single-shot Frequency-Resolved Optical Gating," Opt. Exp. **11**, 68–78 (2003)
25. R. Graf, A. Fernandez, M. Dubov, H.J. Brueckner, B.N. Chichkov, A. Apolonski, "Pearl-chain waveguides written at megahertz repetition rate," Appl. Phys. B-Lasers Opt. **87**, 21–27 (2007)
26. S. Juodkazis, K. Nishimura, S. Tanaka, H. Misawa, E.G. Gamaly, B. Luther-Davies, L. Hallo, P. Nicolai, V.T. Tikhonchuk, "Laser-induced microexplosion confined in the bulk of a sapphire crystal: Evidence of multimegabar pressures," Phys. Rev. Lett. **96**, 166101 (2006)
27. A. Vogel, J. Noack, G. Huttman, G. Paltauf, "Mechanisms of femtosecond laser nanosurgery of cells and tissues," Appl. Phys. B-Lasers Opt. **81**, 1015–1047 (2005)
28. E. Bricchi, P.G. Kazansky, "Extraordinary stability of anisotropic femtosecond direct-written structures embedded in silica glass," Appl. Phys. Lett. **88**, 3 (2006)
29. O. Ogorodnikova, R. Konig, A. Pospieszczyk, B. Schweer, J. Linke, "Thermo-stress analysis of actively cooled diagnostic windows for quasi-continuous operation of the W7-X stellarator," J. Nucl. Mater. **341**, 175–183 (2005)
30. A. Barty, K.A. Nugent, D. Paganin, A. Roberts, "Quantitative optical phase microscopy," Opt. Lett. **23**, 817–819 (1998)
31. A. Podlipensky, A. Abdolvand, G. Seifert, H. Graener, "Femtosecond laser assisted production of dichroitic 3D structures in composite glass containing Ag nanoparticles," Appl. Phys. a-Mater. Sci. Process. **80**, 1647–1652 (2005)
32. L. Gui, B.X. Xu, T.C. Chong, "Microstructure in lithium niobate by use of focused femtosecond laser pulses," IEEE Photonics Technol. Lett. **16**, 1337–1339 (2004)
33. R.R. Thomson, S. Campbell, I.J. Blewett, A.K. Kar, D.T. Reid, "Optical waveguide fabrication in z-cut lithium niobate (LiNbO3) using femtosecond pulses in the low repetition rate regime," Appl. Phys. Lett. **88**, 111109 (2006)
34. B. Sturman, V.M. Fridkin, *The Photovoltaic and Photorefractive Effects in Noncentrosymmetric Materials*, (Gordon&Breach, New York, 1992)
35. E.L. Ivchenko, G.E. Pikus, *Superlattices and Other Heterostructures: Symmetry and Optical Phenomena*, (Springer, Berlin, 1995)
36. A.H. Nejadmalayeri, P.R. Herman, "Rapid thermal annealing in high repetition rate ultrafast laser waveguide writing in lithium niobate," Opt. Exp. **15**, 10842–10854 (2007)
37. G. Rikken, E. Raupach, "Observation of magneto-chiral dichroism," Nature **390**, 493–494 (1997)
38. P.G. Kazansky, Y. Shimotsuma, M. Sakakura, M. Beresna, M. Gecevičius, Y. Svirko, S. Akturk, J. Qiu, K. Miura, K. Hirao, "Photosensitivity control of an isotropic medium through polarization of light pulses with tilted intensity front," Opt. Express **19**, 20657–20664 (2011)

# Part II
# Waveguides and Optical Devices in Glass

# Chapter 7
# Passive Photonic Devices in Glass

**Shane M. Eaton and Peter R. Herman**

**Abstract** Femtosecond laser microfabrication offers the potential for writing passive photonic circuits inside bulk glasses, for use in last-mile photonic networks, sensing, and lab-on-a-chip applications. In this chapter, the fabrication methods for writing low-loss optical waveguides along with waveguide and device characterization techniques are reviewed. The advantages and disadvantages of femtosecond laser writing are analyzed and compared with existing planar lithographic fabrication techniques.

## 7.1 Introduction

Insatiable consumer demand for high-bandwidth Internet services such as high-definition television and voice over internet protocol has driven a rapid development of high-speed optical networks. Dense wavelength division multiplexing, a technology which multiplexes multiple carriers on a single optical fiber to multiply capacity, has enabled bit rates up to 40 Gb/s for longhaul fiber-optic transmission systems. To meet the increasing demand for high-bandwidth last mile fiber-to-the-home (FTTH) access, the passive optical network (PON) architecture was developed in which unpowered optical splitters/routers are used to enable a single optical fiber to serve multiple premises. These advances in optical networking have required improvements not only in semiconductor optoelectronic devices for

---

S.M. Eaton
Istituto di Fotonica e Nanotecnologie – Consiglio Nazionale delle Ricerche (IFN-CNR), Piazza Leonardo da Vinci 32, 20133, Milan, Italy
e-mail: shane.eaton@ifn.cnr.it

P.R. Herman (✉)
Edward S. Rogers Department of Electrical and Computer Engineering, University of Toronto, Toronto, ON, Canada
e-mail: p.herman@utoronto.ca

transmitting and receiving optical signals, but also in passive photonic circuits such as multiplexers and splitters. Further, these robust optical communication devices are finding application in much broader markets for medical instruments, structural sensors, and diagnostic tools.

Fiber-based optical devices are intrinsically limited in their flexibility for dense integration and miniaturization, and have been superseded by two-dimensional planar optical circuits for passive photonic networks. The ideal material for a planar circuit must offer high-optical transparency, good stability, and be flexibly processable for smooth patterning of the refractive index. Polymers are generally not favored due to their strong absorption at telecommunications wavelengths (1.3–1.55 $\mu$m) resulting in high-device insertion loss (IL). High-quality optoelectronic devices are available in semiconductors such as Si or III–V materials but high-cost fabrication facilities and difficult packaging challenges with lower index optical fiber have limited their market success to non-passive applications. Glass therefore remains the favored choice for passive photonic circuits due to its high transmission at visible and near-infrared wavelengths, relatively low cost, ease of manufacture, and compatibility with existing optical fiber technology. The refractive index can be precisely and smoothly controlled and further offer exceptional chemical and physical stability that serve ideally not only in fiber-optic communication networks but also as optical sensors for temperature and strain, and more recently in lab-on-a-chip devices for sensing of biomolecules.

Over the past twenty years, planar lightwave circuit (PLC) technology has become the industry standard for fabricating optical power splitters, wavelength multiplexers and arrayed waveguides gratings (AWGs) in glass for photonic networks [1]. PLCs are typically fabricated by a multi-step flame hydrolysis deposition (FHD), lithographic patterning, and reactive ion etching (RIE) method, as shown in Fig. 7.1. This multi-step process is restricted to fabricating rectangular cross-sectional waveguides in two-dimensional planar geometries with waveguide

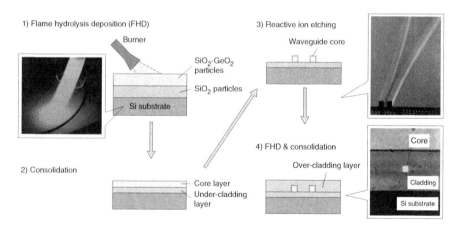

**Fig. 7.1** Fabrication steps in flame hydrolysis method for producing planar lightwave circuits in glass [1]

propagation losses typically below 0.01 dB/cm at telecom wavelengths. High-index contrast core (index difference $\Delta n = 0.025$) waveguides today offer tight bends of 2-mm radius with negligible bend loss to enable circuits with higher integration density such as switches and large-scale AWGs in wavelength division multiplexing (WDM) telecom applications [1]. Nevertheless, the two-dimensional geometry imposes restrictions on cross-connection, tapering, packaging to optical fiber, and further circuit densification.

As highlighted in earlier chapters, femtosecond laser writing is a promising fabrication technology for writing optical waveguides throughout the three-dimensional bulk of silica and other transparent materials. Compared to PLC technology, femtosecond laser writing is a maskless direct-writing fabrication tool, which is ideal for rapid prototyping of integrated optical circuits. Cleanroom conditions are not required for femtosecond laser writing, reducing the complexity, and potential cost compared with PLC fabrication. Also, because femtosecond laser pulses are nonlinearly absorbed to deposit energy within the focal volume, the technique can be applied to tailor the refractive index of a variety of glasses, polymers, and crystalline materials along 3D pathways that greatly expands the architecture and miniaturization geometries possible over that in PLC fabrication, which is limited to planar geometries and specific glass compositions.

Femtosecond laser writing compared with PLCs, is limited to relatively low induced refractive-index change of $\Delta n \sim 0.01$ that imposes large limiting bend radii of $R > 20$ mm, thus preventing the fabrication of highly compact and integrated passive photonic circuits [2, 3]. Furthermore, propagation losses in waveguides fabricated with femtosecond laser pulses are $\sim 0.2$ dB/cm [4], which is significantly higher than the $\sim 0.01$ dB/cm losses in PLC waveguides. Furthermore, femtosecond laser writing is a serial process which is less attractive than PLC fabrication.

In this chapter, Sect. 7.2 reviews the characterization techniques required for assessing performance of passive optical circuits. Section 7.3 reviews the exposure parameters which determine waveguide properties and highlights various milestones reached today in femtosecond laser waveguide writing in glasses. In Sect. 7.4, the important passive photonic devices fabricated by ultrashort laser pulses are reviewed. Finally, the results presented in this chapter are summarized with an outlook given on future research directions which could improve the quality of femtosecond laser-written passive photonic devices in glass.

## 7.2 Characterization of Femtosecond Laser-Written Waveguides

In developing new fabrication processes for optical waveguides, it is important to characterize the waveguide morphology, losses, and mode profile among many other variables to provide important feedback on determining the optimum and most reproducible fabrication conditions.

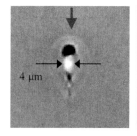 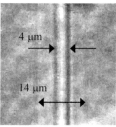

**Fig. 7.2** Cross sectional (*left*) and overhead (*right*) view of a low-loss waveguide written in borosilicate glass. In the cross sectional view, the incident laser was incident from the top as shown by the red arrow [5]

## 7.2.1 Microscope Observation

After waveguide writing and facet polishing, the first characterization step is usually optical microscopy for a qualitative assessment of the laser-induced refractive-index change and waveguide uniformity. In a transverse view of the waveguide, one seeks a uniform axial shape with a brighter central contrast relative to the bulk, indicative of a positive refractive-index change that concentrates the illumination light of the microscope. With the waveguide viewed end-on at the facet, ideal waveguides are those with a brighter circular core, similar to that of optical fibers, which may suggest good optical confinement of light with a circular mode shape, possibly well-matched to optical fiber. The axial and cross sectional microscope image of a typical low-loss waveguide formed by focused femtosecond laser pulses are shown in Fig. 7.2.

## 7.2.2 Insertion Loss

Once reasonable quality waveguides are identified by microscope inspection, a 633-nm wavelength HeNe laser or visible laser diode source can be coupled by end fire with a microscope objective or fiber butt-coupling with single-mode fiber (SMF) to the input facet of the sample to qualitatively test the guiding characteristics of the laser-written tracks. If there is sufficient scattering from laser-induced color centers or scattering points, waveguiding may be confirmed by observing an exponentially decreasing intensity of the scattered guided light overhead by eye. Optimum alignment of the objective or fiber at the input facet is obtained when the scattered streak is brightest. If red scattering is not visible, guiding should be confirmed by observing the guided light projected from the waveguide output facet on a white screen, as shown in Fig. 7.3. The bright central spot and concentric interference rings demonstrated that the laser-written track guided red light. Waveguide insertion loss and mode profiles may be measured at visible wavelengths but typically, this visible light source is used as a starting point for alignment and followed by switching to other light sources for characterization within the low-loss telecom wavelength window (1250–1600 nm).

**Fig. 7.3** Far-field intensity distribution of 633-nm wavelength light output from femtosecond laser-written waveguide

In the telecom band, the waveguide mode profile, insertion loss, and propagation loss are conveniently characterized with a standard SMF butt-coupled to the input waveguide facet and lit with a tunable or broadband laser source. This method offers low IL only if the mode diameter of the laser-formed waveguide matches closely to that (MFD = $10.5\mu$m) of SMF. Input/output fibers and the waveguide sample can be manipulated with independent high-resolution (sub-micron) multi-axis positioning stages while observing positioning with an overhead vision system. The waveguide insertion loss is obtained from the power transmitted $P_{trans}$ through the laser-formed waveguide (detected with an InGaAs photodiode) by butt-coupling SMF fiber at both the input and the output waveguide facet and normalized to the power propagated by directly butt-coupling input and output fibers, $P_{ref}$. Index matching fluid should be applied at the end facets of all fiber-to-fiber and fiber-to-waveguide connections to avoid Fresnel reflection losses and improve measurement accuracy. If the power transmitted through the sample and the reference power are measured in dBm, the IL is given by:

$$IL(dB) = P_{ref} - P_{trans} \quad (7.1)$$

Typically, a monochromatic laser tuned to 1550-nm wavelength is used for IL measurements, but broadband sources such as Erbium-doped fiber amplifiers (1520–1600 nm) or multiple LED sources (1250–1700 nm) may also be coupled to measure the wavelength-dependent transmission. With a broadband source, the power transmitted through the waveguide can be collected by a butt-coupled SMF and transmitted to an optical spectrum analyzer.

### 7.2.3 Mode Profile and Coupling Loss

To profile the optical mode of a femtosecond laser-written waveguide, a microscope objective is placed at the output facet to image and magnify the mode onto a camera sensor, typically mounted together with an empty microscope tube for a fixed magnification ratio and to reject stray room light. To calibrate the pixel spacing of the imaging setup, the outer cladding diameter of SMF or a microscope reticule can be back-illuminated and imaged to provide a size reference. Three different camera technologies are favored for mode profiling: vidicon, phosphor-coated CCD and InGaAs cameras. For a detailed comparison of these technologies, refer to [6].

The coupling loss is defined here as the power loss when the light is coupled from a fiber to the laser-formed waveguide or vice versa. When index matching fluid is applied in the small fiber-substrate gap, Fresnel reflection losses can be neglected. The coupling loss is attributed to a mismatch between the waveguide and fiber mode profiles, which can be estimated from the overlap integral between the waveguide and fiber modes, as described in Chap. 5. The mode field diameter (MFD) describes the size of the guided (fundamental) waveguide mode and is usually reported as the diameter where the intensity has dropped to $1/e^2$ of its peak value. For guided intensity distributions that deviate from Gaussian distributions, other definitions may be applied such as the $4\sigma$ method or 90/10 knife edge [6].

### 7.2.4 Propagation Loss

The waveguide propagation loss resulting from scattering and absorption may be measured several ways. All methods suffer from drawbacks with measurement accuracy typically degrading for shorter (<5 cm) and lower loss (<1 dB/cm) waveguides. In the Fabry–Perot technique, the small Fresnel reflection from the air–glass interface sets up a Fabry–Perot cavity. The fringe contrast observed in the transmission spectrum provides the propagation loss [7]. Since the fringe contrast increases with decreasing propagation loss, the measurement accuracy of the Fabry–Perot technique improves with decreasing waveguide loss, unlike most measurement techniques. However, additional losses from the small angular deviation between the waveguide and end facets will cause the propagation loss to be overestimated. Since it is difficult to align waveguides to within 0.5°, the Fabry–Perot technique is inaccurate for the low (<1 dB/cm) propagation losses available in most glasses.

Another technique for measuring propagation loss is by observing the exponentially decaying light scattered signal along the waveguide from overhead [8] by imaging with a standard CCD camera for visible wavelengths or an InGaAs camera at infrared wavelengths. Due to the weak intensity and noisy nature of the scattered signal, this measurement method is only accurate for moderately high waveguide loss (>1 dB/cm) and long waveguide lengths (>2 cm) [9]. Another difficulty is that the scattering centers that produce the decaying streak are often randomly located along the waveguide length, so that dark regions absent of scattering centers contribute to measurement error.

A commonly employed technique for measuring propagation loss is by recording the IL for a series of identical waveguides with different lengths. This can be accomplished by writing waveguides with identical conditions in different sample lengths or by cutting a long sample into different lengths. This technique requires long sample lengths (>5 cm) for reasonable accuracy and is prone to inaccuracy from the variability in the quality of facet cutting/polishing, in repeated exposures in different sample lengths, and in alignment for each sample.

The least error-prone technique for measuring propagation loss is to use a very long sample, resulting in a large contribution from the propagation loss to the total IL. The propagation loss can then be determined by dividing the IL by the sample length if total coupling loss is relatively small. Therefore, this simple method only provides an upper limit on the propagation loss but can be improved by measuring the waveguide-fiber IL as in (7.1). By then subtracting the coupling loss from both facets as obtained by the overlap integral, the propagation loss can be determined from:

$$\alpha(\text{dB/cm}) = \frac{\text{IL} - 2\text{CL}}{L} \qquad (7.2)$$

where $L$ is the waveguide length in cm.

### 7.2.5 Refracted Near Field Method

An important quantitative method to determine the refractive index distribution of a laser-written waveguide is known as the refracted near field (RNF) method, and is reviewed earlier in Chap. 5.

## 7.3 Femtosecond Laser Microfabrication of Optical Waveguides

The properties of femtosecond laser-written waveguides in bulk glass depend strongly on laser exposure conditions such as repetition rate [10, 11], scan speed, average power, wavelength [12], polarization [13, 14], numerical aperture [10], and focus depth [10, 15]. In this section, the effect of these parameters on waveguide quality will be discussed for the single-pulse (low-repetition rates <100 kHz) and cumulative pulse interaction (high-repetition rates >100 kHz) regimes. The reader is referred to other chapters for a detailed discussion of secondary exposure variables which have been shown to influence waveguide properties. These parameters include pulse duration (Chaps. 4 and 9), spatial-temporal pulse shaping [16] (Chap. 4), multi-scan fabrication [17] (Chap. 5), pulse front tilt [18] (Chap. 6), and nanogratings [19] (Chaps. 6 and 14).

## 7.3.1 Low-Repetition Rate Regime

As described in Chap. 1, at low-repetition rates (<100 kHz), the time between pulses is long enough so that thermal diffusion has carried the heat away from the focus before the next pulse arrives. In this situation, the ensuing pulses may add to the overall modification, but still act independently of one another. Most results in the field of femtosecond laser microfabrication have been carried out at low-repetition rates, and usually at 1-kHz, due to the common use of 1-kHz, 800-nm regeneratively amplified Ti:Sapphire femtosecond lasers. One limitation of waveguide writing in the single-pulse interaction regime is that waveguide cross sections take on a similar shape as the asymmetric focal volume. The resulting waveguides written with the transverse writing scheme are elliptical-like, giving modes that couple poorly to optical fiber. Several methods have been proposed to produce a more symmetric focal volume including astigmatic focusing with a cylindrical lens telescope [20], slit reshaping [21, 22], multi-scan writing [17] and applying a two-dimensional deformable mirror [23]. These methods are reviewed in Chap. 5.

### 7.3.1.1 Fused Silica

Many groups have applied the femtosecond-laser writing method to pure fused silica glass, but few have shown good quality waveguides with operation at both visible and telecom wavelengths. The best result in fused silica to date was obtained with a multi-scan writing scheme to form waveguides with nearly square cross sections (7.4 × 8.2 $\mu$m) with a refractive-index change of $\sim$4 × 10$^{-3}$, suitable for low-loss coupling to SMF [17]. The refractive index profile obtained with RNF for a waveguide written in Ge-doped silica glass for PLCs is shown in Fig. 7.4. The

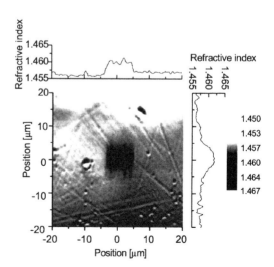

**Fig. 7.4** Refractive index profile of a waveguide written in Ge-doped silica PLC glass by multiple scans [17]

waveguides written in PLC glass were nearly identical to the waveguides written in pure silica. A 775-nm, 150-fs Ti:Sapphire laser with 1-kHz was applied in the transverse writing geometry with a 0.4-NA objective, 182-nJ pulse energy and 10-$\mu$m/s writing speed. The waveguides were fabricated with 20 scans separated transversely by 0.4 $\mu$m. A propagation loss of 0.12 dB/cm at 1550 nm was reported, which is the lowest reported in the field, and is attributed to the gentle refractive index modification enabled by the novel low-fluence, multi-scan fabrication method.

The effect of writing laser polarization on waveguide transmission properties was studied using a 1-kHz, 120-fs Ti:Sapphire laser with a 0.5-mm slit placed before the 0.46-NA focusing objective [13, 24]. With 3-$\mu$J pulse energy measured after the slit and 25-$\mu$m/s writing speed, a refractive-index change of $2.3 \times 10^{-3}$ was measured for circular polarization, which was about twice as high as that obtained with linear polarizations [13]. This enhancement was attributed to the higher photoionization rate for circular polarization compared to linear polarization in the range of intensities studied (42–50 TW/cm$^2$). It was recently discovered that nanogratings are formed oriented orthogonally to the writing laser polarization within fused silica under certain exposure conditions [25, 26]. It is likely that these nanogratings also influence the transmission properties of femtosecond laser-written waveguides in fused silica.

### 7.3.1.2 Phosphate Glasses

Good quality passive optical waveguides have been demonstrated in phosphate glass, which is easily doped with Er and Yb ions for active waveguide applications. For a thorough review of active waveguide devices, please refer to Chap. 10. Using the astigmatic writing method with a cylindrical lens telescope as described in Chap. 5, waveguides with low damping loss (0.25 dB/cm) at 1550-nm wavelength were written at 20 $\mu$m/s with a 1-kHz, 150-fs Ti:Sapphire laser with 0.3-NA microscope objective and 5-$\mu$J pulse energy [20, 27]. The slit shaping method was applied by Withford and coworkers to produce waveguides with similar propagation loss [21].

### 7.3.1.3 Exotic Glasses

Waveguides were written in a highly nonlinear heavy-metal oxide (HMO) glass with a 1-kHz, 800-nm, 100-fs Ti:Sapphire laser [28]. HMO glasses are attractive due to their high optical nonlinearity ($n_2 \sim 10^{-18}$ m$^2$/W), but this presents significant challenges in femtosecond laser writing because of strong self focusing, resulting in a delocalized spatial distribution of the laser energy which is difficult to control. By focusing 1.8-$\mu$J femtosecond laser pulses with a 0.42-NA objective and scanning the sample transversely at 60 $\mu$m/s, elongated damage structures of $\sim$65-$\mu$m vertical length were observed when the sample was viewed from the end facets. These elongated structures were the result of filamentation when self focusing balances against plasma defocusing. The waveguiding regions were found to be adjacent to the filament-induced damage zone, with propagation losses below 0.7 dB/cm demonstrated at 633-nm wavelength. The regions of refractive index increase adjacent to the filament were attributed to compressive stress induced outside the laser-damaged zone, similar to observations during waveguide writing in crystalline materials (Chap. 11).

## 7.3.2 High-Repetition Rate Regime

The development of high-repetition rate femtosecond lasers is opening new avenues for manipulating thermal relaxation effects that control the properties of optical waveguides formed when ultrashort laser pulses are focused inside glasses. At low to moderate repetition rates (1–100 kHz), an increase in laser pulse energy leads to formation of larger modification structures as thermal diffusion extends the laser-heated region far outside the focal volume. As repetition rate increases, the time between laser pulses becomes shorter than the time for the absorbed laser radiation to diffuse out of the focal volume and heat builds up around the focal volume. Schaffer et al. first reported heat accumulation in the bulk of glass using a 25-MHz femtosecond laser oscillator [29]. With increased dwell time, a dramatic increase in the size of laser-modified structures was observed compared to structures formed with single-pulse interactions, where no variation in modification size with dwell time is observed. The combination of high-repetition rate and heat accumulation offers fast writing speeds and cylindrically symmetric waveguides together with benefits of annealing and decreased thermal cycling that are associated with low propagation and coupling loss to standard optical fiber.

### 7.3.2.1 Borosilicate and Phosphate Glasses

Following the early work by Schaffer et al. [29], further insight into heat accumulation effects was provided by Eaton et al. [5, 6]. Using a finite difference thermal diffusion model, the temperature in the focal volume was calculated as a function of pulse number (dwell time) for repetition rates of 0.1, 0.5, and 1 MHz in borosilicate glass. Typical writing conditions of 200 nJ of absorbed energy, 0.55-NA focusing and a melting point of 985°C were assumed. As shown in Fig. 7.5, at 100-kHz repetition rate, the temperature relaxes to below the softening point before the next pulse arrives, resulting in minimal heat accumulation and significant temperature cycling during waveguide writing. At 0.5- and 1-MHz repetition rates, heat accumulation is strongly evident, leading to a larger melted volume which increases with pulse number and repetition rate. Decreased thermal cycling with increased repetition rate is anticipated to lead to smoother waveguides with less propagation loss and birefringent stress.

**Fig. 7.5** Model of glass temperature versus exposure at repetition rates of 100 kHz, 500 kHz, and 1 MHz, at a radial position of 3 $\mu$m from the center of the laser beam. The absorbed pulse energy of 200 nJ was the same at each repetition rate [6]

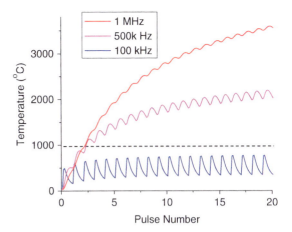

To unravel the contributions of thermal diffusion and heat accumulation to the resulting waveguide morphology, a variable repetition rate (0.1–5 MHz) 300-fs, 1045-nm Yb-doped fiber laser was applied to waveguide writing in Corning EAGLE2000 borosilicate glass [10] with a 0.55-NA focusing lens. Structures were formed at 150 $\mu$m below the surface unless specified otherwise. The onset for heat accumulation was determined by comparing waveguide diameters produced by scanned exposures with that of single-pulse exposures, where the only contribution is from thermal diffusion. The threshold was defined as the minimum pulse energy to increase the waveguide diameter twofold over the diameter produced by diffusion in a single-pulse interaction. Single-pulse diffusion diameters were found to vary from 2 to 6 $\mu$m for pulse energies of 0.1–1 $\mu$J [10].

The threshold pulse energy for driving heat accumulation is shown in Fig. 7.6 as a function of repetition rate (200 kHz to 2 MHz) and for scan speeds of 2, 10, and 40 mm/s. The single pulse modification threshold was 50 nJ (2.5 J/cm$^2$) and invariant with repetition rate. In contrast, the energy threshold for heat accumulation decreased sharply from $\sim$900 nJ at 200 kHz to $\sim$80 nJ at 2 MHz and was only weakly dependent on scan speed. In this 200 kHz to 2 MHz range of repetition rates, the thermal diffusion scale length $\sqrt{D/R}$ decreases from 1.6 to 0.5 $\mu$m, where $D$ is the thermal diffusivity and $R$ is the repetition rate. This decrease in the characteristic thermal diffusion length indicates that the effective laser heating volume decreases dramatically with increasing repetition rate, thereby reducing the threshold pulse energy for heat accumulation. For $R > 1$ MHz, the thermal diffusion scale length of $\sim$0.7 $\mu$m falls inside the laser waist radius of $w_0 = 0.8 \mu$m, resulting in an asymptotic limit of the heat accumulation threshold energy to a minimum value of $\sim$80 nJ at 2-MHz repetition rate in Fig. 7.6. Beyond 2 MHz, the available laser pulse energy was below this value, preventing the observation of heat accumulation effects in EAGLE2000 glass. Similarly, when the laser was operated at the lowest 100-kHz repetition rate, the 2-$\mu$J maximum pulse energy available was insufficient to drive heat accumulation effects beyond a larger thermal diffusion diameter of $\sim$8 $\mu$m.

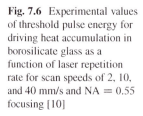

**Fig. 7.6** Experimental values of threshold pulse energy for driving heat accumulation in borosilicate glass as a function of laser repetition rate for scan speeds of 2, 10, and 40 mm/s and NA = 0.55 focusing [10]

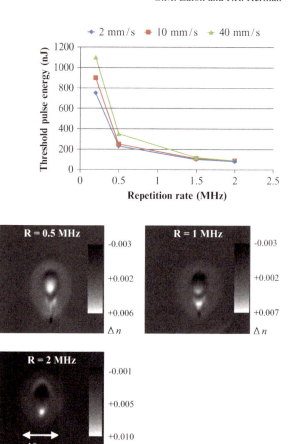

**Fig. 7.7** Cross sectional refractive index profiles of waveguides written with 200-mW power, 25-mm/s scan speed and repetition rates of 0.2, 0.5, 1, 1.5, and 2 MHz. The writing laser was incident from the top and the refractive index scale is shown on the right [10]

In the same study, a new laser processing window was discovered for producing low-loss waveguides across the large 200 kHz to 2 MHz range of repetition rates [10]. By holding the average power constant and delivering the same net fluence exposure at each repetition rate, strong thermal diffusion from high-energy pulses at 200-kHz repetition rate balanced the strong heat accumulation with low pulse energy delivered at high-repetition rates to produce waveguides with similar diameter and strong guiding at 1550-nm wavelength. Figure 7.7 shows cross sectional refractive index profiles measured by RNF for waveguides written with 200-mW average power at repetition rates of 0.2–2 MHz and 25-mm/s scan speed. This average power will be shown later to give the lowest IL at each repetition rate. The core of the waveguides shown in Fig. 7.7 is attributed to the high-temperature spikes induced within the laser spot size by each laser pulse, while the outer lower-contrast cladding is formed by a more slowly evolving near-Gaussian temperature distribution, with the overall size determined by the maximum diameter where the temperature exceeds the melting point. Because of the variable temperature across the modified zones, the cooling rates are highly nonuniform, and therefore are expected to lead to a nonuniform distribution of the final glass density [30].

The small, dark spot at the bottom of the index profiles in Fig. 7.7 is attributed to the focus plane location since its depth was constant with varying exposure conditions. The images show that most of the laser energy was deposited upstream of the focus. At 200-kHz repetition rate, the waveguides showed a vertically elongated central core guiding region with peak $\Delta n = 0.005$. The waveguide is elliptical due to thermal diffusion from a laser heating volume extended vertically at this high-pulse energy of 1 $\mu$J and also from minimal cumulative heating between laser pulses. At 500-kHz, owing to increased heat accumulation, a more circular guiding region was found with a maximum $\Delta n = 0.006$ in the guiding region. A small region of increased refractive index is also observed below the main guiding region. At 1-MHz repetition rate, this region below the core has increased in size and magnitude to a peak $\Delta n = 0.007$. This region is now responsible for guiding of 1550-nm light but the mode also extends into the weaker central core to yield two transverse modes when formed with a higher average power exposure (>250 mW). At 1.5 MHz, the guiding region has clearly transitioned below the central core to form a strong guiding region with peak $\Delta n = 0.008$. At 2-MHz repetition rate, a similar profile is observed but beyond this repetition rate, the pulse energy dropped below the heat accumulation threshold of 90 nJ (Fig. 7.6) and formed weakly guiding structures. The highly nonlinear laser interactions lead to vertical shifts of waveguide position and differing waveguide profiles that must be monitored and accounted for during fabrication with varying exposure conditions and focal depths in the glass.

Figure 7.8 aids in visualizing waveguide properties as a function of average power and scan speed, shown for 1.5-MHz repetition rate. Insertion loss is classified by red, blue, and black squares representing low (<3 dB), medium (3–6 dB) and high (>6 dB) IL, respectively for the 2.5-cm long waveguides. Waveguides exhibiting multiple transverse modes and typically damaged morphology, written with the highest net fluence are found at the top-left. Conversely, the bottom-right corner indicates underexposed waveguides where the index change was too low to efficiently guide 1550-nm light. At 1.5-MHz repetition rate, the lowest IL of ∼1.2 dB for fiber-waveguide-fiber coupling were observed over a large 10–25 mm/s range of scan speeds, but in a narrow 200 mW ± 10 mW average power range (encircled data in Fig. 7.8).

Similar analysis was carried out for waveguides formed with 0.2-, 0.5-, 1-, and 2-MHz repetition rates and in all cases, revealed a similar processing window of 200 mW and 10–25 mm/s for the lowest IL. The minimum IL and MFD at each repetition rate is presented in Fig. 7.9. This constant 200-mW average power exposure window appears consistent with the optimum 250-mW power for generating low-loss waveguides in phosphate glass at repetition rates of 505–885 kHz with a similar femtosecond laser [11]. The decreasing IL with increasing

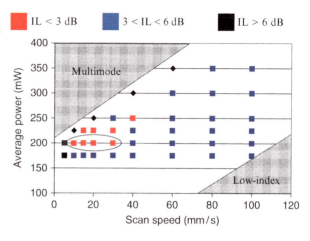

**Fig. 7.8** Processing window map: waveguide properties as a function of average power and scan speed for 1.5-MHz repetition rate [10]

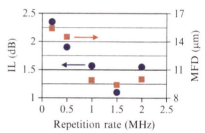

**Fig. 7.9** Insertion loss and mode field diameter versus repetition rate for waveguides written with 200-mW average power and 15-mm/s scan speed [10]

repetition rate in Fig. 7.9 is associated with increasingly stronger heat accumulation that results in higher refractive change and smaller MFD for best coupling to optical fibers at 1.5 MHz. The increased IL and MFD from 1.5 to 2 MHz is attributed to inadequate laser pulse energy (100 nJ) at 2 MHz for driving sufficient laser heating above the ∼90 nJ threshold for heat accumulation shown in Fig. 7.6. Beyond 2-MHz repetition rate, only narrow ∼2 $\mu$m diameter waveguides were formed that were barely guiding and showed no evidence of heat accumulation.

To take advantage of femtosecond laser writing of waveguides in all three dimensions, one must carefully address the problem of spherical aberration at the air–glass interface, which varies dramatically with the focusing depth. The effect of spherical aberration is reduced with oil-immersion lenses [11], objectives with collars for variable depth correction [31] and asymmetric focusing with slit reshaping [15]. It is well known that spherical aberration is less pronounced with lower NA focusing [31, 32]. Further, one can take advantage of strong heat accumulation effects to drive spherically symmetric heat flow which compensates for an axially elongated focal volume. A combination of heat accumulation effect and low numerical aperture of the focusing optic (NA < 0.55) is presented below toward depth-independent, low-loss waveguide formation.

For the cumulative heating regime of 0.55-NA focusing applied above (200 mW, 1.5 MHz, 15 mm/s), waveguides with similar low loss and mode size could only be

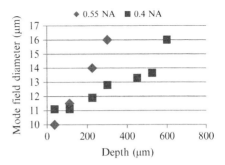

**Fig. 7.10** Mode field diameter versus focusing depth for waveguides formed with 1.5-MHz repetition rate, 15-mm/s scan speed with 0.55 NA (230 mW) and 0.4 NA (200 mW) [10]

obtained in a narrow depth range of $d = 50–200 \mu m$ [10]. Figure 7.10 shows the MFD of these waveguides to increase 60% from 10 $\mu$m at 50-$\mu$m depth to 16 $\mu$m at 300-$\mu$m depth, spherical aberration precluding a deeper waveguide writing range. Much deeper waveguide writing was possible with the 0.25-NA lens, but fivefold lower peak intensity at the maximum available laser power (400 mW) yielded only small diameter waveguides and weak refractive-index change. A better balance was found with the 0.4 NA lens, providing only a small increase in MFD from ∼11.0 to 13.5 $\mu$m as the depth was increased from $d = 50$ to 520 $\mu$m. The measured propagation loss of ∼0.35 dB/cm was nearly independent of focal depth. The ability to write waveguides to depths of 520 $\mu$m is a substantial improvement over the maximum depth of ∼200 $\mu$m reported by other groups employing MHz repetition rate femtosecond lasers with higher NA focusing objectives [33, 34].

### 7.3.2.2 Fused Silica

As demonstrated by Shah et al., the fundamental 1045-nm wavelength led to weak refractive index contrast and irregular morphology in high-repetition rate waveguide writing in pure fused silica [12]. However, processing with the second harmonic wavelength of 522-nm wavelength enabled relatively low loss (∼1 dB/cm) waveguides and moderately high refractive-index change ($\Delta n = 0.01$). The benefits of thermal diffusion acting with heat buildup to form circular waveguide cross sections were not observed in fused silica as reported in lower bandgap silicate glasses [5, 10, 29, 33]. The lack of cumulative heating in fused silica was previously attributed to less absorption in high-bandgap fused silica [12] and later confirmed experimentally with a twofold lower absorption in fused silica measured compared to borosilicate glass for the same laser fluence [35]. Also, the working point temperature of fused silica is ∼1.5-fold higher than that of borosilicate glasses making it more difficult to melt fused silica and drive heat accumulation [5]. As demonstrated by Osellame et al. with a tightly focused 26-MHz repetition rate femtosecond laser, heat accumulation is possible in fused silica with a combination of higher repetition and fluence [34]. However, unlike other glasses [5, 10, 29, 33] where low propagation loss waveguides were reported, the structures defined by heat accumulation in fused silica were nonuniform and unable to guide light [34].

By exploring repetition rates from 0.25 to 2 MHz, a more comprehensive study of waveguide optimization with the second harmonic wavelength was recently performed in fused silica [35] compared to the initial study in 2005 [12]. A repetition rate of 1 MHz yielded the absolute minimum IL of $\sim$1.0 dB (sample length 2.5 cm) with 0.55-NA focusing. Cross sectional profiles taken by optical microscopy and RNF for the optimum waveguide written at 1 MHz, 175 nJ, and 0.2 mm/s are shown in Fig. 7.11a,b, respectively. The modified structures ($\sim$2 × 25$\mu$m) for 0.55-NA focusing were vertically elongated significantly beyond the $\sim$2.2 $\mu$m depth of focus by self focusing and plasma defocusing because the peak power was equal to the 0.8-MW critical power. Good qualitative agreement between the microscope and RNF images was found as shown in Fig. 7.11, with the RNF profile showing a $\sim$2×10$\mu$m guiding region with peak $\Delta n$ = 0.016 formed below an irregular damaged region.

To improve the symmetry of the guiding structures in fused silica, a high 1.25-NA oil-immersion lens was applied in the same study [35] to enable increased laser absorption from higher peak intensity and a more symmetric focal volume. With tighter focusing by the 1.25-NA lens, the best IL of 1.2 dB ($L$ = 1.25 cm) was obtained for 500-kHz repetition rate, 0.2-mm/s speed, and 133-nJ energy, with the corresponding microscope and RNF images shown in Fig. 7.11c,d, respectively. Weaker waveguides were produced at 1-MHz repetition rate, limited by the maximum on-target energy of $\sim$100 nJ. Compared to 0.55-NA, the waveguide formed by 1.25-NA focusing was significantly less elongated, which was attributed to a more symmetric focal volume, and reduced self focusing (0.6-MW peak power). The peak $\Delta n$ = 0.022 represented the highest refractive index increase ever reported for a femtosecond laser-written waveguide in fused silica.

The mode profiles at 1550 nm for the optimum waveguides written by 0.55- and 1.25-NA focusing are shown in Fig. 7.12b and c, respectively, along with the mode for SMF in Fig. 7.12a with 10.5-$\mu$m MFD. The waveguide mode produced by 0.55-NA focusing (MFD = 9.7$\mu$m × 11.9 $\mu$m) becomes significantly smaller and more symmetric (MFD = 7.1 × 7.4$\mu$m) with 1.25-NA focusing, defining the smallest reported mode to date for a laser written waveguide in fused silica with simulations

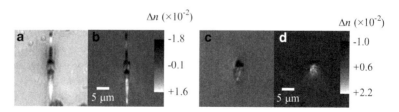

**Fig. 7.11** Cross sectional microscope images (**a,c**) and refracted near filed profiles (**b,d**) of optimum waveguides fabricated with 0.55-NA (**a,b**) and 1.25-NA (**c,d**) objectives. The writing laser was incident from the top [35]

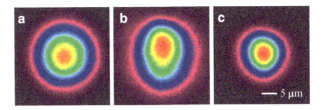

**Fig. 7.12** Mode profile at 1550-nm wavelength for (**a**) standard SMF fiber, (**b**) 0.55-NA waveguide, and (**c**) 1.25-NA waveguide. The writing laser was incident from the top [35]

predicting that small $R = 15$ mm bends are now feasible [35], opening the door for much higher density integrated optical circuits.

It is not immediately evident why processing with the green wavelength offers much better quality waveguides than the fundamental wavelength in pure silica glass. It was previously thought the benefit of the green wavelength was due to an enhanced absorption from a lower order nonlinear process, thus providing stronger index contrast waveguides [12]. However, a similar ~40% absorption for 522- and 1045-nm wavelengths for the same laser fluence was found in a recent study [35]. Schaffer et al. have shown that multiphoton ionization plays a smaller role in large bandgap materials like fused silica. Instead, avalanche ionization dominates, and since this process is relatively independent of wavelength [32], the second harmonic wavelength does not offer an advantage in terms of increased absorption.

The benefit of the green wavelength in previous studies was attributed to its higher on target fluence [35]. After accounting for the 50% conversion efficiency of the SHG process and the fourfold smaller focal area, a twofold higher maximum fluence is possible with 522-nm wavelength. In previous studies, the maximum fluence was required to fabricate the optimum waveguide structures at 522-nm wavelength. It is expected that if higher power lasers were available providing fluences at the fundamental wavelength similar to the optimal value at green wavelength, stronger guiding structures could be formed. This is supported by the recent demonstration by Pospiech et al. of waveguides with 1.2-dB/cm loss in fused silica using a 1-MHz repetition rate femtosecond laser with more energetic pulses (500 nJ) at the 1030-nm fundamental wavelength [36]. However, the reported waveguides showed more irregular morphology compared to the structures demonstrated here with the green wavelength. It is likely that other factors which have yet to be identified may contribute to the green wavelength's advantage in femtosecond laser waveguide formation in fused silica.

### 7.3.2.3 Exotic Glasses

High-quality waveguides were written in an exotic bismuth borate glass (Nippon BZH7) with a 150-fs, 250-kHz regeneratively amplified Ti:Sapphire laser [37]. At this moderate 250-kHz repetition rate, the regime of waveguide fabrication is

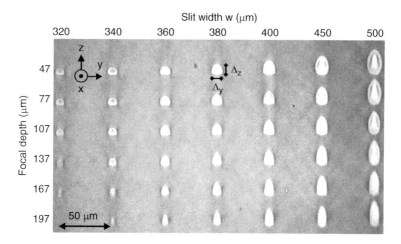

**Fig. 7.13** Cross sectional microscope image of waveguides written in bismuth borate glass showing dependence on focal depth and slit width [37]

likely the result of both thermal diffusion and heat accumulation. Bismuth borate has similar nonlinear properties as the HMO glass described in Sect. 7.3.1 and is therefore a good candidate for nonlinear optics applications but because of its low critical power, presents a challenge for writing symmetric waveguide structures. To overcome the problem of an axially elongated distribution of laser energy at the focus, the slit reshaping method was applied. With 200-nJ pulses focused with a 0.55-NA objective and with a writing speed of 200 $\mu$m/s, a symmetric waveguide cross section was produced with a slit width of 380 $\mu$m, as shown in Fig. 7.13. The symmetric aspect ratio was found to be preserved up to focal depths of about 100 $\mu$m but beyond this value, spherical aberration due to the large refractive index of the glass ($n \sim 2$) led to an increase in the aspect ratio. When probed with 1550-nm light, the waveguide exhibited a circular mode with MFD of 11 $\mu$m, well-matched to the 10.5-$\mu$m MFD of SMF. Using the Fabry–Perot method, a propagation loss of 0.2 dB/cm was measured, which is the lowest loss achieved in a high-index glass waveguide fabricated by femtosecond laser writing.

## 7.4 Devices

Since the first discovery of femtosecond laser fabrication of waveguides by Hirao and co-workers in 1996 [38], various optical devices have been photowritten inside glasses such as Y-junction power splitters, directional couplers and novel three-dimensional circuits that are reviewed in this section.

## 7.4.1 Y-Junctions

Y-junctions are important optical power splitting devices that find use in power dividers and Mach–Zehnder interferometers for sensing and PON. The first Y-junction fabricated by femtosecond laser writing was demonstrated in fused silica by Homoelle et al. in 1999 [39], and was able to divide 514.5-nm light from an argon laser with an approximate 50/50 ratio in the output branches when the splitting angle was 0.5°. More complex 1×N power splitters based on Y-junctions were fabricated in fused silica glass by focusing 125-fs pules from a 150-kHz Ti:Sapphire amplified laser with a 0.4-NA lens [40]. The average power was 52 mW and the writing speed was 50 $\mu$m/s in the transverse plane. The schematic design of a 1×8 power splitter is shown in Fig. 7.14 and consists of seven cascaded Y-junctions with bend radii of 200 mm to avoid bend loss. During fabrication, the 50-$\mu$m long waveguides just before each Y-junction splitting point were overwritten to ensure an equal splitting ratio. Each device was written two times with a sample shift of 1.5 $\mu$m between the first and second writing to avoid excessive insertion loss. Also shown in Fig. 7.14 are the near-field patterns at 1550-nm, which show excellent uniformity among the 8 outputs. The insertion and polarization dependent losses were 16.4 and 0.4 dB, respectively.

The versatility of femtosecond laser writing was demonstrated by an integrated power splitting circuit where both the buried Y-junction waveguide and surface microchannels for aligning input/output fibers were fabricated with the same femtosecond laser [42]. The device was packaged by inserting single-mode fibers in the machined U-grooves, thus avoiding the use of optical fiber array blocks in the alignment or complex post-fabrication etching processes employed to fabricate grooves in PLC optical chips. The integrated optical splitter was fabricated in fused silica using 1-kHz, 100-fs laser pulses focused by a 0.42-NA microscope

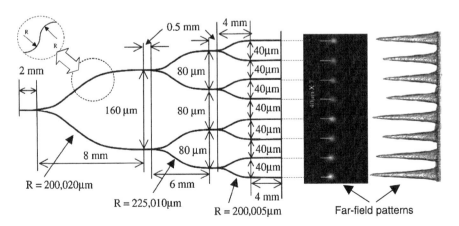

**Fig. 7.14** Schematic design diagrams of a 1–8 optical waveguide power splitter and near-field patterns [40, 41]

objective using the transverse writing method. The waveguides forming the Y-junction were written with 400-nJ pulse energy and a scan speed of 10 $\mu$m/s. The U-grooves for passive alignment were ablated using 30-$\mu$J pulse energy and 500-$\mu$m/s scan speed. The resulting device is shown in Fig. 7.15. The U-grooves were 126-$\mu$m wide, 87-$\mu$m deep with low ~0.3-$\mu$m surface roughness, allowing for SMF (125-$\mu$m diameter) to efficiently couple to the input and output access waveguides. The fiber-aligned splitter had a lower IL of less than 4 dB, including an intrinsic splitting loss of 3 dB and excess loss due to the passive alignment of SMF.

The first demonstration of a 3D optical circuit using femtosecond laser writing was a 1 × 3 Y-junction splitter by Nolte et al. [43]. The 3D Y-junction (Fig. 7.16) was fabricated in fused silica using 500-nJ, 50-fs pulses from a 1-kHz laser focused 200$\mu$m beneath the surface with a 0.45-NA lens. The sample was scanned transversely at 125 $\mu$m/s. The inset in Fig. 7.16 is the near-field output at 1-$\mu$m wavelength and demonstrates a nearly equal 32:33:35 splitting ratio. The propagation losses were estimated to be 0.8 dB/cm for the straight waveguides and the additional losses due to splitting in the Y-junction were approximately 6 dB.

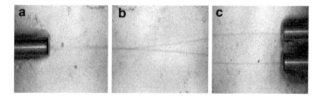

**Fig. 7.15** Fiber aligned one-input (a) and two-output channels (c) of U-grooved Y-junction (b) optical splitter [42]

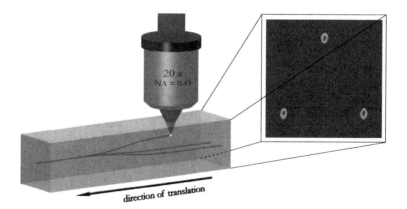

**Fig. 7.16** Schematic of the experimental setup and the 3D Y-junction splitter fabricated. The inset shows the near-field intensity distribution at 1.05 $\mu$m measured at the exit facet. The output ports of the splitter are separated by 100 $\mu$m [41, 43]

## 7.4.2 Directional Couplers

A directional coupler relies on closely spaced waveguides for periodic power transfer between the waveguides through evanescent coupling. The directional coupler is an important building block in optical circuits and enables more complicated designs such as Mach–Zehnder interferometers (MZIs), sensors, power splitters and wavelength multiplexers. The directional coupler is advantageous compared to the Y-junction because it avoids the radiation loss caused by the branching point. However, directional couplers are inherently wavelength dependent so that careful design is needed to engineer broadband spectrally insensitive power splitters.

### 7.4.2.1 Directional Couplers at Visible Wavelengths

The first femtosecond laser-written directional coupler was demonstrated by Streltsov and Borrelli [44]. The coupler had a simple design consisting of a straight waveguide and a curved three-segment waveguide. The latter had 50-mm radius arcs at the ends, with the straight sector of 9-mm length written parallel and 3.5 $\mu$m from (center to center) the straight waveguide arm. The resulting coupler showed single-mode operation and a 1.9-dB coupling ratio at 633-nm wavelength.

An important advance of the Streltsov and Borrelli paper [44] was the first use of the second harmonic wavelength for waveguide writing. Due to the low energy ($\sim$10 nJ) of the Ti:Sapphire oscillator employed in the experiments, the laser was frequency doubled to give 2.8-nJ pulse energy at 400-nm wavelength to increase the photosensitivity of the borosilicate glass (Corning 7980), similar to the experiments described in Sect. 7.3.2 with 522-nm wavelength femtosecond laser writing of fused silica waveguides. The waveguides were inscribed at a relatively slow speed of 10 $\mu$m/s using the longitudinal writing method with a 0.28-NA lens. The RNF method (Chap. 5) was applied for the first time to femtosecond laser-written waveguides to measure a refractive-index change of $\sim$0.0045.

Femtosecond laser-written couplers were extended to the third dimension by Watanabe et al. [45] using a 1-kHz, 130-fs Ti:Sapphire laser with a longitudinal writing scheme. The authors applied a 0.3-NA lens with 0.68-$\mu$J pulse energy producing 40-$\mu$m long filaments in fused silica glass. Two-dimensional translation of the filament led to the formation of a curved waveguide because of bending by the previously induced refractive-index change, allowing the fabrication of a 3D directional coupler as shown in Fig. 7.17a. The near-field patterns from coupling a monochromatic HeNe source are shown in Fig. 7.17b. For the first time, wavelength division was demonstrated in a femtosecond laser-written coupler by launching a broadband supercontinuum source (450–700 nm), as shown by the near-field profiles in Fig. 7.17c.

The first thorough characterization of femtosecond laser-written directional couplers at visible wavelengths was done by Minoshima et al. [46]. In the study, a novel Ti:sapphire multi-pass cavity (MPC) oscillator was applied to write optical

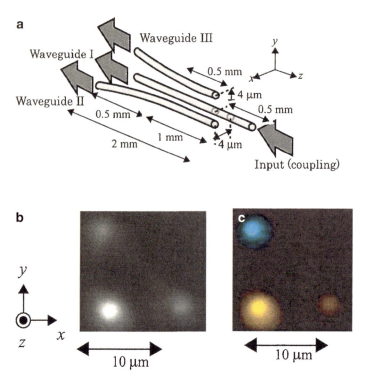

**Fig. 7.17** Schematic of a three-dimensional directional coupler (**a**), near-field patterns of coupler output when (**b**) coupling to 633-nm HeNe, and (**c**) coupling to a broadband source (450–700 nm) showing wavelength filtering [45]

waveguides in standard microscope slide soda lime glass (Corning 0215). The 20-nJ, 4-MHz pulse train was focused with an oil-immersion objective enabling rapid writing speeds of 1 cm/s in the transverse writing geometry. The power coupling behavior of the device was characterized at 633 nm for different interaction lengths and waveguide separations. A sinusoidal power exchange was observed which agreed with coupled mode theory, but a sharp decrease in the peak coupling ratio from 0.6 to 0.08 was reported as the separation distance increased from 8 to 12 $\mu$m. Directional couplers are very sensitive to deviations in the writing conditions due to the exponentially decaying evanescent field responsible for the power exchange between adjacent waveguides. Presumably, some drift in the writing parameters caused a difference in the propagation constants of the adjacent waveguides, manifesting in a non-unity peak coupling ratio.

An important analysis of the effect of orientation angle between coupled waveguides on the coupling strength was carried out by Szameit et al. [47] at visible wavelengths. Waveguides were fabricated using a Ti:Sapphire laser system with a repetition rate of 100 kHz, a pulse duration of 150 fs and 0.3-$\mu$J pulse energy. The beam was focused into a fused-silica sample by a microscope objective with

7 Passive Photonic Devices in Glass

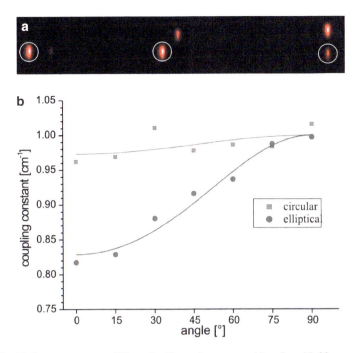

**Fig. 7.18** (a) Output patterns at 633 nm for 10-mm long waveguide pairs with 20-$\mu$m separation at different orientations. The excited waveguide is marked by a white circle. (b) Comparison of the dependence of the coupling on the direction for 18-$\mu$m waveguide separation at 800-nm wavelength [47]

0.45-NA and the sample was scanned transversely relative to the laser at 1.25 mm/s. As shown in Fig. 7.18a, the coupling between waveguides (633-nm wavelength) oriented at different angles varied significantly due to the elliptical refractive index cross section (4 × 13$\mu$m$^2$). In Fig. 7.18b, the coupling coefficient as a function of orientation angle (800-nm characterization wavelength) is shown for the case of nominal focusing and also with slit reshaping to give a more circular waveguide cross section (9 × 14$\mu$m$^2$), as described in Chap. 5. The slit reshaping method gives nearly isotropic coupling as a function of orientation angle.

In most waveguides formed by femtosecond lasers, the cross sectional refractive index profile is asymmetric and nonuniform, regardless of repetition rate regime or beam shaping technique. Waveguides often have damaged regions of depressed refractive index contrast above or below the main guiding region and as a result, show different coupling behavior when the waveguides are coupled transversely or vertically. The present waveguides in fused silica have a uniform, symmetric refractive-index change, but unfortunately show a limited refractive index increase of ∼0.001 [47]. Nevertheless, the waveguides are very well suited to study of linear and nonlinear phenomena in 2D waveguide arrays, a topic covered in Chap. 13.

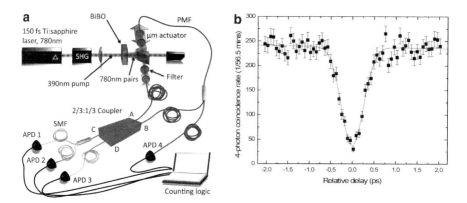

**Fig. 7.19** Quantum interference with three photons. (**a**) Measurement setup with spontaneous parametric down conversion source based on frequency doubled Ti:Sapphire femtosecond oscillator. (**b**) The number of coincident detections are shown as a function of the arrival delay between two interfering photons with an observed interference dip for three photons [48]

An interesting application of femtosecond laser-written directional couplers to quantum information science was recently presented by Marshall and colleagues [48]. The authors wrote 2×2 directional couplers using curved regions with raised-sine curve form to reduce bend loss. With couplers designed to operate at 806-nm wavelength, the authors demonstrated a three-photon entangled state using a 4-photon source shown in Fig. 7.19a. In the quantum interference experiment, the visibility of the interference dip (Fig. 7.19b) surpassed that of an interferometer based on bulk optics, demonstrating the robustness of guided wave optics fabricated by femtosecond lasers.

In a recent article, Sansoni and colleagues applied femtosecond laser writing to form a directional coupler with circular cross sections and low birefringence in Corning EAGLE2000 borosilicate glass [49]. An Yb:KYW cavity-dumped mode-locked oscillator was employed for waveguide writing, which delivered 300-fs, 1-$\mu$J pulses at 1030-nm wavelength at 1-MHz repetition rate. With a pulse energy of 240 nJ focused by 0.6-NA lens and a scan speed of 40 mm/s, waveguides supporting a single mode at 800-nm wavelength were produced. The 8-$\mu$m MFD led to an estimated 0.7-dB coupling loss with a single-mode fiber designed for 800-nm wavelength. Measured propagation losses were 0.5 dB/cm, and using a curvature radius of 30 mm, additional bending losses were lower than 0.3 dB/cm. The birefringence of the waveguides was measured to be $\sim 10^{-5}$, an order of magnitude lower than conventional silica-on-silicon waveguides. In the laser-written directional coupler, quantum interference effects were demonstrated with polarization entangled photons as shown in the inset of Fig. 7.20. The polarization-preserving feature of the directional coupler makes it potentially useful for tasks like cryptography and linear optics quantum computing that are based on the information stored in the polarization of light.

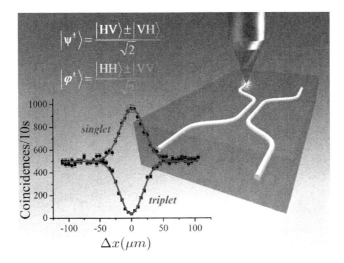

**Fig. 7.20** Schematic of the femtosecond laser-written directional coupler in the bulk of a borosilicate glass. Inset shows characteristic Hong-Ou-Mandel interference for singlet and triplet polarization entangled Bell states [49]

### 7.4.2.2 Directional Couplers at Telecom Wavelengths

The first femtosecond laser-written couplers characterized at the low-loss telecom window were reported by Osellame et al. [50]. The authors fabricated an X-coupler rotated with respect to the transverse plane as shown in Fig. 7.21. The waveguides were written at 1 mm/s in alkali-zinc-silicate glass (Schott IOG-10) using 25-nJ pulses provided by a 26-MHz extended-cavity oscillator and focused by a 1.4-NA oil immersion objective. As shown in Fig. 7.21, the intersection angle, $\alpha$, sets the interaction length of the coupler and the depth displacement, $h$, influences the coupling strength, similar to a standard directional coupler. Several couplers with different values of the two parameters were fabricated and various coupling ratios demonstrated at 1550-nm wavelength. Further, wavelength division multiplexing at near-infrared wavelengths was reported for the first time for laser formed waveguides, showing high rejection of 980 nm wavelength with respect to 1550 nm, a combination useful in laser pumping applications.

Another example of a 3D coupled-mode device was a $3 \times 3$ directional coupler formed in soda-lime glass (Corning 0215) [51]. The waveguides were fabricated by means of a 5.85-MHz repetition rate, 67-fs MPC Ti:sapphire oscillator. The average power of the laser was 200 mW and the laser pulses were focused with a 1.25-NA microscope objective with the sample scanned transversely at 8 mm/s. The $3 \times 3$ directional coupler consisted of three waveguides fabricated on the edges of an equilateral triangle, as shown in Fig. 7.22. The IL of the couplers was 6.6 dB and the largest discrepancy in the coupling ratio for TE and TM polarizations was only

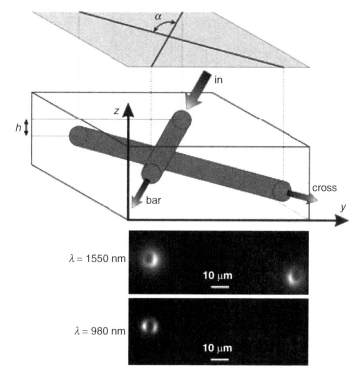

**Fig. 7.21** Schematic of 1×2 splitter with measured near-field mode profiles of output face at 1550 and 980 nm [50]

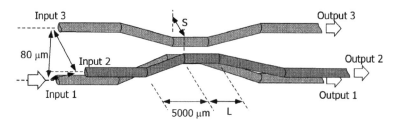

**Fig. 7.22** Schematic configuration of a 3 × 3 directional coupler [51]

11%, suggesting that waveguides fabricated within the cumulative heating regime show little polarization dependence.

Using a femtosecond fiber laser with variable 0.1- to 5-MHz repetition rate, directional couplers were fabricated in borosilicate glass (Corning EAGLE2000) and characterized at the low-loss (1550 nm) and zero-dispersion (1310 nm) telecom wavelengths [2]. The optimum writing conditions for low propagation loss and efficient coupling to SMF were 1-MHz repetition rate, 200-nJ pulse energy, 20-mm/s scan speed, and 0.55-NA focusing in the transverse writing configuration. A MFD

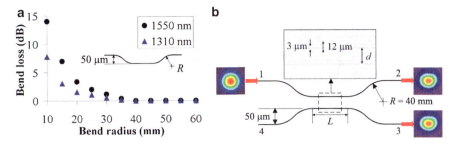

**Fig. 7.23** (a) Bend loss versus radius of curvature at 1310 and 1550 nm. Inset shows one arm of the coupler under test with four bends of radius $R$. (b) Schematic of directional coupler and optical microscope image of waveguides in coupling region (top) together with mode profiles of 1550-nm light from launch fiber at port 1 and from output facets at ports 2 and 3 [2]

of ~12.0 μm at 1550-nm wavelength was found at 20-mm/s scan speed, closely matching that of SMF (10.5 μm). Using a SMF at the output facet, the waveguide exhibited a 2-dB IL over its 5-cm length, which compares favorably to the result by Suzuki et al. [51], perhaps due to the higher quality optical glass employed.

Before writing the directional couplers, the bend loss versus radius of curvature, $R$, was analyzed for single arms of the coupler, yielding the results for 1310 and 1550 nm as shown in Fig. 7.23a. The measured bend loss was defined as the difference in IL between a single coupler arm and a straight waveguide of the same length, and thus included pure bend loss and transition loss at inflection points in the arm. A bend radius of 40 mm (~0.1 dB bend loss) was selected for the directional coupler design because the loss rapidly increased for smaller bend radii.

Directional couplers were then defined by symmetric double S-bend waveguides with 40-mm radius of curvature as shown in Fig. 7.23b. Directional couplers were fabricated with separation distance $d$ (center to center) varied from 7.5 to 22.5 μm and interaction length $L$ varied from 2 to 30 mm. Also shown in Fig. 7.23b are the 1550-nm wavelength near-field mode profiles for the input fiber at port 1 and the waveguides at output ports 2 and 3. A 50/50 coupling ratio was demonstrated in Fig. 7.23b for $d = 10$ μm and $L = 20.6$ mm. The IL of the directional couplers was 2.5 and 2.2 dB at 1310 and 1550 nm, respectively.

The power splitting ratio for couplers was investigated as function of interaction length, $L$, and separation distance, $d$. The power coupling ratio, defined as the power at the cross port, normalized to the total power at the output ports 2 and 3 can be derived from coupled mode theory:

$$r = \frac{P_3}{P_2 + P_3} = \frac{\kappa^2}{\delta^2} \sin^2(\delta L) \qquad (7.3)$$

where $\delta = \sqrt{(\beta_1 - \beta_2)^2 + \kappa^2}$, $\beta_{1,2}$ are the propagation constants of the adjacent waveguides, and $\kappa$ is the coupling coefficient, which increases with separation, $d$. Deviation from zero to one sinusoidal modulation of the coupling ratio is due to a

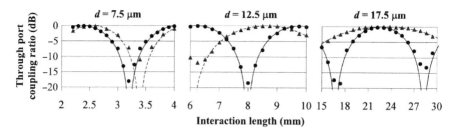

**Fig. 7.24** Through port coupling ratio dependence on interaction length for separations $d = 7.5$, 12.5, and 17.5 $\mu$m at 1310 (*triangles*) and 1550-nm (*circles*) wavelengths. Sinusoidal curve fits are shown as dashed and solid lines for 1310 and 1550 nm, respectively [2]

difference in propagation constants, such as that reported above by Minoshima et al. [46]. Note that a phase term should be included in (7.3) to account for coupling in the curved transition regions, but is not relevant to the beat length for the present discussion.

Data for the through port coupling ratio, $1 - r$, is plotted in Fig. 7.24 versus interaction length for separations of 7.5, 12.5, and 17.5 $\mu$m at wavelengths of 1310 and 1550 nm. Due to weaker coupling between waveguides, the beat length increases with separation distance and decreases with wavelength. The modulation depth consistently reaches near unity and closely follows the expected sinusoidal response. At 1550-nm wavelength, the minimum through port coupling ratio is $-19$ dB (1%) corresponding to $\sim$99% maximum coupling ratio. The modulation range is slightly lower at 1310-nm wavelength, yielding a minimum of $-12$ dB, corresponding to $\sim$93% coupling ratio. Impressively, the coupling performance at $d = 7.5$ $\mu$m is not hindered by the slight overlapping of the outer waveguide cladding zones which have 6-$\mu$m radius.

The present results [2] are an improvement over Minoshima et al. [46], which showed a decrease in peak coupling ratio from 0.6 to 0.08 as the separation distance increased from 8 to 12 $\mu$m. As described above, the evanescently decaying field responsible for power exchange in directional couplers is very sensitive to fabrication errors, and since the modes were more strongly confined in Minoshima's work due to the shorter 633-nm wavelength, the weaker evanescent tail resulted in lower peak coupling ratios. This insight also explains the lower peak coupling ratio at 1310 nm compared to 1550 nm in the present work. Another reason for the lower peak coupling ratio is due to the mismatch in propagation constants arising from different bend radii used in the two arms of the coupler compared with the present work's symmetric S-bend design [2].

#### 7.4.2.3 Wavelength Multi-Demultiplexers

Wavelength multi-demultiplexers play a vital role in WDM optical networks and sensor applications. Wavelength multiplexers have been demonstrated in PLC using directional couplers, offering the potential for miniaturization and mass production [52], but are limited to specific glass compositions with planar geometries. The most common fabrication method for wavelength demultiplexers is using two fibers fused into a single tapered element by heating and drawing. The adiabatic taper at the input and output results in undesirable device lengths of several centimeters [53]. Furthermore, the fiber approach cannot be integrated on a photonic chip for added functionality. In comparison to the commonly employed fused fiber taper technique for fabricating wavelength demultiplexers, femtosecond-laser writing enables the close and accurate placement of adjacent waveguides resulting in small interaction lengths of ∼1 mm in combination with curved transitions of ∼5-mm length, allowing for devices shorter than 1 cm, with well-predicted responses from coupled mode theory, as demonstrated above.

With the aid of an empirical model, the directional couplers described previously [2] were optimized for wavelength demultiplexing of 1.3 and 1.55-$\mu$m wavelengths [54] that may serve in PONs [55] for FTTH applications. Directional couplers with S-bend geometry were written with 1.5-MHz repetition rate, 200-mW (133-nJ) average power (pulse energy) and 12-mm/s scan speed for minimum loss. To characterize the spectral response of the coupling ratio, a broadband light emitting diode (LED) source (1200–1700 nm) was coupled at the input of the couplers. When analyzing the spectral response of a directional coupler, the simplified expression for the power coupling ratio shown in (7.3) must be modified to include a phase term due to the evanescent coupling in the curved transition regions [54, 56]:

$$r(\lambda) = \sin^2 (\kappa(\lambda)L + \phi(\lambda)) \quad (7.4)$$

After accounting for the bending phase, a coupler was designed for demultiplexing at $L = 3.2$ mm and $d = 10\mu$m. Figure 7.25b shows the coupling ratio across the full telecom band (1250–1650 nm) for a separation of $d = 10\mu$m and interaction lengths of $L = 0.8$, 3.2, and 5.2 mm. As expected from (7.4) and Fig. 7.25a,

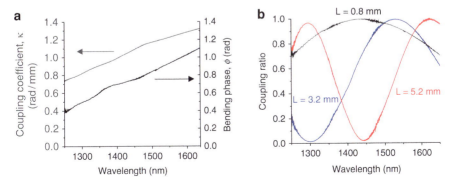

**Fig. 7.25** (a) Coupling coefficient and bending phase versus wavelength for a directional coupler with separation of $d = 10\mu$m. (b) Measured coupling ratio versus wavelength for separation of $d = 10\mu$m and interaction lengths of $L = 0.8$, 3.2, and 5.2 mm [54]

the spectral responses are nearly sinusoidal with period of oscillation decreasing with increasing interaction length. The coupling ratio spectral responses are the first demonstration of 0 to near-unity sinusoidal wavelength response for a femtosecond laser-written coupler, confirming the propagation constants in the adjacent arms were nearly identical. Among the various lengths tested, only $L = 3.2$ mm provided the highest contrast demultiplexing of 1300 and 1550 nm (−19 dB crosstalk). Improvement in crosstalk to compete with typical −30 dB values achieved by conventional fabrication techniques [52, 57] is expected through finer scaling of the separation distance and bend radius as well as improvements in the laser writing to reduce loss and improve waveguide uniformity.

### 7.4.2.4 Wavelength-Flattened Power Splitters

The wavelength sensitivity of directional couplers due to the dispersion in their propagation constants can be exploited for filtering and wavelength multiplexing as described above, but is a hindrance to broadband power splitting needed for power monitoring and also power division in PON. To overcome the wavelength sensitivity, several designs have been proposed which include tapered fiber couplers [58], bent couplers [59], Mach–Zehnder interferometers [60], and asymmetric directional couplers [61]. Y-splitter branching devices may also be used, but are undesirable due to the excess loss introduced by the Y-junction and the increased fabrication complexity in joining the output arms as described in Sect. 7.4.1.

Chen et al. showed that with short separation distances, the wavelength dependence of the straight and curved regions of a directional coupler are compensated, giving a wavelength-flattened broadband response [62]. Wavelength flattening occurs when the total phase $\kappa(\lambda)L + \phi(\lambda)$ is independent of wavelength. Similar writing conditions were applied to the WDM couplers described above [54]. A scan speed of 12 mm/s defined the first arm while the second arm was scanned at 8 (asymmetric), 12 (symmetric), or 20 mm/s (asymmetric). The center-to-center separation distance $d$ ranged from 6 to 10 $\mu$m.

The experimentally measured coupling coefficient and bending phase are presented against wavelength in Fig. 7.26 for separations of $d = 6, 6.5, 7.5,$ and

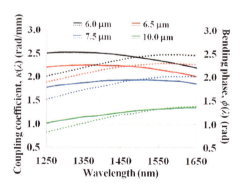

**Fig. 7.26** Coupling coefficient (*solid lines*) and bending phase (*dashed lines*) versus wavelength for symmetric coupler with separation distances of 6.0, 6.5, 7.5, and 10 $\mu$m [62]

10 µm. As the separation distance increases, both the coupling coefficient and bending phase decrease due to the exponentially decaying evanescent tail of the waveguide mode. For 10-µm separation, the coupling coefficient and bending phase increase with wavelength, similar to the trend in Fig. 7.25a, which as mentioned, can be useful for WDM. As the separation distance decreases below 10 µm, the slope of the coupling coefficient is reduced, eventually becoming negative over the entire wavelength range. In this case, broadband coupling behavior is possible through suitable choice of interaction length. A widely held, but incorrect view is that the coupling coefficient increases monotonically with wavelength [63]. The transition of the coupling coefficient from positive to negative slope at small separation can be understood by analyzing the functional form of the coupling coefficient:

$$\kappa \propto \frac{k_0^2}{2\beta} \int_2 E_1(x,y) E_2(x,y) dx dy \tag{7.5}$$

where $\beta$ is the propagation constant in a waveguide arm of the coupler (symmetric case), $k_0$ is the free-space wavenumber and $E_{1,2}$ are the mode fields in waveguides 1,2. Since the integral is taken over waveguide 2, increasing wavelength (less confinement of the modes) results in $E_1$ increasing, but $E_2$ decreasing. The factor outside the integral $k_0^2/2\beta$ varies approximately as $1/\lambda$. Therefore, it is clear that the coupling coefficient can decrease with wavelength at a certain separation distance.

The coupling ratio versus wavelength for 6-µm separation distance and interaction lengths of 0, 0.5, 1, 1.5, 2, and 2.5 mm is shown in Fig. 7.27. With 6-µm separation, the waveguide claddings overlap in the interaction region, yet the waveguides show negligible difference in propagation constant as evidenced by an observed zero to unity sinusoidal variation in coupling ratio with interaction length [62]. This underscores the finesse of the high-repetition rate femtosecond laser writing process, particularly in comparison to UV laser writing of PLCs, where a decrease in photosensitivity contributed to a significant difference in the propagation constants of closely spaced waveguides [61].

Broadband unity coupling with greater than 400-nm bandwidth (±5%) was achieved for $L = 1$ mm as seen in Fig. 7.27. The common view in the literature is that the coupling ratio modulates sinusoidally from zero to unity in the wavelength

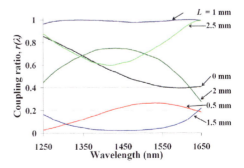

**Fig. 7.27** Coupling ratio versus wavelength for symmetric coupler with separation distance of 6 µm [62]

domain for a symmetric directional coupler [63]. However, for all the interaction lengths tested, the modulation clearly does not follow this trend, but rather is due to the balance between the $\kappa(\lambda)L$ and $\phi(\lambda)$ terms in (7.4) as described above.

To further flatten the wavelength responses, in particular for the 3-dB coupling ratio, which is the most sensitive to wavelength [53], an asymmetry between the adjacent waveguides was introduced by scanning the second arm at a different speed. Coupling ratios from zero to unity were obtained with $\pm 5\%$ bandwidth (BW) ranging from 300 to 400 nm [62], comparable to PLC wavelength-flattened power splitters written by photolithography [63] and direct UV laser writing [61].

### 7.4.3 Mach–Zehnder Interferometers

Y-junction splitters and directional couplers can be cascaded to form more complex optical circuits, such as the MZI commonly used in optical sensors or WDM interleavers. The first demonstration of a femtosecond laser-written MZI consisted of two cascaded X-couplers crossing at an angle of 2°, as shown in Fig. 7.28a. The device was fabricated with the same Ti:Sapphire MPC oscillator used to make the directional couplers presented by the same group [46, 51]. An unbalanced interferometer arrangement was used with a path length difference of 10 $\mu$m between the two arms to allow for wavelength filtering. By coupling a broadband source at the input, the wavelength transfer function in the crossed arm (black line in Fig. 7.28c) showed good agreement with the theoretical wavelength dependence of 9.3 $\mu$m (red line in Fig. 7.28c). The transfer function was constructed by normalizing the output spectrum by the input spectrum (red line in Fig. 7.28b). This small deviation from the design path length difference was attributed to fabrication errors.

Label-free sensing in lab-on-a-chip was implemented by integrating a MZI (see Chap. 14 for details) [64]. The MZI (Fig. 7.29a) was fabricated in fused silica with the second harmonic wavelength of 520 nm, an average power of 90 mW, scan speed of 0.1 mm/s, and repetition rate of 1 MHz, yielding a mode field diameter of 11 $\mu$m at 1550-nm wavelength, similar to previous studies [12, 35]. The MZI was first characterized by launching light into the input port using a fiber-coupled laser with tunable wavelength from $\lambda = 1460$ to 1570 nm. The wavelength-dependent transmission of the unbalanced MZI ($\Delta s = 56$-$\mu$m offset in arm lengths) was found to follow closely the theoretical behavior (Fig. 7.29b), showing an excellent fringe contrast of 13 dB.

### 7.4.4 Other Devices

Multimode interference (MMI) couplers are power dividing circuits with a low sensitivity to manufacturing variations due to their simple design. The MMI coupler consists of a rectangular multimode waveguide, in which the multimode interference

7 Passive Photonic Devices in Glass 187

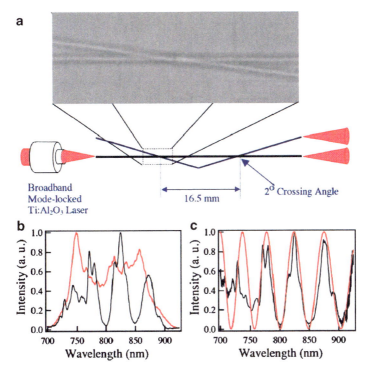

**Fig. 7.28** (a) Schematic of Mach–Zehnder interferometer with phase-contrast microscope images of waveguides. (b) Input (*red line*) and output (*black line*) spectra. (c) Normalized output spectrum (*black line*) compared to the theoretical model (*red line*) [41, 46]

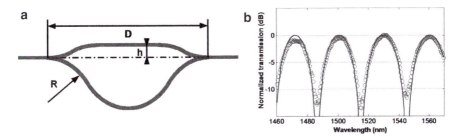

**Fig. 7.29** (a) Schematic of the unbalanced MZI with $h = 50\ \mu\text{m}$, $R = 30$ mm and $D = 17$ mm. (b) *Open circles*: measured transmission spectrum of the MZI. *Solid line*: least-squares fit of the experimental data [64]

leads to self-imaging effects. By placing output waveguides at different image points, it is possible to implement a wide range of splitting and recombining functions with the possibility of a large number of input and output arms [65]. A multimode interference (MMI) device was fabricated in fused silica by scanning a femtosecond-laser induced filament two dimensionally below the surface of fused

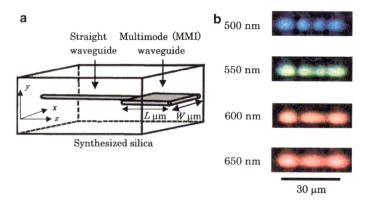

**Fig. 7.30** (a) Schematic of a MMI waveguide device in fused silica. (b) Near-field pattern at the MMI waveguide output for various wavelengths, with $W = 30\mu$m. The central wavelengths of the interference filters are indicated [66]

silica. The 85-fs, 1.5-$\mu$J pulses provided by a 1-kHz amplified Ti:Sapphire laser and focused with a 0.3-NA objective yielded a 100-$\mu$m long filament within the glass. To fabricate the MMI device shown in Fig. 7.30a, the filament was scanned transversely along the x-axis for $W = 30$ $\mu$m producing a 2-$\mu$m thick layer of increased refractive index. The filament region was then translated 50 $\mu$m in the $-z$ direction. By repeating this two-dimensional movement, an MMI waveguide with a width of $W = 30\mu$m, length of $L = 870\mu$m, and a thickness of 2 $\mu$m was produced. Finally, a straight waveguide with 2-$\mu$m diameter was connected to the MMI region by scanning the sample along the optical axis.

To characterize the MMI device, a white light source was end-fire coupled into the straight waveguide and near-field mode profiles were captured at the output of the multimode section. Interference filters with central wavelengths of 500, 550, 600, and 650 nm (FWHM 10 nm) allowed for spectral characterization of the MMI as shown in Fig. 7.30b. The input light is split into three outputs at 600 and 650 nm, and into four outputs for 500 and 550 nm. Comparison of the near-field intensity distribution with a theoretical model based on the beam propagation method showed excellent agreement.

The 3D ring resonator shown in Fig. 7.31 was fabricated at a scanning speed of 10 mm/s in soda lime glass (Corning 0215) using 5.85-MHz, 15-nJ, 43-fs pulses focused by a 0.86-NA objective. Unlike planar ring resonators, the 3D geometry allows the through and drop ports to be located on the same end of the device, with the input at the opposite end. The ring is composed of semicircular arcs of 1-mm bend radius connected by two straight waveguides of 0.5-mm length. The input and output waveguides were fabricated 5 $\mu$m above and below one straight section of the ring. The closed ring was fabricated in a continuous fashion by approaching and diverging the structure at an angle of 10°.

The wavelength-dependent transfer function of the 3D ring resonator was measured using the broadband output from a Ti:Sapphire laser as shown in Fig. 7.32.

# 7 Passive Photonic Devices in Glass

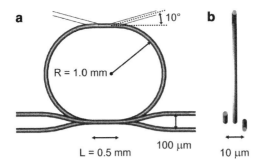

**Fig. 7.31** Schematic of the 3D microring resonator. (**a**) Top view: the ring is fabricated in the plane of the substrate and composed of two semicircular arcs with 1-mm radii connected by 0.5-mm waveguides. (**b**) Side view: the waveguides are separated by 5 $\mu$m with a total depth separation of 10 $\mu$m [67]

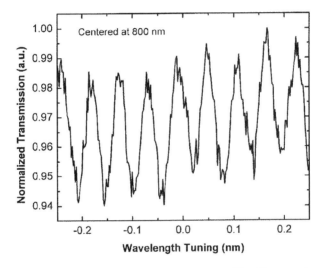

**Fig. 7.32** Normalized transfer function of the ring resonator. The fringe spacing of 57 pm at 800 nm concurs with the 58-pm calculated for a ring length of 7.3 mm [67]

The 57-pm fringe is close to the expected value of 58 pm for a ring of 7.3-mm perimeter. The relatively low fringe visibility of 2% results from a round-trip loss of approximately 37 dB, which is mainly due to bending loss from a combination of the low index contrast ($\sim$0.005) of the written waveguides together with the small 1-mm bend radius chosen for the ring resonator.

In another interesting application of the 3D patterning capability of femtosecond laser writing, a fan-out device was fabricated that allows each core of a 2 × 2 core array multicore optical fiber (MCF) to be addressed individually by one of four single-mode fibers held in 4 × 1 fiber V-groove array (FVA) [68]. Due to the close proximity and geometrical arrangement of the cores, coupling of light to MCFs is a significant technical challenge. A regeneratively amplified Ti:Sapphire laser (5 kHz, 100 fs) was applied to fabricate the fan-out device in silicate glass. A 300-$\mu$m slit was placed before a 0.42-NA objective to form symmetric waveguides as described in Chap. 5. A pulse energy of 1.3 $\mu$J was applied with the sample scanned transversely at 400 $\mu$m/s. A schematic of the device in Fig. 7.33 shows how the square arrangement of the four MFC cores was interfaced with the FVA.

A transmission microscope image of the cross section of the fan-out device at the MFC coupling end is shown together with an end view of the MFC fiber in Fig. 7.34. The fabricated waveguides were found to have MFDs of 15.8 × 11.8$\mu$m in the $x$ and $z$ axis, respectively. For each waveguide, the coupling loss was measured to be $\sim$0.5 dB to SMF with IL of $\sim$3 dB and $\sim$5 dB for waveguides 2,3, and 1,4 shown in Fig. 7.33, respectively. The increased loss for waveguides 1,4 compared to 2,3 was attributed to the larger angular misalignment between the linear sections that form the total waveguide. When the fan-out device was aligned at the input to the MFC, having similar core dimensions as SMF, with the FVA at the output, only slightly higher insertion losses were measured.

Femtosecond laser waveguide writing is effective for writing 3D optical circuits but because it is a serial process, it is slow compared with planar photolithography

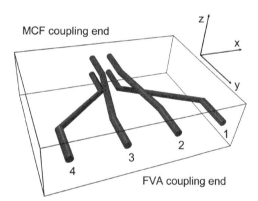

**Fig. 7.33** Graphical representation of the fabricated fan-out device [68]

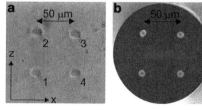

**Fig. 7.34** Transmission mode optical micrographs of (**a**) the MCF coupling end of the fabricated fan-out and (**b**) the end of the MCF fiber for which the fan-out device was designed [68]

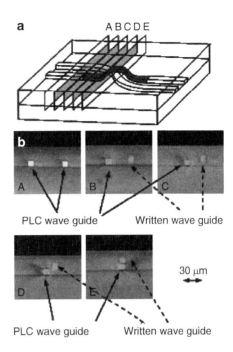

**Fig. 7.35** (a) Schematic diagram of cut back position. (b) Image of the cross-section. (c) Near-field pattern of written and PLC waveguides [69]

for fabricating complex optical circuits. However, the density of integrated optical devices in PLCs are much lower in comparison with that of microelectronic circuits, diminishing the overall benefit of parallel processing for optical circuits. Further, the commercial development of high power femtosecond lasers promise greatly increased (100-fold) serial writing speeds. An interesting proposal for a hybrid laser and PLC manufacturing by Nasu et al. employs conventional planar techniques to fabricate the main waveguides in a PLC, but leaving gaps between waveguides where waveguide crossings would normally occur [69]. Using femtosecond laser technology, 3D waveguide paths could then be written to interconnect the waveguide links and avoid losses due to crossings found in many PLC devices [69]. The authors used a 1-kHz, 150-fs Ti:Sapphire laser with a multi-scan approach to form a waveguide core with similar shape as the PLC waveguides. The sample was cut back at five positions to show the spatial evolution of the 3D waveguide interconnect in the transverse plane (Fig. 7.35). The femtosecond laser-written waveguides appeared very similar in shape compared to the PLC waveguides. However, the mode mismatch between PLC and femtosecond laser-written waveguides resulted in a coupling loss of 1 dB/point. An additional connection loss due to position misalignment was estimated to be less than 0.4 dB/point. The total excess loss was found to be about 2.7 dB at 1550 nm which could be reduced significantly by further tailoring of the waveguide refractive index profile to reduce the high connection losses.

## 7.5 Summary and Future Outlook

In terms of waveguide properties, femtosecond-laser writing now offers low propagation losses of ~0.2 dB/cm, but this is still substantially higher than the ~1 dB/m loss typical with PLC technology [1]. The maximum refractive index contrast induced is $\Delta n = 0.01–0.02$ using high-repetition rate femtosecond lasers, which is suitable for coupling to SMF, but still limits waveguide bends to a radius of 15–40 mm. Higher index contrast would lead to smaller mode sizes for tighter bends allowing for increased density of functions on a photonic chip. However, laser writing in out-of-plane architectures, exploiting the 3D volume of bulk glass, remains the most highly promising opportunity for dense optical integration enabled by femtosecond lasers.

The rapid development of optimized recipes for femtosecond laser fabrication of waveguides now emerging is enabling high-quality passive devices to be demonstrated in a variety of glasses in both 2D and 3D geometries. For three-dimensional coupled-mode devices, future work must address the problem of nonuniform coupling between waveguides at different orientations due to asymmetric refractive index profiles. The goal of writing a cylindrical waveguide with a symmetric, uniform, and high-refractive index contrast has still eluded researchers.

Higher resolution translation stages may result in more uniform structures for reduced propagation losses. However, the relatively high propagation loss of femtosecond laser-written waveguides is mainly attributed to the nonuniform refractive index modification resulting in scattering loss. By exploring a wider range of exposure parameters aided by the recent commercial development of higher power (>20 W), shorter pulse duration, high-repetition rate femtosecond lasers, more extreme temperatures giving increased heat diffusion and accumulation with faster cooling rates may be driven, yielding greater material densification over larger volumes, possibly with smoother refractive index profiles. Further development of real-time monitoring systems to study spectral emissions and infer glass temperatures may in the future be used to control laser exposure to avoid cracks, stress, and other defects that lead to scattering loss.

In the last few years, laser polarization (Chaps. 6, 11, and 14), pulse duration (Chaps. 4 and 9), multiple scan lines (Chap. 5), and titled pulse fronts (Chap. 6) have shown to be play an important role in waveguide writing. The effect of other exposure parameters such as spatial beam profile (e.g., Gaussian, top-hat, square, and donut-shape), temporal pulse shape, bursts of pulses and external heat or stress treatment during writing, have yet to be studied and may offer improvement in waveguide losses and higher refractive index contrast. Lastly, the birefringent properties of laser formed waveguides have been barely examined and suggest rich opportunities for opening another dimension of application in sensing and optical communications, particularly as manifested by the strong form birefringence associated with nanograting structures formed in fused silica glasses [18].

The commercial success of the femtosecond laser waveguide writing in glasses in the context of displacing existing PLC technology is contingent upon addressing the remaining roadblocks of moderate refractive index contrast and waveguide losses described above. The femtosecond laser writing technique is maskless, simple, and flexible, making it ideal for rapid prototyping, refractive index trimming of existing optical circuits, and sensor fabrication. With recent discoveries of easily tailored and predicted waveguide shapes enabled by cumulative heating and thermal diffusion, along with Bragg grating waveguide functionality (Chap. 9) in three dimensional architectures, femtosecond laser fabrication now shows tremendous potential in fabricating tomorrow's integrated passive photonic circuits.

# References

1. Y. Hibino, MRS Bull. **28**(5), 365 (2003)
2. S.M. Eaton, W. Chen, L. Zhang, H. Zhang, R. Iyer, J.S. Aitchison, P.R. Herman, IEEE Photon. Technol. Lett. **18**(17–20), 2174 (2006)
3. L.M. Tong, R.R. Gattass, I. Maxwell, J.B. Ashcom, E. Mazur, Opt. Commun. **259**(2), 626 (2006)
4. R.R. Gattass, E. Mazur, Nat. Photon. **2**(4), 219 (2008)
5. S.M. Eaton, H. Zhang, P.R. Herman, F. Yoshino, L. Shah, J. Bovatsek, A.Y. Arai, Opt. Exp. **13**(12), 4708 (2005)
6. S.M. Eaton, Contrasts in thermal diffusion and heat accumulation effects in the fabrication of waveguides in glasses using variable repetition rate femtosecond laser. Ph.D. thesis (2008)
7. C. Florea, K. Winick, J. Lightwave Technol. **21**(1), 246 (2003)
8. Y. Okamura, S. Yoshinaka, S. Yamamoto, Appl. Opt. **22**(23), 3892 (1983)
9. H. Zhang, S.M. Eaton, P.R. Herman, Opt. Exp. **14**(11), 4826 (2006)
10. S.M. Eaton, H. Zhang, M.L. Ng, J. Li, W.J. Chen, S. Ho, P.R. Herman, Opt. Express **16**(13), 9443 (2008)
11. R. Osellame, N. Chiodo, G. Della Valle, G. Cerullo, R. Ramponi, P. Laporta, A. Killi, U. Morgner, O. Svelto, IEEE J. Sel. Top. Quantum Electron. **12**(2), 277 (2006)
12. L. Shah, A. Arai, S.M. Eaton, P.R. Herman, Opt. Exp. **13**(6), 1999 (2005)
13. D.J. Little, M. Ams, P. Dekker, G.D. Marshall, J.M. Dawes, M.J. Withford, Opt. Express **16**(24), 20029 (2008)
14. W. Gawelda, D. Puerto, J. Siegel, A. Ferrer, A.R. de la Cruz, H. Fernandez, J. Solis, Appl. Phys. Lett. **93**(12), 121109 (2008)
15. A. Ferrer, V. Diez-Blanco, A. Ruiz, J. Siegel, J. Solis, Appl. Surface Sci. **254**(4), 1121 (2007)
16. C. Mauclair, A. Mermillod-Blondin, N. Huot, E. Audouard, R. Stoian, Opt. Exp. **16**(8), 5481 (2008)
17. Y. Nasu, M. Kohtoku, Y. Hibino, Opt. Lett. **30**(7), 723 (2005)
18. W. Yang, P.G. Kazansky, Y. Shimotsuma, M. Sakakura, K. Miura, K. Hirao, Appl. Phys. Lett. **93**(17), 171109 (2008)
19. W.J. Yang, E. Bricchi, P.G. Kazansky, J. Bovatsek, A.Y. Arai, Opt. Exp. **14**(21), 10117 (2006)
20. R. Osellame, S. Taccheo, M. Marangoni, R. Ramponi, P. Laporta, D. Polli, S. De Silvestri, G. Cerullo, J. Opt. Soc. Am. B **20**(7), 1559 (2003)
21. M. Ams, G.D. Marshall, D.J. Spence, M.J. Withford, Opt. Exp. **13**(15), 5676 (2005)
22. Y. Cheng, K. Sugioka, K. Midorikawa, M. Masuda, K. Toyoda, M. Kawachi, K. Shihoyama, Opt. Lett. **28**(1), 55 (2003)
23. R.R. Thomson, A.S. Bockelt, E. Ramsay, S. Beecher, A.H. Greenaway, A.K. Kar, D.T. Reid, Opt. Exp. **16**(17), 12786 (2008)
24. M. Ams, G.D. Marshall, M.J. Withford, Opt. Exp. **14**(26), 13158 (2006)
25. Y. Shimotsuma, P.G. Kazansky, J. Qiu, K. Hirao, Phys. Rev. Lett. **91**(24), 247405 (2003)

26. C. Hnatovsky, R.S. Taylor, E. Simova, V.R. Bhardwaj, D.M. Rayner, P.B. Corkum, Opt. Lett. **30**(14), 1867 (2005)
27. G. Cerullo, R. Osellame, S. Taccheo, M. Marangoni, D. Polli, R. Ramponi, P. Laporta, S. De Silvestri, Opt. Lett. **27**(21), 1938 (2002)
28. J. Siegel, J.M. Fernandez-Navarro, A. Garcia-Navarro, V. Diez-Blanco, O. Sanz, J. Solis, F. Vega, J. Armengol, Appl. Phys. Lett. **86**(12) (2005). 121109
29. C. Schaffer, J. Garcia, E. Mazur, Appl. Phys. A **A76**(3), 351 (2003)
30. J. Chan, T. Huser, S. Risbud, J. Hayden, D. Krol, Appl. Phys. Lett. **82**(15), 2371 (2003)
31. C. Hnatovsky, R.S. Taylor, E. Simova, V.R. Bhardwaj, D.M. Rayner, P.B. Corkum, J. Appl. Phys. **98**(1), 013517 (2005)
32. C. Schaffer, A. Brodeur, E. Mazur, Meas. Sci. Technol. **12**(11), 1784 (2001)
33. K. Minoshima, A. Kowalevicz, I. Hartl, E. Ippen, J. Fujimoto, Opt. Lett. **26**(19), 1516 (2001)
34. R. Osellame, N. Chiodo, V. Maselli, A. Yin, M. Zavelani-Rossi, G. Cerullo, P. Laporta, L. Aiello, S. De Nicola, P. Ferraro, A. Finizio, G. Pierattini, Opt. Exp. **13**(2), 612 (2005)
35. S.M. Eaton, M.L. Ng, R. Osellame, P.R. Herman, J. Non-Cryst. Solids **357**(11–13), 2387 (2011)
36. M. Pospiech, M. Emons, A. Steinmann, G. Palmer, R. Osellame, N. Bellini, G. Cerullo, U. Morgner, Opt. Exp. **17**(5), 3555 (2009)
37. W. Yang, C. Corbari, P.G. Kazansky, K. Sakaguchi, I.C. Carvalho, Opt. Exp. **16**(20), 16215 (2008)
38. K. Davis, K. Miura, N. Sugimoto, K. Hirao, Opt. Lett. **21**(21), 1729 (1996)
39. D. Homoelle, S. Wielandy, A. Gaeta, N. Borrelli, C. Smith, Opt. Lett. **24**(18), 1311 (1999)
40. J.R. Liu, Z.Y. Zhang, S.D. Chang, C. Flueraru, C.P. Grover, Opt. Commun. **253**(4-6), 315 (2005)
41. G. Della Valle, R. Osellame, P. Laporta, J. Opt. A: Pure Appl. Opt. (1), 013001 (2009)
42. S. Ik-Bu, L. Man-Seop, C. Jeong-Yong, IEEE Photon. Technol. Lett. **17**(11), 2349 (2005)
43. S. Nolte, M. Will, J. Burghoff, A. Tuennermann, Appl. Phys. A (Mater. Sci. Process.) **A77**(1), 109 (2003)
44. A. Streltsov, N. Borrelli, Opt. Lett. **26**(1), 42 (2001)
45. W. Watanabe, T. Asano, K. Yamada, K. Itoh, J. Nishii, Opt. Lett. **28**(24), 2491 (2003)
46. K. Minoshima, A. Kowalevicz, E. Ippen, J. Fujimoto, Opt. Exp. **10**(15), 645 (2002)
47. A. Szameit, F. Dreisow, T. Pertsch, S. Nolte, A. Tnnermann, Opt. Exp. **15**(4), 1579 (2007)
48. G.D. Marshall, A. Politi, J.C.F. Matthews, P. Dekker, M. Ams, M.J. Withford, J.L. O'Brien, Opt. Exp. **17**(15), 12546 (2009)
49. L. Sansoni, F. Sciarrino, G. Vallone, P. Mataloni, A. Crespi, R. Ramponi, R. Osellame, Phys. Rev. Lett. **105**(20), 200503 (2010)
50. R. Osellame, V. Maselli, N. Chiodo, D. Polli, R.M. Vazquez, R. Ramponi, G. Cerullo, Electron. Lett. **41**(6), 315 (2005)
51. K. Suzuki, V. Sharma, J.G. Fujimoto, E.P. Ippen, Opt. Exp. **14**(6), 2335 (2006)
52. K. Imoto, H. Sano, M. Miyazaki, Appl. Opt. **26**(19), 4214 (1987)
53. F. Ladouceur, J.D. Love, *Silica-based Buried Channel Waveguides and Devices* (Chapman and Hall, London, 1996)
54. S.M. Eaton, W. Chen, H. Zhang, R. Iyer, M.L. Ng, S. Ho, J. Li, J.S. Aitchison, P.R. Herman, IEEE J. Lightwave Technol. **27**(9) (2009)
55. R.D. Feldman, E.E. Harstead, S. Jiang, T.H. Wood, M. Zirngibl, IEEE J. Lightwave Technol. **16**(9), 1546 (1998)
56. P.L. Auger, S. Iraj Najafi, Opt. Comm. **111**(1-2), 43 (1994)
57. I.P. Januar, A.R. Mickelson, Opt. Lett. **18**(6), 417 (1993)
58. D.B. Mortimore, Electron. Lett. **21**(17), 742 (1985)
59. C.R. Doerr, M. Cappuzzo, E. Chen, A. Wong-Foy, L. Gomez, A. Griffin, L. Buhl, IEEE Photon. Technol. Lett. **17**(6), 1211 (2005)
60. K. Jinguji, N. Takato, A. Sugita, M. Kawachi, Electron. Lett. **26**(17), 1326 (1990)
61. M. Olivero, M. Svalgaard, Opt. Exp. **13**(21), 8390 (2005)
62. W.J. Chen, S.M. Eaton, H. Zhang, P.R. Herman, Opt. Exp. **16**(15), 11470 (2008)

63. A. Takagi, K. Jinguji, M. Kawachi, IEEE J. Lightwave Technol. **10**(6), 735 (1992)
64. A. Crespi, Y. Gu, B. Ngamsom, H.J.W.M. Hoekstra, C. Dongre, M. Pollnau, R. Ramponi, H.H. van den Vlekkert, P. Watts, G. Cerullo, R. Osellame, Lab Chip **10**(9), 1167 (2010)
65. M.R. Poulsen, P.I. Borel, J. Fage-Pedersen, J. Hubner, M. Kristensen, J.H. Povlsen, K. Rottwitt, M. Svalgaard, W. Svendsen, Opt. Eng. **42**(10), 2821 (2003)
66. W. Watanabe, Y. Note, K. Itoh, Opt. Lett. **30**(21), 2888 (2005)
67. A.M. Kowalevicz, V. Sharma, E.P. Ippen, J.G. Fujimoto, K. Minoshima, Opt. Lett. **30**(9), 1060 (2005)
68. R.R. Thomson, H.T. Bookey, N.D. Psaila, A. Fender, S. Campbell, W.N. MacPherson, J.S. Barton, D.T. Reid, A.K. Kar, Opt. Exp. **15**(18), 11691 (2007)
69. Y. Nasu, M. Kohtoku, Y. Hibino, Y. Inoue, Jpn. J. Appl. Phys. **44**, L1446 (2005)

# Chapter 8
# Fibre Grating Inscription and Applications

Nemanja Jovanovic, Alex Fuerbach, Graham D. Marshall, Martin Ams, and Michael J. Withford

**Abstract** The diverse range of opportunities and activity in fibre sensing and fibre lasers has triggered an equally diverse range of research into new grating fabrication methods using femtosecond lasers. Femtosecond laser written fibre gratings can now exhibit similar spectral properties yet superior thermal stability compared to those produced using conventional methods. Foremost of the advantages offered by femtosecond laser fibre grating inscription is the reduced need for photosensitivity. Key examples include demonstrations of high temperature sensors based on sapphire fibres and high power fibre lasers with intra-core optical resonators. Research and development of femtosecond grating inscription methods, properties and devices is reviewed.

## 8.1 Introduction

Fibre gratings have been under continual investigation since 1978 when they were first discovered by Hill et al. [1], during investigations into the non-linear properties of germanium-doped silica fibres. In that seminal experiment, interference inside the fibre core resulted in a resonant periodic modulation of the core's refractive index and partial reflection of the transmitted light. Later studies showed that the index changes were generated by 2-photon excitation of an oxygen vacancy defect site [2]. The next major breakthrough in this field was the demonstration of a robust grating fabrication method using single photon absorption and inscription via transverse illumination of the fibre core with an interference pattern produced by a phase mask [3]. Related pursuits in this field include the development of fibre

N. Jovanovic · A. Fuerbach · G.D. Marshall · M. Ams · M.J. Withford (✉)
Centre for Ultrahigh bandwidth Devices for Optical Systems (CUDOS) and MQ Photonics Research Centre, Department of Physics and Astronomy, Macquarie University, New South Wales, 2109, Australia
e-mail: michael.withford@mq.edu.au

Bragg gratings (FBGs) using point-by-point (PbP) inscription [4] and long period gratings (LPGs) using amplitude masks [5].

UV laser inscribed FBGs played a key role as high fidelity, wavelength filters during the telecommunications boom of the mid to late 90s, when a significant proportion of the world's optical fibre networks were being installed. However, the relatively high fabrication costs of UV laser inscribed FBGs resulted in the emergence of competing technologies such as thin film filters replacing FBGs in many telecommunication applications, even before the end of the boom. Nevertheless, FBG development has still flourished in the last ten years, driven largely by opportunities arising from fibre sensors and fibre lasers. The diverse range of research activity and fibre types used in fibre sensor and fibre laser technology has triggered an equally diverse range of research into new grating fabrication methods that are not limited to conventional single mode optical fibres. The commercial availability of affordable femtosecond lasers has resulted in research aimed at developing new grating fabrication methods using these systems.

Research into femtosecond laser inscribed gratings has evolved in a different fashion to the work in UV laser grating inscription. In particular, for many years following the report of Kondo et al. [6] in 1999, research in this field centred around femtosecond laser inscribed long period gratings, typically fabricated using direct-write techniques. These gratings were reported to have superior resistance to thermal decay compared to conventional UV laser written gratings. The first studies into phase mask inscription methods using femtosecond lasers were undertaken by Mihailov et al. [7] and Dragomir et al. [8] in 2003. Studies into FBG fabrication using direct-write methods, namely PbP inscription, followed soon after in 2004 with consecutive reports by Martinez et al. [9] and Wikszak et al. [10] of femtosecond laser written Type II gratings.

FBGs written using femtosecond laser inscription methods offer many advantages over conventional UV laser written gratings. Foremost of these is the reduced need for photosensitive fibres. A case in point is the recent demonstration of FBGs in sapphire fibre [11], pure silica photonic crystal fibre [12] and double-clad Yb-doped fibre [13]. Femtosecond laser inscribed gratings can also induce relatively high index changes of $6 \times 10^{-3}$ [14] compared to $2 \times 10^{-3}$ for conventional UV photo-inscribed gratings. This field has also opened new avenues for writing through the polymer coating because the commonly used acrylates are transparent to 800 nm light.

This chapter opens with a review of the basic grating types, namely LPGs and FBGs, and their fundamental properties. The chapter includes discussion of both Type I and Type II gratings. Activity in PbP fabrication of Type II gratings is largely discussed in the context of device development. The following section reviews the significant body of research in phase mask inscription which has highlighted numerous similarities and differences with conventional UV laser inscription technology. The chapter concludes with an overview of FBG fabrication in novel fibres and their applications.

## 8.2 Review of Gratings Types

A fibre grating is generated whenever the effective refractive index of the core of an optical fibre is periodically modulated along the length of the fibre. Depending on the length of the period between modulations one can distinguish between fibre Bragg gratings (FBGs, periods of a few hundred nanometres to a few microns) and long period gratings (LPGs, periods of hundreds of microns to a few millimetres).

### 8.2.1 Long Period Gratings

LPGs couple light between different modes that have the same propagation direction [15]. The fundamental mode of a multimode fibre can, for example, be coupled into a certain higher-order mode or, more commonly, a core mode can be coupled into several forward-propagating cladding modes. Cladding modes are attenuated in the coated part of the fibre beyond the grating due to absorption and scattering, and this results in a strong and spectrally narrow resonant loss peak appearing in the transmission spectrum (Fig. 8.1). Although the light that is coupled into the cladding is lost, LPGs can serve as spectrally selective absorbers and are routinely found in telecommunication applications where they are used as band rejection or gain flattening filters (e.g. in erbium-doped fibre amplifiers). Because the resonance condition in LPGs is very sensitive to changes in temperature and mechanical deformation, they can also be used as sensors.

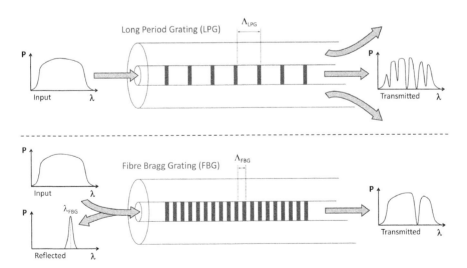

**Fig. 8.1** Schematic of a (*top*) long period grating and (*bottom*) fibre Bragg grating

Efficient coupling of the fundamental fibre mode into one of the cladding modes only happens if the wavelength of the light is equal to the resonant wavelength of the LPG, $\lambda_{LPG}$, given by

$$(n_{core} - n_{clad})\Lambda_{LPG} = \lambda_{LPG} \qquad (8.1)$$

where $n_{core}$ and $n_{clad}$ are effective refractive indices of the core and cladding modes respectively, and $\Lambda_{LPG}$ is the period of the grating. Due to the fact that LPGs couple co-propagating modes that have close propagation constants ($n_{core} \approx n_{clad}$), the period of this type of grating can considerably exceed the wavelength of radiation that is propagating inside the fibre. Thus, LPGs are comparatively simple to manufacture (e.g. by merely pressing the fibre against a plate with periodic grooves [16]) and it is thus unsurprising that the first femtosecond-laser inscribed fibre grating was indeed a LPG [6].

### 8.2.2 Fibre Bragg Gratings

The shorter periods present in fibre Bragg gratings (FBGs) change the coupling condition such that a forward-propagating core mode that is resonant with the grating is coupled into a backward-propagating mode [17, 18]. Since the reflected light can be guided within the core (Fig. 8.1) it is possible to make use of the reflected light and use these structures as mirrors or spectral filters. Indeed, fibre Bragg gratings have been used for just this purpose in both telecommunications systems for routing signals from one place to another and in fibre laser systems as highly reflecting cavity mirrors [19]. Efficient coupling of the forward-propagating core mode into the backward-propagating core mode occurs only at the Bragg wavelength ($\lambda_{FBG}$, the wavelength resonant with the grating structure) and is governed by the Bragg condition

$$\lambda_{FBG} = \frac{2 n_{core} \Lambda_{FBG}}{m} \qquad (8.2)$$

where $\Lambda_{FBG}$ is the period of the grating, $n_{core}$ the effective core index in the region of the grating and $m$ the order of the grating [17, 18].

FBGs are often written interferometrically into photosensitive fibres (fibres typically doped with germanium and loaded with hydrogen) using UV wavelengths [20]. This technique relies on an interference pattern of various orders of diffraction formed by the illumination of a uniform fused silica phase mask with a UV laser. By placing the core of an optically sensitive fibre within the interference pattern, it is possible to create a sinusoidally varying refractive index modulation along the length of the core. The peak linear reflectivity, $R$, (derived from coupled-mode theory) of a first order, sinusoidal refractive index profiled grating at the Bragg wavelength can be determined from

$$R = \tanh^2(\kappa l) \tag{8.3}$$

where $\kappa$ is the coupling coefficient of the grating and $l$ is the grating length [17, 18]. The coupling coefficient is given by

$$\kappa = \frac{\pi \Delta n M}{\lambda} \tag{8.4}$$

where $\Delta n$ is the magnitude of the index modulation and $M$ is the overlap integral between the forward-propagating mode and the index modification [17, 18].

Recently, femtosecond lasers operating in the near infrared (IR, 800 nm) have been exploited to create fibre Bragg gratings [7, 9, 10, 21]. By using a laser operating in the IR it is not possible to access the absorption bands of the glass directly, i.e. via a 1-photon process, but rather energy exchange takes place via multiphoton (5- or 6-photon) processes including absorption and photoionisation. The high peak power of a femtosecond laser make these processes highly efficient. As the absorption is highly non-linear, the volume of modification is not only constrained by the beam waist but also by pulse energy and the amount of which this energy is above the threshold for absorption. In this way it is also possible to obtain a greater spatial confinement of the modifications created using a femtosecond laser.

Three techniques for inscribing gratings in a fibre core using a femtosecond laser have been demonstrated; namely the phase mask (PM) [7] technique, phase mask scanning (PMS) [21] technique and the PbP [9, 10] technique. Gratings written with a phase mask have been shown to have similar characteristics to those written with UV radiation and will be discussed further in Sect. 8.4.

By translating a fibre with constant velocity through the focus of a pulsed femtosecond laser, a periodic refractive index modulation or grating can be created in a PbP fashion. In contrast to gratings fabricated using a phase mask, PbP gratings have refractive index profiles more closely approximated by a square rather than a sinusoidal wave and generally do not extend across the entire core causing increased coupling to cladding modes [9]. Low-order gratings of this sort (gratings which have a period approaching the wavelength of the writing beam) have only become feasible due to the threshold effect outlined above. However, refraction of the inscription beam becomes the major barrier to tight focal confinement and hence strong grating formation. Schemes for rectifying this issue as well as a detailed history and results of PbP FBGs written with femtosecond lasers can be found in Sect. 8.3.

Before discussing in detail the various grating fabrication methods it is important to understand the mechanisms or processes which drive index changes when femtosecond lasers are focussed into transparent materials. The two main mechanisms which underpin energy exchange between the laser beam and the medium are multiphoton absorption and ionisation. For *low pulse energies* both occur and while the former leads to colour centre formation, or the breaking of Si–O bonds in silica based hosts, the latter leads to the melting of the glass and rapid resolidification [22, 23]. This rapid cooling process causes densification and can cause an increase in the local fictive temperature and induced stress-fields [22, 23]. It is widely accepted that a combination of colour centres, densification and induced stress-fields give rise to the index change. Modifications created via these mechanisms are called Type I-IR, where IR denotes the wavelength used to induce the index change and to distinguish from Type I-UV modifications where no melting of the glass occurs [24]. Note that the ratio of multiphoton absorption to photoionisation varies depending on the glass dopants and whether the fibre is pre-sensitised with hydrogen or not. For *high pulse energies* multiphoton and avalanche ionisation lead to plasma formation and once enough energy has been deposited, a Coulombic micro-explosion occurs pushing the material outwards [25]. When the plasma hits the cold outer material, it rapidly cools and resolidifies leaving a void surrounded by a highly densified shell similar in properties to the Type I-IR modification. If a void is formed, the modification is referred to as Type II-IR [24]. These types of modifications occur not only for PbP written gratings but also for all laser written gratings including those written with a phase mask. Although these two pulse energy regimes were first observed in gratings written with a femtosecond laser and a phase-mask, Marshall et al. also demonstrated this pulse energy dependence in PbP gratings by using the ferrule based experimental setup (discussed in Sect. 8.3.1) which offered the appropriate amount of reproducibility [26].

## 8.3 Point-by-Point Inscribed Gratings

### *8.3.1 Fabrication Methods*

The PbP grating writing technique involves writing one index modulation period of a grating at a time, hence the name. This can be done either with high irradiances where only a single pulse is required to create each modification or at moderate irradiances where the overlapping of numerous pulses (up to thousands of pulses) takes place. The technique was first demonstrated by Malo et al. in 1993 where they used single pulses from an excimer laser operating in the UV to inscribe the refractive index modulations of a grating in photosensitive fibre [4]. However the first demonstration of a PbP grating written with a femtosecond laser was by Kondo et al. in 1999, when they focussed a laser with a 120 fs pulse duration, operating at 800 nm and a repetition rate of 200 kHz into the core of a standard germanium-doped single mode fibre to create a LPG [6]. In this case a low numerical aperture (NA) objective, $20 \times 0.46$ NA, produced an estimated 2 $\mu$m spot size in the core of the fibre. Pulses with energies of 750 nJ were focussed at each spot for 10 s at a time. Much like in the experiment of Malo et al., the fibre was suspended between two fixed points and translated by a three axis translation stage which kept the core in focus along the length of the grating. This experimental setup, although simple in principle, is limited in many ways. For example, the grooves of the clamps suspending the fibre must be well aligned with one another in order to minimise bending in the fibre, which would result in the laser focus departing

# 8 Fibre Grating Inscription and Applications

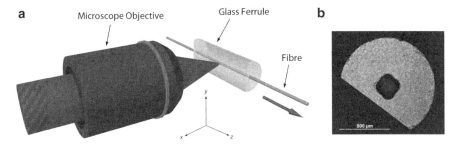

**Fig. 8.2** (**a**) PbP grating fabrication incorporating a fixed glass ferrule and translating fibre configuration. (**b**) Micrograph of a D-shaped glass ferrule

from the core of the fibre during translation. This could be remedied by translating using a catenary function to account for the bend between the start and finish points; however, this introduces an unnecessary complication. The other major issue with this setup is drift in the stages and plates used to support the fibre over time due to environmental changes. These can cause the fibre to drift by only a few microns over a minute, from the focus of the laser, which is enough to make the alignment and grating writing process difficult and irreproducible.

A more elegant solution [27] is to draw the fibre through the focus of an objective, fixed relative to a glass ferrule which is transparent to the femtosecond radiation (Fig. 8.2). The fibre is immersed in index matching oil as it is threaded through the ferrule so that no refraction occurs passing through the ferrule to the fibre. In this case the ferrule, and hence the core of the fibre, remain fixed with respect to the objective while the translation stage pulls the fibre through the ferrule during the writing process. The system is inherently more stable than the former one and results in a high degree of reproducibility [26].

A limitation on the spot size achievable in the core of the optical fibre is imposed by both the objective used as well as the degree of refraction of the femtosecond pulse from the circular shaped cladding of the fibre or the ferrule [9]. Indeed, large spot sizes from weak focussing through a circular cladding limited Malo et al. [4] to 2nd and 3rd order gratings. However, this effect can be overcome or at least minimised by employing high NA oil immersion objectives or in the case of the ferrule based setup, polishing the front face of the ferrule flat and parallel to the output aperture of the objective lens[1] [28]. This technique was employed by Lai et al. to achieve high quality 1st order FBGs in the C-band [29].

One of the key advantages of the PbP technique is the fact that the operator has complete control over the target location of the periods of the grating, this is because PbP gratings rely on one laser pulse to create one period of a grating. For a constant repetition rate, $f_r$, the period, $\Lambda_{FBG}$, of the grating depends on the

---

[1]Shorter working distance microscope objectives can also be used as a result of a ferrule polished close to its centre.

translation velocity, $v_{\text{trans}}$, in the following way

$$\Lambda_{\text{FBG}} = \frac{v_{\text{trans}}}{f_{\text{r}}}, \tag{8.5}$$

which means that the Bragg condition (8.2) can be amended to

$$\lambda_{\text{FBG}} = \frac{2n_{\text{core}}v_{\text{trans}}}{mf_{\text{r}}}. \tag{8.6}$$

Hence by changing the translation velocity it is possible to inscribe gratings at any arbitrary wavelength which is not possible using other methods. By varying the velocity of the stage during the grating writing process, chirped gratings can also be realised. Furthermore, careful positional control of modifications of a PbP FBG can result in many different types of gratings being created including, but not limited to, super-structure, apodised and phase shifted gratings. These types of gratings will be discussed in Sect. 8.3.3.

### 8.3.2 Development of Femtosecond Laser Direct-Write LPGs

As already mentioned, Kondo et al. [6] were the first to use a femtosecond laser to inscribe a periodic grating structure directly into the core of an optical fibre. A 30 mm long LPG with a grating period of 460 μm was produced that exhibited a series of transmission dips in the wavelength range between 1.3 and 1.7 μm. The peak transmission loss of the grating (which was most likely a Type II grating) was a relatively modest 8 dB whereas the off resonance loss, approximately 4 dB, was much higher than conventional UV laser inscribed LPGs (typically 0.2 dB).

Type I PbP written LPGs have been fabricated in $H_2$-loaded and Ge-doped fibres using unamplified frequency-doubled (400 nm) Ti:Sapphire laser pulses [30] (16 dB loss peak) and frequency-quadrupled (264 nm) pulses from an Nd:glass femtosecond laser [31] (28 dB loss peak). Hindle et al. later demonstrated fabrication of LPGs with near-IR (800 nm) femtosecond laser pulses [32]. Keeping the intensity below the actual damage threshold of the glass, they could inscribe a 16 dB strong grating exhibiting only 0.3 dB excess losses in a germanium-doped fibre. In addition, they also demonstrated the first (albeit weak) laser-written LPG in a pure silica fibre. In the same work a study of alignment issues was conducted and the grating spectra were found to be extremely sensitive even to small changes in alignment between the centre of the fibre core and the focus of the writing beam.

In the above-mentioned cases the incident laser fluence was kept orders of magnitude below that used by Kondo et al. (hundreds of J/cm$^2$ vs. 0.5 MJ/cm$^2$). A systematic study by Kryukov et al. [33] showed that the refractive index change induced by high-energy femtosecond laser pulses is the result of non-photochemical mechanisms (local melting, densification and void formation) whereas low-energy femtosecond laser pulses induce photochemical changes via multiphoton processes with an efficiency depending on the wavelength of the laser [33]. Based on these findings, high-quality gratings with resonant losses of up to 30 dB have recently been fabricated using the third [34] and the fifth [35] harmonic of a femtosecond Nd:glass laser (352 nm and 211 nm, respectively).

### 8.3.3 Development of Femtosecond Laser Direct-Write FBGs

The first demonstrations of PbP FBGs inscribed with a femtosecond laser were reported almost simultaneously by Martinez et al. [9] and Wikszak et al. [10]. Martinez et al. showed that by using a high NA 100× objective, 1st, 2nd and 3rd order Type II-IR gratings could successfully be fabricated. However, the 1st order resonances were unexpectedly weaker than the 2nd order resonances. This was due to defocussing of the laser pulse by the circular shaped cladding of the optical fibre causing subsequent grating periods to overlap. They later remedied this by employing the ferrule based setup as described in Sect. 8.3.1 and managed to write strong 1st order FBGs [29]. In the initial publication by Martinez et al., the authors demonstrated that the gratings coupled strongly to cladding modes due to the highly localised, strong index modifications that were non-uniform across the core of the fibre. The gratings also exhibited a strong birefringence of the order of $3.6 \times 10^{-5}$ which will be discussed further in Sect. 8.5.2. Using the birefringence of Type II-IR gratings, Jovanovic et al. were able to determine that the effective indices of the two orthogonal polarised modes were lower than that of the pristine fibre and hence indirectly confirm the presence of a micro-void at the core of the index modifications [28]. Martinez et al. also demonstrated that these Type II-IR gratings show high thermal stability which opens up new opportunities for their application [36]. Thermal stability issues will be elucidated further in Sect. 8.5.1.

It was not long before the ability to inscribe gratings into non-photosensitive active fibres was exploited. A PbP grating was inscribed by Lai et al. into an Er:Yb-doped phosphosilicate laser fibre and a distributed Bragg reflector (DBR) fibre laser was realised [37]. This compact source was highly thermally stable and operated off a fully linearly polarised single-longitudinal-mode, but more importantly, demonstrated for the first time one of the key advantages of femtosecond inscribed FBGs, namely the ability to fabricate fibre lasers with an intra-core optical resonator.

Marshall et al. took advantage of the ability to control the position of each modification in order to realise both amplitude and phase sampled gratings as well as chirped gratings [38]. Figure 8.3 shows a grating spectrum for a phase modulated super-structure grating fabricated in SMF-28 fibre using PbP inscription. The 20-mm long grating consisted of a square wave modulation of the phase by $\pi/2$ shifts which completely suppressed the Bragg resonance peak. A microscope image of a single phase shift is also shown.

Jovanovic et al. exploited the high thermal stability and direct-write capability of these gratings to realise high power fibre lasers operating >100 W with minimal (5%) degradation to the performance of the gratings over time [13, 39]. The highly narrow linewidth of these PbP gratings resulted in highly narrow linewidth laser

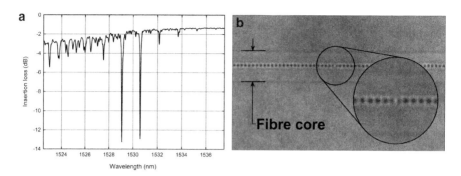

**Fig. 8.3** (a) Graph showing the spectral characteristics of a phase-modulated sampled grating fabricated in SMF-28 fibre using PbP inscription [38]. (b) A micrograph of one of the $\pi/2$ phase modulations of the sampled grating

sources which were applied to non-linear frequency doubling experiments outlined in Sect. 8.6.1.

Another advantage of the technique demonstrated in the work by Jovanovic et al. was the ability to inscribe gratings into non-standard, non-circular fibres. Indeed the fibre used in the high power laser experiments was a 300 μm diameter hexagonal shaped double-clad laser fibre. Further to this, Jovanovic et al. also demonstrated the successful inscription of PbP gratings in polarisation maintaining (PM) fibres [40].

## 8.4 Phase Mask Inscribed Gratings

### 8.4.1 Fabrication Method

A diverse range of beam splitting interferometric configurations are used routinely, with nanosecond laser systems, to produce the desired modulated intensity pattern needed to photoinscribe a grating into the core of an optical fibre. However, the relatively short coherence length of femtosecond lasers prescribes the use of phase mask writing geometries, which place the fibre in close proximity to the beam splitting element and minimise path length mismatches. A typical phase mask writing configuration is shown in Fig. 8.4. A cylindrical lens is used to produce a line focus within the fibre's core. The phase mask consists of a surface relief grating, which splits the incoming laser beam into numerous diffractive orders. The majority of the diffracted energy is present in the $+1$ and $-1$ orders.[2] The fibre is placed close enough to the phase mask, typically within a few millimetres, so that it

---

[2]Phase masks are often designed to be zero order nulled in order to minimise a DC index contribution.

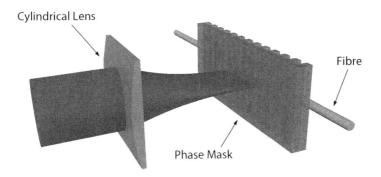

**Fig. 8.4** Phase mask writing technique

is illuminated by the interference pattern produced by the +1 and −1 orders, yet far enough away so that higher order contributions (i.e. Talbot fringes) are minimised.

Phase mask grating writing places limitations on the range of incident pulse energies available for grating inscription. This is particularly true of femtosecond laser grating writing. For example, at high pulse energies it is possible to exceed the damage threshold of the phase mask due to its close proximity to the intense line focus produced by the cylindrical lens. One method for extending the range of pulse energies is to use double phase mask interferometers which can relax the requirement to have the fibre in close proximity to the phase mask [41].

## 8.4.2 Development of Femtosecond Laser-Phase Mask Inscription

The first demonstration of gratings inscribed using a femtosecond laser and phase mask was by Mihailov et al. in 2003 [7,42]. In this case IR radiation at 800 nm was used to photoinscribe FBGs in Ge-doped single mode fibre via a highly multiphoton absorption process. The peak transmission loss of the resulting FBGs was 45 dB and optical microscopy showed that the photoinduced index change extended across the entire fibre core[3] (Fig. 8.5). Further, the FBGs were reported to be stable over a period of 2 weeks during continual exposure to an elevated temperature of 200°C. The authors summarised these observations with the statement that the "gratings have spectral quality similar to Type I, UV-induced FBGs but with the ultrastable temperature characteristics of Type II gratings". In the same year Dragomir et al. [8] reported FBG inscription using an ultraviolet (264 nm) femtosecond laser and phase

---

[3]It is important to note that cladding index modification can also occur under different writing conditions.

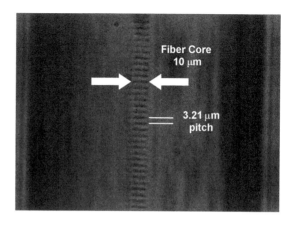

Fig. 8.5 Microscope image showing the photoinduced index changes produced in Ge-doped single mode fibre using an infrared femtosecond laser and the phase mask technique [7]

mask writing techniques. In this case pre-sensitised, $H_2$ loaded Ge-doped single mode fibres were used. Once again relatively strong grating reflectivities were observed, with peak band rejections of up to 30 dB. A study of the dependence of the refractive index modulation on the intensity of the laser implied that UV femtosecond laser grating inscription was driven by a 2-photon absorption mechanism. The non-linear nature of femtosecond laser photo-inscription results in a deviation from the standard sinusoidal index modulation usually induced in the fibre core using single photon inscription mechanisms. In particular, the index modulation has a reduced DC index contribution and complex spatial structure in the case of femtosecond laser written gratings, which in turn influences the Bragg wavelength and increases the grating bandwidth [43].

The groundbreaking studies by Mihailov et al. and Dragomir et al. were rapidly followed by a number of reports comparing and contrasting both femtosecond laser FBG inscription (IR and UV wavelengths) with conventional nanosecond and continuous wave (CW) laser grating writing (UV only), and fibres with different core dopants against those with pure silica cores. For example, Mihailov et al. [44] showed that IR femtosecond laser grating inscription could produce gratings in all silica core fibre that were comparable in strength to those written in Ge-doped fibres. Similarly, Zagorulko et al. [45] showed that gratings of a similar reflectivity could be written in a $P_2O_5$-doped core, Ge-doped core and all silica core fibre types using a 267 nm femtosecond laser and a phase mask. They also directly compared UV femtosecond laser written gratings with UV nanosecond laser written ones. In general, the fibres tested showed a higher susceptibility to index changes when irradiated with femtosecond laser UV pulses compared to their nanosecond laser counterparts. In addition, the femtosecond laser inscribed gratings were found to have similar spectral and annealing properties to conventional Type I-UV gratings. This is in contrast to previous reports of IR femtosecond laser written gratings [7,44] that exhibit Type II annealing characteristics. However, this discrepancy can be explained by the influence of the laser writing intensity on the type of material change photoinduced in the fibre core. In particular, as mentioned before, relatively low laser writing intensities will result in the formation of a Type I grating via colour centre mechanisms whereas high laser writing intensities (typically above the threshold for supercontinuum generation within the fibre) result in Type II gratings associated with void formation and peripheral compaction within the core [24].

Although the inherent advantage of femtosecond laser grating inscription is the reduced requirement for a photosensitised fibre core, a comparative study by Li et al. [46] of $H_2$-free and $H_2$-loaded fibres showed that the grating strength of a femtosecond laser written grating could be enhanced by the use of a $H_2$-loaded Ge-doped core fibre. This effect was attributed to $H_2$ pre-sensitisation restricting the index modification to the core. However, $H_2$ loading was also found to reduce the thermal stability of both Type I and Type II FBGs written in the Ge-doped core fibre compared to those produced under $H_2$-free conditions. This was attributed to the higher concentration of colour centres formed in the $H_2$ loading case.

Several studies investigating novel forms of femtosecond laser grating inscription with phase masks have also been reported. Fu et al. [12] showed that this writing method could be used to produce FBGs in all silica photonic crystal fibres. In this case the rotational alignment of the fibre with respect to the focussed laser beam was found to be critical if the incident radiation was to circumnavigate the internal cladding microstructure and reach the core [47]. Whereas all the previous reports had used static beam delivery configurations resulting in grating lengths of only a few millimetres, Thomas et al. [21] showed that long gratings, up to 40 mm, could be written using phase mask scanning methods. Furthermore, the phase mask scanning method was used to produce a 20-mm long continuously chirped FBG, with a bandwidth of 6 nm, using femtosecond laser photoinscription [48].

## 8.5 Properties of Femtosecond Laser Written Gratings

### 8.5.1 Thermal Stability

The thermal stability of gratings written with femtosecond lasers is highly dependent on the fluence used and therefore the type of grating formed, i.e. Type I or Type II. The thermal characteristics of both Type I-IR and Type II-IR gratings written via the PbP technique have been characterised up to 1050°C and are summarised in Figs. 8.6 [36] and 8.7 [26].

Figure 8.6a highlights the evolution of the reflectivity of a Type II-IR PbP grating at various annealing temperatures for 20 h at each step. It can be seen that there is a negligible change in the grating reflectivity for temperatures as high as 900°C and at temperatures of 1000°C the grating reflectivity, after initially decaying, stabilises to a 25% lower value. As outlined by Chan et al. [23] this is a result of modifications inscribed with femtosecond lasers possessing a higher fictive temperature than the surrounding glass, resulting in the gratings being less susceptible to annealing at high temperatures. Upon further annealing the gratings can be seen to decay rapidly

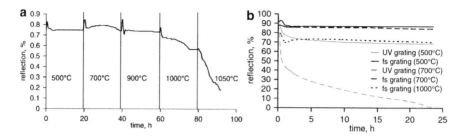

**Fig. 8.6** (**a**) Evolution of reflectivity of a single PbP FBG after annealing at various temperatures for 20 h. (**b**) Evolution of grating reflectivity for a single UV-ns phase mask written and single IR-fs PbP written FBG over 25 h for various temperatures [36]

which is due to the annealing temperature approaching the softening point of the glass and hence the onset of major restructuring of the glass matrix.

Figure 8.6b compares the annealing properties of a Type II-IR PbP grating written with a femtosecond laser and a grating written with a UV nanosecond laser and a phase mask. The 2 gratings were taken to successively higher temperatures for 25 h periods. It can be seen that the Type II-IR femtosecond laser written grating is considerably more stable than the UV grating which is completely erased at 700°C. The superior thermal stability of the Type II-IR femtosecond laser written grating over the UV written grating creates opportunities for these gratings to be applied in high power fibre lasers and high temperature sensors, both of which will be discussed in Sect. 8.6.

In comparison Marshall et al. [26] annealed gratings written with pulse energies corresponding to both the Type I-IR (110 nJ) and Type II-IR (140–160 nJ) regimes and the results are displayed in Fig. 8.7. It can be seen that the annealing results for the Type II-IR gratings are consistent with Martinez et al. in that at 1000°C the gratings remain but have a reduced grating depth. However, the Type I-IR gratings were erased at similar temperatures (700°C) as UV-written gratings. This is understood to be a result of colour centres being annealed out and the reconnection of bonds in the glass matrix.

Interestingly, it can be seen in Fig. 8.7 that there is a small range of pulse energies between the Type I-IR and Type II-IR regimes, 120 nJ and 130 nJ respectively, where the grating depth increases as a function of temperature. This is attributed to a strong, thermally resistant, Type II-IR central refractive index modification that is surrounded by a weak Type I-IR modification. As the Type I-IR region anneals, the grating contrast and hence the grating depth increases.

Results for phase mask written ultrafast gratings are shown in Fig. 8.8. It can be seen that Type I-IR gratings written with thousands of pulses considerably degrade in performance upon heating to 1000°C while Type II-IR gratings written with 20 or so pulses, which have observable damage when viewed under a microscope, are thermally stable at temperatures of 1000°C [24, 49].

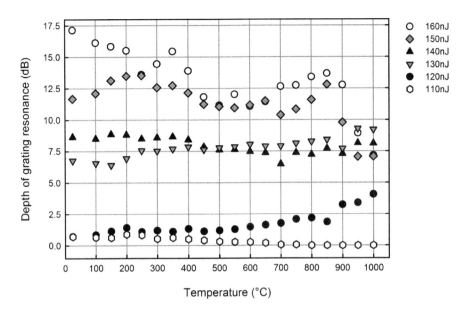

**Fig. 8.7** Graph showing the evolution of the grating depth resonance for PbP gratings written with various pulse energies as a function of temperature [26]

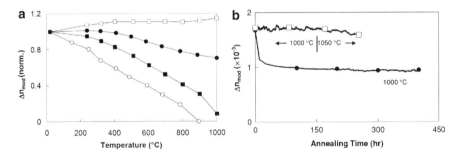

**Fig. 8.8** (a) Evolution of the refractive index contrast of Type I-IR-fs (*black squares*), Type II-IR-fs (*white squares*), Type II-IR-ps (*black circles*) and Type I-UV-ns (*white circles*) written gratings [24]. (b) Evolution of the refractive index contrast of Type II-IR-fs (*white squares*) and Type II-IR-ps (*black circles*) written gratings as a function of time at constant temperatures [49]

It can be seen from Fig. 8.8 that unlike Type II-IR PbP FBGs which were observed to decay rapidly at 1050°C, gratings written using phase masks have minimal reduction in the index contrast at these temperatures. This could be a result of the modified regions of the two types of gratings having different fictive temperatures causing them to anneal differently [23]. In fact, Smelser et al. [50] showed that by exposing a Type I-IR grating for a period twenty times longer than that experienced by the grating shown in Fig. 8.8, it is possible to maintain 60% of the refractive index contrast after isochronal annealing at 1000°C, which again is understood to be due to a greater fictive temperature in the grating.

Grobnic et al. [11] also demonstrated ultrahigh temperature stability of Type II-IR FBGs written into a sapphire fibre (a fibre which has a higher melting point then silica) with a phase mask. These gratings were tested up to 2000°C and will be discussed further in Sect. 8.6.2. The ability to process transparent materials with higher melting points extends the range of ultrafast written gratings beyond that of conventional UV-ns laser inscription and therefore opens up new applications for these types of gratings.

## 8.5.2 Stress and Birefringence

The stress and birefringence induced during grating writing are strongly dependent on the focussing geometry used and the pulse energy or type of modification created. The axial stress profile for a Type I-IR LPG modification by a femtosecond pulse in SMF-28 (germanosilicate) fibre is mapped tomographically in Fig. 8.9a [51]. It was determined that an asymmetric increase in the axial core stress of up to 5.2 kg/mm$^2$ was produced as a result of the LPG inscription process. This increase in stress led to a change in the $x$ and $y$ components of the refractive index profiles for the core (Fig. 8.9b) constituting the index contrast of the grating. Since the magnitudes of these changes were different, a birefringence was formed. The similarities seen by Fonjallaz et al. [52] between the change in the glass properties in this case and the case for a UV-ns irradiated SMF-28 fibre, led to the conclusion that the underlying mechanisms may be similar. Unfortunately this experiment has not been repeated for Type II-IR modifications, which as outlined in the previous section have stronger structural changes and would therefore have greater stress-fields associated with them.

Limberger et al. also showed that Type I gratings written with a UV (264 nm) femtosecond laser and a phase mask into H$_2$-loaded SMF-28 fibre, via a 2-photon

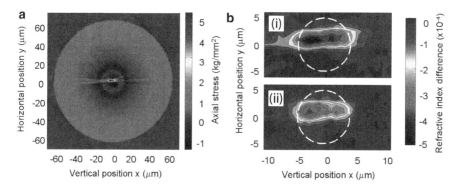

**Fig. 8.9** (a) Axial stress profile of a femtosecond modification in SMF-28 fibre. (b) Stress-induced change for the x (i) and y (ii) component of the refractive index contrast. The dotted circles indicate the core of the fibre [51]

absorption process, did not induce any stress to the core or the cladding while gratings written into non-$H_2$-loaded SMF-28 fibre did [22]. They proposed that the mechanism for index change in the non-$H_2$-loaded case was compaction, stress and colour centre formation while for $H_2$-loaded gratings the only mechanism left, as there was no stress observed, was colour centre formation.

The birefringence induced upon inscription of a Type I-IR FBG using the phase mask technique has been demonstrated to be low for SMF-28 fibres that are loaded with $H_2$ ($\sim 10^{-6}$, similar to that of a pristine fibre) and can be annealed at room temperature [53]. This is consistent with the conclusion of Limberger et al. that the refractive index modification of a Type I femtosecond grating written into a $H_2$-loaded fibre is due to colour centre formation, which would not affect the birefringence of the grating [22]. However non-$H_2$-loaded SMF-28 fibres exhibit a greater birefringence ($\sim 10^{-5}$) that strongly depends on the polarisation of the writing beam and which can only be annealed out with the grating itself. This is also in agreement with Limberger et al. in that the index modulation of a Type I femtosecond grating written into a non-$H_2$-loaded fibre is not only due to colour centre formation but also compaction and stress which induce a birefringence [22]. In addition, Type II-IR gratings inscribed into $H_2$-loaded SMF-28 fibre with either a linear writing polarisation aligned parallel or perpendicular to the fibre axis and those inscribed into non-$H_2$-loaded fibre with the S-polarisation (light polarised normal the fibre axis) have shown birefringent values of $\sim 10^{-4}$, while gratings written into non-$H_2$-loaded fibre with the P-polarisation (light polarised parallel to the fibre axis) have birefringent values of $\sim 10^{-5}$ [53].

It has been shown that even higher levels of birefringence (up to $4 \times 10^{-4}$) can be achieved by inscribing a Type I grating into the core of SMF-28 optical fibre, using the phase mask method, and by subsequently post-inscribing a Type II grating into the cladding close to the core [54]. It is important that a modulated exposure is used for the grating in the cladding as uniform exposures did not work, and it must be placed far enough from the core so as to not interfere with the core mode. Further increases in birefringence to $\sim 8 \times 10^{-4}$ were observed when a second cladding grating was placed on the opposite side of the core and the writing parameters were optimised. The mechanism causing the birefringence in this case is understood to be due to an asymmetric stress-field similar to the one created by the Stress Applying Parts (SAPs) in polarisation maintaining (PM) fibres. The technique was also demonstrated in a pure silica core fibre and a low cutoff wavelength fibre with similar values of birefringence being produced. This approach presents an opportunity for the creation of highly birefringent gratings that can be used in fibre laser systems. In fact, the birefringence levels achieved by this technique are a factor of 2 greater than those observed in standard PM fibres.

The birefringence values induced upon inscription of a Type II-IR FBG via the PbP technique have been observed to range from $10^{-5}$ to $10^{-4}$ which is consistent with observations of the phase mask written gratings outlined above [28, 29]. It has been determined that the majority of the birefringence in PbP FBGs occurs as a result of the elliptical morphology of the refractive index modifications which is a result of the focussing geometry used during fabrication. The elliptical shape and different boundary conditions for the two orthogonal states of linear polarisation between the densified shell and the micro-void, which is presumed to have an index of 1, generate different effective indices for the two orthogonal states of linear polarisation, manifesting in birefringence [28]. The elliptical shape of the modifications means that there is also a difference in coupling coefficients for the two orthogonal states of linear polarisation and therefore the grating exhibits different gratings strengths for orthogonal polarisations. For these gratings, the magnitude of the birefringence could be explained excluding stress-induced birefringence although the authors believe it to be still present. No measurements of the birefringence of Type I-IR gratings written via the PbP technique have been reported.

Finally, the mechanisms that cause birefringence in both femtosecond written Type I and Type II gratings share some similarities with those written by nanosecond sources. However, there are mechanisms which differ between the two regimes with Caucheteur et al. showing that Type I gratings written with femtosecond lasers operating at 264 nm in the UV, induce significantly larger polarisation dependant losses and hence have greater birefringences than those written with a 244 nm nanosecond laser system even though the absorption is a 1- or 2-photon process for both wavelengths [55].

### 8.5.3 Photoattenuation

Photoattenuation refers to the attenuation of light as it propagates through a medium as a result of exposing the medium to another light source, or in this case a femtosecond laser. This was first observed for femtosecond written gratings by Kondo et al. when they inscribed a LPG using the PbP technique [6]. The refractive index modifications were created by exposing each point for 10 s at a time with weak focussing and high pulse energy which resulted in Type II modifications. The spectrum of the LPG can be seen in Fig. 8.10a showing an out-of-band insertion loss of the order of 2 dB at 1750 nm increasing towards the shorter wavelengths with a maximum value of 8 dB at 700 nm. This example of photoattenuation has been explained by scattering from the strong, highly localised refractive index modulations as well as stronger coupling to the cladding due to a non-uniform index profile across the core. Strong photoattenuation such as this would limit the application of such a LPG as a gain flattening filter in telecommunication applications where the losses must be kept low, but for suppressing lasing in fibre lasers, where there is a high gain, the losses can be tolerable. This technique was used by Himei et al. [56] to demonstrate a fixed optical attenuator where the attenuation was a function of the number of modifications inscribed in the fibre core (Fig. 8.10b).

Mihailov et al. showed that the losses associated with Type I-IR gratings written with a phase mask could be reduced to values of 0.3–0.57 dB for grating strengths

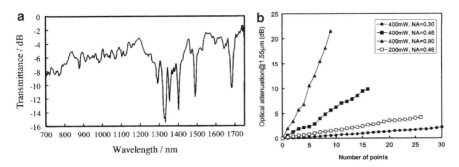

**Fig. 8.10** (a) Transmission spectrum of a Type I-IR PbP LPG [6]. (b) Optical attenuation as a function of the number of refractive index modifications inscribed into the core of a fibre [56]

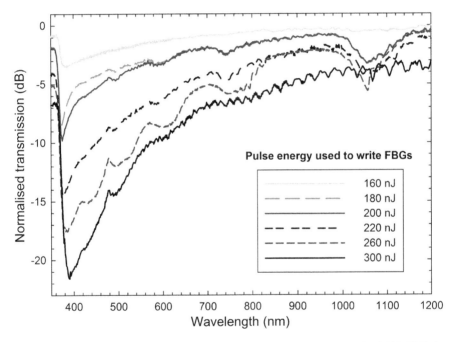

**Fig. 8.11** Normalised transmission spectra for Type II-IR femtosecond inscribed PbP FBGs in SMF-28 [57]

between 15 dB and 45 dB respectively; however, this is still ~2–3 times greater than the typical 0.1–0.2 dB observed for UV-written gratings [7].

By moving to the Type II-IR regime, gratings are not only stronger but the out of band insertion losses increase as well, with typical values of the order of 1.5 dB for a 25 dB deep grating in the C-band [9]. Aslund et al. [57] have shown that the optical losses associated with Type II-IR PbP FBGs increase in magnitude towards the shorter wavelengths (Fig. 8.11) but with a much greater magnitude relative to Type I-IR gratings. Also the magnitude of the out-of-band insertion losses can be seen to be a strong function of the pulse energy of the laser used to write the gratings. The increasing pulse energy is understood to increase both the physical size and the index contrast of the laser-induced modifications, which in turn increases the amount of insertion loss. In this case, the losses are attributed to diffractive Mie scattering from the high index contrast densified regions surrounding the microvoids. By thermally annealing the gratings at 400°C no change was observed from the gratings, which once again indicates that colour centres do not play a significant role in the index contrast of Type II-IR gratings. The strong scattering is a result of the index modifications being highly localised and non-uniform across the core of the fibre [57]. The large insertion losses for Type II-IR gratings rule them out for use in telecommunications systems where losses should be kept to a minimum, but are tolerable in fibre laser systems and also fibre sensors as these FBGs offer other advantages such as great thermal stability.

## 8.6 Applications

### 8.6.1 Fibre Lasers

Gratings written via multiphoton techniques, directly into transparent materials such as glasses, are attractive for use in applications such as high reflectors and output couplers in fibre laser systems as they can potentially simplify these architectures. They also have many interesting properties such as thermal robustness and birefringence which can be exploited. The first demonstration of a directly written grating with a femtosecond laser applied to a fibre laser was by Lai et al. in 2006, who realised a distributed Bragg reflector fibre laser that was 31 mm long in a Er:Yb-codoped phosphosilicate glass fibre [37]. In this laser two 8 mm long Type II-IR gratings written via the PbP technique were spaced by 15 mm. The laser operated off a single-longitudinal-mode, demonstrated an output power of 45 $\mu$W, and was very stable over the 12 h period of test in both wavelength and output power. Most interestingly, the fibre laser was fully linearly polarised with an extinction ratio of 40 dB. This was a result of the different grating strengths for the two orthogonal polarisations which has been shown to be a result of the ellipticity of refractive index modifications for PbP gratings [28].

This was shortly followed by Wikszak et al. who demonstrated power scaling to 38 mW with a modest slope efficiency of 21%. They achieved this by inscribing a highly reflecting Type I-IR grating via the phase mask scanning technique, into a 0.85 m long $Er^{3+}$-doped fibre laser [58]. Jovanovic et al. continued the power scaling of this type of fibre laser to 5 W, in an $Yb^{3+}$-doped double-clad fibre laser [39]. The cavity contained a highly reflecting Type II FBG inscribed directly into one end of the 20 m long double-clad fibre via the PbP technique, whereas a Fresnel reflection from the other end of the fibre was used as an output coupler. This laser exhibited a highly narrow linewidth of the order of 15 pm and was extremely

# 8 Fibre Grating Inscription and Applications

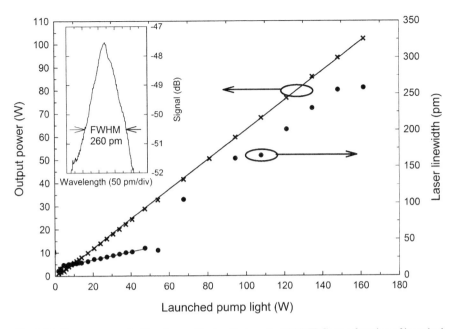

**Fig. 8.12** Output power of a fibre laser with directly inscribed PbP FBGs as a function of launched pump power [13]

stable (<2 pm wavelength fluctuations and 0.3% variations in output power) and showed no signs of power saturation. This was quickly followed up by the same group, with further power scaling of a similar fibre laser to >100 W, as shown in Fig. 8.12 [13]. In this case, the laser had a near Stokes limited efficiency of 64%. At 50 W output power a degradation in the lasers output of the order of 5% was observed after 8 h of operation, beyond which the laser was stable [59]. This is believed to be due to photo-annealing rather than thermal-annealing of the grating in the laser as the temperatures in the core were not sufficiently high to cause grating decay. The laser exhibited a narrow linewidth at all output powers with a maximum linewidth at 103 W of 260 pm. The high output power and narrow linewidth made the laser attractive for non-linear frequency doubling and resulted in the generation of 2.1 W of green light (from 20 W of IR light) using a MgO-doped PPLN crystal, which was limited by the laser's linewidth.

So far, all the fibre lasers outlined above were unpolarised. Linearly polarised fibre lasers are also of major interest for many applications and the first medium power demonstration came from Wikszak et al. [60]. They demonstrated that by inscribing a highly reflecting Type I-IR grating via the PMS technique into a $Yb^{3+}$-doped polarisation maintaining (PM) fibre, it was possible to realise a 100 mW linearly polarised fibre laser. The output coupler in this case was a bulk mirror and the mechanism of polarisation selection came from rotating a polarising beam cube in the cavity between the mirror and the fibre. Jovanovic et al. followed this by creating an all-fibre based version of the laser. Type II-IR gratings were written directly into the active PM-fibre via the PbP technique and a splice, orienting the stressors in the PM-fibre of the output coupler orthogonal to that of the high reflector, acted as the polarisation discriminator. Wavelength and polarisation switchability was achieved via temperature or strain tuning of the output coupler [40]. Finally, it has been shown that by optimising the difference in grating strength for orthogonal states of polarisation, it is possible to create highly linearly polarised fibre lasers utilising low birefringence, long length (up to 10 m) Er-doped fibres [28]. The technique demonstrated the potential for simplifying linearly polarised fibre lasers by removing the need for polarisation discrimination through expensive PM-fibres or external polarisation controllers.

There have been numerous other demonstrations of gratings inscribed directly into fibres with either different glass hosts or geometries. For example Bernier et al. demonstrated the direct inscription of Type I-IR gratings via a phase mask into Tm-doped ZBLAN fibres [61]. Although the gratings were strong, they were not thermally stable and decayed by 50% at 150°C, but nonetheless demonstrated a potential avenue for achieving 2 micron laser sources. Grobnic et al. demonstrated the direct inscription of strong Type I-IR gratings written via the phase mask technique into heavily doped Er:Yb-phosphate glass as a potential means to miniaturising laser sources [62]. Groothoff et al. realised the direct inscription of Type II-IR gratings written via the PbP technique into Yb-doped air-clad fibre for high power applications [63]. And finally, Aslund et al. showed that it was possible to create a highly simplified laser by butt-coupling an unfocussed diode bar up to a double-clad fibre laser, if a Type II-IR grating was inscribed into the core of the fibre as the high reflector [64].

### 8.6.2 Sensors

The resonant wavelength of any fibre grating (a FBG or a LPG) is defined by the physical period of the grating and by the effective refractive index of the core and/or cladding modes as outlined in Sect. 8.2. Both of these properties can be influenced by external factors, especially temperature and strain. As a result, fibre gratings have a long history of being used for sensing applications and constitute highly versatile and simple intrinsic sensing elements [65].

Compared to fibre sensors that are fabricated using conventional UV lasers, femtosecond laser written gratings can offer superior performance for specific applications due to their unique properties that have already been discussed in Sect. 8.5. The most notable property of direct written gratings is the extremely high temperature stability of Type II-IR gratings that makes them ideal candidates for use in harsh environments. Other factors include potentially higher robustness, as it is possible to write the gratings directly through the polymer coating and the increased flexibility in sensor design that results from the fact that the gratings can be inscribed in virtually any kind of optical fibre [66, 67]. Grobnic et al. have

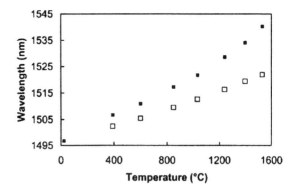

**Fig. 8.13** Variation of Bragg resonance in sapphire fibre grating with temperature. *Black squares* are the measured $\lambda_{FBG}$ as a function of temperature while the *white squares* are the calculated $\lambda_{FBG}$ based on pitch variation resulting from fibre elongation [11]

utilised these advantages and have fabricated a temperature sensor by inscribing a Type II-IR FBG with a femtosecond-laser and a phase-mask into the core of a single crystal sapphire fibre [11]. The device displayed a wavelength-dependent thermal sensitivity of 25 pm/°C up to 1500°C, with no observed degradation of the grating strength at those temperatures (Fig. 8.13). Based on this technology distributed optical sensor arrays that can operate to temperatures of up to 2000°C are feasible. Later they demonstrated that by inscribing a Type I-IR FBG into the core and the cladding region of a tapered single mode fibre, more precisely within the waist of a biconical tapered fibre and a tapered fibre tip, it was possible to realise a highly sensitive refractive index sensor [68]. The sensor exploits the fact that the effective refractive index of the fibre mode is in this case strongly influenced by the interaction of the highly evanescent field with the surrounding environment, which makes it extremely sensitive to refractive index changes and has a resolution of about $2.4 \times 10^{-5}$.

PbP written FBGs offer the additional freedom that they can be offset from the centre of the core, which stems from the fact that the physical dimensions of the individual grating periods are typically only a fraction of the core diameter. This results in a reduced overlap between the grating and the fibre mode, and thus in a reduced grating strength [32]. Normally this is a nuisance in applications were a high wavelength-selective-reflectivity is desired (e.g. in fibre lasers) and great care is taken to ensure that the entire FBG is well aligned in the centre of the core. However, Martinez et al. have shown that by inscribing an off-axis Type II-IR FBG into an untreated single mode fibre using a femtosecond laser, a novel direction-sensitive bending sensor could be realised [69]. Recently, this approach has been extended to the fabrication of asymmetric LPGs in photonic crystal fibres (PCFs) [70]. These PCF devices are characterised by a very low spectral temperature sensitivity and consequently can be used as high resolution bend sensors.

Gratings inscribed into multimode fibres, instead of single mode fibres, can be used for multi-parameter sensors. These gratings have multiple resonant dips in their transmission and reflection spectra, respectively, and the change of the spectral shape is different for perturbations due to temperature and pressure. While this theoretically offers a high degree of flexibility in sensing applications, it has the practical problem that measurement data can be distorted by changing excitation conditions. Polarisation-maintaining (PM) single mode fibres have also been used in sensing applications [71]. In this case the wavelength shift between the 2 linearly polarised modes in a Type II-IR grating were used to sense temperature and pressure.

Recently, temperature sensors based on femtosecond laser written gratings in PCFs have also been demonstrated. Fotiadi et al. have photochemically recorded a LPG into an endlessly single mode PCF using a femtosecond UV laser and have as such realised a temperature sensor with a sensitivity of as high as 300 pm/°C [72]. Other applications of femtosecond lasers for the fabrication of fibre based sensor systems include the selective removal of the cladding of (conventionally fabricated) FBGs in order to enhance their sensitivity for the simultaneous monitoring of the temperature and concentration of chemical and biological solutions [73] and the fabrication of Micro-Fabry-Perot interferometers in SMFs and PCFs that constitute ultra-compact sensors with a high strain and a low temperature sensitivity [74].

### 8.6.3 Other Applications

Westbrook et al. have shown that supercontinuum generation in a fibre containing a Bragg-grating exhibits a more than ten times enhancement near the Bragg resonance wavelength [75]. While their experimental setup was based on a conventional, UV-written FBG in a highly non-linear fibre, femtosecond laser based fabrication methods have the potential to greatly enhance the flexibility of this approach. Kulishov et al. have shown that LPGs incorporating $N - 1$ $\pi$-phase shifts can serve as an $N$-th order temporal differentiator that operates in transmission [76]. Compared to phase-mask written gratings where phase shifts are relatively difficult to implement, phase shifts can be easily implemented if the PbP technique is used.

## 8.7 Novel Fibre Types and Challenges

So far we have reviewed the femtosecond laser inscription of grating structures in common single- and multi-mode fibre forms along with more exotic geometries such as fibre tapers, polarisation-maintaining and double-clad fibres. We conclude this section with a summary of activities performed in other fibre types such as MOFs and fibres fabricated from more unusual materials such as polymers and chalcogenide glasses.

### 8.7.1 Microstructured Optical-Fibres (MOFs)

Photonic crystal fibres (PCFs) enable the unique opportunity of custom engineering of the linear, non-linear and modal properties of a waveguiding structure which can be hundreds of metres long (budget permitting). It is often desirable to combine these fibres with grating structures in order to enhance the functionality of the fibre or perform mode investigations; however the densely packed air-hole photonic crystal region of the fibre cladding presents unique challenges for grating inscription. The air-hole region of a MOF presents a scattering layer that obscures the core from the transversely launched laser beams used in all modern grating inscription techniques (Fig. 8.14). To complicate matters further many of the commercially available PCFs are fabricated from pure fused silica and hence the core lacks photosensitiser ions such as germanium. The sensitivity of fused silica to UV radiation can be increased via commonly used $H_2$-loading processes however the air-structured regions of the cladding again compound fabrication difficulties due to rapid effusion of the loaded hydrogen from the core, and if hydrogen loading is to be used it is common practice to pre-seal the PCF facets through fusion splicing to fibre-pigtails. Another important consideration regarding grating inscription in PCFs is the orientation of the grating writing beam with respect to the hexagonal lattice of the crystal. Studies of the effect of fibre rotational angle on a common PCF have shown that a 2.5× increase in the core coupled light is achievable by selecting an approach direction that is close to $\Gamma K$ [47]. Despite these apparent difficulties phase mask fabricated gratings have been written in hydrogen loaded pure fused silica PCFs using a 267 nm femtosecond laser [12], and in fused silica core fibre using an 800 nm femtosecond laser. Using custom fabricated PCFs with photosensitisers added to the core removes many of the difficulties associated with grating fabrication and this technique is equally applicable to femtosecond laser inscription of gratings as with other laser sources [77]. However, approaches of improving through-cladding coupling to the fibre core such as using an index matching oil to remove the scattering effects of the air-cladding region [78] are not well suited to phase mask or PbP femtosecond laser inscription of gratings in PCFs due to the fact that the high peak laser intensities lead to rapid vaporisation of the oil. In air-clad fibre geometries, where there are no air voids in close proximity to the

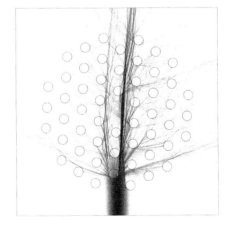

**Fig. 8.14** Computer model of the scattering of a UV grating writing laser beam in the microstructured region of a PCF at the angle of maximum coupling to the core [47]

core, index matching oils can however be used to alleviate scattering in sufficiently high numerical aperture focussing arrangements [63].

### 8.7.2 Polymer and Non-linear Fibres

The low cost, convenience of manufacture and low Young's modulus of polymer materials has lead to the development of polymer optical fibre (POF) platforms that are single moded and have propagation losses suitable for short-haul communication links and sensor networks. Grating inscription schemes for POF are presently based on UV nanosecond lasers [79] however IR femtosecond laser radiation has been demonstrated to create modulations in the refractive index of bulk polymethyl methacrylate (PMMA) of $5 \times 10^{-4}$, an amount eminently suitable to produce strong Bragg gratings in a fibre [80]. The advantage of using a multiphoton process to modify PMMA lies in the low absorption of the polymer material at this wavelength. Thus it is possible to write localised gratings deep inside a bulk or fibre sample, in contrast to the short penetration length of UV lasers in polymers.

Of significant interest to the IR and non-linear waveguide community are the recent developments in chalcogenide optical fibres. These glasses, fabricated from materials such as arsenic and selenium, have high third-order non-linearity and long wavelength IR transmission. Gratings have been fabricated on the surface of chalcogenide rib-waveguides and in chalcogenide fibre using the materials photosensitivity to visible light wavelengths. However these gratings are, by nature, erased by exposure to ambient light. A solution to this problem of grating permanency is to employ a Type II grating writing regime, and PbP gratings have been written in both bulk and fibre samples using the 800 nm output from a femtosecond laser system [81].

### 8.7.3 Through Jacket Grating Writing

One of the key advantages of using IR femtosecond laser light for the fabrication of gratings in fibres is that the polymer, silica or more exotic glass materials are transparent to these wavelengths and allow deep grating writing. An excellent exploitation of this is to write fibre gratings through the polymer fibre jacket. Stripping this protective layer from glass fibres in particular severely impacts their robustness. Both phase mask written [82] and PbP gratings [66] have been written through the polymer fibre jacket with little impact on the fibres' integrity.

## 8.8 Summary

Femtosecond laser inscription is now capable of producing Type I fibre gratings that exhibit excellent spectral linewidths (<100 pm) and reflectivities (>30 dB). In addition, there is renewed interest in Type II gratings because the off resonant losses of femtosecond inscribed gratings can be relatively low (<1 dB) and they have superior thermal stability compared to conventional fibre gratings. This field continues to make inroads into fibre grating devices in a range of novel fibre types, unrestrained by the need for photosensitivity.

# References

1. K.O. Hill, Y. Fujii, D.C. Johnson, B.S. Kawasaki, Appl. Phys. Lett. **32**(10), 647 (1978)
2. D.K.W. Lam, B.K. Garside, Appl. Opt. **20**(3), 440 (1981)
3. G. Meltz, W.W. Morey, W.H. Glenn, Opt. Lett. **14**(15), 823 (1989)
4. B. Malo, K.O. Hill, F. Bilodeau, D.C. Johnson, J. Albert, Electron. Lett. **29**(18), 1668 (1993)
5. A.M. Vengsarkar, P.J. Lemaire, J.B. Judkins, V. Bhatia, T. Erdogan, J.E. Sipe, J. Lightwave Technol. **14**(1), 58 (1996)
6. Y. Kondo, K. Nouchi, T. Mitsuyu, M. Watanabe, P.G. Kazansky, K. Hirao, Opt. Lett. **24**(10), 646 (1999)
7. S.J. Mihailov, C.W. Smelser, P. Lu, R.B. Walker, D. Grobnic, H.M. Ding, G. Henderson, J. Unruh, Opt. Lett. **28**(12), 995 (2003)
8. A. Dragomir, D.N. Nikogosyan, K.A. Zagorulko, P.G. Kryukov, E.M. Dianov, Opt. Lett. **28**(22), 2171 (2003)
9. A. Martinez, M. Dubov, I. Khrushchev, I. Bennion, Electron. Lett. **40**(19), 1170 (2004)
10. E. Wikszak, J. Burghoff, M. Will, S. Nolte, A. Tunnermann, T. Gabler, in *Conference on Lasers and Electro-Optics (CLEO)* (2004), p. CThM7
11. D. Grobnic, S.J. Mihailov, C.W. Smelser, H.M. Ding, IEEE Photon. Technol. Lett. **16**(11), 2505 (2004)
12. L.B. Fu, G.D. Marshall, J.A. Bolger, P. Steinvurzel, E.C. Magi, M.J. Withford, B.J. Eggleton, Electron. Lett. **41**(11), 638 (2005)
13. N. Jovanovic, M. Aslund, A. Fuerbach, S.D. Jackson, G.D. Marshall, M.J. Withford, Opt. Lett. **32**(19), 2804 (2007)
14. E. Fertein, C. Przygodzki, H. Delbarre, A. Hidayat, M. Douay, P. Niay, Appl. Opt. **40**(21), 3506 (2001)
15. A.M. Vengsarkar, P.J. Lemaire, J.B. Judkins, V. Bhatia, J.E. Sipe, T. Erdogan, in *Optical Fiber Communication Conference (OFC)* (1995), pp. PD4–2
16. S. Savin, M.J.F. Digonnet, G.S. Kino, H.J. Shaw, Opt. Lett. **25**(10), 710 (2000)
17. A. Othonos, K. Kalli, *Fiber Bragg Gratings: Fundamentals and Applications in Telecommunications and Sensing* (Artech House, Inc., Norwood, MA, 1999)
18. T. Erdogan, J. Lightwave Technol. **15**(8), 1277 (1997)
19. J. Canning, S.D. Jackson, M. Aslund, N. Groothoff, B. Ashton, K. Lyytikainen, Electron. Lett. **41**(20), 1103 (2005)
20. K.O. Hill, B. Malo, F. Bilodeau, D.C. Johnson, J. Albert, Appl. Phys. Lett. **62**(10), 1035 (1993)
21. J. Thomas, E. Wikszak, T. Clausnitzer, U. Fuchs, U. Zeitner, S. Nolte, A. Tunnermann, Appl. Phys. Mater. Sci. Process. **86**(2), 153 (2007)
22. H.G. Limberger, C. Ban, R.P. Salathe, S.A. Slattery, D.N. Nikogosyan, Opt. Exp. **15**(9), 5610 (2007)
23. J.W. Chan, T. Huser, S. Risbud, D.M. Krol, Opt. Lett. **26**(21), 1726 (2001)
24. C.W. Smelser, S.J. Mihailov, D. Grobnic, Opt. Exp. **13**(14), 5377 (2005)
25. E.N. Glezer, E. Mazur, Appl. Phys. Lett. **71**(7), 882 (1997)
26. G.D. Marshall, M.J. Withford, in *Conference on Bragg gratings, photosensitivity and periodic poling (BGPP)* (Quebec, Canada, 2007), p. JWA30
27. Intellectual property, Macquarie University

28. N. Jovanovic, J. Thomas, R.J. Williams, M.J. Steel, G.D. Marshall, A. Fuerbach, S. Nolte, A. Tnnermann, M.J. Withford, Opt. Exp. **17**(8), 6082 (2009)
29. Y. Lai, K. Zhou, K. Sugden, I. Bennion, Opt. Exp. **15**(26), 18318 (2007)
30. K.A. Zagorul'ko, P.G. Kryukov, Y.V. Larionov, A.A. Rybaltovskii, E.M. Dianov, N.S. Vorob'ev, A.V. Smirnov, M.Y. Shchelev, A.M. Prokhorov, Quantum Electron. **31**(11), 999 (2001)
31. A. Dragomir, D.N. Nikogosyan, A.A. Ruth, K.A. Zagorul'ko, P.G. Kryukov, Electron. Lett. **38**(6), 269 (2002)
32. F. Hindle, E. Fertein, C. Przygodzki, F. Durr, L. Paccou, R. Bocquet, P. Niay, H.G. Limberger, M. Douay, IEEE Photon. Technol. Lett. **16**(8), 1861 (2004)
33. P.G. Kryukov, Y.V. Larionov, A.A. Rybaltovskii, K.A. Zagorul'ko, A. Dragomir, D.N. Nikogosyan, A.A. Ruth, Microelectronic Eng. **69**(2–4), 248 (2003)
34. S.A. Slattery, D.N. Nikogosyan, Opt. Commun. **255**(1–3), 81 (2005)
35. A.I. Kalachev, D.N. Nikogosyan, G. Brambilla, J. Lightwave Technol. **23**(8), 2568 (2005)
36. A. Martinez, I.Y. Khrushchev, I. Bennion, Electron. Lett. **41**(4), 176 (2005)
37. Y. Lai, A. Martinez, I. Khrushchev, I. Bennion, Opt. Lett. **31**(11), 1672 (2006)
38. G.D. Marshall, R.J. Williams, N. Jovanovic, M.J. Steel and M.J. Withford, Opt. Exp., **18**(19), 19844 (2010)
39. N. Jovanovic, A. Fuerbach, G.D. Marshall, M.J. Withford, S.D. Jackson, Opt. Lett. **32**(11), 1486 (2007)
40. N. Jovanovic, G.D. Marshall, A. Fuerbach, G.E. Town, S. Bennetts, D.G. Lancaster, M.J. Withford, IEEE Photon. Technol. Lett. **20**(9-12), 809 (2008)
41. M. Livitziis, S. Pissadakis, Opt. Lett. **33**(13), 1449 (2008)
42. S.J. Mihailov, C.W. Smelser, P. Lu, R.B. Walker, D. Grobnic, H.M. Ding, J. Unruh, in *Opt. Fiber Commun. Conf. (OFC)*, vol. 86 (2003), p. Postdeadline Paper PD30
43. C.W. Smelser, S.J. Mihailov, D. Grobnic, J. Optical Soc. Am. B Optical Phys. **23**(10), 2011 (2006)
44. S.J. Mihailov, C.W. Smelser, D. Grobnic, R.B. Walker, P. Lu, H.M. Ding, J. Unruh, J. Lightwave Technol. **22**(1), 94 (2004)
45. K.A. Zagorulko, P.G. Kryukov, Y.V. Larionov, A.A. Rybaltovsky, E.M. Dianov, Opt. Exp. **12**(24), 5996 (2004)
46. Y.H. Li, C.R. Liao, D.N. Wang, T. Sun, K.T.V. Grattan, Opt. Exp. **16**(26), 21239 (2008)
47. G.D. Marshall, D.J. Kan, A.A. Asatryan, L.C. Botten, M.J. Withford, Opt. Exp. **15**(12), 7876 (2007)
48. J. Thomas, C. Voigtlander, D. Schimpf, F. Stutzki, E. Wikszak, J. Limpert, S. Nolte, A. Tunnermann, Opt. Lett. **33**(14), 1560 (2008)
49. D. Grobnic, C.W. Smelser, S.J. Mihailov, R.B. Walker, Meas. Sci. Technol. **17**(5), 1009 (2006)
50. C.W. Smelser, D. Grobnic, P. Lu, S.J. Mihailov, in 34th *Eur. Conf. Opt. Commun. (ECOC)* (2008), article no. 4729336
51. F. Durr, H.G. Limberger, R.P. Salathe, F. Hindle, M. Douay, E. Fertein, C. Przygodzki, Appl. Phys. Lett. **84**(24), 4983 (2004)
52. P.Y. Fonjallaz, H.G. Limberger, R.P. Salathe, F. Cochet, B. Leuenberger, Opt. Lett. **20**(11), 1346 (1995)
53. P. Lu, D. Grobnic, S.J. Mihailov, J. Lightwave Technol. **25**(3), 779 (2007)
54. D. Grobnic, S.J. Mihailov, C.W. Smelser, J. Lightwave Technol. **25**(8), 1996 (2007)
55. C. Caucheteur, P. Megret, T. Ernst, D.N. Nikogosyan, Opt. Commun. **271**(2), 303 (2007)
56. Y.S. Himei, J.R. Qiu, S. Nakajima, A. Sakamoto, K. Hirao, Opt. Lett. **29**(23), 2728 (2004)
57. M.L. Aslund, N. Jovanovic, N. Groothoff, J. Canning, G.D. Marshall, S.D. Jackson, A. Fuerbach, M.J. Withford, Opt. Exp. **16**(18), 14248 (2008)
58. E. Wikszak, J. Thomas, J. Burghoff, B. Ortac, J. Limpert, S. Nolte, U. Fuchs, A. Tunnermann, Opt. Lett. **31**(16), 2390 (2006)
59. N. Jovanovic, M. Aslund, A. Fuerbach, S.D. Jackson, G.D. Marshall, M.J. Withford, in *Conference on Bragg gratings, photosensitivity and periodic poling (BGPP)* (Quebec, Canada, 2007), p. BWB7

60. E. Wikszak, J. Thomas, S. Klingebiel, B. Ortac, J. Limpert, S. Nolte, A. Tunnermann, Opt. Lett. **32**(18), 2756 (2007)
61. M. Bernier, D. Faucher, R. Vallee, A. Saliminia, G. Androz, Y. Sheng, S.L. Chin, Opt. Lett. **32**(5), 454 (2007)
62. D. Grobnic, S.J. Mihailov, R.B. Walker, C.W. Smelser, C. Lafond, A. Croteau, IEEE Photon. Technol. Lett. **19**(9-12), 943 (2007)
63. N. Groothoff, N. Jovanovic, G.D. Marshall, J. Canning, M.J. Withford, Glass Technol. Eur. J. Glass Sci. Technol. A **50**(1), 75 (2009)
64. M.L. Aslund, S.D. Jackson, N. Jovanovic, M.J. Withford, Opt. Commun. **281**(15-16), 4092 (2008)
65. A.D. Kersey, M.A. Davis, H.J. Patrick, M. LeBlanc, K.P. Koo, C.G. Askins, M.A. Putnam, E.J. Friebele, J. Lightwave Technol. **15**(8), 1442 (1997)
66. A. Martinez, I.Y. Khrushchev, I. Bennion, Opt. Lett. **31**(11), 1603 (2006)
67. D. Grobnic, S.J. Mihailov, C.W. Smelser, R.T. Ramos, IEEE Photon. Technol. Lett. **20**(9-12), 973 (2008)
68. D. Grobnic, S.J. Mihailov, H.M. Ding, C.W. Smelser, IEEE Photon. Technol. Lett. **18**(1-4), 160 (2006)
69. A. Martinez, Y. Lai, M. Dubov, I.Y. Khrushchev, I. Bennion, Electron. Lett. **41**(8), 472 (2005)
70. T. Allsop, K. Kalli, K. Zhou, Y. Lai, G. Smith, M. Dubov, D.J. Webb, I. Bennion, Opt. Commun. **281**(20), 5092 (2008)
71. C. Zhan, Y. Zhu, S. Yin, P. Ruffin, Optical Fiber Technol. **13**(2), 98 (2007)
72. A.A. Fotiadi, G. Brambilla, T. Ernst, S.A. Slattery, D.N. Nikogosyan, J. Optical Soc. Am. B Optical Phys. **24**(7), 1475 (2007)
73. H. Alemohammad, E. Toyserkani, A.J. Pinkerton, J. Phys. D Appl. Phys. **41**(18), (2008)
74. Y.J. Rao, M. Deng, D.W. Duan, X.C. Yang, T. Zhu, G.H. Cheng, Opt. Exp. **15**(21), 14123 (2007)
75. P.S. Westbrook, J.W. Nicholson, K.S. Feder, Y. Li, T. Brown, Appl. Phys. Lett. **85**(20), 4600 (2004)
76. M. Kulishov, D. Krcmarik, R. Slavik, Opt. Lett. **32**(20), 2978 (2007)
77. B.J. Eggleton, P.S. Westbrook, C.A. White, C. Kerbage, R.S. Windeler, G.L. Burdge, J. Lightwave Technol. **18**(8), 1084 (2000)
78. H.R. Sørensen, J. Canning, J. Lgsgaard, K. Hansen, Opt. Exp. **14**(14), 6428 (2006)
79. H.Y. Liu, H.B. Liu, G.D. Peng, P.L. Chu, Opt. Commun. **266**(1), 132 (2006)
80. P.J. Scully, D. Jones, D.A. Jaroszynski, J. Opt. a Pure Appl. Opt. **5**(4), S92 (2003)
81. C. Florea, J.S. Sanghera, I.D. Aggarwal, Opt. Mater. **30**(10), 1603 (2008)
82. S.J. Mihailov, D. Grobnic, C.W. Smelser, Electron. Lett. **43**(8), 442 (2007)

# Chapter 9
# 3D Bragg Grating Waveguide Devices

**Haibin Zhang and Peter R. Herman**

**Abstract** Over the past decade, ultrashort pulse laser processing has opened a large suite of photonic devices that can be formed inside bulk optical glasses by direct writing. Such processes promise rapid and seamless integration into novel three-dimensional optical circuits. One obstacle towards commercial application of this technology has been finding an effective means for inscribing high-quality grating devices. Such gratings, when embedded within the laser written waveguides, enable multi-functional spectral filters to be tailored to specific applications required in optical sensing, fiber lasers, and telecommunications. In this chapter, a new Bragg grating waveguide device is introduced that can be fabricated directly inside transparent glass materials by ultrashort laser direct writing. These Bragg grating waveguide devices are composed of arrays of partially overlapped refractive index voxels (volume pixels), defining a finely pitched segmented waveguide which simultaneously offers low-loss light guiding and strong Bragg filter resonances. Two approaches, a single-pulse writing method and a burst writing method, are introduced for inscribing the grating waveguide devices with respective low-and high-repetition rate ultrashort laser systems. Optimal laser exposure parameters are presented for fabricating high rejection notch filters (>35 dB) with narrow spectral bandwidth (0.2 nm) in the 1550-nm telecom band. Examples of Bragg grating waveguide circuits are presented for filter and sensor applications.

## 9.1 Introduction

Optical waveguides typically rely on formation of high refractive index structures in transparent materials that guide electromagnetic fields in the optical spectrum by total internal reflection. The most common example of such waveguides is the

---

H. Zhang · P.R. Herman (✉)
Department of Electrical and Computer Engineering and Institute for Optical Science,
University of Toronto, Toronto, ON, M5S 3G4, Canada
e-mail: p.herman@utoronto.ca

optical fiber which delivers light signals over long distances with extremely low loss of $\sim$0.2 dB km$^{-1}$. With careful design and integration, waveguides are the key building block for connecting photonic devices such as power splitters, modulators, multiplexers, detectors, and lasers into multi-functional optical systems.

Photolithography and etching, the same technology used for large scale fabrication of microelectronic chips in the semiconductor industry, have also been adopted in fabrication of integrated photonic circuits for telecommunication application. This approach relies on ultraviolet light to replicate patterns of optical circuits from pre-defined photomasks to substrates through multiple and complex steps of optical exposure and chemical reaction processes (etching, ion exchange, or diffusion) inside a clean room environment. Despite its advantages in repeatability and parallel processing, the overall process is time-consuming, multi-step, and inflexible, and leads only to two-dimensional designs of integrated optical waveguide circuits.

The advent of ultrashort-laser material processing has opened a new realm of opportunities for writing integrated optical devices directly inside transparent materials. In 1996, Davis and coworkers first demonstrated waveguide writing inside transparent glass materials [1]. When laser pulses of ultrashort duration are focused into a transparent substrate, the normally transparent material will turn absorptive due to non-linear optical interactions [2], and the refractive index of the material will change only locally inside or near the laser focus [3]. Translating the material with respect to the laser focus will then form a track of refractive index change which can serve under appropriate conditions as a low-loss optical waveguide. Because the changes only happen locally in the laser focal volume, complex photonic circuits can be fabricated with pre-programmed motion stage routines in three dimensions (3D). Compared with the traditional photolithography method, this direct write technology is fast, maskless, flexible, and insensitive to environment (no clean room needed). Although the minimum loss of such laser-written waveguides ($\sim$0.2 dB cm$^{-1}$) is still not comparable with that of the single-mode fibers (0.2 dB km$^{-1}$) or planar lightwave circuits (PLCs, <0.02 dB cm$^{-1}$), this laser technology provides novel means for writing of compact photonic circuits over all three dimensions of the glass substrate promising to greatly enhance the functionality and integration capacity of photonic circuits.

Ultrashort laser written waveguides have been demonstrated in a broad range of materials including various types of glasses [1, 4, 5], semiconductors [6], crystals [7, 8], polymers [9], and optically active materials [10]. In the mean time, numerous optical circuit devices such as couplers [11, 12], waveguide amplifiers [10], and multi-mode interference waveguides [13] have also been demonstrated. A rapidly expanding research community is now actively pushing forward the applications of the ultrashort laser direct write technology as described in the many chapters in this book.

Despite these advances in ultrashort laser direct fabrication of photonic devices, a significant roadblock exists towards commercialization of such technology. While advanced Bragg grating filters can be tailor designed and fabricated inside the guiding core of optical fiber and PLCs, no practical means for direct formation of strong Bragg gratings have been demonstrated within such laser-written waveguides until recently.

Bragg grating waveguides rely on periodic perturbation of refractive index along the guiding core of a waveguide to create spectrally controlled filter or reflection features in the transmitted and reflected light. Responses can be strong at very narrowly defined resonant wavelengths that satisfy the Bragg condition [14]. The Bragg resonance wavelength is extremely sensitive to environmental parameters such as temperature and strain and thus is further attractive for strain or temperature gauging. Fiber Bragg gratings (FBGs) [15, 16] have been the most intensively studied grating-waveguide device over the past 30 years and FBGs are found today in numerous commercial applications in telecom, sensing, and fiber laser markets.

The formation of strong and low-loss Bragg gratings inside the ultrashort laser written waveguides has been challenging, possibly owing to a saturated photosensitivity response following the strong ultrashort-laser material interaction. Such strong exposure may wash out any further fine (0.53 μm) periodic grating modulation. Nevertheless, such gratings are highly desirable for providing narrow-spectrum filters, mirrors, couplers, multiplexers, taps, and sensing elements that define the basic functions required, for example, in sensing, passive optical networks (PON), and optical communications.

Ultraviolet [17] and ultrafast lasers [18] readily generate Bragg gratings in optical fibers by holographic, phase mask, or point-by-point writing means [19, 20], where the pre-existing waveguide core often facilitates strong photosensitivity enhancement from rare-earth element dopants and hydrogen soaking. In contrast, only weak or high-order Bragg gratings have been reported in waveguides first formed in bulk transparent materials with ultrashort pulse lasers. Kamata and Obara [21] proposed using a second laser exposure step to overlay gratings on the previously formed waveguides and generated weak (∼6 dB) tenth-order gratings. Marshall et al. [22] used a two-step exposure method to produce a weak (∼3 dB) second-order Bragg gratings at ∼1550-nm telecom wavelength. Chung et al. also demonstrated ∼3-dB first-order Bragg gratings inside a 65-mm long soda-lime glass waveguide by using a similar two-step process [23]. However, in order to inscribe the ∼500-nm period grating features, a 1.25-NA, 100× oil immersion objective lens was used for high optical resolution. The short working distance and large spherical aberration of such oil-immersion lenses limits the applicable depth of this writing geometry.

This chapter summarizes the recent effects by our group to bypass the multi-step fabrication methods used previously to form continuous waveguides and periodic refractive index modulation in sequential exposure steps. Two methods for single-step waveguide writing are presented that simultaneously fabricate segmented waveguides with low-loss and high-strength Bragg resonances. Such Bragg grating waveguides (BGWs) are composed of an array of laser induced refractive index voxels (volume pixels) that provide sufficiently strong average refractive index change for efficient light guiding while also undulating with moderated refractive index modulation for a strong Bragg resonance reflection. The two BGW writing methods center on creation of each BGW voxel with either a single laser pulse or a burst of ultrashort laser pulses. The Bragg resonant wavelength of such BGW devices are controlled by the voxel separation, whereas the waveguide losses and grating strengths are to be determined by the relative modulation contrast and blending of neighboring refractive index voxels.

The single pulse and burst BGW writing techniques rely on low- and high-repetition rate ultrashort lasers, respectively. Both methods yield similar BGWs with strong >35 dB transmission strength, >90% peak reflection, and <0.2-nm spectral bandwidth (3 dB bandwidth) in borosilicate and fused silica glasses for the infrared telecommunication band from 1200 to 1600 nm. These strong waveguide grating responses compare closely with the responses available today in FBGs. However, the new direct writing method of BGWs is single-step and maskless, permitting flexible, rapid, and alignment-insensitive processing that enables facile wavelength tuning and device cascading into BGW optical circuits. Advanced devices such as chirped BGWs and 3D BGW sensor networks are presented. Because of the strong laser interactions, the present BGW devices are shown to have high temperature stability up to 500°C.

## 9.2 Bragg Grating Waveguide Fabrication

Ultrashort laser waveguide writing involves a sequence of intense physical processes that may include microexplosions triggered by the extremely high peak intensity of the focused laser light inside the bulk glasses. The available material photosensitivity response for refractive index change is most likely depleted after the first laser exposure for forming waveguides, making difficult the further inscription of refractive index modulation on which high-strength Bragg responses rely. This consumed photosensitivity is likely responsible for the generation of only weak gratings of <3 dB strength by the two-step laser fabrication approach as reported by other groups [21–23].

The key to high-strength Bragg response is to form both high refractive index contrast for strong guiding while retaining a moderate AC refractive index modulation along the waveguide for strong Bragg reflection. Figure 9.1 presents a conceptual image of high contrast BGW device. The BGW is composed of an array of partially overlapped refractive index voxels each formed by single or multiple ultrashort laser pulses. By carefully exploring the laser exposure and writing

**Fig. 9.1** Bragg grating waveguide composed of an array of partially overlapped refractive index voxels of period $\Lambda = d$. (Isolated voxels were drawn here for clarity of the concept.)

conditions, BGWs have been generated in a single exposure step having strong refractive index contrast for both strong guiding and low propagation loss while also retaining a moderate AC index modulation for a high-strength Bragg resonance. The Bragg condition is defined by, $\lambda_B = 2n_{eff}\Lambda$, where $\lambda_B$ is the Bragg wavelength and $n_{eff}$ is the effective refractive index of the waveguide. For glass materials with $n_{eff}$ close to 1.5, the grating period needs to be approximately $\Lambda = 0.5\,\mu\text{m}$ for Bragg responses near the telecom wavelength of $\lambda_B = 1550$ nm. Therefore, laser interaction volumes must be confined to very challenging dimensions as small as $\Lambda/2 = 250$ nm.

The formation of periodically modulated or segmented waveguides with very fine pitch of $\Lambda = 0.5\,\mu\text{m}$ requires comprehensive optimization effort and control of laser exposure parameters to submicron focusing spots and 200 nm sample positioning that must be controlled over very long exposure length (~25 mm). The two ultrafast laser and exposure systems are presented here that center on low- (Ti:sapphire CPA; 1 kHz) and high-repetition rate (fiber CPA; 100 kHz to 5 MHz) laser systems. A wide range of laser exposure conditions were examined to optimize the BGW writing process for combination of low loss guiding and strong BGW response. In this section, results of the BGW device optimization, fabrication, and characterization are presented, first for the single-pulse BGW writing and followed by the burst laser BGW writing method. Thermal properties of the BGWs and several demonstrations of BGW devices are given at the end of the section.

## 9.2.1 BGW Fabrication Method 1: Single-pulse Writing

In nearly all cases, ultrafast laser waveguide writing involves the presentation of multiple overlapping pulses at a laser focus to change the refractive index in a continuous line in the material. In 2006, a new domain of rapid waveguide writing technique was demonstrated with a 1-kHz ultrashort laser, where non-overlapping exposure of single laser pulses led to segmented but strongly confining waveguides [24]. Under this domain, the waveguide consisted of an array of nearly isolated voxels, each formed by a single-pulse interaction that sharply contrasts with the much higher exposures of tens to thousands of overlapping laser pulses applied along a slowly moving laser focus. Because strong refractive index modification was formed with a single-pulse laser interaction, the photosensitivity responses was labeled "type II" to follow the damaged gratings (type II) reported for FBGs formed by high fluence single-pulse exposure of UV laser light [14].

Figure 9.2 illustrates the formation of the segmented waveguides with voxels created by single laser pulses. The voxels, shown enlarged in Fig. 9.2b, have separations defined by the sample scan speed, $v$, and the laser repetition rate, $R$, according to the period $\Lambda = v/R$. At exposures of $R = 1$ kHz and $v = 1\,\text{mm s}^{-1}$, the spacing $\Lambda = 1\,\mu\text{m}$ is comparable with the 1-$\mu$m laser spot size which marks the boundary of partial overlapping ($v < 1\,\text{mm s}^{-1}$) and non-overlapping

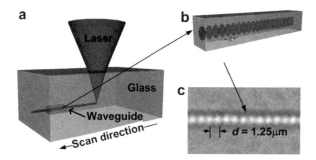

**Fig. 9.2** Schematic illustration (**a**) of the waveguide fabrication process with the sample translated perpendicularly to the focused laser beam direction, and leading to an array of isolated focal volumes (**b**) under high (>1 mm s$^{-1}$) scan velocities. An overhead microscope image (**c**) of a waveguide showing the isolated modification volumes written with 1.25 mm s$^{-1}$ scan speed

($v > 1$ mm s$^{-1}$) voxel arrays. In Fig. 9.2c, the overhead microscope image resolves the individual voxels of a waveguide formed in fused silica with $\Lambda = 1.25\,\mu$m.

Single-pulse segmented waveguides formed in fused silica offered low propagation loss when the index voxels were partially overlapping. This new regime of laser exposure was distinguishable in two ways from traditional approaches in laser waveguide writing. First, an examination of a wide 50-fs to 5-ps range of pulse duration showed the lowest loss waveguides to form in a narrow $1.0 \pm 0.2$ ps window that significantly exceeded the 50–200 fs duration reported as optimal in other studies. Second, an unusually high scan speed of $1.0 \pm 0.2$ mm s$^{-1}$ indicated a strong type-II photosensitivity mechanism in forming nearly isolated refractive index voxels. A minimum propagation loss of ~0.2 dB cm$^{-1}$ and a slightly asymmetric mode diameter of ~9 μm was demonstrated for 633-nm light. However, Bragg grating responses in these waveguides could not be confirmed without any broadband sources available in the visible spectrum in our lab.

The single-pulse waveguide writing process was successfully extended to the borosilicate glass (Corning EAGLE2000$^{TM}$), which offered much stronger photosensitivity response than fused silica for this method. A good processing window for low-loss light guiding was reported in the infrared spectrum (~1550 nm) over a large range of scan speeds. With this flexibility, Bragg grating periods of ~500 nm were successfully written with sufficient AC index contrast to generate narrow-bandwidth first-order Bragg responses for the first time in bulk glasses. The Bragg wavelength of these BGWs was tunable over the whole telecom spectrum (1200–1600 nm) by simply tuning the scan speed during laser exposure.

Figure 9.3 shows the waveguide fabrication arrangement based on a commercial 1-kHz laser system (Spectra-Physics: *Spitfire Pro 40F*). The system delivered 800-nm laser pulses at 1-kHz repetition rate with 2.5-mJ maximum pulse energy and $M^2$ beam quality close to 1. The laser pulse duration could be adjusted from 40 fs to 2 ps by tuning the compressor grating in the amplifier and verified with an autocorrelator. The laser power was adjusted with a variable neutral density (ND) filter while a

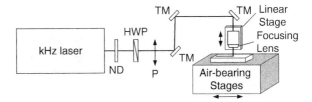

**Fig. 9.3** The kHz laser system beam delivery arrangement for BGW writing. ND: neutral density filter; HWP: half-wave plate; P: linear polarizer; TM: turning mirror

combination of half-wave plate and a linear polarizer controlled both the laser power and the orientation of laser polarization relative to the scan direction of the target. An aspheric lens of 0.55 NA was used to produce a ∼1-μm focal diameter ($1/e^2$ intensity) at a position 200 μm below the surface of a borosilicate glass sample (Corning EAGLE2000, 50 mm × 10 mm × 1 mm). The sample was fixed on two-axis air-bearing motion stages (Aerotech ABL1000, 2-nm resolution and 50-nm repeatability) and scanned perpendicular to the laser polarization to yield 10-mm long waveguides. For scan speed fixed at 0.52 mm s$^{-1}$, first-order Bragg reflection at 1550 nm was observed for the 1-kHz laser repetition rate. The laser exposure energy was adjusted from 0.5 to 10 μJ by the half-wave plate and linear polarizers.

After laser exposure, the waveguides were characterized for their appearance and guided mode profiles at 1.55 μm telecom bandwidth. These BGWs comprised of a periodic array of refractive index voxels with 0.52-μm centre-to-centre separation, as defined by 0.52-mm s$^{-1}$ scan speed and 1-kHz laser repetition rate. [24]. Similar to the narrow processing windows noted in the previous example, continuous and homogenous tracks were observed only for pulse durations in the range of 100 fs to 1.5 ps and pulse energy in the range of 2 to 7 μJ. The laser tracks appeared faint, discontinuous, or invisible for lower pulse energy of <2 μJ, or appeared inhomogeneous and damaged above 7 μJ. For all pulse durations tested, the lowest propagation losses were found at pulse energies only near 3 μJ, slightly above the ∼2-μJ threshold for generating guiding tracks.

Figure 9.4a shows the measured waveguide propagation loss as a function of pulse duration for 1560-nm wavelength guiding in BGWs written with 3-μJ pulse energy and pulse durations from 100 fs to 1.5 ps. Two low-loss processing windows of ∼0.5 dB cm$^{-1}$ can be identified at short 100-fs and long 1-ps pulse duration. Outside these two windows, the propagation loss increased to above ∼3 dB cm$^{-1}$.

Propagation losses of BGWs written with pulse durations of 100 fs, 300 fs, and 1 ps are plotted in Fig. 9.4b against the incoming laser pulse energy. In all cases, insufficient refractive index change leads to high insertion losses for energies lower than 3 μJ. For 1-ps pulses, only pulse energy near 3 μJ generated low propagation losses at around 0.5 dB cm$^{-1}$, whereas the BGWs written with energies greater than 7 μJ become increasingly inhomogeneous, yielding high >3-dB cm$^{-1}$ propagation losses. In contrast, a more forgiving low-loss processing window exists for the short 100-fs pulse duration. At this duration, the BGW propagation loss is less than 1 dB cm$^{-1}$ for all the energies above 2 μJ, which compares favorably to the long 1-ps window duration for low-loss waveguide fabrication. It is unclear why higher

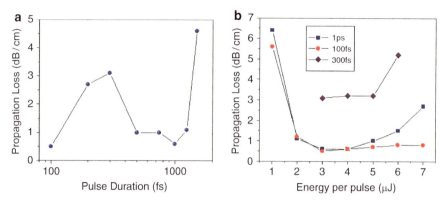

**Fig. 9.4** Propagation loss of BGWs written with (**a**) 3-μJ pulse energy for increasing the pulse duration and with (**b**) 100 fs, 300 fs, and 1 ps pulse durations with increasing pulse energy. These devices were written inside borosilicate glasses with the kHz laser system

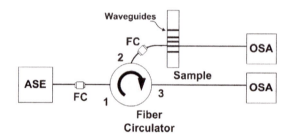

**Fig. 9.5** Spectral response arrangement of BGWs. The broadband light from an ASE is guided to the waveguide by the fiber circulator which also routed the waveguide reflection signal. An OSA was used to analyze both the transmitted and reflected light. FC: fiber connector; OSA: optical spectrum analyzer; and ASE: broadband light source

losses occurred for writing laser durations between 100 fs and 1 ps, warranting further analysis of the structural changes induced in the laser focal volume.

Bragg grating responses were systematically characterized for BGWs written with 1 to 7 μJ laser pulse energy and 100- to 1500-fs pulse duration. Figure 9.5 illustrates the experimental setup for characterizing BGW spectral responses in both transmission and reflection. Infrared light from a randomly polarized broadband source (Thorlabs ASE-FL7002, 1530 nm to 1610 nm) was routed through a fiber circulator (transmit signal from port 1 to 2, 2 to 3 with minimum loss) and butt coupled into the BGWs with a single-mode fiber. Another single-mode fiber was aligned to the exit facet of the waveguides to guide the transmitted light to an optical spectrum analyzer (OSA, Ando AQ6317B with 0.01-nm resolution) for recording transmission spectra. The reflected light from the BGWs was passed back to port 2 of the optical circulator and guided to port 3, which was connected to the same

# 9 3D Bragg Grating Waveguide Devices

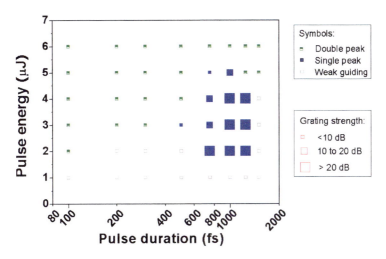

**Fig. 9.6** Classification of BGWs properties as functions of laser pulse duration and energy. The devices were fabricated inside borosilicate glasses with the kHz ultrashort laser system

OSA for reflection responses. Index matching fluid was used for all the waveguide-fiber interfaces to eliminate Fresnel reflections and Fabry–Perot spectral effects.

Optimized Bragg transmission responses varied from relatively weak (<5 dB) dual-peaked lines that indicated waveguide birefringence at short 100-fs pulse duration to strong >35-dB single-peaked resonances at long 1.0-ps pulse duration. In Fig. 9.6, the BGWs were classified by their grating strength, propagation loss, and number of peaks according to laser pulse energy and duration. The symbol size represents the grating strength in transmission, with small, medium, and large corresponding to <10 dB, 10 to 20 dB, and >20 dB response, respectively. Half-filled squares represent dual-peak BGWs and solid squares correspond to BGWs with single peaks. Open squares represent BGWs with large propagation loss (>3 dB cm$^{-1}$). For pulse duration less than 500 fs, effective guiding was observed above a 1 to 2 μJ energy threshold, but only weak dual-peak gratings (<10 dB) were observed. An optimum window for generating strong (>20 dB) and low-loss (~0.5 dB cm$^{-1}$) BGWs was identified for 0.8 to 1.2 ps duration and 3 to 6 μJ pulse energy. Higher pulse energy yielded weaker and dual-peaked gratings while longer pulse duration (1.5 ps) required higher pulse energy (5 μJ) for guiding and provided only weak and dual-peak BGWs.

In Fig. 9.7, the reflection and transmission spectra of the highest strength BGW observed are plotted. The BGWs were written using 1-ps pulse duration and 3-μJ pulse energy. Strong responses of 35 dB in transmission and 95% in reflection are available with 0.2-nm bandwidth. Considering the short 10-mm sample length, this Bragg response is three orders of magnitude stronger than the 11-dB transmission and ~40% reflection responses in 50-mm long BGWs written with weaker focusing condition (NA = 0.25), presented in the previous section. Radiation mode losses of ~5 dB are apparent on the short wavelength side of the Bragg resonance that

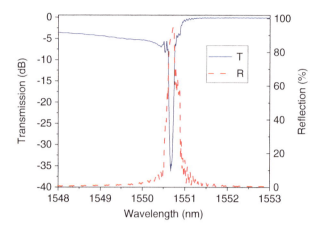

**Fig. 9.7** Transmission (*solid line*) and reflection (*dashed line*) spectral response of the borosilicate BGW written with 3-μJ pulse energy and 1-ps pulse duration for the kHz laser system

indicates a non-uniform refractive index modulation transverse to the guiding direction.

The BGW refractive index modulation $\Delta n_{AC}$ can be inferred from [16]

$$\Delta n_{AC} = \frac{\lambda}{\pi L \eta} \tanh^{-1}\left(\sqrt{R}\right), \qquad (9.1)$$

where $\lambda$ is the Bragg wavelength, $L$ is the grating length, and $R = 1 - T$ is the reflectance derived from the grating transmittance, $T$, at the Bragg resonance. $\eta$ is the modal overlap factor defined by the fraction of modal power inside the waveguide core, and was approximated as the modal power inside a 2-μm diameter waveguide, yielding $\eta = 0.06$ for the ∼11-μm diameter mode measured for this BGW. For the 10-mm long grating waveguide in Fig. 9.7, one estimates a strong $\Delta n_{AC}$ of ∼4 × 10$^{-3}$, which represents a large ∼40% component of the average refractive index $\Delta n_{DC} = $ ∼0.01 inferred from the 2-μm waveguide core and 11-μm mode diameter. Such index modulation is an order of magnitude larger than typically found in strong FBGs, and is consistent with formation of optically isolated index voxels during the single-pulse waveguide writing process.

The transmission and reflection spectra of a typical dual-peak BGW written with 100-fs duration and 3-μJ pulse energy are shown in Fig. 9.8. Two sharply resolved (0.1-nm wide FWHM) peaks at 1550.9 nm and 1551.1 nm wavelength are present, but only weak 1.9 dB and 2.6 dB transmission and 31% and 38% reflection resonances, respectively, were generated that indicate waveguide birefringence.

It is clear from Fig. 9.6 that for laser pulse energy of 6 μJ, BGWs written with durations from 100 fs to 1.5 ps are all birefringent. The birefringence, $\Delta n_B$, of these BGWs were inferred from $\Delta n_B = n_{TM} - n_{TE} = (\lambda_{TM} - \lambda_{TE})/2\Lambda$, where $\Lambda = 0.52$ μm is the grating period, and plotted in Fig. 9.9 against the pulse duration.

# 9  3D Bragg Grating Waveguide Devices

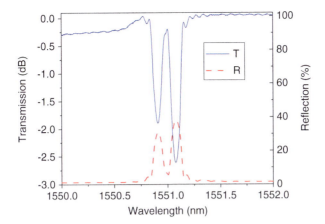

**Fig. 9.8** Transmission (*solid*) and reflection (*dash*) spectral response of the borosilicate BGW written with 3-µJ pulse energy and 100-fs pulse duration for the kHz laser system

**Fig. 9.9** Waveguide birefringence dependence on pulse duration for the borosilicate BGWs written with 6-µJ pulse energy with the kHz laser system

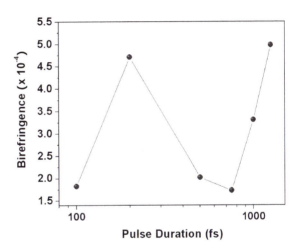

The birefringence strongly correlates with the propagation loss data in Fig. 9.4a, suggesting laser damage induced birefringence possibly related to asymmetric waveguide stresses. The lowest measurable birefringence by using this method corresponds to ∼0.1-nm spectral separation where the TE and TM resonances would merge into unresolved lines, corresponding to an lower bound of $\Delta n_B = {\sim}1 \times 10^{-4}$. The present results suggest new possibilities for fabricating polarization-dependent components in 3D optical circuits or for trimming birefringence-free waveguides in telecom systems.

As we show above, by using the kHz system (low repetition rate of 1 kHz), BGW structures could be fabricated by using single-pulse exposures to define individual modification voxels inside transparent glasses. By carefully exploring laser exposure parameters including pulse energy, pulse duration, and focusing geometry, it is possible to find processing windows that form such BGW devices with both low-loss light guiding and high-strength grating resonances. However, the single-pulse writing method has limitations. For certain transparent materials, this single-pulse exposure might not generate sufficient refractive index change for efficient guiding at 1550-nm, greatly restricting such BGW writing technique for telecom applications. A second BGW fabrication approach, namely, the burst writing approach, is adopted here to ensure enough refractive index modification for infrared light-guiding. Multiple ultrashort laser pulses are used to form each refractive index voxel in the BGW structure.

## 9.2.2 BGW Fabrication Method 2: Burst Writing

In Sect. 9.2.1, a novel BGW device was demonstrated to have both low-loss light guiding and high-strength Bragg resonance, formed simultaneously in a single exposure step. By using a low-repetition rate (1 kHz) laser, a one-step single-pulse method was applied to shape refractive index voxel arrays in borosilicate glass under fast scan speed. However, for materials like fused silica, this single-pulse interaction failed to generate sufficiently large refractive index change for infrared guiding, which poses a limitation of the single-pulse method for sensing and filtering device applications in the telecom band and nearby spectral regions.

In contrast with single-pulse BGW writing, a highly overlapping method is preferred in waveguide writing as reported by most research groups, where hundreds or thousands of laser pulses overlap in the laser focal volume to induce large refractive index changes. Typically, one uses high-repetition rate ultrashort lasers, slow sample scan speed, or a combination of both for this high fluence exposure. This latter method has been applied to generate low-loss waveguides in a much broader range of materials including various types of glasses [1, 4, 25], crystal materials [7, 26, 27], semiconductors [6], and polymers [28, 29]. With exposure recipes available for low-loss waveguide writing in these materials, it is highly desirable to form Bragg gratings in them for enhanced spectral filtering applications in 3D integrated optical circuits.

This section introduces a burst method to form BGWs with highly overlapping pulse exposures. First, an optimum waveguide writing recipe that generates low-loss waveguides was found in the highly overlapping domain. Secondly, modulation of the laser source was developed to generate a burst of laser pulses. When applied for waveguide writing, each burst generated a single segment in an array of refractive index voxels that forms an optical waveguide with periodic index modulation. The result is a similar BGW structure as presented in Fig. 9.1, but each voxel is now formed by multiple laser pulses to ensure enough refractive index change for infrared guiding. The voxel size and separation can be further optimized by adjusting the modulation duty cycle for high refractive index contrasts targeting strong grating response. This method was practiced on fused silica and borosilicate glasses, but only optimized for fused silica due to low thermal diffusion and low heat accumulation effects.

An ultrashort fiber laser systems (*IMRA America: FCPA μJewel D400-VR*) delivered laser pulses at 1045 nm with adjustable repetition rate from 0.1 to 5 MHz and $M^2$ value of ∼1.3. The laser layout and its beam delivery system for waveguide writing are shown in Fig. 9.10. The uncompressed pulses from the laser system were compressed by an external grating compressor, yielding ∼5-mm diameter output laser beam with ∼400-fs pulse duration and ∼400 mW maximum power. A half-wave plate rotated the linear laser polarization which was passed through a linear polarizer for laser power adjustment. An acousto-optic modulator (AOM; NEOS 23080–3–1.06-LTD) was introduced in the beam path to modulate the incoming laser beam. The beam was reduced to ∼1-mm diameter with a telescope (T1) to allow fast (∼150-ns) rise time response of the AOM while the following telescope (T2) re-collimated the beam back to ∼5-mm diameter.

Shah et al. reported that the second harmonic generation (SHG) of the MHz laser system at 523 nm is more efficient for inscribing low-loss waveguides inside fused silica glasses [5]. In contrast, Eaton and coworkers demonstrated effective waveguide writing using the fundamental wavelength of the MHz laser inside other types of glasses such as borosilicate [30]. Hence, an optional SHG beam line (dotted box in Fig. 9.10) was used to frequency-double the AOM output beam when needed. The SHG setup was composed of a temperature-controlled 3-mm thick lithium triborate (LBO) crystal, two convex lenses (L1 and L2, focal lengths 200 mm) to achieve high focusing for efficient SHG conversion, and a short-pass filter (SP) placed after convex lens L2 to filter out the fundamental laser wavelength. When the fundamental wavelength was used, the SHG beam line was bypassed. Further, the laser power could be adjusted with a tuneable ND filter that follows. Similar to the setup for kHz laser system (previous section), the MHz laser beam was also directed by a set of turning mirrors (TMs), and then focused down to a few microns diameter beneath the surface of the glass samples. The focusing lens was fixed on the same linear stage and adjusted vertically to focus the laser beam to a pre-defined depth beneath the glass substrate. The samples were mounted on 2D air bearing motion stages and translated transversely to the laser beam direction. A quarter-

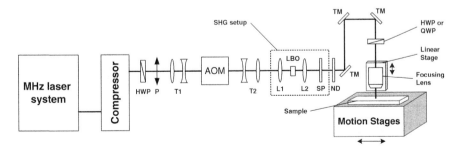

**Fig. 9.10** MHz laser system and the waveguide writing arrangement. See text for definitions of symbols

wave plate (QWP) or a half-wave plate (HWP) was placed before the focusing lens to control the laser polarization to be either circular or linear (either parallel or perpendicular to the sample scan direction).

With the AOM turned off, a non-modulated laser beam was delivered to the fused silica samples and continuous waveguides were fabricated and optimized as the basic building block of burst writing of BGWs. Propagation losses and guided mode profiles were measured and optimum processing parameters for low-loss waveguides were identified for laser repetition rate of 250 kHz to 1 MHz. Repetition rates higher than 1 MHz resulted in low SHG power that was insufficient to yield low-loss waveguides in fused silica. The SHG laser power was adjusted from 10 mW to the maximum available power at each repetition rate, and the sample scan speed was tested from 0.05 mm s$^{-1}$ to 5 mm s$^{-1}$ for all repetition rates. The waveguides were fabricated ~75 μm below the surface of a 25-mm long fused silica sample (Corning 7980, 50 mm × 25 mm × 1 mm). Laser polarization was kept parallel to the sample scan direction. Table 9.1 shows the optimum laser conditions found for low-loss waveguide writing at various laser repetition rates under the conditions of continuous waveguide writing.

With the low-loss waveguide writing recipe in hand, AOM modulation was used to introduce pulse bursts aiming for BGW structures with periodic refractive index modulations inside the fused silica glass. Due to power losses induced by various optics, the AOM, and the SHG conversion in the beam deliver system, the maximum SHG average power reaching the sample was about 80 mW for the first-order AOM diffracted beam. Compare this value with the data presented in Table 9.1, 500 kHz laser repetition rate was selected as the highest repetition rate for ample power delivery in low-loss BGW fabrication.

The first-order diffracted beam from the AOM setup was applied to give 0 to 100% intensity modulation (on/off) of the incoming laser beam for maximum possible refractive index contrast among the voxels. Figure 9.11 illustrates the modulation output of the beam. The vertical bars in the top row represent the SHG pulses from the high-repetition rate laser when the AOM is off, which is equivalent to 100% modulation duty cycle (ratio of on-time compared to the modulation period). When turned on, the AOM was fed with square waveforms from a function generator with duty cycles adjustable from 20% to 100%. The middle and bottom

**Table 9.1** Optimum waveguide writing conditions for the MHz laser system

| Laser repetition rate (kHz) | Optimum laser conditions | | Waveguide properties | |
|---|---|---|---|---|
| | SHG power (mW)/pulse energy (μJ) | Scan speed (mm s$^{-1}$) | $L_p$ (dB cm$^{-1}$) | MFD (μm) |
| 250 | 50/0.2 | 0.12 | 0.8 | 11.2 |
| 500 | 75/0.15 | 0.25 | 0.5 | 11.0 |
| 750 | 100/0.13 | 0.35 | 0.6 | 11.5 |
| 1000 | 150/0.15 | 0.5 | 0.3 | 11.3 |

MFD: mode field diameter; $L_p$: propagation loss

# 9 3D Bragg Grating Waveguide Devices

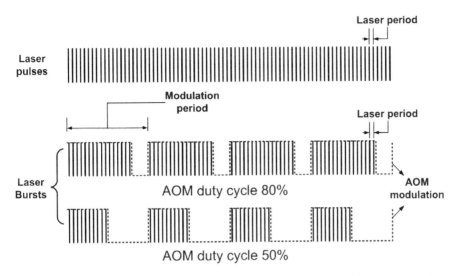

**Fig. 9.11** AOM modulation for burst generation. The continuous high-repetition rate laser pulses (*top*) were modulated by the AOM with a square waveform (*dotted envelopes in middle and bottom rows*) generating a series of laser pulse bursts. The modulation frequency and duty cycle were controlled by an external function generator

row of the figure illustrate the generated laser bursts with duty cycles of 80% and 50%, respectively. The modulation envelope is represented by the dashed square wave. The AOM modulation frequency and modulation duty cycle were controlled by an external function generator.

The modulated laser burst was used to form 25-mm long BGWs inside fused silica glasses. In this case, the index modulation periodicity, $\Lambda$, was determined collectively by the modulation frequency, $f$, and the scan velocity, $v$, by $\Lambda = v/f$. The Bragg condition can be expressed as

$$\lambda_B = 2n_{\text{eff}} \Lambda = \frac{2n_{\text{eff}} v}{f}, \tag{9.2}$$

where $\lambda_B$ is the BGW Bragg wavelength where the maximum reflection occurs and $n_{\text{eff}} \approx 1.445$ is the effective refractive index of the waveguide which is close to the refractive index of fused silica as expected. A modulation frequency of $f = 500$ Hz and scan speed of $v = 0.2678$ were applied for an index modulation period of $\Lambda = 535.6$ nm, targeting Bragg resonance near the 1550 nm telecom band.

With such a modulated laser beam, each burst of pulses formed a single refractive index voxel inside the glass sample during the laser scan process. The top row of Fig. 9.12 shows the overhead microscope images of the BGWs written with 20, 50, 80, and 100% AOM duty cycles, recorded with a 40× objective. With the decrease of the AOM duty cycle, the waveguide appearance changed gradually from uniform at 100% to inhomogeneous at 20%, caused by the increasing discontinuity of

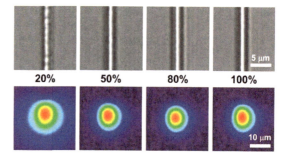

**Fig. 9.12** Overhead microscope images of the BGWs written with 20, 50, 80, and 100% AOM duty cycles (*top row*) inside fused silica glass, and near-field guided modes for these waveguides at 1560 nm shown in the *bottom row*. The laser repetition rate was 500 kHz

the refractive index along the waveguide path. Near-field guided modes for these waveguides at 1560 nm are shown in the bottom row of Fig. 9.12, which can be well-represented by Gaussian profiles in both transverse directions with ∼1.1 aspect ratios (slightly larger vertically). For lower modulation duty cycles, the smaller voxel sizes and larger voxel gaps resulted in less optical confinement due to lower average refractive index changes, which is confirmed by the mode diameters that increased from ∼12 µm at 100% duty cycle to ∼18 µm at 20%. By using the Lumerical MODE Solutions software and matching the simulated mode size with measurement, the refractive index change of the BGW written with 100% duty cycle was estimated to be ∼0.01 for the 2-µm diameter of the waveguide as inferred from Fig. 9.12.

Attenuation of the laser beam power by the AOM modulation lowered the waveguide index contrast. Nevertheless, an AC index modulation was demonstrated to induce strong Bragg effects and low-loss waveguiding simultaneously over a range of AOM duty cycles adjustable from 20 to 100%. As shown in Fig. 9.13, a high transmission notch of 35 dB with a strong reflection of ∼90% was obtained at a Bragg wavelength of 1547.95 nm for 60% duty cycle, matching the expected Bragg wavelength of $\lambda_B = 2n_{\text{eff}}\Lambda = 2n_{\text{eff}}v/f = 1547.96$ nm. The bandwidth of this BGW is 0.2-nm (3-dB bandwidth) in both transmission and reflection spectra. The ∼2 dB radiation mode loss that appears at shorter wavelength side is also much less than the >5 dB loss for the BGWs formed by single-pulse method, e.g., in Fig. 9.7, which is a significant improvement for telecom applications where such losses might induce crosstalk in closely separated optical channels.

The effective refractive indices ($n_{\text{eff}}$) of the BGWs follow the changes in the BGW voxel size, separation, and individual material modifications that are controlled by the laser exposure condition, which can be directly inferred from the waveguide Bragg wavelength according to $\lambda_B = 2n_{\text{eff}}\Lambda$, where $\Lambda = 535.6$ nm is the grating period (see (9.2)). Figure 9.14 shows the observed Bragg wavelength and the calculated effective index as a function of the AOM duty cycle. Below 50% duty cycle, the increase of the Bragg wavelength is linear with duty cycle indicating a linear change in the waveguide effective index. The data extrapolate to intersect the wavelength axis at 1547.34 nm, corresponding to the pristine index of the fused silica glass of 1.444 [31]. At duty cycles higher than 50%, the effective

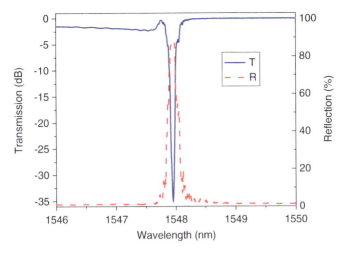

**Fig. 9.13** The transmission and reflection spectra of the BGW written in fused silica with 60% AOM modulation duty cycle. The laser repetition rate was 500 kHz and AOM modulation frequency was 500 Hz. The grating strength is 35 dB and the maximum reflection is ∼90%

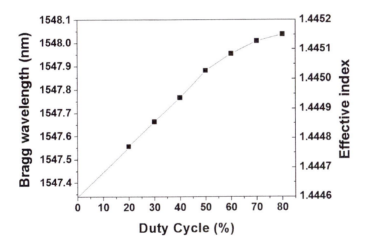

**Fig. 9.14** Bragg wavelength dependence of the grating waveguides on the AOM modulation duty cycle. The BGWs were written inside fused silica with the 500-kHz laser system

index deviates from the linear increase due to more overlapping of the individual voxels and a saturation of the available refractive index change.

Figure 9.15 illustrates the change of grating strength (top graph), peak reflectivity (top graph), and the propagation loss (bottom graph) of the BGWs with respect to the AOM duty cycle. The propagation loss increases significantly from 0.5 dB cm$^{-1}$ for smooth waveguides (100%) to 2.5 dB cm$^{-1}$ for the discontinuous waveguide track at 20% duty cycle. On the other hand, the grating strength grows rapidly as the duty cycle decreases from 100% to 60%, owing to larger AC refractive index

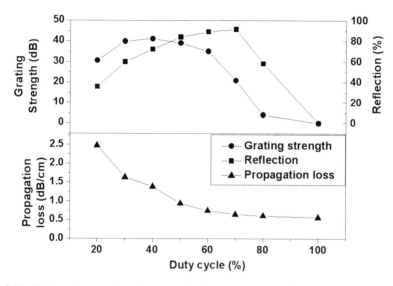

**Fig. 9.15** Change of grating strength, peak reflection percentage, and the propagation loss of the fused silica BGWs with respect to the AOM duty cycle

modulation induced by smaller voxel sizes and larger voxel gaps for smaller duty cycles. The grating strength peaks at 41.2 dB with ∼70% reflection for 40% duty cycle and drops thereafter due to optical losses induced by weaker guiding. The BGW reflectivity follows a similar trend but peaking at 91% reflectivity for a higher 70% duty cycle, indicating a higher sensitivity to waveguide losses for the reflection peak in comparison with the transmission peak as the duty cycle decreases.

The duty cycle of the modulation is an important control parameter that affects every aspect of the BGW spectral response. As Fig. 9.15 demonstrates, several trade-offs exist when optimizing BGW grating strength, reflectivity, and the propagation loss. As a result, the AOM duty cycle should be chosen according to the application requirement. The BGW written with 60% modulation offers a practically useful combination of strong Bragg resonance (35-dB transmission, 89% reflection) and moderately low-loss ($\sim$0.6 dB cm$^{-1}$) guiding, with additional improvements expected with further optimization of other laser exposure conditions.

### 9.2.3 BGW Thermal Stability

Bragg grating waveguides have strong potential applications in telecom and sensing, where high thermal stability and high temperature endurance are required. The single-pulse written borosilicate BGWs (Sect. 9.2.1) and the burst written fused silica BGWs (Sect. 9.2.2) were both characterized in several heating cycles beginning at 250°C for 1, 1, 2, and 4 h (i.e., total accumulated 1, 2, 4, and 8 h), then for 1 h at 500°C, and then for one 1 h at 750°C. The fused silica BGWs

# 9 3D Bragg Grating Waveguide Devices

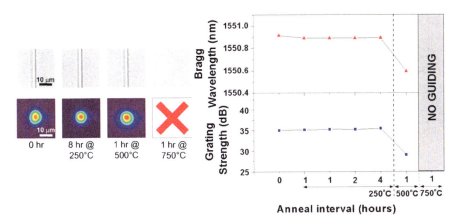

**Fig. 9.16** Microscope images (*upper left corner*) and mode profiles (*lower left corner*) of the single-pulse written borosilicate glass BGW together with (*right*) plots of Bragg wavelength shift and transmission grating strength under various heat cycles

were further annealed for an additional 1-hour period at 1000°C. After each thermal cycle, the samples were cooled to room temperature, observed under an optical microscope, and characterized for their mode profiles, Bragg responses, and waveguide birefringence.

Figure 9.16 shows the images and the guided modes of the single-pulse written BGW discussed in Fig. 9.7, with the annealing cycle noted at the bottom. No change in the waveguide morphology (top figures) was found following the 250°C and 500°C heating steps. However, strong fading is noted after the 750°C annealing step that exceeds the 666°C strain point for the borosilicate glass. Waveguiding was no longer observable after this 750°C cycle. Inspection of mode profiles revealed strong degradation after the 500°C heat cycle, with mode diameter increasing from 10 μm to 18 μm. Figure 9.16 illustrates the shifts of Bragg wavelength and grating strength following each thermal cycle. The BGW remained highly stable with no degradation of grating strength following each of the four annealing steps at 250°C. A small 0.015-nm shift of Bragg wavelength occurred only for the first 250°C heat cycle, representing a small $\sim 1 \times 10^{-5}$ decrease in refractive index that may be attributed to a thermo-optic response due to ±1°C room temperature fluctuation. The high-strength BGW appeared most stable, and generated very little overall degradation especially compared with FBGs fabricated with traditional ultraviolet lasers [32], suggesting good prospects for meeting Telecordia standards for 20-year product lifetime.

Figure 9.17 shows the response of the burst-written high-strength BGW in fused silica (as shown in Fig. 9.13) after the same thermal annealing processes. There are no visual changes in BGW morphology after all the heating cycles, even for temperature of 1000°C that is higher than the 893°C strain point for fused silica glass. The mode profiles remained almost identical with only a slight increase in diameter until 500°C, but expanded to a poorly confined mode at 750°C as shown in

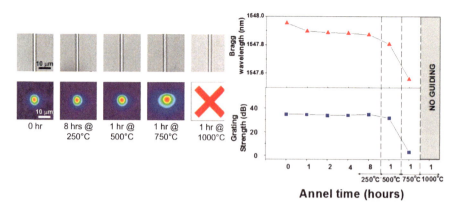

**Fig. 9.17** Microscope images (*upper left corner*) and mode profiles (*lower left corner*) of BGWs written in fused silica glass with burst method of 60% AOM duty cycle, under various heat cycles of annealing together with (*right*) plots of Bragg wavelength and grating strength under various heat cycles

the bottom row of Fig. 9.17. No guiding was observed after the annealing procedure at 1000°C.

A slight increase of the mode diameter from ~12 μm to ~14 μm was observed after the heating cycles for 8 h at 250°C. The mode diameter increased slightly to ~15 μm after baking at 500°C for 1 h, but expanded quickly to ~17 μm after 1-hour of annealing at 750°C, indicating less light confinement caused by reduced average waveguide index. The waveguide propagation loss remained stable at ~0.6 dB cm$^{-1}$ after baking cycles at 250°C and 500°C, but increased to 2.2 dB cm$^{-1}$ loss after annealing at 750°C, indicating dramatic structural changes of the BGW under such high temperature. The BGW grating strength stayed stable (~30 dB) for annealing cycles under 250°C and decayed only slightly by ~3 dB at 500°C. After the 750°C heating, the grating transmission notch decreased dramatically to only ~3 dB representing a larger reduction in the refractive index modulation contrast. The Bragg wavelength change of 0.246 nm after the 750°C annealing points to a relaxation of the effective refractive index change of $2.3 \times 10^{-4}$, representing ~2% of the total refractive index change of ~0.01 as obtained in Sect. 9.2.2. These results show that the BGWs written inside the fused silica glass can endure higher temperature with better stability compared to the BGWs written inside borosilicate glass, which is in agreement with their high and low strain point of 893°C (fused silica) and 666°C (borosilicate), respectively. As a result, fused silica BGWs are preferred for high-temperature applications.

## 9.3 BGW Devices

Bragg grating devices have broad applications. Because of the sharp transmission attenuation notch at the Bragg wavelength, these devices can be used as narrow-band filters in optical communications [33, 34]. The near-unity reflections of such devices at the Bragg resonance also render them extremely useful in add/drop devices [35]. Chirped gratings, i.e., gratings with gradually changed periodicity, can be used not only as broad-bandwidth filters or reflectors but also as dispersion compensation devices essential to high-speed long-distance optical networks [33]. Further, because the grating Bragg wavelength is very sensitive to environmental variables such as temperature, stress, bending, and pressure, they have been widely applied as high-precision sensors [36].

While numerous publications address the application of FBG devices, we present here the first application of the new BGW devices. The purpose here is not to encompass as many directions as possible, but rather to demonstrate the feasibility of the BGWs as practical devices in telecom and sensing applications, to stimulate research interests among the photonics community, and finally to initiate development efforts to make the BGW a competitive technology serving unique possibilities in 3D optical circuits.

### 9.3.1 Multi-wavelength BGWs

Wavelength tuning in traditional FBG fabrication is expensive and time consuming. In practice, a separate phase mask is fabricated with e-beam or other expensive means for each wavelength channel needed. When cascading, multiple fabrication steps must be taken in serial, with each step requiring tedious alignment.

On the other hand, wavelength tuning and device cascading for the new BGW devices are shown to be facile and fast processes. As demonstrated in Sect. 9.2, the Bragg wavelength of a BGW depends on the voxel array periodicity which is in turn determined by the sample scan speed and the laser repetition rate (single-pulse writing) or the modulation frequency (burst writing). One can easily tune the BGW wavelength by adjusting either one of these parameters. In this section, fine scan speed adjustment during laser exposure was demonstrated to effectively tune the BGW wavelength over the whole 1200 to 1600-nm telecom band.

Since wavelength tuning was easily obtained, cascading of multi-wavelength BGW devices is demonstrated next by changing the scan speed during the same one-step BGW writing process. One advantage of the BGW cascading is that the BGW segments could be closely connected, which is challenging in traditional interference-based FBG fabrication techniques.

Figure 9.18 illustrates the dependence of the measured Bragg reflection wavelength verses the sample scan speed, for the BGWs written inside borosilicate glasses (EAGLE2000) with 320-fs duration, 3 µJ pulse energy, and 0.25-NA

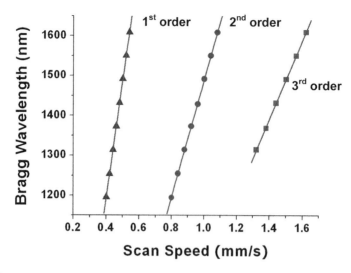

**Fig. 9.18** Controllable selection of Bragg resonance wavelength versus scan velocities of the BGWs written in borosilicate glasses with the single-pulse technique, in the first, second, and third order

aspheric lens focusing, using the single-pulse fabrication technique as described in Sect. 9.2. Scan velocities around 0.5, 1.0, and 1.5 mm s$^{-1}$ yielded Bragg wavelengths in the telecom band. The data are clustered in three groups representing first, second, and third-order Bragg reflections, respectively. Each cluster is accurately represented (solid lines) by the Bragg condition, $\lambda_B = 2n_{eff}v/mR$, where $m$ is an integer representing the grating order, with decreasing line slopes yielding $n_{eff}$ values of 1.479, 1.477, and 1.473, respectively, as expected because increasing scan speed (increasing order) reduces the average refractive index changes. Scan speeds from 0.42 to 0.56 mm s$^{-1}$ provided first-order Bragg resonances from 1200 nm to 1600 nm, fully covering the telecom band.

Cascading of multiple BGW devices was simply demonstrated by abruptly changing the sample scan speed during a one-step laser scan process. This method was implemented for writing a 50-mm long BGW inside a borosilicate glass (EAGLE2000) with four 12.5 mm long segments with scan speeds of 0.5168, 0.5201, 0.5235, and 0.5268 mm s$^{-1}$. Figure 9.19 illustrates the spectra of this cascaded BGW where four distinct resonances are sharply defined both in reflection and transmission. The 1571-nm Bragg segment closest to the input facet provided the strongest reflection amplitude of 41% compared with 18% for the most distant segment at 1542.2 nm. The 0.6-dB cm$^{-1}$ waveguide loss contributes 1.2, 2.4, and 3.6 dB additional attenuation to the second, third, and fourth BGW segments, yielding corrected reflectance values of 44%, 49%, and 41%, respectively, that better match the first segment reflection. Each resonance faces similar transmission loss of 3 dB to yield comparable Bragg strengths of $\sim$10 dB.

# 9 3D Bragg Grating Waveguide Devices

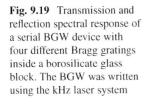

**Fig. 9.19** Transmission and reflection spectral response of a serial BGW device with four different Bragg gratings inside a borosilicate glass block. The BGW was written using the kHz laser system

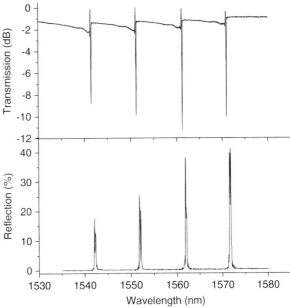

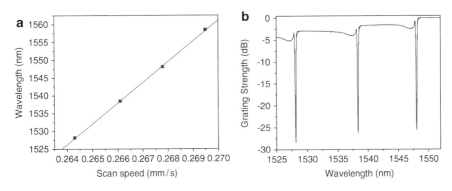

**Fig. 9.20** First-order Bragg wavelength tuning by velocity change (**a**) and the transmission spectrum of the cascaded BGW (**b**). These BGWs were fabricated in fused silica glass with the burst fabrication method, using the MHz laser system

For the burst fabrication method, AOM modulation frequency and scan speed adjustment are both effective in BGW wavelength tuning and cascading. Laser parameters and focusing conditions used in Sect. 9.2.2 were employed for this set of experiments with 60% AOM modulation duty cycle. In Fig. 9.20(a), scan speeds of 0.2643, 0.2661, 0.2678, and 0.2695 mm s$^{-1}$ yielded Bragg resonances at 1528.11, 1538.37, 1548.04, and 1558.42 nm, respectively. A linear fit (thin solid line) confirmed the linear dependence of the Bragg wavelength on the sample scan speed. Figure 9.20(b) shows the transmission spectrum of the cascaded BGW devices with three distinct wavelengths at 1528.11, 1538.37, and 1548.04 nm, written with a similar one-step process as used in the previous section by applying two abrupt scan speed changes during laser writing.

## 9.3.2 Chirped BGWs

Chirped Bragg gratings are Bragg gratings with non-uniform grating period along their lengths [37]. The chirp in gratings could take many forms including linear chirp, quadratic chirp, and other spatial dependencies. Linearly chirped Bragg grating, especially, has broad applications in telecom applications in dispersion compensation [38], chirped pulse amplification [39], and ASE rejection [40]. The chirped gratings can be fabricated by using gradually varied temperature [41], strain [42], or refractive index profiles [33], or by tilting or curving the fiber during the laser exposures [43, 44]. These methods are somewhat limited in the addressable wavelength range, and lack adequate control or flexibility of the chirping. For this reason, point-by-point writing with tightly focused laser light is a promising alternate method that provides more control and flexibility for complex spatial profiling and chirping of the grating response.

Since the period of the BGWs can be easily adjusted by the sample scan speed, it is also possible to use acceleration in the motion stages to introduce aperiodic gratings with linear or higher order chirps. Compared to the methods mentioned in the previous paragraph that involve complex procedures and difficult alignments, this method provides dramatically improved controllability and flexibility with a simple one-step scan. In this section, a linearly chirped BGW was demonstrated by varying the laser scan speed linearly during the fabrication process.

The chirped BGWs were fabricated with the burst writing technique inside a 25-mm long fused silica glass block. Laser parameters and focusing conditions were the same as in Sect. 9.2.2, with the AOM set at 60% duty cycle and 500 Hz frequency. The sample was scanned near the velocity $v = 0.2678$ mm s$^{-1}$ targeting center Bragg wavelength at $\sim$1550 nm. During the sample scan, the acceleration was varied targeting BGW bandwidth, $\Delta\lambda$, of 0.2 to 20 nm.

From (9.2), the scan speed for a Bragg wavelength $\lambda_B$ can be expressed as $v = \lambda_B f / 2n_{\text{eff}}$. For a BGW designed for bandwidth of $\Delta\lambda$, the starting scan speed ($v_1$), finishing scan speed ($v_2$), and the scan acceleration ($a$) should satisfy, respectively,

$$v_1 = \frac{\left(\lambda_B - \frac{\Delta\lambda}{2}\right) f}{2n_{\text{eff}}}; \quad v_2 = \frac{\left(\lambda_B + \frac{\Delta\lambda}{2}\right) f}{2n_{\text{eff}}}; \quad a = \frac{v_2^2 - v_1^2}{2L}. \tag{9.3}$$

Chirped BGWs were fabricated according to velocities and accelerations calculated from (9.3). Figure 9.21 illustrates the measured reflection spectra of the laser-written chirped BGWs with designed bandwidth of 0.2, 1, 2, 5, 10, and 20 nm. Slightly deviated from flat-top reflection spectra, the reflection curves show oscillations and

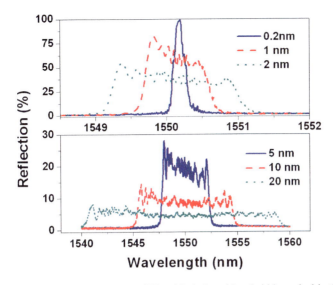

**Fig. 9.21** Reflection spectra of chirped BGWs with designed bandwidth marked in the spectrum graphs. Burst laser writing was applied to fused silica glass

**Fig. 9.22** Peak transmission strength and reflection of the chirped BGWs in Fig. 9.21

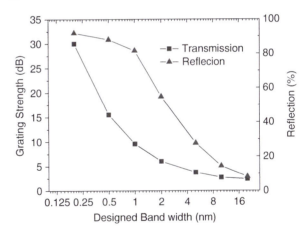

a slight slope. Fluctuation in the reflectivity from different BGW segments likely arises from non-uniform acceleration of the target motion stages. The stronger reflection at short wavelength is possibly due to the increased refractive index contrast on the corresponding short-period side of the BGW, which received higher laser exposure due to slower writing speed.

The peak reflection and transmission strength of the chirped BGWs are plotted in Fig. 9.22. Because all the chirped BGWs have the same length (25 mm), the effective grating length per unit bandwidth decreases with increasing bandwidth (0.2 to 20 nm), causing a rapid decrease of the peak reflection from 95% to 8% and a similar drop of grating transmission strength from 30 dB to 2 dB.

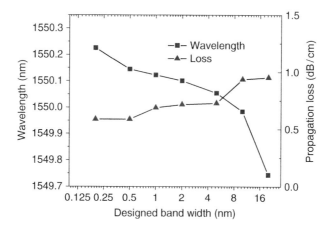

**Fig. 9.23** Center Bragg wavelength and waveguide propagation losses at 1560 nm, of the BGWs shown in Fig. 9.21

**Table 9.2** Calculated dispersion properties of the BGWs in Fig. 9.21

| BGW bandwidth (nm) | Dispersion (ps nm$^{-1}$) |
|---|---|
| 0.2 | 1204.17 |
| 0.5 | 481.67 |
| 1 | 240.83 |
| 2 | 120.42 |
| 5 | 48.17 |
| 10 | 24.08 |
| 20 | 12.04 |

As shown in Fig. 9.23, the BGW center wavelength also decreased as the bandwidth increases, caused by lower net exposure introduced by higher velocity after a period of acceleration. The BGW propagation loss dependence on the chirp bandwidth is shown in the same graph, which increased from 0.5 dB cm$^{-1}$ at 0.2-nm bandwidth to 0.95 dB cm$^{-1}$ for 20-nm bandwidth. More optical scattering from increased waveguide roughness or discontinuities for larger acceleration values is the possible cause of this increased loss.

The time delay of the light reflected from the long-wavelength side of the BGW compared to the short wavelength side, $\tau$, can be estimated from $\tau = 2Ln_{\text{eff}}/C$ [45], where $L$ is the BGW length, $n_{\text{eff}} = 1.445$ is the waveguide effective index, and $C$ is the speed of light in vacuum. For the 25-mm BGW, $\tau$ is calculated to be 240 ps. The dispersion properties of the chirped BGWs are estimated as $D = \tau/\Delta\lambda$ [45,46] and are listed in Table 9.2. Dispersion measurements were not carried out here.

**Fig. 9.24** A 3D sensor network in a fused silica plate (50 mm × 50 mm × 1 mm) consisting of multi-wavelength BGW segments (*red lines*) laid out as shown in the schematic of (**a**) and the photograph of (**b**). Single-mode fibers are shown epoxied for butt-coupling to six of the BGW segments here

### 9.3.3 3D BGW Sensor Network

Bragg gratings have broad applications in sensing of various physical quantities such as temperature, strain, pressure, and acceleration [47–49], due to their modest fabrication cost and high sensitivity to the environmental parameters. As an example, a 3D sensor network is presented in this section representing the versatile sensing capabilities of the BGW devices. Due to the high thermal stability, fused silica BGWs were fabricated and tested for 3D temperature and strain detection.

Figure 9.24 illustrates a 3D optical sensor network which was directly fabricated inside a transparent fused silica glass substrate (Corning 7980) of 50 mm × 50 mm × 1 mm dimension. The waveguides were written with the femtosecond fiber laser (IMRA FCPA μJewel D-400-VR) providing ∼300-fs duration pulses at 522-nm wavelength with 500-kHz repetition rate. The laser beam was modulated by an acousto-optic modulator (AOM; NEOS 23080–3–1.06-LTD) to form laser bursts with 60% duty cycle and 500-Hz modulation frequency. The laser bursts were focused by a 0.55 numeric aperture (NA) aspheric lens to ∼1-μm spot size ($1/e^2$ diameter) below the surface of the glass substrate. The glass sample was mounted on two-dimensional (2D) air bearing motion stages (Aerotech ABL1000 with 2-nm resolution and 50-nm repeatability) and translated transversely to the laser propagation direction with the laser polarization parallel to the translation direction. During the waveguide writing process, the laser bursts formed a series of refractive index voxels that simultaneously provided low-loss waveguiding and high-strength Bragg resonance as first demonstrated in [50] and [51].

For each waveguide in Fig. 9.24(a), three 16.7-mm long BGW segments with different Bragg resonant wavelengths, $\lambda_B$, were cascaded in series, by scanning each segment with a different scan speed, $v$ [52]. The Bragg wavelength was given by:

$$\lambda_B = 2n_{\text{eff}}\Lambda = 2n_{\text{eff}}\, v/f \qquad (9.4)$$

where $\Lambda$ is the grating period, $f = 500\,\text{Hz}$ is the AOM modulation frequency, and $n_\text{eff} \approx 1.445$ is the effective index of the BGW as obtained previously [50] for the same exposure parameters. Two identical layers of BGW networks were laser-written at 75-µm beneath the top and above the bottom surfaces of the glass chip. For each layer, two rows and two columns of cascaded BGWs divided the glass chip into 9 sensing zones of 16.7 mm × 16.7 mm in size. The Bragg wavelength of each BGW sensor segment, as labeled in Fig. 9.24(a), is 1530.0 nm for H1, H4, V1, and V4, 1540.0 nm for H2, H5, V2, and V5, and 1550 nm for H3, H6, V3, and V6. After laser fabrication, the BGW waveguides were butt-coupled with standard SMF28 fibers and bonded with UV curing polymers as shown in Fig. 9.24(b).

As described in [51], light from a broadband source (Thorlabs ASE-FL7002, 1530 nm to 1610 nm) was passed through a fiber circulator and launched into the BGW segments through the single-mode fibers for spectral analysis. Back-reflected light from the sample was routed by the circulator and recorded by an optical spectrum analyzer (OSA, Ando AQ6317B) with 0.01-nm resolution. The reflection spectra were normalized against the power returned by a high reflectivity fiber reflector ($R = 96\%$). The Bragg resonance wavelength of the various BGW segments were recorded over a 25 to 125°C temperature range and 0 to ~500 µε strain range by heating and bending the glass substrate, respectively.

By differentiating (9.4), the shift in the Bragg grating wavelength, $\Delta\lambda_\text{B}$, separates into strain-optic and thermo-optic components given by the induced strain and temperature changes, $\varepsilon_z = \Delta l/l$ and $\Delta T$, respectively, and following [53]:

$$\Delta\lambda_\text{B} = 2\left(\Lambda\frac{\partial n}{\partial l} + n\frac{\partial\Lambda}{\partial l}\right)\Delta l + 2\left(\Lambda\frac{\partial n}{\partial T} + n\frac{\partial\Lambda}{\partial T}\right)\Delta T. \quad (9.5)$$

The first term in (9.5) accounts for changes in the grating spacing and refractive index, $n$, induced by the strain-optic effect, and can be further expressed as [54]:

$$\Delta\lambda_\text{B-st} = \lambda_\text{B}(1 - p_\text{e})\varepsilon_z, \quad (9.6)$$

where the effective strain-optic term is given by $p_\text{e} = n^2[p_{12}-\nu(p_{11} + p_{12})]/2$, and $p_{11}$, $p_{12}$ are components of the strain-optic tensor and $\nu$ is Poisson's ratio. For fused silica, Borrelli and Miller applied ultrasonic methods [55] to report $p_{11} = 0.126$, $p_{12} = 0.26$, $\nu = 0.168$, and $p_\text{e} = 0.204$. Using these values and $n_\text{eff} = 1.445$, the expected strain-optic response near 1550 nm is estimated as $\Delta\lambda_\text{B}/\varepsilon_z \approx 1.23\,\text{pm}/\mu\varepsilon$.

The second term in (9.2) follows changes in grating spacing and waveguide refractive index due to thermal expansion and can be simplified to [54]:

$$\Delta\lambda_\text{B-th} = \lambda_\text{B}(\alpha + \zeta)\Delta T, \quad (9.7)$$

where $\alpha = 0.55 \times 10^{-6}/°\text{C}$ is the thermal expansion coefficient for fused silica [17] and $\zeta$ is the thermo-optic coefficient of the waveguide core. For a single-mode fiber, $\zeta$ was measured to be $8.6 \times 10^{-6}/°\text{C}$ [53], leading to an expected temperature

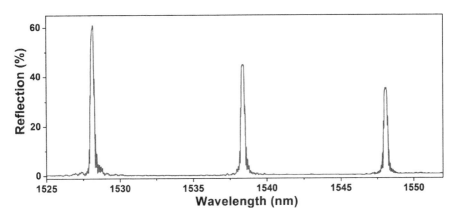

**Fig. 9.25** Reflection spectrum of BGW segments H1, H2, and H3 (*left to right*)

sensitivity of $\Delta \lambda_B / \Delta T = 13.7$ pm/°C at 1550-nm wavelength for a FBG. A similar value is anticipated for the bulk BGWs formed in fused silica.

Figure 9.25 shows a typical reflection spectrum collected from BGW segments H1, H2, and H3, collected from the probe fiber directly bonded to segment H1. The Bragg wavelengths for the three reflection peaks at 1528.2 nm, 1538.4 nm, and 1548.1 nm slightly underestimate (~0.1%) the design values of 1530 nm, 1540 nm, and 1550 nm, respectively. These small shifts arise from experimental variations in laser exposure and indicate an effective index of $n_{\text{eff}} \approx 1.4435$ as opposed to 1.445 obtained previously [50]. The maximum reflectivity of the three peaks decreases from 60.2% (2.2 dB) to 44.7% (3.5 dB) and to 35.3% (4.5 dB) due to higher waveguide propagation loss for BGW segments H2 and H3 which have ~33 mm and ~66 mm extra travel distance from the measurement facet than for the segment H1. This attenuation indicates a ~0.3 to 0.4 dB cm$^{-1}$ propagation loss which is slightly lower than the 0.6 dB cm$^{-1}$ propagation loss reported in [50] for BGWs written with the same exposure conditions. The ~0.3 nm spectral bandwidth (full width half maximum: FWHM) of all three peaks slightly exceeds the ~0.2-nm width reported for the 25-mm long BGWs in [50] due to shorter grating length.

The thermal response of the BGW devices was tested in the 25 to 125°C range by monitoring the Bragg wavelength shifts of the sensor during physical contact with a hotplate. Because the glass sample was thin (1-mm thick by 50-mm wide) and uniformly heated, a one-dimensional heat transfer model [18] could be used to estimate the temperature at each BGW layer ~75 μm from the glass surfaces. Given a 1.3 W m$^{-1}$ K$^{-1}$ thermal conductivity for the fused silica and a convective heat transfer coefficient of ~10 W m$^{-2}$ K$^{-1}$ for air [19], a maximum temperature drop of only 0.07°C at 35°C to 0.71°C at 125°C was expected across the sample thickness (bottom to top surface) and, thus, a uniform substrate temperature could be assumed.

The Bragg wavelength of a BGW with room temperature Bragg wavelength $\lambda_B = 1550.2$ nm is plotted in Fig. 9.26 as a function of hotplate temperature,

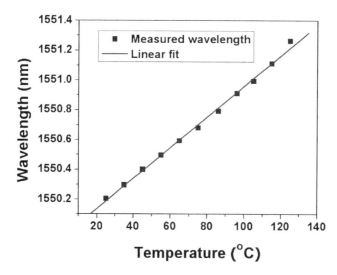

**Fig. 9.26** BGW resonance wavelength plotted with respect to temperature of the hotplate

revealing a linear thermal response (solid line) as expected from (9.4). The slope of the line is 10.4 pm/°C, yielding a thermal response that is 24% lower than the 13.7 pm/°C value for the standard SMF28 fiber. Similarly, the thermo-optic coefficient of $\zeta = 6.16 \times 10^{-6}$ for the fused silica BGW is 28% lower than the optical fiber value ($\zeta = 8.6 \times 10^{-6}$ [53]), a difference possibly arising from the 3 mol% germanium dopant in the fiber core.

The distributed temperature sensing is demonstrated in Fig. 9.27, where non-uniform heating of the glass block was provided by heating only the centre of the fused silica sensing chip via a small aluminum block (10 mm × 10 mm × 5 mm) placed between the hotplate and the 3D sensor chip. The Bragg wavelength shift (with respect to room temperature) of BGW segments H1, H2, and H3 is plotted in Fig. 9.27(a) and segments V1, V2, and V3 are plotted in Fig. 9.27(b). The wavelength shifts were converted to temperature changes using the 10.4 pm/°C thermal response obtained in Fig. 9.26 to define the right axis of the graph. It is clear from Fig. 9.27 that the center sensors, H2 and V2, are hotter than the respective peripheral sensors, H1, H3, V1, and V3, thus confirming a higher temperature in the center of the plate. Such distributed temperature sensing thus permits one to pinpoint the heat source and obtain temperature gradients in bulk glass substrates.

The 3D BGW network will respond to strain induced by either parallel or perpendicular pressure applied to the BGW. Here, only parallel pressure was studied by bending the sensor plate as shown in Fig. 9.28(a). Because of the symmetric positioning of the two BGW layers 75 μm from both surfaces, the top and bottom BGW segments will experience equal amounts of tensile and compressive strain, respectively, that shift the Bragg wavelengths by identical magnitudes but in opposite spectral directions. Further, since each Bragg grating has identical

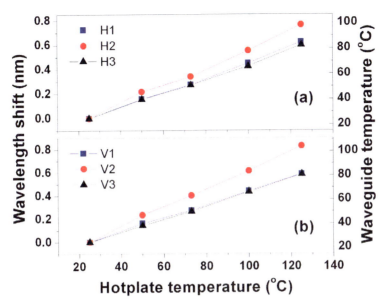

**Fig. 9.27** Observed Bragg wavelength shift and the calculated local temperature of BGW segments (**a**) H1, H2, and H3 and (**b**) V1, V2, V3, in the 3D sensor network, for hotplate temperature from 25 to 125°C

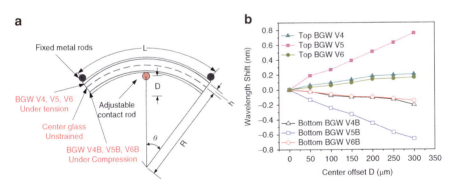

**Fig. 9.28** Beam bending arrangement (**a**) for optical strain sensing in a 3D BGW network and correspond Bragg wavelength shifts (**b**) under increasing beam displacement

temperature response, this strain-gauge arrangement is insensitive to environmental temperature fluctuation – an important advantage for field deployment.

In Fig. 9.28(a), the central rod was pushed with a micrometer against the glass plate to bend the plate between the two fixed cylindrical rods with displacements up to $D = 300\,\mu\text{m}$. The strain of the glass plate at the top and bottom waveguide can be expressed as $\varepsilon = \Delta L/L = h/R$, where $L = 50\,\text{mm}$ is the top rod-to-rod arc length subtending twice the half angle, $2\theta = L/R$, $R$ is the sample bend radius under displacement $D$, and $h = 425\,\mu\text{m}$ is the waveguide to the glass center plane distance. Using the small angle approximation, one obtains a simple relationship between the glass displacement and bend radius:

$$D = R - R\cos\theta = 2R\sin^2(\theta/2) = R\theta^2/2 = L^2/8R \quad (9.8)$$

which then gives the following relationship for strain at the BGW:

$$\varepsilon = \frac{h}{R} = \frac{8Dh}{L^2} \quad (9.9)$$

Figure 9.28(b) plots the measured Bragg wavelength shift with respect to the center beam offset, $D$, for BGW segments V4, V5, V6, V4B, V5B, and V6B, where $B$ indicates BGWs in the bottom plane. The Bragg wavelength on the top and bottom BGWs shift to longer and shorter wavelength, respectively, as expected by the respective tensile and compressive strains induced for bending as in Fig. 9.28(a). It is also clear in Fig. 9.28(b) that much smaller Bragg shifts were induced in the peripheral BGW segments (V4, V6, V4B, and V6B). These smaller shifts were accompanied by broader wavelength-chirped lineshapes that overall reflect a range of smaller bending radii and incomplete bending at the end points of the beam. In this way, 3D distributed optical strain sensing has been demonstrated for the first time in bulk transparent media.

The center BGW segments, V5 and V5B, underwent more uniform bending stress, providing good linear response of wavelength shift with beam displacement as seen in Fig. 9.28(b). From the slopes of this data, we find a BGW strain-optic sensitivity $s = \Delta\lambda_B/\Delta D$ of $2.32\,\text{pm}/\mu\text{m}$ and $-2.13\,\text{pm}/\mu\text{m}$ for the top and bottom BGW pairs, respectively. Combining $s$ with (9.9) yields the Bragg wavelength-strain coefficient, $\Delta\lambda_B/\Delta\varepsilon = sL^2/8h$, from which were inferred values of $1.38\,\text{pm}/\mu\varepsilon$ and $-1.27\,\text{pm}/\mu\varepsilon$, respectively, showing good agreement with the calculated value of $\Delta\lambda_B/\Delta\varepsilon = 1.23\,\text{pm}/\mu\varepsilon$. The sensitivities of the top and bottom BGWs were 20% and 10% greater, respectively, than the $1.15\,\text{pm}/\mu\varepsilon$ reported for standard fused silica FBGs for 1,550-nm radiation [54].

The symmetric placement of BGWs in the present 3D sensor network provides a unique and highly desirable sensing capability that can locally sample the beam bending radius and bending moments from differential strain measurement as in Fig. 9.28 and provide distributed 2D and 3D optical strain sensing. At the same time, this symmetric pairing of BGWs greatly improves the sensor precision by separating out the temperature and other environmental drifting factors which shift the BGW pairs identically in the same spectral direction.

The sensor chip can also survive exposure to very high temperature up to 500°C, according to data shown in Sect. 9.2.3, which opens sensing to a broad range of high temperature applications. The 16.7 mm spatial resolution can be further improved by introducing shorter BGW segments, which could lead to high resolution strain and temperature sensing in rigid structures such as building or bridges through to biochips and microreactors. Further, each BGW device (24 in present sample) can be written to have different Bragg wavelengths and be conveniently monitored with a single low-cost broadband source. To our best knowledge, this is the first demonstration of a 3D distributed sensor network fabricated directly inside a transparent material. Extension of this technology for sensing of other environmental quantities is expected.

## 9.4 Summary and Future Outlook

In this chapter, the Bragg grating waveguide device was introduced, optimized, and designed for applications in sensing and telecom. The BGW devices were composed of arrays of isolated or partially overlapped refractive index voxels which provide low-loss guiding and strong Bragg resonances simultaneously. Two novel methods, the single-pulse writing method and the burst writing method, were presented for forming such devices inside various types of glasses. In these methods, each waveguide voxel is defined in a single-scan step by either single-pulse or a burst of ultrashort pulses from commercial laser systems with low (<5 kHz) and high (>200 kHz) repetition rate, respectively, that drive different mechanisms for controlling voxel sizes and separations.

By optimizing the laser exposure conditions and focusing conditions, strong Bragg gratings with >35 dB transmission notches were both demonstrated using these two methods, inside borosilicate glass (single-pulse method) and fused silica (burst method) glass. The peak reflection of these BGWs exceeds 90% at the Bragg resonances showing great potential as integrated reflection mirrors in telecom and laser applications. These new results mark the highest quality 3D Bragg grating waveguides with first-order Bragg responses of >35 dB strength in transmission and >90% in reflection.

For multi-wavelength applications, both writing approaches offer maskless fabrication of BGWs with facile wavelength tuning. Sample scan velocity was proven successful in tuning the Bragg wavelength over the entire optical communication band. Other exposure or laser parameters such as laser repetition rate (single-pulse method) or the AOM modulation frequency (burst method) are also anticipated to adjust the wavelength just as effectively. Cascading of the BGW devices was demonstrated using both writing methods targeting distributed device applications, by simply applying different scan velocities at different segments. These properties all compare favorably with the traditional methods in multi-wavelength FBG fabrication in both fabrication cost and processing time. Chirped BGWs were also fabricated by applying acceleration during the sample scan process and broadband gratings with up to 20-nm response were demonstrated targeting applications in telecom such as dispersion compensation and power taps.

The BGW devices were also proven to have superior thermal stability compared with UV-written FBGs. Both the borosilicate BGWs and the fused silica BGWs showed extremely stable operation at 250°C with very minor changes in appearances, guided modes, and grating responses. The fused silica BGWs further demonstrated steady performance at high 500°C temperature. These results all surpass the performance of UV-written FBGs indicating long life time of the BGW devices under room temperature. BGW application in 3D sensing was demonstrated by forming a two-layer distributed sensor network inside fused silica glass. Theoretical expectations of the device responses to temperature and strain were calculated and compared to experimental measurements with good agreements.

The BGW devices promise to enable novel 3D applications in several directions. New challenges occur as the BGW technology quickly develops, demanding more efforts, both theoretical and experimental, for better understanding of the BGW operation. In the mean time, integration of BGW devices into optical systems can trigger revolutionary new designs of photonic devices with rich spectral filtering or sensing capabilities.

With two novel fabrication methods using the low- and high-repetition rate ultrashort lasers, respectively, this chapter successfully provides effective means of fabricating sharp first-order BGWs inside two different glass materials: borosilicate glass for the single-pulse method and fused silica for the burst method. It is highly desirable to extend such BGW fabrication methods to other transparent materials, such as other types of glasses, crystal materials, polymers, and active materials, which will open new application directions in telecom, sensing, lasing, and biophysics. Further, it is also attractive to incorporate the BGW devices with other optical devices such as directional couplers, ring resonators, power splitters, or waveguide lasers for added device functionalities.

Using a variation on the burst method described here, the Withford group have recently demonstrated a monolithic distributed feedback waveguide laser [56] in doped phosphate glass which was further optimized to achieve high 100-mW output power [57] (see Chap. 10 for more details). Using waveguide Bragg gratings as a diagnostic tool, the same group demonstrated that for doped phosphate glasses, color centers are induced by the femtosecond laser irradiation accounting for ∼15% of the refractive index change [58]. Due to the visible light generated from co-operative luminescence between neighboring Yb ions, the waveguide Bragg gratings are photo-annealed, so that thermal aging treatments are needed to obtain long-lifetime active BGWs in Yb-doped phosphate glasses [59].

Further optimizations of the two BGW fabrication methods are also needed in order to produce devices with much improved qualities. Lower propagation losses and circular mode profiles matching fiber mode could be realized through careful exploration of exposure conditions. It is important to minimize the radiation losses currently present on the short wavelength side of the BGW transmission spectrum which poses limitations of BGWs in dense wavelength division multiplexing (DWDM) applications, potentially by optimizing the overlap between individual index voxels. This could be dramatically improved by applying tighter focusing or beam shaping methods such as spatial filtering or astigmatic focusing, and targeting more circular and expanded BGW cross-sections that would reduced the strength of power coupling to the radiation modes.

The fact that the BGWs exhibit high birefringence under certain conditions could be useful in exploring the polarization properties of ultrashort laser written waveguides. It would be helpful to study the mechanisms of such birefringence formation for better control of the device birefringence targeting various application needs.

It is also very attractive to combine the 3D sensing capability of the BGWs with the recently developed microfluidic systems for spectral analysis of liquid solvents targeting novel lab-on-a-chip devices or integrated spectrometers that could replace the current bulky, expensive, and multi-component spectrometers. A recent paper applied a BGW device to accurately sense using evanescent probing, the refractive index of a liquid in a closely positioned microfluidic channel [59] (see Chap. 14 for more details).

With novel designs and integration, many more applications could be foreseen with these newly demonstrated BGW photonic devices.

# References

1. K.M. Davis, K. Miura, N. Sugimoto, K. Hirao, Writing waveguides in glass with a femtosecond laser. Opt. Lett. **21**, 1729–1731 (1996)
2. A.P. Joglekar, H.H. Liu, E. Meyhofer, G. Mourou, A.J. Hunt, Optics at critical intensity: applications to nanomorphing. Proc. Natl. Acad. Sci. U.S.A. **101**, 5856–5861 (2004)
3. K. Itoh, W. Watanabe, S. Nolte, C.B. Schaffer, Ultrafast processes for bulk modification of transparent materials. MRS Bull. **31**, 620–625 (2006)
4. K. Miura, J. Qiu, H. Inouye, T. Mitsuyu, K. Hirao, Photowritten optical waveguides in various glasses with ultrashort pulse laser. Appl. Phys. Lett. **71**, 3329 (1997)
5. L. Shah, A.Y. Arai, S. Eaton, P. Herman, Waveguide writing in fused silica with a femtosecond fiber laser at 522 nm and 1 MHz repetition rate. Opt. Express **13**, 1999–2006 (2005)
6. A.H. Nejadmalayeri, P.R. Herman, J. Burghoff, M. Will, S. Nolte, A. Tunnermann, Inscription of optical waveguides in crystalline silicon by mid-infrared femtosecond laser pulses. Opt. Lett. **30**, 964–966 (2005)
7. S. Nolte, M. Will, B.N. Chichkov, A. Tunnermann, in *Photonics West*, Waveguides Produced by Ultra-short Laser Pulses Inside Glasses and Crystals (The International Society for Optical Engineering, San Jose, 2002), pp. 188–196
8. A.H. Nejadmalayeri, P.R. Herman, Ultrafast laser waveguide writing: lithium niobate and the role of circular polarization and picosecond pulse width. Opt. Lett. **31**, 2987–2989 (2006)
9. S. Klein, A. Barsella, H. Leblond, H. Bulou, A. Fort, C. Andraud, G. Lemercier, J.C. Mulatier, K. Dorkenoo, One-step waveguide and optical circuit writing in photopolymerizable materials processed by two-photon absorption. Appl. Phys. Lett. **86**, 211118 (2005)
10. R. Osellame, S. Taccheo, M. Marangoni, R. Ramponi, P. Laporta, D. Polli, S. De Silvestri, G. Cerullo, Femtosecond writing of active optical waveguides with astigmatically shaped beams. J. Opt. Soc. Am. B **20**, 1559–1567 (2003)
11. A.M. Streltsov, N.F. Borrelli, Fabrication and analysis of a directional coupler written in glass by nanojoule femtosecond laser pulses. Opt. Lett. **26**, 42–43 (2001)
12. S.M. Eaton, W. Chen, L. Zhang, H. Zhang, R. Iyer, J.S. Aitchison, P.R. Herman, Telecomband directional coupler written with femtosecond fiber laser. Photon. Technol. Lett., IEEE **18**, 2174–2176 (2006)
13. W. Watanabe, Y. Note, K. Itoh, Fabrication of multimode interference waveguides in glass by use of a femtosecond laser. Opt. Lett. **30**, 2888–2890 (2005)
14. R. Kashyap, *Fiber Bragg gratings* (Academic, San Diego, 1999))
15. K.O. Hill, Y. Fujii, D.C. Johnson, B.S. Kawasaki, Photosensitivity in optical fiber waveguides – application to reflection filter fabrication. Appl. Phys. Lett. **32**, 647–649 (1978)

16. K.O. Hill, G. Meltz, Fiber Bragg grating technology fundamentals and overview. J. Lightwave Technol. **15**, 1263–1276 (1997)
17. K.O. Hill, B. Malo, F. Bilodeau, D.C. Johnson, J. Albert, Bragg gratings fabricated in monomode photosensitive optical fiber by UV exposure through a phase mask. Appl. Phys. Lett. **62**, 1035–1037 (1993)
18. S.J. Mihailov, C.W. Smelser, P. Lu, R.B. Walker, D. Grobnic, H. Ding, G. Henderson, J. Unruh, Fiber Bragg gratings made with a phase mask and 800-nm femtosecond radiation. Opt. Lett. **28**, 995–997 (2003)
19. E. Wikszak, J. Burghoff, M. Will, S. Nolte, A. Tunnermann, T. Gabler, in *Conference on Lasers and Electro-Optics (CLEO)*, vol. 2. Recording of fiber Bragg Gratings with Femtosecond Pulses Using a "point by point" Technique (IEEE, San Francisco, 2004), p. 2
20. A. Martinez, M. Dubov, I. Khrushchev, I. Bennion, Direct writing of fibre Bragg gratings by femtosecond laser. Electron. Lett. **40**, 1170–1172 (2004)
21. M. Kamata, M. Obara, in *Conference on Lasers and Electro-Optics Europe (CLEO Europe)*. Waveguide-based Bragg Filters Inside Bulk Glasses Integrated by Femtosecond Laser Processing (IEEE, Munich, 2005), p. 492
22. G.D. Marshall, M. Ams, M.J. Withford, Direct laser written waveguide-Bragg gratings in bulk fused silica. Opt. Lett. **31**, 2690–2691 (2006)
23. J.H. Chung, Y. Gu, J.G. Fujimoto, in *Conference on Lasers and Electro-Optics (CLEO07)*. Submicron-Period Waveguide Bragg Gratings Direct Written by an 800-nm Femtosecond Oscillator (Baltimore, Maryland, 2007)
24. H. Zhang, S.M. Eaton, P.R. Herman, Low-loss Type II waveguide writing in fused silica with single picosecond laser pulses. Opt. Express **14**, 4826–4834 (2006)
25. E.N. Glezer, M. Milosavljevic, L. Huang, R.J. Finlay, T.H. Her, J.P. Callan, E. Mazur, Three-dimensional optical storage inside transparent materials. Opt. Lett. **21**, 2023 (1996)
26. T. Gorelik, M. Will, S. Nolte, A. Tuennermann, U. Glatzel, Transmission electron microscopy studies of femtosecond laser induced modifications in quartz. Appl. Phys. A **76**, 309–311 (2003)
27. R.R. Thomson, S. Campbell, I.J. Blewett, A.K. Kar, D.T. Reid, Optical waveguide fabrication in z-cut lithium niobate (LiNbO$_3$) using femtosecond pulses in the low repetition rate regime. Appl. Phys. Lett. **88**, 111109–111101 (2006)
28. A. Zoubir, C. Lopez, M. Richardson, K. Richardson, Femtosecond laser fabrication of tubular waveguides in poly(methyl methacrylate). Opt. Lett. **29**, 1840–1842 (2004)
29. S. Sowa, W. Watanabe, T. Tamaki, J. Nishii, K. Itoh, Symmetric waveguides in poly(methyl methacrylate) fabricated by femtosecond laser pulses. Opt. Express **14**, 291–297 (2006)
30. S.M. Eaton, H. Zhang, P.R. Herman, Heat accumulation effects in femtosecond laser-written waveguides with variable repetition rate. Opt. Express **13**, 4708–4716 (2005)
31. Data sheet of fused silica glass http://www.corning.com/docs/specialtymaterials/pisheets/H0607_hpfs_Standard_ProductSheet.pdf
32. T. Erdogan, V. Mizrahi, P.J. Lemaire, D. Monroe, Decay of ultraviolet-induced fiber Bragg gratings. J. Appl. Phys. **76**, 73–80 (1994)
33. K.O. Hill, F. Bilodeau, B. Malo, T. Kitagawa, S. Theriault, D.C. Johnson, J. Albert, K. Takiguchi, Chirped in-fiber Bragg gratings for compensation of optical-fiber dispersion. Opt. Lett. **19**, 1314–1316 (1994)
34. F. Bilodeau, D.C. Johnson, S. Theriault, B. Malo, J. Albert, K.O. Hill, All-fiber dense-wavelength-division multiplexer/demultiplexer using photoimprinted Bragg gratings. IEEE Photon. Technol. Lett. **7**, 388–390 (1995)

35. L. Dong, P. Hua, T.A. Birks, L. Reekie, P.S.J. Russell, Novel add/drop filters for wavelength-division-multiplexing optical fiber systems using a Bragg grating assisted mismatched coupler. IEEE Photon. Technol. Lett. **8**, 1656–1658 (1996)
36. A.D. Kersey, A review of recent developments in fiber optic sensor technology. Opt. Fiber Technol.: Mater., Devices Syst. **2**, 291–317 (1996)
37. K.C. Byron, K. Sugden, T. Bricheno, I. Bennion, Fabrication of chirped Bragg gratings in photosensitive fiber. Electron. Lett. **29**, 1659–1660 (1993)
38. W.H. Loh, R.I. Laming, X. Gu, M.N. Zervas, M.J. Cole, T. Widdowson, A.D. Ellis, 10 cm chirped fibre Bragg grating for dispersion compensation at 10 Gbit/s over 400 km of non-dispersion shifted fibre. Electron. Lett. **31**, 2203–2204 (1995)
39. A. Boskovic, M.J. Guy, S.V. Chernikov, J.R. Taylor, R. Kashyap, All-fiber diode-pumped, femtosecond chirped pulse amplification system. Electron. Lett. **31**, 877–879 (1995)
40. M.C. Farries, C.M. Ragdale, D.C.J. Reid, Broadband chirped fibre Bragg filters for pump rejection and recycling in erbium doped fibre amplifiers. Electron. Lett. **28**, 487–489 (1992)
41. J. Lauzon, S. Thibault, J. Martin, F. Ouellette, Implementation and characterization of fiber Bragg gratings linearly chirped by a temperature gradient. Opt. Lett. **19**, 2027–2029 (1994)
42. I.C. Byron, H.N. Rourke, Fabrication of chirped fibre gratings by novel stretch and write technique. Electron. Lett. **31**, 60–61 (1995)
43. Y. Painchaud, A. Chandonnet, J. Lauzon, Chirped fibre gratings produced by tilting the fibre. Electron. Lett. **31**, 171–172 (1995)
44. K. Sugden, I. Bennion, A. Molony, N.J. Copner, Chirped gratings produced in photosensitive optical fibres by fibre deformation during exposure. Electron. Lett. **30**, 440–442 (1994)
45. R. Kashyap, P.F. McKee, R.J. Campbell, D.L. Williams, Novel method of producing all fibre photoinduced chirped gratings. Electron. Lett. **30**, 996–998 (1994)
46. I. Bennion, J.A.R. Williams, L. Zhang, K. Sugden, N.J. Doran, UV-written in-fibre Bragg gratings. Opt. Quant. Electron. **28**, 93–135 (1996)
47. W.W. Morey, J.R. Dunphy, G. Meltz, in *Distributed and Multiplexed Fiber Optic Sensors*, Multiplexing Fiber Bragg Grating Sensors (SPIE, Boston, 1992), pp. 216–224
48. M.G. Xu, L. Reekie, Y.T. Chow, J.P. Dakin, Optical in-fiber grating high-pressure sensor. Electron. Lett. **29**, 398–399 (1993)
49. S. Theriault, K.O. Hill, D.C. Johnson, J. Albert, F. Bilodeau, G. Drouin, A. Beliveau, in *1998 International Conference on Applications of Photonic Technology III: Closing the Gap between Theory, Development, and Applications*. High-g Accelerometer Based On In-fiber Bragg Grating: a Novel Detection Scheme (SPIE, Bellingham, 1998), pp. 926–930
50. H. Zhang, S.M. Eaton, P.R. Herman, Single-step writing of Bragg grating waveguides in fused silica with an externally modulated femtosecond fiber laser. Opt. Lett. **32**, 2559–2561 (2007)
51. H. Zhang, S.M. Eaton, J. Li, A.H. Nejadmalayeri, P.R. Herman, Type II high-strength Bragg grating waveguides photowritten with ultrashort laser pulses. Opt. Express **15**, 4182–4191 (2007)
52. H. Zhang, S.M. Eaton, J. Li, P.R. Herman, Femtosecond laser direct-writing of multi-wavelength Bragg grating waveguides in bulk glass. Opt. Lett. **31**, 3495–3497 (2006)
53. A. Othonos, Fiber Bragg gratings. Rev. Sci. Instrum. **68**, 4309 (1997)
54. W.W. Morey, G. Meltz, W.H. Glenn, in *Optical Fiber Sensors. Proceedings of the 6th International Conference. OFS '89*. Bragg-grating temperature and strain sensors (Springer, Paris, 1989), pp. 526–531
55. N.F. Borrelli, R.A. Miller, Determination of individual strain-optic coefficients of glass by an ultrasonic technique. Appl. Opt. **7**, 745 (1968)
56. G.D. Marshall, P. Dekker, M. Ams, J.A. Piper, M.J. Withford, Directly written monolithic waveguide laser incorporating a distributed feedback waveguide-Bragg grating. Opt. Lett. **33**, 956–958 (2008)

57. M. Ams, P. Dekker, G.D. Marshall, M.J. Withford, Monolithic 100 mW Yb waveguide laser fabricated using the femtosecond-laser direct-write technique. Opt. Lett. **34**, 247–249 (2009)
58. P. Dekker, M. Ams, G.D. Marshall, J.D. Little, M.J. Withford, Annealing dynamics of waveguide Bragg gratings: evidence of femtosecond laser induced colour centres. Opt. Express **18**, 3274–3283 (2010)
59. V. Maselli, J. Grenier, S. Ho, P.R. Herman, Femtosecond laser written optofluidic sensor: Bragg grating waveguide evanescent probing of microfluidic channel. Opt. Express **17**, 11719–11729 (2009)

# Chapter 10
# Active Photonic Devices

**Giuseppe Della Valle and Roberto Osellame**

**Abstract** The chapter is devoted to active photonic devices fabricated by fs-laser writing. After a brief introduction focused on the role played by fs-laser written active devices, Sect. 10.2 briefly reviews the spectroscopical properties of the most interesting active ions so far exploited, namely erbium, ytterbium, neodimium, and bismuth. In Sect. 10.3 the main figures of merit for an active waveguide, namely the internal gain, the insertion loss, the net gain, and the noise figure are introduced and the experimental procedure for accurate gain measurement is also detailed. A thorough review of the active photonic devices demonstrated with the femtosecond laser microfabrication technique is presented in Sects. 10.4, 10.5, and 10.6, where several active waveguides and amplifiers, prototypal lasers, as well as more functionalized laser devices (operating under single longitudinal mode or stable mode-locking regime) are illustrated, respectively. Finally, conclusions and future perspectives of femtosecond-laser micromachining of active photonic devices are provided.

## 10.1 Introduction

Photonic devices based on ion-doped glass substrates have become increasingly interesting for a variety of applications in spectroscopy, optical communications, and optical sensing. In many cases, such as metro-local optical communications [1], free-space optics [2], compact laser radar and remote sensing [3], and distribute fiber-optic sensors [4], the requirements may become extremely demanding in terms of power efficiency, compactness, insensitivity to environmental disturbance, and

G. Della Valle · R. Osellame (✉)
Istituto di Fotonica e Nanotecnologie - Consiglio Nazionale delle Ricerche (IFN-CNR), and Department of Physics - Politecnico di Milano, Piazza Leonardo da Vinci 32, I-20133 Milano, Italy
e-mail: giuseppe.dellavalle@polimi.it; roberto.osellame@ifn.cnr.it

low cost production. Bulk active devices exploiting large modal volumes are able to provide high gains or laser output powers, but generally are quite sensitive to technical noise, are not compact devices, and typically result in complex and expensive systems. Short fiber-based [5,6] and waveguide-based active devices [7,8] are compact and monolithic structures which have the potential to satisfy all the requirements previously outlined.

In the last decade, a strong effort has been devoted to the development of fs-laser written active as well as passive waveguide devices (see [9] and references therein) in view of the small footprint capability and ease of experimentation on new glass materials offered by the fs-laser writing technique. In particular, the first demonstration in 2004 of a fs-laser written waveguide laser [10] was a breakthrough for fs-laser written active devices, a result that also strongly contributed to the assessment of the fs-laser micromachining as a fabrication process that can compete with standard techniques, such as silica-on-silicon, plasma-enhanced chemical vapor deposition, ion-exchange, and solgel in terms of waveguide quality. More recently, Er-doped and Er:Yb-doped waveguide amplifiers operating in the whole C-band (1530–1565 nm) of optical communications, as well as tunable lasers and single-mode or mode-locked lasers have been demonstrated. Also, very efficient femtosecond laser written Nd-based active devices have been reported, and the experimentation on new exciting active materials is ongoing.

In this chapter we review the state of the art of fs-laser written active photonic devices. Section 10.2 is devoted to the spectroscopy of the most representative active ions successfully exploited or exhibiting new potentialities for future development in the field, namely erbium, ytterbium, neodymium, and bismuth. In Sect. 10.3 the main figures of an active waveguide, namely the internal gain, the insertion loss, the net gain, and the noise figure are rigorously introduced starting from the waveguide propagation equations. A typical experimental setup for the characterization of active waveguides and the *ON–OFF* procedure for accurate gain measurement are also illustrated in detail. A thorough review of the active photonic devices demonstrated with the fs-laser micromachining is presented in Sects. 10.4, 10.5, and 10.6, where several active waveguides and waveguide amplifiers, prototypical lasers, as well as advanced laser devices operating under single longitudinal mode or stable mode-locking regime are illustrated, respectively. Finally, conclusions and future perspectives of femtosecond-laser micromachining of active photonic devices are provided.

## 10.2 Active Ions for Waveguide Devices

Active ions embedded in dielectric media have been extensively studied in the past decades and successfully employed for the development of optical amplifiers and lasers. In particular, comprehensive research activity has been focused on several rare-earth ions in many different hosts [11]. Actually, rare-earth energy levels arises from the $4f$ or $5f$ inner-shells which are effectively screened by the complete

8-electrons outer shells of Xenon and Radon for Lanthanides and Actinides, respectively. This results in sharp intense absorption and emission lines of rare-earth doped crystals. On the contrary, strong inhomogeneous line-broadening in glasses determines broad absorption and emission bands. Rather than being a limit, this fact reduces the sensitivity of active glasses to the pump wavelength as compared to crystals and, most importantly, allows broad-band amplification or tunability of laser devices. Low thermal conductivity certainly may be a limit of glass hosts, a minor drawback almost rewarded with the low cost and high quality manufacturing possibilities offered by glasses. However, newly developed glass bases present an increased thermal conductivity which is sufficient for most low-power applications, like optical communications and sensing.

In this section we briefly review the main properties of those active ions that have been so far exploited in the development of fs-laser written active photonic devices in crystals and glasses, namely, erbium, ytterbium, neodymium as well as bismuth which has recently attracted much attention for the ultra-broad gain bandwidth exhibited in the infrared.

### 10.2.0.1 Neodymium

One of the most popular active ion is neodymium $Nd^{3+}$. Typical hosts are $Y_3Al_5O_{12}$ (YAG), $YLiF_4$, and $YVO_4$ crystals, as well as silicate and phosphate glasses. Typical doping level is limited to $\sim 1\%$ in crystals and $\sim 4\%$ in glasses to prevent fluorescent quenching or defect formation. A simplified scheme of $Nd^{3+}$ energy levels is reported in Fig. 10.1a (labelled according to Russell–Saunders nomenclature) aside with the principal electronic transitions. Four main pumping bands occur at 0.52, 0.58, 0.73, and 0.80 $\mu$m. These bands are coupled by fast non-radiative decay to the metastable $^4F_{3/2}$ energy level ($\sim$230 $\mu$s lifetime in YAG) from where radiative transition to the $^4I_{11/2}$ allows stimulated emission to take place at around 1.05 $\mu$m wavelength. Since the $^4I_{11/2}$ level is coupled to the $^4I_{9/2}$ ground level by fast non-radiative decay, amplification from transition $^4F_{3/2} \rightarrow {}^4I_{11/2}$ corresponds to a 4-level scheme [12]. The exact emission peak wavelength of the transition depends on the particular host, ranging from 1.049 $\mu$m (ZBLAN) to 1.064 $\mu$m (YAG). The emission linewidth (full-width at half maximum of the emission cross-section) is typically less than 1 nm for crystals but turns to be much wider for glasses (10–15 nm) because of the inhomogeneous broadening, also resulting in a much lower peak value (e.g., $\sigma_e = 2.8 \times 10^{-19}$ cm$^2$ in YAG$_4$ against $2.5 \times 10^{-20}$ cm$^2$ in silicate glass) [11]. Due to the four-level scheme preventing from any re-absorption of the signal radiation, Nd-based active systems are intrinsically very efficient and thus particularly suitable for the development of high power bulk lasers to be employed in material processing (drilling and welding) and medical surgery. Nevertheless, Nd-doped optical waveguides can possibly find new interesting application as efficient and compact active media in low-power miniaturized lasers for optical ranging and remote sensing[3].

**Fig. 10.1** Typical energy levels for (**a**) $Nd^{3+}$, (**b**) $Er^{3+}$:$Yb^{3+}$, (**c**) $Yb^{3+}$ and (**d**) $Bi^+$ active ions in glasses. Principal electronic transitions are also shown: pump absorptions and gain (or laser) emission (*solid line*); fast non-radiative decays (*wavy lines*); energy-transfer (*dotted lines*); cooperative up-conversion (*dashed lines*)

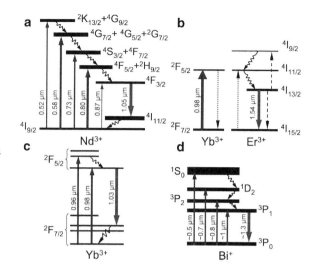

### 10.2.0.2 Erbium-Ytterbium

Another very popular and widely exploited active system is the erbium ion, allowing optical gain and laser action at 1.5 μm, i.e., in the third window of optical communications. Sensitization with ytterbium is often used and the resulting Er:Yb active system is sketched in Fig. 10.1b. Pumping at 0.98 μm provides inversion of Yb ions and Er inversion is thus obtained through an indirect pumping process based on a resonant energy-transfer mechanism [13] from ytterbium to erbium (dotted line in Fig. 10.1b). From the pumping level $^4I_{11/2}$ Er ions decay to the metastable state $^4I_{13/2}$ by non-radiative processes and the radiative transition $^4I_{13/2} \rightarrow {}^4I_{15/2}$ provides optical amplification at 1.5 μm by means of stimulated emission. The ytterbium codoping turns out to be particularly suitable for the development of compact waveguide devices because Yb absorption cross section at 0.98 μm is much higher than erbium one (by a factor of 10 or even more [13]) and thus Yb-codoping offers the advantage of increasing the absorption efficiency of the pump radiation without acting on the Er concentration, which can be independently optimized. Actually, contrary to a 4-level system, in a three-level system like erbium at 1.5 μm, doping concentration ought to be carefully optimized to achieve maximum gain, because those ions that are not excited to the metastable state absorbs signal radiation rather than amplifying it. To achieve an effective energy transfer process from ytterbium to erbium the back energy transfer (from erbium to ytterbium) needs to be minimized by fast non-radiative multi-phonon decay from the erbium pump level $^4I_{11/2}$ to the metastable state $^4I_{13/2}$. Therefore, phosphates are particularly suitable hosts among glass substrates in view of their high-phonon energy allowing higher non-radiative decay rate from $^4I_{11/2}$ [11]. Phosphates also exhibit a particularly long lifetime of the metastable state of about 8 ms. Another benefit of Yb-codoping is the possibility to obtain less sensitivity to the pump wavelength (though at the expense of longer absorption lengths) by exploiting the lower shoulder exhibited by Yb absorption cross section below 0.97 $\mu$m [13].

Finally, it is worth noting that with increasing doping concentration quenching effects take place, the most relevant channel of energy loss being represented by the up-conversion process [14] between two excited erbium ions (dashed lines in Fig. 10.1b): the interaction between two erbium ions, initially excited to the $^4I_{13/2}$ level, results in one ion quenched to the $^4I_{15/2}$ ground level and the other raised to the $^4I_{9/2}$ excited level. This latter ion then relaxes to its original $^4I_{13/2}$ state through a fast non-radiative decay. Therefore the global effect of the up-conversion is the loss of one $^4I_{13/2}$ excited ion with a rate proportional to the square of the $^4I_{13/2}$ level population. Multiple up-conversion processes involving higher energy levels of erbium and subsequent radiative decay to the ground state are responsible for the typical green fluorescence exhibited by heavily-doped erbium active systems under strong inversion [15].

### 10.2.0.3 Ytterbium

Ytterbium ion is itself a very effective active ion for the development of near infrared amplifiers and lasers. The ytterbium energy structure is very simple, consisting of only two electronic manifolds, sketched in Fig. 10.1c aside from the most relevant electronic transitions [16]. The two most exploited absorption lines occurs at around 0.96 $\mu$m and at 0.98 $\mu$m (the exact peak wavelength depending on the particular host) and the main gain line occurs at 1.03 $\mu$m. Therefore, Yb is a direct competitor to Nd but note that according to the level scheme above reported Yb-based active systems operate on a quasi-three-level scheme, and thus typically exhibit an higher threshold for inversion as compared to Nd-based systems. Despite such limitation, that in recent years has been overcome by the availability of commercial high-power InGaAs laser diodes, ytterbium offers several advantages: the simple energy level structure prevents several parasitic effects like excited-state absorption and up-conversion and also allows heavy doping of the host material (up to 20% in weight in glass without significant fluorescence quenching); a broad emission bandwidth (of the order of 100 nm in glass) allows laser wavelength tunability or fs-mode-locking operation; the *in-band* pumping results in a very high quantum efficiency ($\lambda_{Pump}/\lambda_{Signal} > 90\%$). The lifetime of the metastable state is also much longer than Nd metastable state lifetime (about 1.16 ms against 0.23 ms in YAG) thus making Yb a better active ion for laser Q-switching.

### 10.2.0.4 Bismuth

After the first demonstration of optical amplification in bismuth-doped silica glass [17, 18], bismuth has been recently attracted much attention, offering the potentiality of an ultra-broad-band amplification at 1.3 $\mu$m telecom wavelength [19], which is the natural zero-dispersion region of silica glass fiber.

Differently from rare-earth ions, where energy levels are only weakly influenced by the crystal field, bismuth ions exhibit very broad absorption and emission lines with peak wavelengths that are strongly sensitive to the host material. Furthermore, bismuth always exhibits several kinds of valence states simultaneously existing in the same crystal or glass. Under such a complex scenario, it is worth saying that the infrared luminescence mechanism of Bi-doped materials is still not completely clear. Nevertheless recent studies revealed that infrared luminescence most probably arises from monovalent ions $Bi^+$ [20, 21]. According to this picture, a tentative energy band diagram for $Bi^+$ should be sketched as in Fig. 10.1d (after [20]). Four main absorption bands occur at around 0.5, 0.7, 0.8, and 1 $\mu$m, and the $^3P_1 \rightarrow {}^3P_0$ transition is responsible for optical gain. Being the $^3P_1 \rightarrow {}^3P_0$ transition electric-dipole forbidden, the lifetime $\tau$ of the $^3P_1$ metastable state of $Bi^+$ is several hundred microseconds long. A typical value for peak emission cross-section at 1.3 $\mu$m is of the order of $1 \times 10^{-20}$ cm$^2$ in glass, thus resulting in a $\sigma\tau$ product that is about three times larger than that of Ti-sapphire. This indicates that Bi-doped glasses are really promising material for optical amplification at 1.3 $\mu$m and in fact several efficient bulk as well as fiber laser devices have already been demonstrated [22, 23].

## 10.3 Gain Definitions and Measurement Technique

The performance of an active waveguide is defined by its active and passive figures, namely: the internal gain, the insertion loss, the net gain and the noise figure. In this section we provide a rigorous definition of such figures based on the waveguide propagation equations, explaining their dependence on the constitutive parameters of the waveguide (i.e., dopant concentration, waveguide length, absorption, and emission cross-sections). Finally, we illustrate the typical setup for active characterization of channel waveguides and provide a brief review of the *ON/OFF* technique for accurate gain measurement.

### 10.3.1 *Definition of the Main Figures of An Active Waveguide*

#### 10.3.1.1 Internal Gain

Consider an active channel waveguide of length $L$. Under steady state pumping condition, the evolution of the signal optical power $P$ at signal wavelength $\lambda$ along the waveguide axis $z$ obeys the following propagation equation [24–26]:

$$\frac{dP(z,\lambda)}{dz} = \left[\alpha_{21}(z,\lambda) - \alpha_{12}(z,\lambda) - \gamma(z,\lambda)\right] P(z,\lambda) \qquad (10.1)$$

with

$$\alpha_{12}(z,\lambda) = \iint_A n_1(r,\phi,z)\sigma_a(\lambda)\varphi_s(r,\phi)rdrd\phi \quad (10.2)$$

$$\alpha_{21}(z,\lambda) = \iint_A n_2(r,\phi,z)\sigma_e(\lambda)\varphi_s(r,\phi)rdrd\phi \quad (10.3)$$

where integration is performed in the whole $r, \phi$ transverse plane of the waveguide, $n_1(r,\phi,z)$ and $n_2(r,\phi,z)$ are the local population densities of active ions in the lower and upper (metastable) state, respectively, $\sigma_a(\lambda)$ and $\sigma_e(\lambda)$ are the absorption cross-section and the emission cross-section of the active ions, respectively, $\varphi(r,\phi)$ is the normalized intensity mode profile of the signal, and $\gamma(z,\lambda)$ is the local propagation loss coefficient.

If we integrate (10.1) from 0 to $L$ and introduce the *effective (average) population densities* $N_1$ and $N_2$ (ions·m$^{-3}$) for the lower and higher laser levels, respectively

$$N_1 = \frac{1}{L}\int_0^L \left(\iint_A \varphi_s(r,\phi)n_1(z,\phi,r)rdrd\phi\right)dz \quad (10.4)$$

$$N_2 = \frac{1}{L}\int_0^L \left(\iint_A \varphi_s(r,\phi)n_2(z,\phi,r)rdrd\phi\right)dz \quad (10.5)$$

we can define the so-called *internal gain* of the active waveguide as:

$$G_{\text{int}}(\lambda) = \frac{P(L,\lambda)\Gamma_{\text{PL}}(\lambda)}{P(0,\lambda)} = e^{[\sigma_e(\lambda)N_2 - \sigma_a(\lambda)N_1]L} \quad (10.6)$$

where $\Gamma_{\text{PL}}(\lambda) = \exp[\int_0^L \gamma(z,\lambda)dz]$ is the overall propagation loss attenuation (i.e., $10\log_{10}[\Gamma_{\text{PL}}(\lambda)]/L$ representing the average propagation loss per unit length in dB unit).

Equation (10.6) shows that $G_{\text{int}}$ only depends on the average inversion and is independent of the population spatial profiles $n_1(r,\phi,z)$ and $n_2(r,\phi,z)$. Also, it is worth noting that according to (10.6) the internal gain of a 4-level system (where $N_1 \simeq 0$) turns to be $G_{\text{int}} \simeq e^{\sigma_e N_2 L} > 1$. On the contrary, for a 3-level system, the internal gain can be greater, equal, or even lower than 1 depending on the actual inversion and on the interplay between stimulated emission and absorption.

### 10.3.1.2 Absorption

Under no pumping (pump OFF), the signal launched into an active waveguide can eventually be absorbed rather than amplified. The signal *absorption* is defined as follows:

**Fig. 10.2** Schematic of an optical waveguide inserted in an external fiber circuit.

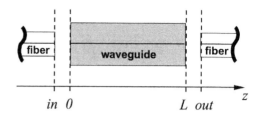

$$A(\lambda) = \frac{P(0, \lambda)}{P^{\text{OFF}}(L, \lambda) \Gamma_{\text{PL}}(\lambda)} \quad (10.7)$$

Note that a 4-level system, being transparent to signal radiation, does not exhibit any absorption and $A(\lambda) = 1$. On the contrary, in a 3-level system, the absorption is $A(\lambda) = e^{\sigma_a(\lambda) n_t L} > 1$, which can be obtained from (10.1) and (10.2) by setting $n_1 = n_t$ (the total concentration of the active ions) and thus $n_2 = 0$.

### 10.3.1.3 Insertion Loss

The passive performance of the waveguide is accounted by its *insertion loss* which is defined as the excess loss introduced when the waveguide is inserted in an optical fiber circuit (Fig. 10.2).

More precisely we can define the *insertion loss* of the waveguide as:

$$\Gamma_{\text{IL}}(\lambda) = \frac{P_{\text{in}}(\lambda)}{P_{\text{out}}^{\text{OFF}}(\lambda)} \quad (10.8)$$

where $P_{\text{in}}$ is the incident signal power provided by the input fiber and $P_{\text{out}}^{\text{OFF}}$ is the signal power measured at the output fiber with no pumping. In the operating bandwidth $\Delta\lambda$ of the active optical waveguide, being $\Delta\lambda \ll \lambda$, $\Gamma_{\text{IL}}$ can be assumed as wavelength independent. To prevent spurious contribution due to signal absorption, the $\Gamma_{\text{IL}}$ of an active waveguide should be measured at a wavelength where the absorption cross-section of the active ion is expected to be negligible.

The insertion loss arises from two different contributions: the *coupling loss* $\Gamma_{\text{CL}}$ and the *propagation loss* $\Gamma_{\text{PL}}$. The former takes into account the mismatch between the fiber and waveguide mode (accounted twice, once for each facet), and depends on the transverse refractive index profile of the waveguide; the latter takes into account the light attenuation due to scattering (and passive absorption) along the waveguide, and thus depends on the quality of the writing process (in terms of uniformity achieved by the refractive index modification along the waveguide axis) and is ultimately limited by the scattering losses in the bulk substrate. Therefore we have $\Gamma_{\text{IL}} = \Gamma_{\text{CL}}^2 \Gamma_{\text{PL}}$.

### 10.3.1.4 Net Gain

The internal gain, as defined by (10.6), accounts for the amplification achieved by the launched signal power $P(0)$ after propagating along a scattering free waveguide of length $L$. Therefore, $G_{int}$ is aimed at specifying the *pure* active performance of the waveguide medium. More practically, we can define a *net gain* or *external gain* of an active optical waveguide as follows:

$$G_{net}(\lambda) = \frac{P_{out}^{ON}(\lambda)}{P_{in}(\lambda)} \qquad (10.9)$$

Note that being $P_{out}^{ON} = P(L)/\Gamma_{CL}$, $P_{in} = P(0)\Gamma_{CL}$, and remembering that $\Gamma_{IL} = \Gamma_{PL}\Gamma_{CL}^2$, we have the following relation between the net gain and the internal gain:

$$G_{net}(\lambda) = G_{int}(\lambda)/\Gamma_{IL}(\lambda) \qquad (10.10)$$

The net gain is thus aimed at specifying the capability of an active optical waveguide to really behave as an optical amplifier. Note that in dB (i.e., logarithmic) units the latter equation becomes $G_{net}|_{dB} = G_{int}|_{dB} - IL|_{dB}$, where $IL|_{dB} = 10\log_{10}(\Gamma_{IL})$.

### 10.3.1.5 Noise Figure

The origin of amplifier noise in an active-ion doped waveguide is the spontaneous emission generated in a wavelength range corresponding to the emission bandwidth of the active ions, and subsequently amplified by the waveguide itself. The resulting amplified spontaneous emission (ASE) is thus added as a background noise to the signal radiation and not only affects the noise performance but it can also limit the maximum achievable gain by contributing to the phenomenon of gain saturation.

The noise performance of an amplifier is usually expressed by means of its *noise figure* ($NF$), which is simply defined as the ratio

$$NF = \frac{(SNR)_{in}}{(SNR)_{out}}. \qquad (10.11)$$

where $(SNR)_{in}$ and $(SNR)_{out}$ are the input and output signal to noise ratio, respectively. Obviously, because any amplifier degrades the signal to noise ratio of the input signal, the noise figure is always grater than 1.

A simple and very useful expression of the NF can be obtained if we consider an ideal signal source and an ideal photodetector whose noise performance are limited by shot noise. In the presence of a narrow pass-band integrating optical filter or when the monochromatic signal is sufficiently high in power, the noise figure takes the simplified expression [26]:

$$NF(\lambda) = \frac{P_{\text{out,ASE}}(\lambda)}{h\nu B_0 G_{\text{net}}(\lambda)} + \frac{1}{G_{\text{net}}(\lambda)} \quad (10.12)$$

where

- $P_{\text{out,ASE}}(\lambda)$ is the total ASE power measured in the bandwidth $B_0$ at the waveguide output,
- $h = 6.626 \times 10^{-34}$ J s is Planck constant,
- $\nu$ (Hz) is the signal optical frequency,
- $G_{\text{net}}(\lambda)$ is the (net) gain of the optical waveguide,
- $B_0$ (Hz) is the integration bandwidth of the optical filter.

### 10.3.2 The On/Off Technique for Gain Measurement

#### 10.3.2.1 Experimental Set-Up

Figure 10.3 shows the schematic of a typical *all-in-fiber* set-up for active waveguide characterization. The active channel waveguide is mounted on a 4-axes micro-positioning stage and the waveguide facets are butt-coupled to standard optical fibers by means of 5-axes micro-positioning stages.

An index matching fluid (designed for high power operation) is inserted between waveguide and fiber ends. The index matching fluid, by reducing detrimental Fresnel

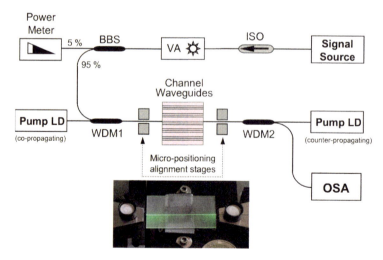

**Fig. 10.3** A typical *all-in-fiber* experimental set-up for active waveguide characterization. *ISO*: optical isolator. *VA*: variable attenuator. *BBS*: broad-band splitter. *WDM1 and WDM2*: wavelength-division multiplexer. *OSA*: Optical spectrum analyzer. Inset shows an Er:Yb-doped channel waveguide under bi-propagating pumping scheme exhibiting the characteristic green up-conversion fluorescence from erbium ions

reflections, not only allows minimization of coupling losses, but also improves the reliability of the measurement by preventing the Fabry–Perot effects in the fiber-waveguide coupling region. The probe signal (typically from a fiber-pigtailed single-frequency laser diode or broad-band stabilized ASE source), is properly attenuated by means of a variable attenuator and is then split in two arms, with 5% of the signal feeding a high precision power meter in order to monitor the input power level. A couple of fiber-pigtailed laser diodes, connected to the input and output arm of the setup by means of wavelength-division-multiplexer (WDM) devices, allows to exploit single (i.e., *co-* or *counter-propagating*) as well as double (i.e., *bi-propagating*) pumping schemes. The WDM2 signal arm is connected to an optical spectrum analyzer (OSA) allowing accurate investigation of the output signal in the spectral domain. All fiber connectors are FC/APC type to prevent detrimental back reflections, and the source is isolated by means of an optical isolator.

The *all-in-fiber* setup described above is particularly representative as a standard setup for the characterization of photonic devices operating at 1.3 or 1.5 $\mu$m telecom wavelengths, where standard fiber-based components like WDM, BBS, ISO, VA, etc. (see Fig. 10.3) are commercially available. Unfortunately, for Yb- and Nd-based active waveguides operating at around 1 $\mu$m some of these components are not available as fiber-based or fiber-coupled standard products and a complete *all-in-fiber* setup is not possible. This results in a less standardized setup that eventually employs bulk dichroic mirrors instead of fiber-based WDMs and microscope objective focusing instead of fiber butt-coupling to couple light into and out of the waveguide.

### 10.3.2.2 Experimental Procedure

The internal gain of an active optical waveguide can be accurately measured according to the so-called *ON–OFF* method, which is independent from the knowledge of the coupling losses and the wavelength-dependent response of the measurement set-up.

The method is based on the measurement of the *enhancement* $E(\lambda)$ exhibited by the signal output power level due to the pumping process, which is simply given by the ratio between the output power level with pump ON and the output power level with pump OFF:

$$E(\lambda) = \frac{P_{out}^{ON}(\lambda)}{P_{out}^{OFF}(\lambda)} \tag{10.13}$$

During the measurement, the input signal power level is carefully monitored (by means of the in-line power meter, see Fig. 10.3) and kept at a constant value. Note that, under the hypothesis that the pumping process does not affect the coupling losses (see the following section for an accurate discussion on this point), being $P_{out}(\lambda) = P(L, \lambda)/\Gamma_{CL}$, we have $E(\lambda) = P^{ON}(L, \lambda)/P^{OFF}(L, \lambda)$, and the internal gain can thus be expressed as the ratio between the enhancement and the absorption:

$$G_{\text{int}}(\lambda) = \frac{E(\lambda)}{A(\lambda)} \qquad (10.14)$$

The absorption spectrum $A(\lambda)$ can be accurately measured from bulk spectroscopical investigations, and thus the internal gain measurement is reduced to the measurement of the enhancement, which is intrinsically accurate because coupling losses and the wavelength-dependent response of the measurement set-up are common to $P_{\text{out}}^{\text{ON}}(\lambda)$ and $P_{\text{out}}^{\text{OFF}}(\lambda)$ and thus rejected in their ratio. As the insertion loss is known, the *net gain* is simply given by (10.10).

It is worth noting that when an optical amplifier is fed by a monochromatic input signal, the spectral power at the output exhibits a peak at the signal wavelength emerging from a background noise. In active-ion doped amplifiers such noise is dominated by the amplified spontaneous emission (ASE), and should be properly estimated and taken into account in the gain measurement. More precisely, $P(L,\lambda)$ should be replaced everywhere in the previous formulae by $P(L,\lambda) - P_{\text{ASE}}(L,\lambda)$, where $P_{\text{ASE}}(L,\lambda)$ is the ASE power at the signal wavelength (recorded with the same resolution bandwidth). However, in the case of low noise amplifiers, $P_{\text{ASE}}(L,\lambda)$ can be practically disregarded being the loss of accuracy within the experimental error, i.e., typically about 0.5 dB (ultimately due to the polarization dependent spectral response of the measurement setup). This is particularly true for compact active waveguides, which are relatively low gain and well-inverted devices and thus exhibit a low noise figure [26].

For the noise figure measurement, (10.12) can be followed. Note that once $G_{\text{net}}(\lambda)$ is measured, the $NF$ measurement reduces to the measurement of the ASE output power comprised in the frequency bandwidth $B_0$ (typically corresponding to a 1 nm bandwidth). Note that though (10.12) provides a simple recipe for NF measurement, the wavelength-dependent transfer function of the measurement set-up from the OSA back to the waveguide output is needed to properly estimate the ASE output power.

### 10.3.2.3 Thermal Contribution

The *ON–OFF* method for the gain measurement assumes that the pumping process does not affect the fiber-waveguide coupling. Actually, under few hundreds mW pump power, a local temperature rise of the order of 100 degrees is expected at the waveguide facet where the pump radiation is launched, and this results in the formation of a thermal lens. It has been reported that in a typical commercial laser glass, one can obtain a thermally induced refractive index change of $5 \times 10^{-4}$ at around 1 $\mu$m [27]. As the refractive index increases, the numerical aperture of the waveguide also increases, and for those active waveguides exhibiting a very low numerical aperture as compared to standard optical fibers (in view of the low fs-laser-generated refractive index change) the pump-induced thermal lensing results in a better coupling efficiency to the fiber, enhancing the output signal power level. For a correct estimation of the internal gain by *ON–OFF* measurements, this thermally induced contribution should be removed from the *in-band* measured enhancement $E(\lambda)$. The spurious contribution to the enhancement can be easily estimated by measuring the (unexpected) signal enhancement achieved at those wavelengths where the emission cross-section of the active ion is negligible. As an example, 3 dB thermally induced enhancement has been reported in Nd-doped glass waveguides with a fs-laser induced refractive index change of $3 \times 10^{-4}$ [27]. On the contrary, for those waveguides that demonstrate coupling losses to standard fibers < 1 dB because of a much higher refractive index contrast (typically above $\sim 3 \times 10^{-3}$) the spurious thermally induced enhancement is practically negligible, and the *ON–OFF* method illustrated in the previous paragraph requires no correction. Finally, note that in any case the eventual correction to the internal gain due to thermal contribution is not crucial for the estimation of the net gain, which is defined as the ratio between the output and the input signal power level regardless of the nature of the internal mechanism being responsible for the observed signal gain. The main drawback of a thermally induced enhancement is the sensitivity exhibited by the net gain to environmental temperature variations.

## 10.4 Active Waveguides and Amplifiers

### *10.4.1 Internal Gain in Nd-Doped Active Waveguides*

The first active device manufactured by fs-laser writing was an active Nd-doped waveguide. The waveguide was fabricated in a commercially available Nd-doped silicate glass rod ($2 \times 10^{20}$ ion/cm$^3$ doping concentration) by means of a low-repetition rate Ti:sapphire laser [28]. According to the *ON–OFF* method, 1.5 dB/cm enhancement (equivalent to internal gain for four-level systems such as Nd, in which signal absorption is absent) was measured at 1054 nm for an incident pump power of 350 mW at 514 nm (provided by an Ar ion laser). Nd-doped systems are very efficient 4-level active media, but are not suitable for applications at telecom wavelengths. Therefore, much effort was devoted to fs-laser writing in other active glasses, with particular interest to Er and Er:Yb doped glasses allowing operation in the C-band.

### *10.4.2 Waveguide Amplifier in Er:Yb-Doped Phosphate Glass*

A phosphate glass (Kigre QX) suitable for fs-laser writing with a low repetition rate system allowed the first demonstration of an active waveguide exhibiting internal gain at 1.5 $\mu$m [29]. The waveguide was 25-mm long and diode-pumped at 980 nm under a bi-propagating scheme (similar to the one reported in Fig. 10.3). A maximum internal gain of 0.7 dB was measured at 1557 nm. The single-mode guided-field was Gaussian but significantly larger than the mode-field supported by a standard single mode fiber, resulting in coupling losses as high as 4.3 dB/facet and thus high overall insertion losses (even though the device showed remarkably low propagation loss of 0.3 dB/cm). By exploiting an astigmatic shaping of the writing laser beam, active waveguides with a more symmetric refractive index profile were obtained in the same substrate, allowing lower coupling losses to single mode fiber (2 dB/facet) and consequently higher gain efficiency: internal gain in the whole C-band from a compact 9-mm long active waveguide was reported [30]. However, a real waveguide amplifier able to provide net gain was still lacking.

A significant improvement was demonstrated by exploiting an innovative writing system based on a cavity-dumped Yb-oscillator [31], operating in a repetition rate regime that is intermediate between regeneratively amplified systems and high repetition rate oscillators [9]. The same phosphate glass was used, but doping concentrations were optimized in order to obtain higher gain per unit length and longer devices ($1.8 \times 10^{20}$ Er ions/cm$^3$ and $3.6 \times 10^{20}$ Yb ions/cm$^3$). The versatility of this cavity-dumped system led to the definition of an optimized set of writing parameters, namely optimal pulse energy (200–500 nJ), repetition rate (500 kHz–1 MHz), and writing speed (50–300 $\mu$m/s), allowing high optical quality of the waveguides. Coupling losses as low as 0.2 dB/facet and propagation loss of about 0.3 dB/cm were demonstrated. Thanks to the strongly reduced insertion loss and to the improved doping levels, net gain in femtosecond laser written waveguides was achieved for the first time [32, 33]. Also, as a full demonstration of feasibility of a real device, a 37-mm long waveguide amplifier operating in the whole C-band of optical communications was fabricated and characterized in terms of insertion losses, spectral gain, and noise figure [34]. A couple of 975-nm wavelength InGaAs laser diodes in a bi-propagating pumping scheme were adopted to supply up to 460 mW (260 mW + 200 mW) incident pump power through 980/1550 nm wavelength-division-multiplexers (WDMs) (cf. Fig. 10.3). Overall insertion loss was as low as 1.9 dB and net gain was demonstrated in the whole C-band as shown in Fig. 10.4a, with 3.6 dB residual gain at the upper edge (1565 nm) and peak net gain of 7.3 dB at 1535 nm. The peak internal gain per unit length of $\sim$2.5 dB/cm, is comparable to the value of 3 dB/cm reported in state-of-the-art erbium-ytterbium doped waveguide amplifiers fabricated by ion-exchange in a phosphate glass with the same doping concentration [35]. The measured noise figure (Fig. 10.4c) was in good agreement with the theoretical estimation [26], and was below the 3 dB level for wavelengths longer than 1560 nm, where the gain also decreased, as expected in short, heavily-doped erbium amplifiers, which are well inverted and relatively low gain devices [26].

### 10.4.3 Waveguide Amplifier in Er:Yb-Doped Oxyfluoride Silicate Glass

Phosphate glasses are very promising hosts for rare earth doping in view of their excellent spectroscopic properties, namely, broadband emission spectrum around 1.5 $\mu$m, long lifetime of the Er metastable state, and high phonon energies allowing

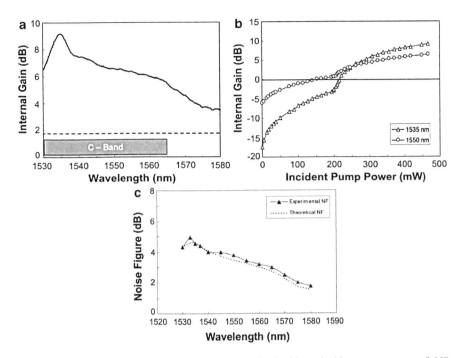

**Fig. 10.4** (a) Measured internal gain spectrum obtained with an incident pump power of 460 mW in bi-directional pumping configuration. The dashed line indicates the total insertion losses. (b) Internal gain of the 37-mm long Er-Yb waveguide amplifier as a function of the incident pump power for two different signal wavelengths: 1535 nm (*triangles*) and 1550 nm (*circles*). (c) Experimental (*triangles*) and theoretical (*dotted line*) noise figure. Reprinted with permission from [34]

very efficient pumping. Erbium-doped oxyfluoride silicate glass is also a particularly promising host for rare earth ions because it combines the attractive spectroscopic properties of fluoride glass (similar to phosphate) with the structural compatibility and stability of silica [36]. A 19-mm long active waveguide in oxyfluoride silicate glass was demonstrated by Thomson and co-workers [37]. The glass was doped with erbium (1% of $ErF_3$ in weight) and the waveguide was fabricated by means of a Ti:sapphire laser system at 5 kHz repetition rate, operated at very low writing speed (2 $\mu$m/s). The waveguides exhibited high propagation losses (about 3.2 dB/cm), dominated by Rayleigh scattering, and 1.6 dB/facet coupling loss were measured at 1550 nm, thus preventing an efficient pumping. A limited enhancement of few dB was achieved under 140 mW pump power at 1480 nm, but no internal gain was reported. Improved results were obtained by exploiting the multiscan technique [38], in which the desired waveguide cross section is obtained by writing many lines of modified material slightly shifted to each other. Total insertion loss of 4.3 dB was achieved in a 18.5 mm long waveguide and a maximum internal gain of 1.7 dB at 1537 under double-wavelength pumping (412 mW at 980 nm and 243 mW at 1480 nm) was also demonstrated. Recently, feasibility of a waveguide amplifier in this oxyfluoride silicate glass (codoped with Yb, 2% in weight of $Yb_2O_3$, to enhance pumping efficiency at 980 nm) was ascertained by the demonstration of net gain in a 10-mm long waveguide [39]. The low repetition rate Ti:sapphire laser was replaced by the same cavity-dumped Yb:glass laser that demonstrated the best results obtained in phosphate glass [31, 32]. Multiscan technique was adopted to produce waveguides with a square cross section by means of 40 individual scans each with 0.2 μm separation or 20 individual scans with 0.4 μm separation (Fig. 10.5a). The mode-field profile was quite symmetric (Fig. 10.5b) and exhibited good matching with the mode-field of standard single mode fiber, allowing coupling losses as low as 0.4 dB/facet. The propagation loss was 0.34 dB/cm, a remarkably low value with respect to those previously reported in oxyfluorides and also comparable to the propagation loss reported in phosphate

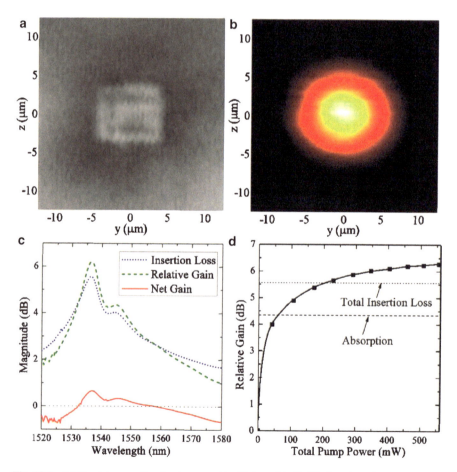

**Fig. 10.5** (a) Facet image of optimal waveguide fabricated in Er:Yb-doped oxyfluoride silicate glass. (b) Near-field mode image of the waveguide. (c) Relative gain, insertion loss, and net gain spectra. (d) Relative gain vs pump power at the peak of the gain spectrum using dual wavelength pumping (62%, 980 nm; 38%, 1480 nm). Reprinted with permission from [39]

glass. The improvement of waveguide quality and Yb codoping allowed significant population inversion (about 72%) resulting in a peak net gain of 0.72 dB under double-wavelength pumping (Fig. 10.5c–d). Improvement of amplifier performance is achievable by further optimization of doping concentrations.

### 10.4.4 Active Waveguides in New Glass Materials

It is worth pointing out that the cavity-dumped Yb:glass laser also allowed fabrication of high quality fs-laser written waveguides in another promising active substrate, a Bi-doped silica glass [40], which is known to exhibit broadband emission around 1.3 $\mu$m. Under optimized writing conditions, both single-scan and multi-scan fabricated waveguides demonstrated propagation loss as low as 0.4 dB/cm and coupling loss to standard telecom fiber of about 0.6 dB/facet. At 980 nm pump wavelength a fluorescence bandwidth as wide as 500-nm was reported.

Finally, micromachining with a low repetition rate Ti:sapphire laser was successfully applied to a new Er-doped tellurite glass modified by phosphate and aluminum [41]. Tellurite glasses doped with rare earth elements, though exhibiting broadband emission at telecom wavelengths [42], suffer from relatively high up-conversion effects due to low phonon energies. Phosphate modification induced higher phonon energies and also led to positive index change by fs-laser writing. These phospho-tellurite glasses are therefore extremely interesting for direct writing of optical waveguides with the potential to exhibit ultra-broad-band amplification in the S-C-L telecom bands (1460–1625 nm) [41]. So far, active waveguides showing internal gain over the entire C+L telecom bands and net gain at 1534 nm have been demonstrated [43].

## 10.5 Waveguide Lasers

### 10.5.1 Waveguide Lasers in Er:Yb-Doped Phosphate Glass

#### 10.5.1.1 First Demonstration of Laser Action

The first waveguide laser written by a fs-laser was fabricated in an Er:Yb-doped phosphate glass (Kigre QX) by means of a frequency-doubled cavity-dumped Yb:glass laser, operated at 166 kHz repetition rate [10]. The doping concentration was 2% in weight of erbium and 4% in weight of ytterbium and 20-mm long waveguides were fabricated. The waveguides provided up to 4.4 dB peak internal gain at 1533.8 nm under 420 mW pump power at 980 nm wavelength in a bi-propagating pumping scheme. Insertion losses to standard single mode fiber were 2.1 dB, thus allowing 2.3 dB peak net gain. The laser cavity was obtained by means of two fiber Bragg gratings with peak reflectivity at 1533.5 nm butt-coupled to the waveguide: a broad-band flat-top FBG with 1-nm bandwidth (FWHM) providing

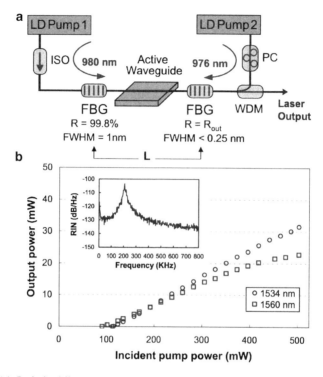

**Fig. 10.6** (**a**) Optimized linear cavity setup for the Er:Yb-doped phosphate waveguide laser in the C-band. *ISO*: Optical Isolator. *PC*: Polarization Controller at $\lambda/2$. *WDM*: Wavelength-Division Multiplexer. (**b**) Input–output characteristics of the laser at 1534 nm and at 1560 nm. Inset shows a typical RIN spectrum of the laser at 1534 nm. Reprinted with permission from [44]

high reflectivity (99.8%) at one side, and a 68% reflectivity 0.25-nm bandwidth FBG used as output coupler. A maximum output power of 1.7 mW with 2% slope efficiency was reported.

### 10.5.1.2 Tunable Lasers in the C-Band

A significant improvement in the waveguide quality was achieved by increasing the repetition rate of the cavity-dumped Yb:glass oscillator. 20-mm long waveguides fabricated at 505 kHz repetition rate, 436 nJ energy per pulse with 50–100 $\mu$m/s writing speed exhibited insertion loss as low as 1 dB. Also, the laser cavity, reported in Fig. 10.6a, was optimized with respect to the first laser experiment: in order to remove the parasitic interaction between the counter-propagating pumping beams, a single-stage optical isolator was connected to pump 1, and, in addition, a half-wave fiber polarization controller was inserted along the fiber connecting pump 2 to the waveguide.

Note that the latter configuration, with respect to discrete mirrors or dielectric coatings, provides output power already coupled into a standard SMF with no additional losses, and allows for a simple change of the laser wavelength and output coupling by substitution of the gratings. This feature of the setup and the high gain exhibited by the waveguides allowed demonstration of laser action at different wavelengths [33]. A maximum output power of 30 mW with 8.4% slope efficiency was obtained at 1533.5 nm and 23 mW output power with 6.6% slope efficiency was also reported at 1560 nm (i.e., at the upper edge of the C-band), thus ascertaining wavelength tunability in the whole C-band of optical communications (Fig. 10.6b).

## 10.5.2 Waveguide Laser in Er:Yb-Doped Oxyfluoride Silicate Glass

A laser device was also recently reported in Er:Yb-doped oxyfluoride silicate glass by exploiting the high quality waveguides previously demonstrated to exhibit net gain (see Sect. 10.4.3) [45]. The laser cavity was based on FBGs (a high reflectivity 0.38 nm bandwidth and a 99.5% reflectivity FBGs with 0.44 nm bandwidth as output coupler) and a dual-wavelength bi-propagating pumping scheme was adopted as sketched in Fig. 10.7a. Total physical length of the cavity was 207.5 cm. A maximum output power of 30 $\mu$W under 342 mW pump power at 980 nm plus 213 mW pump power at 1480 nm was demonstrated in multimode operation (see Fig. 10.7b).

## 10.6 Advanced Waveguide Lasers

Besides the lasers illustrated above and operating in continuous wave multi-mode regime, more advanced waveguide lasers capable of stable single-longitudinal-mode or even mode-locking operation have been demonstrated in different active media.

### 10.6.1 Single-Longitudinal-Mode Operation

#### 10.6.1.1 Er:Yb-Phosphate Glass Laser

A single-longitudinal-mode Er:Yb-glass laser has been demonstrated by exploiting an optimized and ultra-compact Fabry–Perot linear laser cavity based on FBGs [46]. Actually, in previous experiments by the same authors (see Sect. 10.5.1), the laser cavity was longer than 50 cm and the output coupler had a bandwidth of 0.25 nm, thus causing a spectral broadening of the source due to multimode oscillation. To achieve single mode operation, the output coupler was replaced by a narrow bandwidth (64 pm FWHM) FBG and the cavity length was reduced by cutting the FBG fibers. A cavity length of $L = 5.5$ cm was obtained. Such a compact cavity

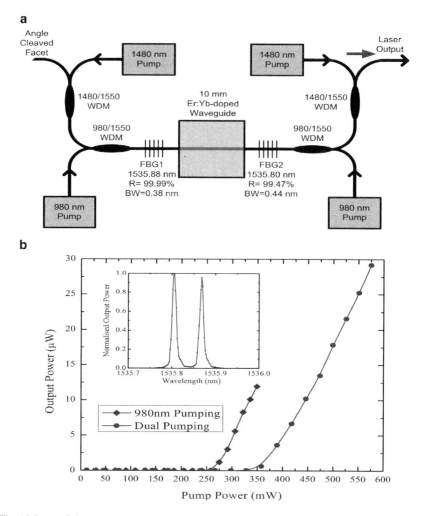

**Fig. 10.7** (a) Schematic diagram of the waveguide laser in oxyfluoride silicate glass. (b) Laser output power versus pump power. Inset graph shows laser output spectrum recorded by an optical spectrum analyzer with 0.01 nm resolution bandwidth. Reprinted with permission from [45]

resulted in a large free spectral range of 1.82 GHz thus consistently reducing the number of modes falling within the FWHM bandwidth of the FBG. Figure 10.8a shows the input–output characteristic of the 5.5-cm long laser cavity. The threshold pump power was 124 mW with 21% slope efficiency. To ascertain single mode operation the laser power spectrum was monitored by means of a scanning confocal Fabry–Perot interferometer. Single mode operation was maintained up to a maximum pump power level of about 400 mW, corresponding to an output power of about 55 mW. At higher pump power a slightly multimodal regime appeared, with weak side peaks 1.8-GHz apart from the central mode, thus corresponding to

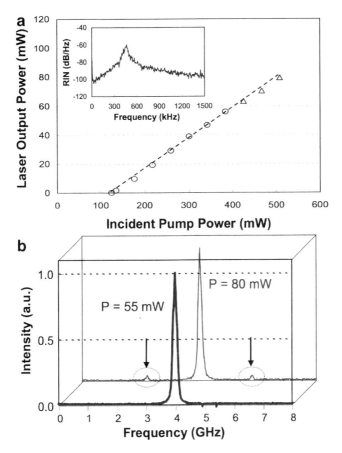

**Fig. 10.8** (a) Input–output characteristic of the Er:Yb-phosphate glass ultra-compact cavity. Inset shows the RIN spectrum of the laser in single mode operation. (b) Fabry–Perot spectra at maximum output power in single mode (*circles in panel (a)*) and slightly multimodal regime (*triangles in panel (a)*). Reprinted with permission from [44]

adjacent longitudinal modes. As compared to previously reported waveguide lasers fabricated with the same technology, this cavity suffers from a higher intensity noise. Actually, the use of a narrow bandwidth FBG makes the single-mode laser intensity noise quite sensitive to cavity length fluctuations induced either by mechanical vibrations at the fiber-waveguide butt-coupling interface or by environmental temperature variations. Nevertheless, it is worth noting that the performance reported above compares very well, in terms of maximum output power, slope efficiency, and spectral quality, with state of the art Er-Yb-based single-mode waveguide lasers fabricated in phosphate glasses by Ag-Na ion-exchange technique [47, 48]. An improvement in power stability could be achieved by using fiber-pigtailing technique to permanently connect the waveguide to FBGs, or eventually by fabricating the Bragg gratings in the active substrate (possibly by means of the

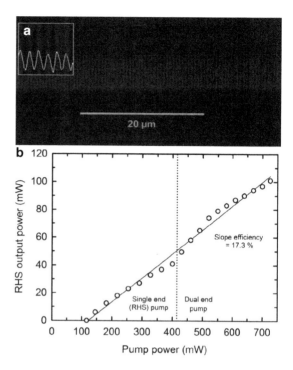

Fig. 10.9 (a) Transmission DIC micrograph of a typical first-order waveguide Bragg grating fabricated by the single-step writing process with a low repetition rate Ti:sapphire system [49]. The inset shows a section of the structure at 3× magnification. Overlaid on the inset is a graph of the grating image intensity. (b) Output power of the DFB Yb-laser, at room temperature, as a function of pump power for single- and dual-end pump geometries using a 9.5-mm long waveguide Bragg grating. Reprinted with permission from [49, 50]

same fs-laser writing process) leading to an integrated Fabry–Perot cavity or to a distributed feedback (DFB) laser configuration. Preliminary results on the feasibility of a DFB laser device have already been reported in Er-Yb-doped phosphate glass [49]. The active waveguide and a superimposed Bragg grating were fabricated in a single-step writing process (see Fig. 10.9a) by using a low repetition rate Ti:sapphire laser. Even though laser efficiency was still very low (0.36 mW maximum output power under 710 mW incident pump power) and pump power threshold very high (639 mW), this result is important for the future development of monolithic fs-laser written waveguide lasers.

### 10.6.1.2 Yb-Phosphate Glass Laser

Very recently, Ams and co-workers also demonstrated an efficient DFB laser in a commercial Yb-doped phosphate glass [50]. Similarly to the previous work on Er:Yb-doped DFB lasers [49], the active waveguide and a superimposed Bragg grating were fabricated in a single-step writing process by using a low repetition rate Ti:sapphire system (1 kHz repetition rate, 120 fs pulse duration, 20× focusing objective). The resulting ultra-compact 9.5-mm long monolithic laser provided up to 100 mW single-end output power at 1032.59 nm under dual pumping at 700 mW with 17.3% slope efficiency (Fig. 10.9b). The laser linewidth was measured (by scanning Michelson interferometry) to be below 2 pm with aside-mode suppression ratio of at least 20 dB. Pumping with unpolarized light (via a non-polarization-maintaining WDM) led the laser to oscillate on two frequency degenerate polarization modes (within the 2 pm linewidth) with polarizations parallel and orthogonal to the direction of the fs-laser writing beam. This indicates that the waveguide Bragg grating exhibits a birefringence as low as $2 \times 10^{-6}$. In any case, single-polarization operation was imposed by removing the WDM and pumping with a polarized pump diode with pump polarization axis aligned with the horizontal or vertical axis of the active waveguide.

This was the first demonstration of a high power waveguide laser created entirely by means of fs-laser micromachining, a result that possibly paves the way to the development of a new generation of monolithic waveguide lasers with high temporal coherence.

### 10.6.1.3 Nd:YAG Ceramic Laser

Recently an efficient fs-laser written waveguide laser has been also demonstrated in a Nd:YAG ceramic material [51]. YAG ceramics have recently attracted much interest because of the lower production cost and the possibility of higher doping concentration with respect to YAG crystals.

Again, the waveguide was fabricated by means of a low repetition rate Ti:sapphire laser by exploiting the negative refractive index change induced in this ceramic material with a *two-line* writing method: two lines of negative refractive index modification were written with a separation of 20 $\mu$m by translating the sample at 50 $\mu$m/s speed; therefore light confinement is achieved in the unmodified central region (Fig. 10.10a). A flat dichroic dielectric mirror (99% reflectivity

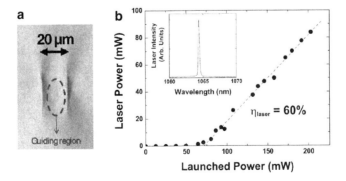

**Fig. 10.10** (a) Microscope image of the end face of the active waveguide fabricated in Nd-doped YAG ceramic by two-line fs-laser writing technique. (b) Output power as a function of the launched pump power of the fs-laser written Nd waveguide laser. Inset shows the spectral dependence of the laser radiation. Reprinted with permission from [51]

around 1.06 μm and 80% transmission in the 600–850 nm range) was attached at one end-facet of the waveguide chip and 8% Fresnel reflection at the other end-facet provided optical feedback. Under 200 mW launched pump power at 748 nm, this Fabry–Perot waveguide cavity delivered up to 80 mW output power in stable single mode operation at 1064 nm (Fig. 10.10b). A remarkable 60% slope efficiency was also demonstrated, which is comparable to state of the art Nd-based waveguide lasers so far reported by other fabrication techniques.

### 10.6.2 Mode-Locking Regime

Another demonstration of advanced functionality in a fs-laser written waveguide laser was the feasibility of passive mode-locking operation. Classically, the technique of passive mode-locking employing semiconductor saturable absorber mirrors has been widely adopted to generate ps and fs laser pulses, but recently a new technology based on carbon nanotubes (CNTs) has emerged as an alternative for the realization of saturable absorbers. CNTs show in fact strong and tunable saturable absorption in the near infrared, ultrashort recovery time, and can be cheaply assembled into polymer composites and easily integrated into optical fiber communication systems. This enabled the first demonstration of a mode-locked laser source based on a carbon nanotubes saturable absorber and a femtosecond laser written waveguide [52]. CNTs specially designed to have efficient saturable absorption at 1.5 μm were embedded in free standing polyvinyl alcohol (PVA) films of 50 μm thickness. By sandwiching such a film between a couple of ferrule plane connectors (FC/PC), with index matching fluid at both fiber ends, a fiber-pigtailed CNT-PVA mode-locker was packaged. Figure 10.11a shows a schematic of the mode-locked waveguide laser in a ring cavity configuration integrating a 36-mm long Er:Yb-doped waveguide amplifier fabricated by femtosecond laser writing [34] and the fiber-pigtailed CNT-PVA mode-locker. Two 976-nm pump laser diodes, providing 480 mW total incident power, were coupled to the waveguide by means of WDMs. A fiber coupler was used to couple 5% of the intracavity radiation out of the ring resonator. Uni-directionality of the ring was imposed by an optical isolator. Figure 10.11b shows the laser output spectrum recorded by an optical spectrum analyzer. Continuous-wave laser action started at 450 mW incident pump power and self-starting single-pulse stable mode-locking was observed just above laser threshold. The laser central wavelength in the mode-locking regime was 1535 nm, with an oscillating bandwidth (at −3 dB) of 1.6 nm. By means of standard autocorrelation techniques (see trace in Fig. 10.11c) the pulse duration was estimated to be about 1.6 ps, resulting in transform limited $sech^2$ pulses. The repetition rate of this cavity was relatively low (16.7 MHz), but much higher repetition rates could be obtained using an ultra-compact linear cavity where the saturable absorber is directly placed on the waveguide facet.

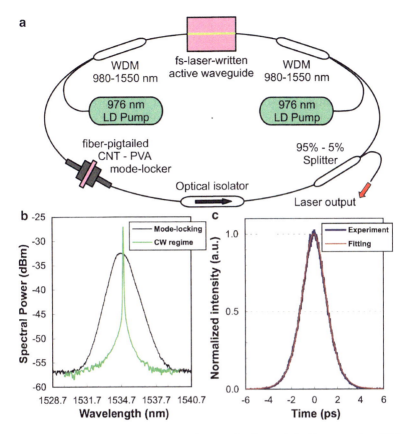

**Fig. 10.11** (a) Schematic of the Er:Yb ring laser cavity for mode-locking operation. (b) Laser power spectra in continuous-wave and passive mode-locking regimes. (c) Autocorrelation trace of the laser pulses. Reprinted with permission from [44]

## 10.7 Outlook and Conclusions

Femtosecond lasers have proved to be very valuable tools for the fabrication of active waveguide devices, as amplifiers and lasers. It should be noted that the fabrication of active devices has been the first test bench where this new microfabrication technology showed the capability to compete with conventional ones. Waveguide amplifiers and lasers, fabricated by ultrafast laser inscription, provided performances comparable to those exhibited by devices fabricated by ion exchange or plasma-enhanced chemical vapor deposition (PECVD); this was a very important result to assess the reliability of femtosecond lasers as a microfabrication tool in transparent materials. However, telecom applications typically require mass production of active devices with a strong drive toward cost reduction, thus the flexibility and rapid prototyping capabilities of femtosecond laser waveguide writing are not seen as key elements. This is probably why, notwithstanding the significant results achieved so far, a real implementation in an industrial environment is still to come. However, the application of femtosecond laser waveguide writing to active device fabrication is still raising interest. In fact, this fabrication technology turned out to be extremely valuable to test new active materials. The nonlinear absorption mechanism at the base of the modification process is flexible enough to allow waveguide writing with the same laser source in almost any transparent material, after a suitable optimization of the irradiation parameters is performed. In this way, material scientists developing new and improved active materials can rapidly test their performances and potentials as substrates for active devices, by direct laser writing waveguides and characterizing them as amplifiers. Fabrication of active devices in new active materials will probably be one of the main fields of application of femtosecond laser waveguide writing also in the near future and every time a small series of prototypes is to be produced.

# References

1. D.R. Zimmermann, L.H. Spiekman, J. Lightwave Technol. **22**, 63 (2004)
2. H. Willebrand, B.S. Ghuman, *Free-Space Optics: Enabling Optical Connectivity in Today's Networks*, (Sams Publishing, Indianapolis, 2002)
3. C.J. Karlsson, F.A. Olsson, D. Letalick, M. Harris, Appl. Opt. **39**, 3716–3726 (2000)
4. F.T.S. Yu, S. Yin, *Fiber Optic Sensors*, (Marcel Dekker, Inc., New York, 2002)
5. C. Spiegelberg, J. Geng, Y. Hu, Y. Kaneda, S. Jiang, N. Peyghambarian, J. Lightwave Technol. **22**, 57–62 (2004)
6. A. Schulzgen, L. Li, V.L. Temyanko, S. Suzuki, J.V. Moloney, N. Peyghambarian, Opt. Exp. **14**, 7087–7092 (2006)
7. S.S. Saini, Y. Hu, F.G. Johnson, D.R. Stone, H. Shen, W. Zhou, J. Pamulapati, M.N. Ott, H.C. Shaw, M. Dagenais, Photon. Technol. Lett. **12**, 840 (2000)
8. G. Della Valle, S. Taccheo, G. Sorbello, E. Cianci, V. Foglietti, P. Laporta, Electron. Lett. **42**, 632 (2006)
9. G. Della Valle, R. Osellame, P. Laporta, J. Opt. A Pure Appl. Opt. **11**, 013001 (2009)
10. S. Taccheo, G. Della Valle, R. Osellame, G. Cerullo, N. Chiodo, P. Laporta, O. Svelto, A. Killi, U. Morgner, M. Lederer, D. Kopf, Opt. Lett. **29** 2626 (2004)
11. W. J. Miniscalco, in *Rare-Earth-Doped Fiber Lasers and Amplifiers*, ed. by M.J.F. Digonnet (Marcel Dekker, Inc., New York, 2001) pp. 19–133
12. O. Svelto, *Principles of Lasers*, (Springer Science, Inc., New York, 2010) pp. 6–8
13. B.-C. Hwang, S. Jiang, T. Luo, G. Sorbello, N. Peyghambarian, J. Opt. Soc. Am. B **17**, 833–839 (2000)
14. V.P. Gapontsev, S.M. Matitsin, A.A. Isineev, Opt. Commun. **46**, 226–230 (1983)
15. Y. Lu, N. Ming, J. Mater. Sci. **30**, 5705 (1995)
16. K. Lu, N.K. Duttaa, J. Appl. Phys. **91**, 576–581 (2002)
17. Y. Fujimoto, M. Nakatsuka, Jpn. J. Appl. Phys. **40**, L279 (2001)
18. Y. Fujimoto, M. Nakatsuka, Appl. Phys. Lett. **82**, 3325 (2003)
19. T. Suzuki, Y. Ohishi, Appl. Phys. Lett. **88**, 191912 (2006)
20. X.-G. Meng, J.-R. Qiu, M.-Y. Peng, D.-P. Chen, Q.-Z. Zhao, X.-W. Jiang, C. S. Zhu, Opt. Exp. **13**, 1628 (2005)

21. J. Ren, G. Dong, S. Xu, R. Bao, J. Qiu, J. Phys. Chem. A **112**, 3036–3039 (2008)
22. I. Razdobreev, L. Bigot, V. Pureur, A. Favre, G. Bouwmans, M. Douay, Appl. Phys. Lett. **90**, 031103 (2007)
23. I.A. Bufetov, K.M. Golant, S.V. Firstov, A.V. Kholodkov, A.V. Shubin, E.M. Dianov, Appl. Opt. **47**, 4940 (2008)
24. G. Sorbello, S. Taccheo, P. Laporta, Opt. Quant. Electron. **33**, 599–619 (2001)
25. E. Desurvire, *Erbium-Doped Fiber Amplifiers*, (Wiley, New York, 1994)
26. P.C. Becker, N.A. Olsson, J.R. Simpson, *Erbium-Doped Fiber Amplifiers: Fundamentals and Technology*, (Academic Press, San Diego, 1999)
27. G. Matthäus, J. Burghoff, W. Will, S. Nolte, A. Tünnermann, Appl. Phys. A **83**, 347–350 (2006)
28. Y. Sikorski, A.A. Said, P. Bado, R. Maynard, C. Florea, K.A. Winick, Electron. Lett. **36**, 226 (2000)
29. R. Osellame, S. Taccheo, G. Cerullo, M. Marangoni, D. Polli, R. Ramponi, P. Laporta, S. De Silvestri, Electron. Lett. **38**, 964 (2002)
30. R. Osellame, S. Taccheo, M. Marangoni, R. Ramponi, P. Laporta, D. Polli, S. De Silvestri, G. Cerullo, J. Opt. Soc. Am. B **20**, 1559 (2003)
31. A. Killi, U. Morgner, M.J. Lederer, D. Kopf, Opt. Lett. **29**, 1288 (2004)
32. R. Osellame, N. Chiodo, G. Della Valle, S. Taccheo, R. Ramponi, G. Cerullo, A. Killi, U. Morgner, M. Lederer, D. Kopf, Opt. Lett. **29**, 1900 (2004)
33. R. Osellame, N. Chiodo, G. Della Valle, G. Cerullo, R. Ramponi, P. Laporta, A. Killi, U. Morgner, M. Lederer, D. Kopf, O. Svelto, IEEE J. Sel. Top. Quant. Electron. **12**, 277 (2006)
34. G. Della Valle, R. Osellame, N. Chiodo, S. Taccheo, G. Cerullo, P. Laporta, A. Killi, U. Morgner, M. Lederer, D. Kopf, Opt. Exp. **13**, 5976 (2005)
35. Y. Jaouën, L. du Mouza, D. Barbier, J.-M. Delavaux, P. Bruno, Photon. Tech. Lett. **11**, 1105 (1999)
36. S.X. Shen, A. Jha, Opt. Mater. **25**, 321 (2004)
37. R.R. Thomson, S. Campbell, I.J. Blewett, A.K. Kar, D.T. Reid, S. Shen, A. Jha, Appl. Phys. Lett. **87**, 121102 (2005)
38. R.R. Thomson, H.T. Bookey, N. Psaila, S. Campbell, D.T. Reid, S. Shen, A. Jha, A.K. Kar, Photon. Technol. Lett. **18**, 1515 (2006)
39. N.D. Psaila, R.R. Thomson, H.T. Bookey, A.K. Kar, N. Chiodo, R. Osellame, G. Cerullo, A. Jha, S. Shen, Appl. Phys. Lett. **90**, 131102 (2007)
40. N.D. Psaila, R.R. Thomson, H.T. Bookey, A.K. Kar, N. Chiodo, R. Osellame, G. Cerullo, G. Brown, A. Jha, S. Shen, Opt. Exp. **14**, 10452 (2006)
41. P. Nandi, G. Jose, C. Jayakrishnan, S. Debbarma, K. Chalapathi, K. Alti, A.K. Dharmadhikari, J.A. Dharmadhikari, D. Mathur, Opt. Exp. **14**, 12145 (2006)
42. L. Huang, A. Jha, S. Shen, X. Liu, Opt. Exp. **12**, 2429 (2004)
43. T.T. Fernandez, S.M. Eaton, G. Della Valle, R. Martinez Vazquez, M. Irannejad, G. Jose, A. Jha, G. Cerullo, R. Osellame, P. Laporta, Opt. Exp. **18**, 20289 (2010)
44. R. Osellame, G. Della Valle, N. Chiodo, S. Taccheo, P. Laporta, O. Svelto, G. Cerullo, Appl. Phys. A **93**, 17–26 (2008)
45. N.D. Psaila, R.R. Thomson, H.T. Bookey, N. Chiodo, S. Shen, R. Osellame, G. Cerullo, A. Jha, A.K. Kar, Photon. Tech. Lett. **20**, 126 (2008)
46. G. Della Valle, S. Taccheo, R. Osellame, A. Festa, G. Cerullo, P. Laporta, Opt. Exp. **15**, 3190 (2007)
47. D.L. Veasey, D.S. Funk, P.M. Peters, N.A. Sanford, G.E. Obarski, N. Fontaine, M. Young, A.P. Peskin, W.-C. Liu, S.N. Houde-Walter, J.S. Hayden, J. Non-Cryst. Solids **263-264**, 369–381 (2000)
48. G. Della Valle, A. Festa, G. Sorbello, K. Ennser, C. Cassagnetes, D. Barbier, S. Taccheo, Opt. Exp. **16**, 12334 (2008)
49. G.D. Marshall, P. Dekker, M. Ams, J.A. Piper, M.J. Withford, Opt. Lett. **33**, 956 (2008)

50. M. Ams, P. Dekker, G.D. Marshall, M.J. Withford, Opt. Lett. **34**, 247 (2009)
51. G.A. Torchia, A. Rodenas, A. Benayas, E. Cantelar, L. Roso, D. Jaque, Appl. Phys. Lett. **92**, 111103 (2008)
52. G. Della Valle, R. Osellame, G. Galzerano, N. Chiodo, G. Cerullo, P. Laporta, O. Svelto, U. Morgner, A.G. Rozhin, V. Scardaci, A.C. Ferrari, Appl. Phys. Lett. **89**, 231115 (2006)

# Part III
# Waveguides and Optical Devices in Other Transparent Materials

# Chapter 11
# Waveguides in Crystalline Materials

Matthias Heinrich, Katja Rademaker, and Stefan Nolte

**Abstract** Crystalline media offer a variety of unique properties, such as even order nonlinearities, birefringence, and broad transparency ranges, which make them of great interest for applications in integrated optics. However, the extension of femtosecond laser direct inscription techniques [Nolte et al., Appl. Phys. A **77**, 109 (2003), J. Mod. Opt. **51**, 2533 (2004); Itoh et al. MRS Bull. **31**, 620 (2006); Gattass, Mazur, Nature **2**, 219–225 (2008); Della Valle et al., J. Opt. A: Pure Appl. Opt. **11** (2009)] with their vast potential for the fabrication of buried three-dimensional structures of arbitrary shapes to crystals has proven to be a challenging task. Results achieved so far in direct writing of optical devices in crystals will be reviewed.

## 11.1 Origins of Refractive Index Changes

Generally, the behavior of crystals under fs-laser irradiation differs significantly from that observed in fused silica. While in the latter, modifications feature an increased refractive index due to densification [1, 2], destruction of a regular crystalline lattice generally reduces the density within the modified region. Similar to phosphate glasses [3], a reduction of the refractive index follows. Thus, considerably more effort is necessary to achieve guiding of light.

One of the first crystalline materials investigated regarding structural changes due to femtosecond laser irradiation was $\alpha$-quartz [4]. In contrast to the amorphous phase (fused silica glass), only a decrease of the refractive index was observed in the modified regions. Cross-sectional transmission electron microscopy investigations for various pulse energy regimes revealed an amorphous core for single nano-Joule pulses, while partial recrystallization in the inner part of the modification

M. Heinrich (✉) · K. Rademaker · S. Nolte
Institute of Applied Physics, Friedrich-Schiller-Universität Jena, Max Wien Platz 1,
07743 Jena, Germany
e-mail: heinrich@iap.uni-jena.de; stefan.nolte@uni-jena.de

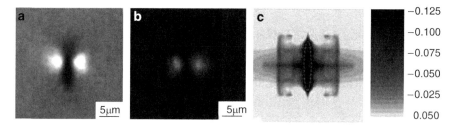

**Fig. 11.1** Lines written with a pulse energy of 14 μJ in a quartz crystal: (**a**) polarization-contrast optical microscope image (end view) of the modifications; (**b**) near-field distribution of 514-nm radiation guided in the produced structure; (**c**) hydrostatic pressure distribution in a quartz matrix after a single-shot exposure according to FEM calculations (the "irradiated" ellipsoid region is marked) [4]

occurred for overlapping pulses. The use of micro-Joule pulse energy generally resulted in an amorphous phase and thus a decrease in density and refractive index. Outside a small transition zone of only some ten nanometers, the material in the vicinity of the modifications still exhibits a crystalline structure. Interestingly, here an increased refractive index (up to $\Delta n \approx 0.01$) was discovered symmetrically on both sides of the irradiated region (Fig. 11.1). This index increase was attributed to the stress caused by expansion of the amorphous core. Consequently, waveguiding with propagation losses $< 5\mathrm{dBcm}^{-1}$ were observed.

Apart from quartz, a crystalline material of considerable more importance for integrated as well as nonlinear optics is lithium niobate ($LiNbO_3$). The characteristics and origin of modifications in $LiNbO_3$ have been thoroughly investigated e.g. in [5, 6, 49].

Exposure at high pulse energies (>1 μJ) results in significant damage of the crystalline structure, leading to the formation of so-called type II modifications (Fig. 11.2d–f) similar to the results in quartz. The crystal lattice is completely destroyed in the focal region, where both refractive indices (ordinary $n_o$ and extraordinary $n_e$) are decreased. Stress fields caused by the expansion of the modified volume extend across the surrounding regions, inducing local changes to the refractive indices. Depending on the direction of stress, either the ordinary or extraordinary index can be increased here.

In addition, so-called type I structures (Fig. 11.2a–c) are formed at pulse energies just above the modification threshold. They are characterized by an increased extraordinary refractive index, while the ordinary index as well as the nonlinear response is decreased.

As high energy protons backscattering experiments have revealed, type II structures can clearly be attributed to a destruction of the crystalline lattice, decreasing the density and thus both refractive indices in the modified region. Further contributions to the index change may arise from local changes in stoichiometry after thermal diffusion of ions during the formation process. However, this contribution is expected to be small due to the short time scales associated

11 Waveguides in Crystalline Materials

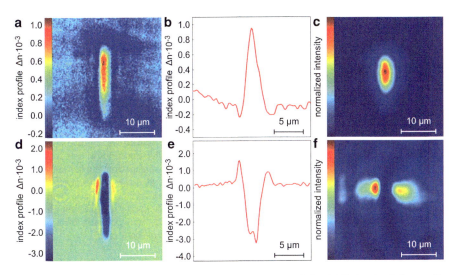

**Fig. 11.2** Femtosecond laser-induced modifications in x-cut LiNbO$_3$. (**a**) Refractive index profile ($n_e$) of a type I structure, (**b**) respective horizontal cross-section, (**c**) observed intensity distribution of the guided mode at 633 nm (TE polarization). (**d**) Refractive index profile ($n_e$) of a type II structure, (**e**) respective horizontal cross-section, (**f**) observed intensity distribution of the guided mode at 633 nm (TE polarization) [5]

with fs-laser irradiation. Photorefractive processes have only minor influence on the laser-induced refractive index change. For type I modifications, this holds true since they can also be fabricated in MgO-doped LiNbO$_3$, which is much less photorefractive than pure LiNbO$_3$. For type II structures, this is obvious since they lack the crystalline properties responsible for photorefractivity altogether.

The occurrence of type I modifications can instead by explained by taking into account the ferroelectric properties of LiNbO$_3$. Its spontaneous polarization leads to a decrease of both refractive indices by means of the quadratic electrooptical effect. In LiNbO$_3$, $n_e$ is influenced more strongly than $n_o$; in fact, a decrease of the polarization due to distortions of the lattice mainly leads to an increase of the extraordinary index. The much smaller increase of $n_o$ is easily overcompensated by the defects inevitably created during the fabrication process [6]. This model is supported by the fact that to this date, no type I modifications have been found in nonferroelectric crystals.

In lithium niobate, the shape of the fs-laser induced structures strongly depends on the energy and length of the pulses. If the critical power $P_{cr} = 3.77\lambda^2/8\pi n_0 n_2$ is exceeded (for 400 fs pulses at 800 nm, this threshold corresponds to a pulse energy of 0.12 µJ), filamentation occurs due to a dynamic balance between self-focusing and plasma defocusing. This effect can extend the modifications far beyond the Rayleigh length of the focus (Fig. 11.3a and b).

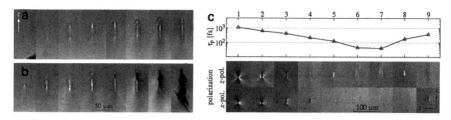

**Fig. 11.3** Transmission microscope images of the end facet of z-cut LiNbO$_3$ illuminated with z-polarized light. The pulse duration was (**a**) 130 fs and (**b**) 420 fs. Pulse energies were 0.2, 0.3, 0.4, 0.5, 0.6, 1, 3, and 4.5 µJ from left to right. The writing laser was incident from the bottom. Bright and dark regions correspond to extraordinary index increases and decreases, respectively [6]. (**c**) Influence of the pulse duration $\tau_p$ on the waveguide properties in x-cut LiNbO$_3$. Shown are transmission microscope images for z- or x-polarized light. The minimum pulse duration was 40 fs while the pulse energy was 0.2 µJ for all structures; the same results were obtained for negatively and positively chirped pulses. The writing laser was incident from the bottom [6]

Even more pronounced is the influence of the pulse duration (Fig. 11.3c). For a fixed pulse energy, the modifications decrease dramatically in size as well as strength if the pulse length is decreased. While pulses with a length of 1 ps cause severe destruction of the material, hardly any modification occurs if the same energy is delivered by a 40 fs pulse. This behavior is due to the strong nonlinear absorption of LiNbO$_3$, which depletes the energy of short pulses significantly before the actual focus. Although the total absorption of a 50 fs pulse (Fig. 11.4a, b) is twice as large as that of a 500 fs pulse (Fig. 11.4c, d), the energy is deposited in a significantly larger volume. Consequently, the peak fluence of the focused 500 fs pulse is about six times that of the 50 fs pulse. Optimum pulse lengths of around 1 ps for the fabrication of waveguides have been reported in [7].

## 11.2 Waveguides Characteristics in Various Crystals

Depending on the material, various approaches to the creation of waveguides via femtosecond laser processing have been developed. Once fabricated, the guides are typically evaluated according to their propagation loss. As opposed to the so-called insertion loss, which depends strongly on the excitation efficiency (i.e., the overlap of the injected light with the waveguide mode) as well as on the bulk refractive index of the host material itself (Fresnel reflections), the loss that light suffers during propagation constitutes an actual measure for guiding properties of a waveguide structure. Other characteristics to be considered include the shape of the mode fields, thermal stability, possible deterioration over time, and finally nonlinear properties of the waveguide, which will be discussed in Sect. 11.3 of this chapter. In actively doped materials, further information can be gained from the observation of changes in the spectral characteristics.

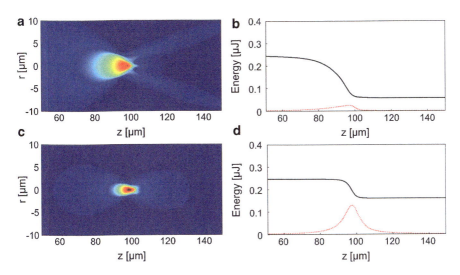

**Fig. 11.4** Simulated propagation of a 0.25 μJ pulse in LiNbO$_3$ [5]. (**a**) Spatial fluence distribution for a pulse length of 50 fs (radial coordinate r, propagation distance z). (**b**) Evolution of the overall energy of the same pulse during propagation (*solid black line*) and energy within a radius of 1 μm around the optical axis (*dotted red line*). (**c** and **d**) Same figures for a pulse length of 500 fs. Note, that the peak fluence in (**c**) is six times larger than in (**a**)

## 11.2.1 Waveguide Fabrication in LiNbO$_3$

### 11.2.1.1 Type I

The first waveguides created via femtosecond laser-processing of LiNbO$_3$ were based on Type I modifications [8]. In such waveguides, propagation losses as low as 0.6 dB cm$^{-1}$ have been reported [9]. Unfortunately, however, type I waveguides have proven to be susceptible to thermal deterioration. Heating of the structures to 150°C results in decreased light confinement [9] and can even cure the modifications completely [10]. As a result, they are not suitable for high average power applications, which require temperatures around 200°C in order to prevent photorefractive damage. Furthermore, a gradual decay of type I structures has been observed even at room temperature [10]. Increased writing energies allow the fabrication of stable type I waveguides, however, at the cost of strongly decreased nonlinear coefficients [9].

### 11.2.1.2 Type II

Type II modifications (exemplary microscope image shown in Fig. 11.5a) feature refractive index increases in the vicinity of the actual structure (Fig. 11.5b). Consequently, light is weakly confined and the modes extend far into the bulk medium

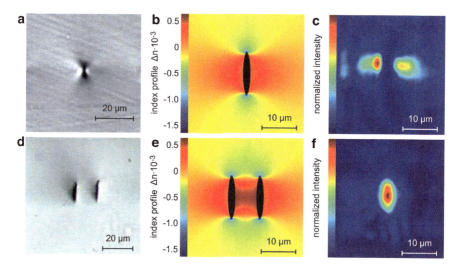

**Fig. 11.5** Single type II modification in z-cut LiNbO$_3$: (**a**) Microscope image, (**b**) simulated stress-induced index-increase ($n_e$) and (**c**) exemplary guided mode at 633 nm [5]. Dual-line waveguide based on type II modifications: (**d**) microscope image [11], (**e**) simulated stress-induced index-increase ($n_e$), and (**f**) exemplary guided mode at 633 nm

(Fig. 11.5c). However, symmetrical waveguides supporting well-confined mode profiles can be fabricated by overlapping the stress fields of two adjacent type II structures (Fig. 11.5d–f). For such dual-line waveguides, propagation losses as low as 1.0 dB cm$^{-1}$ have been reported [11]. Although type I waveguides with lower losses have been demonstrated, type II waveguides have the significant advantage of thermal stability up to more than 300°C [5].

## 11.2.2 Other Crystals

Beside quartz and LiNbO$_3$, many other types of crystalline media are of interest for applications in integrated optics. This section will present an overview of materials, in which femtosecond laser-written waveguides have been realized in the past.

### 11.2.2.1 Silicon

Integrated optical components in silicon, the material of choice for electronic components, will allow the synthesis of optics and electronics on minute scales. To enable the formation of waveguides by means of ultrafast laser processing techniques, however, a carrier wavelength within the transparency range of the material has to be employed. Also, the large refractive index step between air

11 Waveguides in Crystalline Materials 301

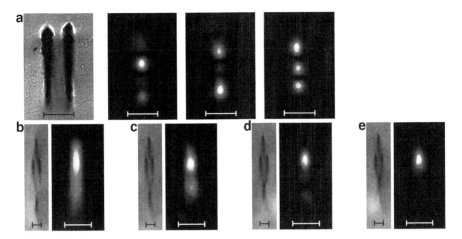

**Fig. 11.6** Type II waveguides in BBO [14]. (**a**) Dual line geometry (*left*: microscope image) supporting multiple vertical modes (*right*) at a wavelength of 633 nm. (**b–e**) Improved vertical confinement in various quad line geometries. All scale bars indicate 20 μm

and silicon and resulting aberrations need to be overcome. To this end, an optical parametric oscillator (OPA) was utilized to generate 1.7 μJ pulses at a wavelength of 2.5 μm. A thin oxide layer covering the silicon wafer served as the index matching medium, allowing the fabrication of waveguiding structures within the crystalline silicon beneath [12]. Despite a broad range of focusing depths, guiding always was observed near the $SiO_2$–Si interface, giving rise to asymmetric mode profiles. Within the 1.3–1.6 μm telecommunication band, losses between 0.7 and 1.2 dBcm$^{-1}$ were reported for light polarized orthogonally to the interface, while the parallel polarization was subject to significantly higher losses.

### 11.2.2.2 β Barium Borate

For nonlinear optical applications in the visible spectrum, β barium borate (BBO) is the material of choice due to its strong nonlinearity and transparency down to the UV range. Beside the fabrication of gratings by femtosecond laser irradiation [13], waveguides based on type II structures have also realized in this material [14]. However, experiments have shown that the usual dual-line approach yields insufficient confinement of light along the direction of the writing beam. Consequently, the waveguides support several transverse modes in the vertical direction (see Fig. 11.6a). This difficulty can be addressed by introducing another pair of modifications to increase vertical confinement. At a sufficiently narrow separation, single mode behavior can be enforced (see Fig. 11.6b–e) [14]. Unfortunately, the additional structures also increase the overlap of light with modified regions of the crystal, causing higher scattering losses. Consequently, the lowest propagation losses achieved to date are in the range of 5 dB cm$^{-1}$.

#### 11.2.2.3 Lithium Tantalate

The fabrication of waveguides in lithium tantalate ($LiTaO_3$) has been reported in [15]. While the modified volume itself showed a decreased refractive index, guiding at a wavelength of 1.5 μm was observed in the region above. Micro-Raman investigations of the guiding regions revealed lattice compression as the dominant guiding mechanism in $LiTaO_3$ [16].

#### 11.2.2.4 Lithium Fluoride

Lithium fluoride (LiF) is one of the most chemically stable alkali halide crystals. The visible range laser active $F_2$ and $F_3^+$ color centers in this material feature high photothermal stability and consequently are of interest for laser applications.

Direct inscription of femtosecond laser-induced waveguides and holographic microgratings in LiF were reported in [17]. Measuring the critical coupling angle of the waveguides at 633 nm, the refractive index increase was estimated as $\sim 10^{-2}$ and ascribed to the presence of color centers. Based on those findings, a distributed feedback $F_2$ color center laser was realized, as discussed in Sect. 11.4.

## 11.2.3 Actively Doped Crystals and Ceramics

Doped crystals are essential materials for solid state lasers. Doping leads to modifications of the lattice and thus may have an influence on the general properties of the host crystal as well as the response to fs-laser exposure. Consequently, the choice and concentration of dopants provide additional degrees of freedom. Waveguides in such doped crystals are of great interest, e.g., for waveguide laser applications, as will be discussed in Sect. 11.4.

#### 11.2.3.1 Neodymium Doped MgO:LiNbO$_3$

Waveguides in neodymium doped $MgO:LiNbO_3$ have been fabricated by ablation of trenches on the crystal surface as presented in [18]. Micro-Raman and microluminescence investigations showed a localized red shift of the $^4F_{3/2} \to {}^4I_{9/2}$ fluorescence band of the $Nd^{3+}$ ions, which the authors attributed to stress induced changes of the crystal field. While the shape of the fluorescence bands remained similar to the bulk, an increase of the luminescence intensity was reported. A permanent densification, extending $\approx 3$ μm around the modified volume, was observed. From the measured Raman shifts, induced stress of up to 2 GPa was determined.

11 Waveguides in Crystalline Materials 303

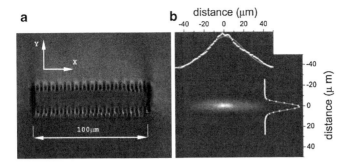

**Fig. 11.7** (a) End view of a depressed cladding waveguide inscribed in $Nd^{3+}$:YAG; (b) near-field image of the laser beam. Field profiles and Gaussian fitting curves are shown as *scatter graphs* and *solid curves*, respectively [20]

### 11.2.3.2 Yttrium Aluminum Garnet

Yttrium aluminum garnet (YAG) crystals and ceramics doped with various rare earth ions have been shown to support laser induced structures which can serve as waveguides. Studies of $Nd^{3+}$ and $Cr^{4+}$ doped YAG crystals have been presented in [19, 20], where a number of parallel structures were combined to form a depressed index cladding surrounding an unexposed waveguide core region (Fig. 11.7a). The index step between core and depressed cladding was reported as $4 \times 10^{-4}$, and propagation losses of $0.05\,cm^{-1}$ ($0.2\,dB\,cm^{-1}$) were estimated. Based on this structure, a waveguide laser was implemented (see Sect. 11.4 for details).

Waveguides have been fabricated in Nd:YAG also following the standard dual-line approach (see Sect. 11.2.1) [21], featuring index steps on the order of $10^{-3}$ [22] and propagation losses of $0.4\,dB\,cm^{-1}$ [23] at a wavelength of $1.06\,\mu m$. The stress-induced nature of the index increase was confirmed by selective etching of the inscribed structures, after which no more guiding was observed in the surrounding unmodified regions. Similar spectra and no changes in fluorescence lifetime compared to the bulk material were reported.

Dual line structures have also been applied to create waveguides in Nd:YAG ceramics [24]. Detailed investigations using scanning confocal Raman and luminescence imaging [25] revealed a slight red shift, reduced luminescence intensity as well as an increase of the width of $Nd^{3+}$ emission bands due to lattice defects. The permanent refractive index decrease (up to $9 \times 10^{-2}$) located at the cores of the modifications is surrounded by defects that can be cured by annealing. Additionally, a slight index decrease ($8 \times 10^{-4}$, stable up to $600°C$ at the upper and lower part of the filaments, was reported. However, the stress induced index increase ($5 \times 10^{-3}$, thermally stable up to $1,000°C$) yields the main contribution to the guiding between the lines [25].

Waveguides in Nd:YAG ceramics have also been fabricated by ablation of a trench on the sample surface [26]. In the region below the modification, single mode guiding at 660 nm with propagation losses of $1\,dB\,cm^{-1}$ were achieved by this technique.

### 11.2.3.3 Titanium Doped Sapphire (Ti:Al$_2$O$_3$)

Ti:sapphire stands out due to its exceptional properties such as low lasing threshold and broad amplification range. Consequently, it is the material of choice for femtosecond laser oscillators and could also serve as broadband light source, e.g., in optical coherence tomography.

Laser induced damage structures have been investigated in Ti:sapphire with a dopant concentration of 0.21 at% [27], revealing a greatly decreased damage threshold compared to the pure crystal. Index changes between $-2 \times 10^{-4}$ within the irradiated spot and $+1 \times 10^{-4}$ in the surrounding regions were observed. Multimode guiding for a wavelength of 633 nm of both TE and TM polarizations took place in the zones of increased index above and below the amorphous damage. Propagation losses of 2.3–2.5 dB cm$^{-1}$ were measured. Despite slightly decreased fluorescence efficiency, the spectra obtained from the waveguides were identical to the bulk material.

### 11.2.3.4 Ytterbium Doped KY(WO$_4$)$_2$(Yb:KYW)

Yb:KYW is of great interest for room temperature high power laser applications due to its favorable properties such as excellent thermal conductivity and high refractive index. Stress-induced ultrafast laser-written waveguides inscribed perpendicular to the [010] direction of Yb(2%):KYW crystals were reported in [28]. Single mode guidance was observed at 670 nm for a line separation of 15 μm; propagation losses at a wavelength of 1 μm were estimated as 2 dB cm$^{-1}$.

## 11.3 Nonlinear Properties

For optical applications, the nonlinear response is one of the most important properties of crystals. It allows energy transfer between different spectral components and the generation of new wavelengths in the signal. Since nonlinear processes are intensity-dependent, the focusing of light to small cross-sections increases their efficiency. Waveguides allow light to stay narrowly confined for arbitrary distances, thereby offering new opportunities for highly efficient nonlinear optical components.

### *11.3.1 Lithium Niobate*

#### 11.3.1.1 Type I

In order to estimate the impact of type I modifications on the nonlinear properties of LiNbO$_3$, measurements of the electrooptic coefficient of such waveguides were conducted in x-cut LiNbO$_3$ [6]. The electrodes in the setup shown in Fig. 11.8a were fabricated by local laser ablation of a thin gold film previously deposited on the LiNbO₃ substrate. Subsequently, the waveguides were inscribed in the gaps. The interference pattern for simultaneous excitation of both waveguides (Fig. 11.8b) was then observed for varying electrode voltages. From the half wave voltage (Fig. 11.8c), the effective electrooptical coefficient of the waveguides was measured to decrease to $r_{33} = 2.9 \times 10^{-11}$ m V$^{-1}$, which is about 52% of the bulk coefficient for the wavelength of 633 nm. However, the effective coefficient constitutes the average value experienced by the guided mode and thus contains significant contributions of the surrounding unmodified material. A rough estimate shows that the actual coefficient in the modified region has to be much more strongly decreased: assuming the coefficient to be zero within the modified volume and intact otherwise, the effective coefficient would still be about 45% of the bulk value.

Nevertheless, efficient frequency doubling (SHG) at 1567 nm has been demonstrated with quasi-phase-matched type I waveguides in periodically poled LiNbO₃ (PPLN) (Fig. 11.9) [9]. Here, a special multiscan writing approach was used to minimize the impact on the nonlinear properties. With optimized processing parameters, virtually no change in the nonlinear properties was reported and a normalized conversion efficiency of 6.5% W$^{-1}$ cm$^{-2}$ was obtained.

However, the lack of thermal stability excludes type I waveguides from high average power applications, since typically temperatures around 200°C are needed to suppress the photorefractive effect. Increased thermal resistance can be achieved at higher pulse energy, however, at the cost of decreased nonlinear properties [9].

#### 11.3.1.2 Type II

Waveguides based on type II modifications promise the opportunity to circumvent these problems. Initial experiments with temperature phase-matched SHG already

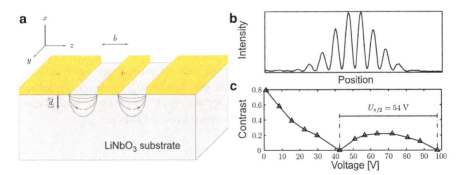

**Fig. 11.8** Investigation of the nonlinear properties of type I structures in x-cut LiNbO₃ [6]. (**a**) Schematic of the experimental setup. Waveguides are represented by *white ellipses*, gold electrodes are marked *yellow*. (**b**) Interference pattern of two neighboring waveguides, recorded with a CCD line camera. (**c**) Fringe contrast as a function of the applied electrode voltage. The distance between the minima corresponds to a phase shift of $\pm\pi/2$ in both waveguides

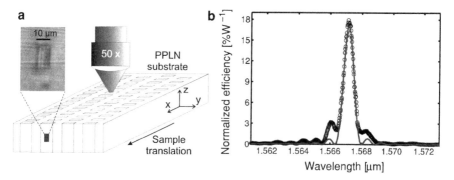

**Fig. 11.9** (a) Schematic of the multiscan writing setup, *inset*: microscope image of the waveguide end face [9]. (b) Normalized SHG efficiency as a function of pump wavelength: experimental data (*circles*) and theoretical fit (*solid line*) [9]

yielded a total conversion efficiency of 49% at 1064 nm [29]. However, the second harmonic was guided in a higher transverse mode of the waveguide (Fig. 11.10a and b). Due to the resulting small overlap, the normalized conversion efficiency only reached the low value of 0.6% $W^{-1}$ $cm^{-2}$. By ensuring single mode guiding at both fundamental and second harmonic wavelength (Fig. 11.10c–e), a normalized efficiency of 2.5% $W^{-1}cm^{-2}$ was achieved in PPLN [11]. In contrast to type I waveguides, the thermal resilience of the type II structures allowed operation of the crystal at the elevated temperature of 195.4°C to avoid photorefractive effects. Since guiding took place in unmodified regions of the crystal, no reduction of the nonlinear coefficient was reported. Consequently, a total conversion efficiency of up to 58% (Fig. 11.10f and g) was achieved.

## 11.4 Integrated Optical Devices

The material of choice for active optical components is lithium niobate with its large electrooptic and nonlinear coefficients as well as its availability in good optical quality. Commonly, devices like optical switches and modulators are produced by microfabrication techniques involving lithography [30, 31]. However, ultrashort laser pulse writing is a promising approach for the rapid fabrication of fully three-dimensional functional integrated optical elements in bulk $LiNbO_3$.

### 11.4.1 Mach–Zehnder Interferometer

Mach–Zehnder interferometers (MZIs) are key components in integrated optics. The first demonstration of such structures in $LiNbO_3$ based on femtosecond laser processing was reported in [32], featuring single mode guiding at 632.8 nm, 808 nm,

11 Waveguides in Crystalline Materials 307

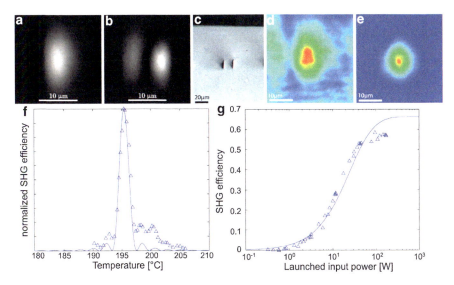

**Fig. 11.10** Mode fields of a type II waveguide in LiNbO$_3$ [29]; (**a**) fundamental transverse mode at pump wavelength 1064 nm; (**b**) higher transverse mode at second harmonic 532 nm. Type II waveguide in PPLN [11]; (**c**) microscope image of end facet. Mode fields for the (**d**) fundamental wavelength 1064 nm, and (**e**) second harmonic 532 nm. (**f**) Thermal phase matching curve of the second harmonic output. (**g**) SHG efficiency as a function of the input peak power

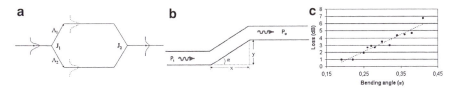

**Fig. 11.11** (**a**) Schematic diagram of a simple MZI with single input and output waveguides. (**b**) Schematic diagram showing the offset transition with sharp corner bend design utilized in this approach. (**c**) Losses for the configuration shown in (**b**) as a function of the bending angle [32]

and 1550 nm (Fig. 11.11). The maximum separation angle $2\alpha$ of the Y junctions to guide light with acceptable losses was determined to be 0.65°.

## 11.4.2 Electrooptic Modulator

The simplest active integrated optical device is the electrooptic modulator, consisting of a MZI and electrodes for the application of an electrical field to introduce an additional phase shift between the arms, and consequently the modulation of the output intensity. Since its functionality is based on the nonlinear properties of the material, type II waveguides are best suited for this kind of device.

Fig. 11.12 (a) Top and (b) end views of optical micrographs of the embedded electrodes and optical waveguides; (c) contour plots of the equipotential contour of the embedded electrodes [33]

A modulator featuring embedded gold electrodes (Fig. 11.12) in x-cut LiNbO$_3$ was presented in [33]. The electrodes were fabricated by fs-laser assisted selective electroless plating in grooves previously excavated by fs-laser ablation. This technique allows the waveguides to be fabricated inside the regions where the electric field is strongest, yielding a half wave voltage of $U_\pi \approx 19$ V and modulation depths around 9 dB for TE polarized light at 632.8 nm. The waveguides were buried about 20 μm below the surface, with a Y junction angle of 1.2° and a separation of 65 μm between the 2.6 mm long branches.

Another approach was presented in [34], where after the inscription of the type II MZI structure, the surface electrodes were fabricated by sputter deposition of a thin gold layer and subsequent laser ablation. Since no lithographic processing steps are necessary, this method (as well as the previous one) features a significant reduction of the time needed for fabrication. Modulators for TE and TM polarized light were inscribed in x-cut lithium niobate samples (Fig. 11.13), achieving half wave voltages of $U_\pi = 8$ V (TE) and $U_\pi = 23$ V (TM), respectively, at a wavelength of 532 nm. Modulation depths up to 11 dB were reported. Here, the waveguides were buried 30–50 μm beneath the surface with a separation of 80 μm. The junctions were fabricated in an optimized S bend design, allowing a branching angle of 1.5°. The length of the electrodes was 9.5 mm and their separation was 20 μm. To enforce single mode behavior for TM polarization, the output waveguide was narrowed to a width of 4 μm over a distance of 1 mm after the combining junction. Calculations of the electrical field based on the measured half wave voltage revealed that practically no change of the electrooptic coefficient occurred in the waveguides compared to the bulk material.

### 11.4.3 Waveguide Lasers

Waveguide lasers fabricated, e.g., in Nd-doped silicate [35] and Er,Yb doped phosphate glasses [36] have already been discussed in Chap. 10. To date, the realization of such lasers on crystalline active materials has been mainly achieved by techniques involving lithographic process steps [37–40].

Using several parallel lines of femtosecond laser induced modifications, a waveguide laser consisting of an unexposed core region surrounded by a depressed

11 Waveguides in Crystalline Materials

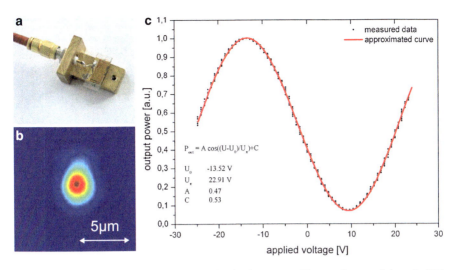

**Fig. 11.13** (a) X-cut LiNbO$_3$ sample containing the femtosecond laser written modulator for TM polarized light. (b) Guided mode in the maximum of transmission. (c) Modulation of the output intensity depending on the applied voltage [34]

index cladding has been reported in Nd(0.8 mol%):YAG [20] (see Fig. 11.7a). The core dimensions were chosen to match the mode profiles of typical pumping diodes. Laser performance studies were conducted with a highly reflective coating on one side and flat mirrors of different output coupling ratios attached to the other side. With an output coupling of 24% a laser threshold of 30 mW and output powers above 150 mW were reported. The corresponding laser mode profile is shown in Fig. 11.7b. The laser predominately oscillated in the fundamental mode, and no degradation of output power was observed over a time period of several tens of hours.

A stress-induced channel waveguide laser in Nd(1%):YAG supported by a dual line structure with a separation of 20 μm was demonstrated in [21–23]. Optical feedback was provided by the Fresnel reflections at the end faces of the crystal. When pumped by a Ti:sapphire source at 808 nm, slope efficiencies up to 54% and output powers as high as 336 mW are achieved at a wavelength of 1.06 μm with a nearly Gaussian beam profile (Fig. 11.14). Recently, improved output powers of up to 1.17 W were reported [41].

Continuous lasing at 1.06 μm has also been demonstrated for waveguides in Nd(2%):YAG ceramics pumped by a Ti:sapphire source [24]. The stress-induced guiding took place between two lines separated by 20 μm. Again, Fresnel reflections provided the necessary feedback, yielding slope efficiencies of 60% and output power in excess of 80 mW for single mode stable laser operation.

A slightly different approach has been demonstrated in LiF [17], where a distributed-feedback (DFB) laser based on writing-induced color centers was fabricated. At ambient temperature, the laser emission occurred at a wavelength of 707 nm with a slope efficiency of ≈10% and a beam divergence of ≈20 mrad.

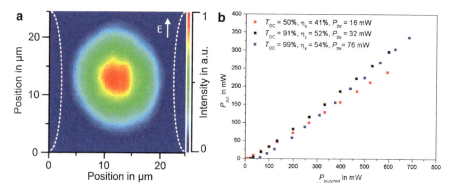

**Fig. 11.14** (a) Near-field image of the guided mode of a Nd(1%):YAG waveguide laser at 330 mW laser power. (b) Efficiency curve for different output coupling percentages [23]

A narrow line width below 1 nm was achieved with a micrograting array induced using the two-beam interference pattern of fs pulses. The concentration of $F_2$ color centers induced by fs-laser irradiation was estimated to be $2 \times 10^{18}$ cm$^{-3}$. However, this concentration had to be increased by an additional X-ray exposure in order to facilitate lasing action.

## 11.5 Conclusion

Femtosecond laser processing has proven to be a viable method for the fabrication of not only waveguides but also nonlinear components and active devices in crystalline materials. Consequently, it provides new possibilities for integrated optics. Nevertheless, intensive research is still necessary to optimize the fabrication parameters in order to generate structures with lower propagation losses.

In addition to the results summarized in this chapter, evanescent coupling is a key component for many integrated-optical devices such as directional couplers, and due to its wavelength dependence offers new degrees of freedom for the design of advanced components. However, the typical dual line type II structures used in processing crystals have the disadvantage of suppressing the evanescent fields of the guided modes due to their strongly confined guiding. Thus, significant coupling between type II waveguides can only be achieved for sufficiently weak structures. Following this approach, at the cost of slightly increased losses (4 dB cm$^{-1}$), planar arrays of evanescently coupled type II waveguides (see Fig. 11.15) have been manufactured recently in x-cut LiNbO$_3$ [42].

Interestingly, even non-waveguiding structures have the potential for applications, as illustrated by the direct inscription of diffractive elements [49] photonic crystals [47, 51] in various crystalline host materials.

11 Waveguides in Crystalline Materials

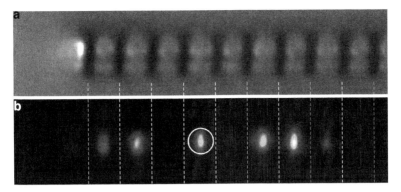

**Fig. 11.15** Planar array of evanescently coupled type II waveguides [42]. (**a**) Micrograph of the end facet. (**b**) Discrete diffraction pattern for excitation at 633 nm (excited site is marked by the *circle*)

Even more possibilities are provided by the influence of fs-laser structuring on the nonlinear properties of the crystal. Hence, this technique might be applied for the fabrication of structures with specifically designed nonlinearity.

## References

1. J.W. Chan, T. Huser, S. Risbud, D.M. Krol, "Structural changes in fused silica after exposure to focused femtosecond laser pulses", Opt. Lett. **26**, 1726 (2001)
2. J.W. Chan, T. Huser, S. Risbud, D.M. Krol, "Modification of the fused silica glass network associated with waveguide fabrication using femtosecond laser pulses", Appl. Phys. A **76**, 367 (2003)
3. J.W. Chan, T. Huser, S. Risbud, J.S. Hayden, D.M. Krol, "Waveguide fabrication in phosphate glasses using femtosecond laser pulses", Appl. Phys. Lett. **82**, 2371 (2003)
4. T. Gorelik, M. Will, J. Burghoff, A. Tünnermann, "Transmission electron microscopy studies of femtosecond laser induced modifications in quartz", Appl. Phys. A: Mater. Sci. process., **76**, 309 (2003)
5. J. Burghoff, "Volumenwellenleiter in kristallinen Medien", Dissertation, Institut für Angewandte Physik, Friedrich-Schiller-Universität Jena, Jena (2007) Braunschweig Verlag; 1.Auflage: 1 (August 2007), ISBN 3981166515
6. J. Burghoff, H. Hartung, S. Nolte, A. Tünnermann, "Structural properties of femtosecond laser-induced modifications in LiNbO$_3$", Appl. Phys. A **86**, 165 (2007)
7. A. Nejadmalayeri, P. Herman, "Ultrafast laser waveguide writing: lithium niobate and the role of circular polarization and picosecond pulse width," Opt. Lett. **31**, 2987 (2006)
8. L. Gui, B. Xu, T.C. Chong, "Microstructure in Lithium Niobate by Use of Focused Femtosecond Laser Pulses" IEEE Photonics Technol. Lett. **16**, 1337 (2004)
9. R. Osellame, M. Lobino, N. Chiodo, M. Marangoni, G. Cerullo, R. Ramponi, H.T. Bookey, R.R. Thomson, N.D. Psaila, A.K. Kar, "Femtosecond laser writing of waveguides in periodically poled lithium niobate preserving the nonlinear coefficient", Appl. Phys. Lett. **90**, 241107 (2007)
10. J. Burghoff, C. Grebing, S. Nolte, A. Tünnermann, "Waveguides in lithium niobate fabricated by focused ultrashort laser pulses", Appl. Surf. Sci. **253**, 7899 (2007)

11. J. Thomas, J. Burghoff, M. Heinrich, S. Nolte, A. Ancona, A. Tünnermann, "Femtosecond laser-written quasi-phase-matched waveguides in lithium niobate", Appl. Phys. Lett. **91**, 151108 (2007)
12. A. Nejadmalayeri, P. Herman, J. Burghoff, M. Will, S. Nolte, A. Tünnermann, "Inscription of optical waveguides in crystalline silicon by mid-infrared femtosecond laser pulses", Opt. Lett. **30**, 964 (2005)
13. Y. Li, P. Lu, N. Dai, et al. Surface relief diffraction gratings written on beta $-$ $BaB_2O_4$ crystal by femtosecond pulses, Appl. Phys. B-Lasers Opt. **88**(2), 227–230 (2007)
14. F. Dreisow, J. Thomas, J. Burghoff, A. Ancona, M. Heinrich, S. Nolte, A. Tuennermann, "Efficient frequency doubling in fs-laser written waveguides in PPLN and BBO," SPIE Photonics West/LASE 2008, paper 6879A-34
15. B. McMillen, K.P. Chen, H. An, S. Fleming, Vincent Hartwell, and David Snoke, "Waveguiding and nonlinear optical properties of three-dimensional waveguides in $LiTaO_3$ written by high-repetition rate ultrafast laser," Appl. Phys. Lett. **93**, 111106 (2008)
16. B. McMillen, K.P. Chen, D. Jaque, "Microstructural imaging of high repetition rate ultrafast laser written $LiTaO_3$ waveguides," Appl. Phys. Lett. **94**, 081106 (2009)
17. K.-I. Kawamura, M. Hirano, T. Kamiya, H. Hosono, Femtosecond-laser-encoded distributed-feedback color center laser in lithium fluoride single crystal, J. Non-Cryst. Solids **352**, 2347–2350 (2006)
18. A. Ródenas, J.A. Sanz García, D. Jaque, G.A. Torchia, C. Mendez, I. Arias, L. Roso, F. Agulló-Rueda, "Optical investigation of femtosecond laser induced microstress in neodymium doped lithium niobate crystals", J. Appl. Phys. **100**, 033521 (2006)
19. M.V. Dubov, L. Krushchev, I. Bennion, A.G. Okhrimchuk, A.V. Shectakov, "Waveguide inscription in YAG:$Cr^{4+}$ crystals by femtosecond laser irradiation, paper CWA49, CLEO (2004)
20. A.G. Okhrimchuk, A.V. Shestakov, I. Khrushchev, J. Mitchell, "Depressed cladding, buried waveguide laser formed in a YAG:$Nd^{3+}$ crystal by femtosecond laser writing," Opt. Lett. **30**, 2248 (2005)
21. J. Siebenmorgen, K. Petermann, G. Huber, K. Rademaker, S. Nolte, A. Tünnermann, Femtosecond laser written stress-induced Nd:$Y_3Al_5O_{12}$ (Nd:YAG) channel waveguide laser, Special Issue Appl. Phys. B (2009)
22. J. Siebenmorgen, K. Rademaker, S. Nolte, G. Huber, Femtosecond laser written stress-induced Nd:YAG channel waveguide laser, THoD.7, Europhoton, Paris (2008)
23. J. Siebenmorgen, T. Calmano, K. Petermann, G. Huber, "Fabrication of a stress-induced Nd:YAG channel waveguide laser using fs-Laser pulses", Advanced Solid-State Photonics, MB29, Denver, USA (2009)
24. G.A. Torchia, A. Rodenas, A. Benayas, E. Canelar, L. Roso, D. Jaque, "Highly efficient laser action in femtosecond-written Nd:yttrium aluminium garnet ceramic waveguides", Appl. Phys. Lett. **92**, 111103 (2008)
25. A. Rodénas, G.A. Torchia, G. Lifante, E. Cantelar, J. Lamela, F. Jaque, L. Roso, D. Jaque, "Refractive index change mechanisms in femtosecond laser written ceramic Nd:YAG waveguides: micro-spectroscopy experiments and beam propagation calculations", Appl. Phys. B **95**(1), 85–96 (2009)
26. G.A. Torchia, P.F. Mailán, A. Rodenas, D. Jaque, C. Mendez, L. Roso, "Femtosecond laser written surface waveguides fabricated in Nd:YAG ceramics", Opt. Exp. **15**, 13266–13271 (2007)
27. V. Apostolopoulos, L. Laversenne, T. Colomb, C. Depeursinge, R.P. Salathé, M. Pollnau, R. Osellame, G. Cerullo, P. Laporta, "Femtosecond-irradiation-induced refractive-index changes and channel waveguiding in bulk Ti3+:Sapphire," Appl. Phys. Lett. **85**, 1122 (2004)
28. C.N. Borca, V. Apostolopoulos, F. Gardillou, H.G. Limberger, M. Pollnau, R.-P. Salathé, "Buried channel waveguides in Yb-doped KY($WO_4)_2$ crystals", Appl. Surf. Sci. **253**, 8300–8303 (2007)
29. J. Burghoff, C. Grebing, S. Nolte, A. Tünnermann, "Efficient frequency doubling in femtosecond laser-written waveguides in lithium niobate", Appl. Phys. Lett. **89**, 081108 (2006)

30. G.L. Destefanis, J.P. Gailliard, E.L. Ligeon, S. Valette, B.W. Farmery, P.D. Townsend, A. Perez, The formation of waveguides and modulators in LiNbO$_3$ by ion implantation, J. Appl. Phys. **50**(12), 7898–7905 (1979)
31. J.L. Jackel, C.E. Rice, J.J. Veselka: Proton-Exchange for high-index waveguides in LiNbO$_3$, Appl. Phys. Lett. **41**(7), 607–608 (1982)
32. C. Méndez, G.A. Torchia, D. Delgado, I. Arias, L. Roso, "Fabrication and characterization of Mach-Zehnder devices in LiNbO$_3$ written with femtosecond pulses", Proc. Of IEEE/LEOS workshop on fibers and optical passive components, pp. 131 (2005)
33. Y. Liao, J. Xu, Y. Cheng, Z. Zhou, F. He, H. Sun, J. Song, X. Wang, Z. Xu, K. Sugioka, K. Midorikawa, "Electrooptic integration of embedded electrodes and waveguides in LiNbO$_3$ using a femtosecond laser", Opt. Lett. **33**(19), 2281 (2008)
34. S. Ringleb, K. Rademaker, S. Nolte, A. Tünnermann, *"Optical waveguide devices and electrooptic modulation in ultrashort-pulse laser written lithium niobate crystals"* (DPG Tagung, Hamburg, 2009)
35. Y. Sikorski, A.A. Said, P. Bado, R. Maynard, C. Florea, K.A. Winick, Electron. Lett. **36**, 226 (2000)
36. S. Taccheo, G. Della Valle, R. Osellame, G. Cerullo, N. Chiodo, P. Laporta, O. Svelto, A. Killi, U. Morgner, M. Lederer, D. Kopf, "Er:Yb-doped waveguide laser fabricated by femtosecond laser pulses," Opt. Lett. **29**, 2626 (2004)
37. L. Laversenne, P. Hoffmann, M. Pollnau, P. Moretti, J. Mugnier, Designable buried waveguides in sapphire by proton implantation, Appl. Phys. Lett. **85**, 22, 5167–5169 (2004)
38. C. Grivas, D.P. Sheperd, R.W. Eason, L. Laversenne, P. Moretti, C.N. Borca, M. Pollnau, "Room-temperature continuos-wave operation of Ti:sapphire buried channel-waveguide lasers fabricated via proton implantation", Opt. Lett. **31**(23), 3450–3452 (2006)
39. Y.E. Romanyuk, C.N. Borca, M. Pollnau, S. Rivier, V. Petrov, U. Griebner, "Yb-doped KY(WO$_4$)$_2$ planar waveguide laser", Opt. Lett. **31**(1), 53–55 (2006)
40. F. Gardillou, Y.E. Romanyuk, C.N. Borca, R.-P. Salathé, M. Pollnau, "Lu, Gd codoped KY(WO$_4$)$_2$:Yb epitaxial layers: towards integrated optics based on KY(WO$_4$)$_2$", Opt. Lett. **32**(5), 488–490 (2007)
41. J. Siebenmorgen, T. Calmano, O. Hellmig, K. Petermann, G. Huber, "Efficient Femtosecond Laser Written Nd:YAG Channel Waveguide Laser with an Output Power of more than 1 W", CLEO Europe Conference, paper CJ7.1 (2009)
42. M. Heinrich, A. Szameit, F. Dreisow, S. Döring, J. Thomas, A. Ancona, S. Nolte, A. Tünnermann, "Evanescent coupling in arrays of type II femtosecond laser-written waveguides in bulk x-cut lithium niobate," Appl. Phys. Lett. **93**, 101111 (2008)
43. R.R. Gattass, E. Mazur, "Femtosecond laser micromachining in transparent materials", Nature **2**, 219–225 (2008)
44. K. Itoh, W. Watanabe, S. Nolte, C.B. Schaffer, "Ultrafast Processes for Bulk Modification of Transparent Materials", MRS Bull. **31**, 620 (2006)
45. S. Nolte, M. Will, J. Burghoff, A. Tünnermann, "Femtosecond waveguide writing: a new avenue to three-dimensional integrated optics", Appl. Phys. A **77**, 109 (2003)
46. S. Nolte, M. Will, J. Burghoff, A. Tünnermann, "Ultrafast laser processing: New options for 3D photonic structures", J. Mod. Opt. **51**, 2533 (2004)
47. A. Ródenas, G. Zhou, D. Jaque, M. Gu, "Direct laser writing of three-dimensional photonic structures in Nd:yttrium aluminium garnet laser ceramics", Appl. Phys. Lett. **93**, 151104 (2008)
48. R.R. Thomson, S. Campbell, I.J. Blewett, A.K. Kar, D.T. Reid, "Optical waveguide fabrication in z-cut lithium niobate (LiNbO$_3$) using femtosecond pulses in the low repetition rate regime," Appl. Phys. Lett. **88**, 111109 (2006)
49. G.A. Torchia, C. Mendez, I. Arias, L. Roso, A. Rodenas, D. Jaque, "Laser gain in femtosecond microstructured Nd:MgO:LiNbO$_3$ crystals", Appl. Phys. B **83**, 559 (2006)
50. G. Della Valle, R. Osellame, P. Laporta, "Micromachining of photonic devices by femtosecond laser pulses", J. Opt. A Pure Appl. Opt. **11**, (2009) – Review article
51. G. Zhou, M. Gu, "Direct optical fabrication of three-dimensional photonic crystals in a high refractive index LiNBO$_3$ crystal", Opt. Lett. **31**(18), 2783 (2006)

# Chapter 12
# Refractive Index Structures in Polymers

**Patricia J. Scully, Alexandra Baum, Dun Liu, and Walter Perrie**

**Abstract** Refractive index structuring of poly(methyl methacrylate) (PMMA) by femtosecond (fs) laser irradiation is discussed, including writing conditions defined by wavelength, pulse duration, and associated photochemistry. The aim is to determine optimal conditions for refractive index modification, $\Delta n$ without doping for photosensitivity. The work presented here forms a generic methodology for other polymers. Nanostructuring using holographic optics and precise control of beam parameters has versatile application for three-dimensional (3D) photonic devices. Self-focusing and filamentation at various depths below the surface of bulk PMMA are discussed together with parallel processing using a spatial light modulator. Applications of refractive index structures in polymers include microfluidics, lab-on-a-chip, organic optoelectronic devices, and gratings in polymer optical fibres.

## 12.1 Introduction

This chapter presents refractive index (RI) structuring of the commonly used polymer, poly(methyl methacrylate) (PMMA), by laser irradiation, where writing conditions defined by wavelength, pulse duration, and associated photochemistry are described.

---

P.J. Scully (✉) · A. Baum
Photon Science Institute, Alan Turing Building, The University of Manchester, Oxford Road, Manchester, M13 9PL, UK

Centre for Instrumentation and Analytical Science, The University of Manchester, CEAS, PO Box 88, Sackville Street, Manchester, M60 1QD, UK
e-mail: p.scully@manchester.ac.uk

D. Liu · W. Perrie
University of Liverpool, Liverpool Lairdside Laser Engineering Centre (LLEC), Wirral, CH41 9HP, UK

The authors present their own work on infrared and ultraviolet femtosecond (fs) laser photomodification at 775, 800, and 387 nm wavelength, with pulse durations between 200 and 40 fs in context with work by other researchers. Their aim is to determine optimal conditions for RI modification $\Delta n$ without doping for photosensitivity.

PMMA represents a commonly available polymer and the work presented here forms a generic methodology for other polymers. The chapter considers conditions for sub-ablation threshold fs laser interaction with PMMA including fluence, wavelength, pulse duration, and elucidation of the photochemical process as a function of writing parameters. Laser-induced RI modification can be related to material density changes and when exposed to laser irradiation, transparent polymers can undergo thermal expansion and rapid cooling, causing reduced density, and cross-linking creating increased density, depending on exact writing conditions.

The nanostructuring techniques developed via holographic optics and precise control of beam parameters have generic application for three dimensional (3D) photonic devices for a wide range of un-photosensitized optical materials.

Direct and holographic techniques are described, together with initial attempts to write waveguides into bulk PMMA and the effect of pulse duration, wavelength, and bandgap on $\Delta n$. The effect of self-focusing, filamentation (FL), and control of these effects to achieve RI modification at a range of depths below the surface of bulk PMMA are discussed. The use of a spatial light modulator (SLM) for parallel processing is demonstrated for increased speed and full use of available laser energy. Finally, applications of RI structures in polymers are considered such as microfluidics, lab-on-a-chip, organic optoelectronic devices, and gratings in polymer optical fibres (POFs).

## 12.2 Motivation for Refractive Index Structures in Polymers

Structured optical materials including waveguides and gratings form important building blocks for photonic devices for telecommunications, sensing, and optical data storage. Ultra-short pulses provided by fs lasers create highly localized RI changes deep within transparent media, forming sub-micron feature sizes due to non-linearity at high intensities, without requiring photosensitization by doping or hydrogen loading. PMMA is of interest because it is inexpensive, rugged, and thus ideal for disposable devices for clinical, biological, and chemical applications. Low processing temperatures permit organic dopants (unsuited to high $T_G$ dielectrics such as glass) that impart optical activity and lasing properties to PMMA and other polymers. Thus amplifiers, fibre-lasers, and electro-optic modulators as well as LEDs can be formed on a common polymer substrate, to be combined with fs written 3D waveguide and stacked optical structures such as photonic crystals. Since fs irradiation can form microchannels and microvoids, integrated photonic circuits for use as lab-on-the chip devices can be created without etching, lithography, or deposition, facilitating polymer based sensing and microfluidic devices. PMMA has high elasticity and thermal expansion coefficients when compared with glass materials and forms the base material for POFs. The ability to write periodic RI structures (Bragg gratings or long period gratings) into the core of single mode optical fibre can create filters and reflectors that are more sensitive to strain or temperature than equivalent structures written into glass optical fibres.

However, there is a knowledge gap about the photochemical processes mechanisms involved in low dose fs laser modification, and the stability and longevity of the photonic structures produced.

## 12.3 Laser Photomodification of PMMA

This section presents a review of laser modification of PMMA, categorized in three groups, including continuous wave (cw) light from broad band UV sources or He:Cd lasers, long pulses (ns, ps) from excimer lasers, and ultrashort pulses from Ti:sapphire lasers. Writing conditions are defined by wavelength, NA, and pulse duration related to the photochemical processes taking place. The authors' own work on infrared and ultraviolet fs laser photomodification with pulse durations between 200 and 40 fs is discussed in context with work by other researchers to indicate the optimal conditions for RI control to create waveguiding structures.

### 12.3.1 Continuous Wave UV Laser Sources

Refractive index increases in PMMA by cw UV laser radiation (He:Cd, 325 nm) were studied as early as 1970 [1]. The RI modification was measured to be $3 \times 10^{-3}$ corresponding to a densification of the material (density increase by 0.8% and thickness reduction observed), explained as a cross-linking effect [1]. This observation of densification combined with a positive RI change is contradictory to earlier work on the photodegradation of PMMA by Alison in 1966 [2]. However, the bulk writing process described by Tomlinson in 1970 [1] and, Moran and Kaminov in 1973 [3] requires long developing times of about 200 h, and the fabrication of 3D structures is constrained due to the close proximity to the UV absorption edge of the material.

Mitsuoka, Torikai and co-workers demonstrated the wavelength sensitivity of PMMA photodegradation, using cw light at a range of UV wavelengths [4–6]. No effect was observed for wavelengths above 340 nm, whereas for 300 nm, photoinduced side chain scission initiated maximum main chain scission.

### 12.3.2 Long Pulse (ns, ps) Laser Sources

A large number of experiments have been conducted with dye-doped PMMA. In 1999, Peng and Xiong et al. were the first to inscribe effective tunable FBGs in doped, single mode POF by excimer laser radiation (248, 280, and 325 nm), exhibiting a RI change of about $10^{-4}$ [7, 8]. For PMMA, containing residual monomer and UV absorbing photoinitiator, the light induced polymerization of the monomer at 514 nm was shown by Marotz in 1985 [9]. The holographically produced index gratings showed self-development times of several weeks or required thermal fixing.

However, material doping is a time and cost consuming step, whereas commercial grade, undoped PMMA is widely available. Several studies were made at excimer laser wavelength to photoinduce RI changes in pure PMMA thin films [10–12].

A RI change of 0.5% at the readout wavelength of 632 nm was measured using a prism coupler by Baker and Dyer in 1993 [12]. It was suggested that, although the VIS and near infrared (NIR) transmission was found nearly unaffected, the new absorption bands in the UV can change the RI in the visible region.

In 2005, a comprehensive study about low dose excimer laser degradation of PMMA over various wavelengths and exposure ranges emphasised the importance of writing conditions on the balance of the photoreaction, and involved positive and negative $\Delta n$ [13]. This study evaluated different wavelengths (193, 248, and 308 nm) and wide exposure ranges at 5 Hz repetition rate, but the pulse duration was not stated [13, 14]. The RI modification was shown to be strongly dependent on the irradiation conditions, showing increased as well as decreased RI (measured by Abbe refractometer and Mach–Zehnder interferometer). This was explained by simultaneous photochemical cross-linking and degradation mechanisms, with shifting balance regarding the reaction products, and was confirmed by in situ volatile analysis and spectroscopy. For wavelengths above 300 nm, no modification was detected, but the effect at shorter wavelengths was strongly dose dependent. For low irradiation doses, cross-linking of ester side chains was observed. This caused an increased RI, contradicting the classical literature on polymer degradation [15]. Medium doses caused main chain breakdown via direct unzipping or main chain scission, which followed the cross-linking step and was accompanied by RI decrease. High dose irradiation was found to be dominated by direct ester side chain cleavage. At 193 nm wavelength, complete cleavage occurred, but at 248 nm cleavage was incomplete. The RI increase was related to densification due to increased dipole–dipole interactions. Absorbing incubation centres in the form of C=C bonds were found, which, together with the destabilized main chain, finally led to photothermal defragmentation and ablation, in tandem with RI decrease at very high doses. Wochnowski's study was the first to indicate the importance of writing conditions on the balance of the photoreaction and consider positive and negative $\Delta n$.

### 12.3.3 Ultrashort fs Laser Sources

The photomodification of commercial grade bulk PMMA was studied at a range of UV pulsed laser wavelengths as well as with fs laser radiation by the authors from 2000 onwards [16–19].

The photomodification of commercial grade bulk PMMA was studied at a range of UV excimer laser wavelengths as well as with fs laser radiation (40 fs, 1 kHz,

800 nm) by Scully et al. in 2000 and 2003, respectively [16, 18]. It was shown that the material could be efficiently modified, requiring neither pre- nor post-processing. Refractive index changes were first estimated by an online Fabry–Perot interferometer measurement, but later, diffraction gratings were fabricated and their diffraction efficiency measured. The resulting high RI change of $\Delta n = 5 \times 10^{-4}$ was estimated using the equation for Bragg diffraction in reflection. Work by Baum et al. [19] uses the methodology, which is based on the original work, carried out by Scully et al. [18]; this is described in more detail in Sect. 12.5. The magnitude of $\Delta n$ and the cumulative behavior of the process were confirmed and the RI change was directly measured to be $\Delta n = 4 \times 10^{-3}$. Photodegradation was monitored via polymer molecular weight distribution (MWD) (random main chain scission detected) and volatile analysis (showing monomer as sole reaction product).

The first use of second harmonic fs laser radiation (200 fs, 70 MHz, 385 nm) for PMMA structuring in optical recording was carried out by Bityurin et al. in 1999 [20]. The fundamental wavelength alone could not induce structures but required combination with the second harmonic. The modification was described as bulk damage that was concurrent with light emission, which probably indicated plasma generation. Therefore, it can be assumed the process was above the ablation threshold.

A single fs laser pulse (130 fs, 800 nm) was used by Li et al. to introduce holographic surface relief as well as bulk RI structures in commercial grade PMMA with a periodicity of 1.5 µm [21]. The phase gratings were about 100 µm thick and showed a first order diffraction efficiency of about 0.8%, corresponding to $\Delta n = 2 \times 10^{-4}$. The holographic arrangement consisted of two separated beam paths, focused by two $f = 500$ mm lenses at an angle of $\theta = 17°$.

## 12.4 Waveguiding and Positive/Negative Refractive Index

Refractive index modification of PMMA using fs laser irradiation is of interest because different researchers observed either positive or negative $\Delta n$ under different conditions. Studies by Zoubir et al., using a high repetition rate fs oscillator at 800 nm (25 MHz, 30 fs), found negative RI changes in the focal volume [22]. Other groups, including Wochnowski et al., who recently started planar structuring of PMMA with 1 kHz repetition rate, 800 nm fs laser radiation, assume positive $\Delta n$ due to densification [13, 21, 23].

Waveguides were first produced in PMMA by longitudinal direct writing in the bulk material [22] by Zoubir (2004), as a depressed waveguide with a tubular structure surrounding the centre, as the central region did not guide light. A commercial waveguide optical analyzer based on the refractive near-field technique was used to measure the index profile in the guiding region of commercial grade PMMA, which showed a maximum RI change of $2 \times 10^{-3}$. The measured profile of the RI modification was consistent with relative measurements on the basis of selective etching. The exposed PMMA was found to show different etch rates in methyl isobutyl ketone (MIBK), linearly depending on its RI. The revealed topographic structure was studied, using white light interferometry. The central region, which was exposed to the focal volume, showed the lowest etch rate and was assumed to have a lower RI than the primary material. This behavior is rather unusual as the solubility generally shows inverse proportionality to the material density [24].

Light guiding structures, produced by fs laser radiation in the laser focal volume in PMMA, were first reported by Ohta et al. in 2004 [23], using 100 fs, 1 kHz, and 800 nm laser pulses. Measurement of the waveguide numerical aperture inferred $\Delta n$ between $7.8 \times 10^{-5}$ and $1.5 \times 10^{-4}$, the sign of $\Delta n$ was not directly measured.

In 2006, Sowa et al. were the first to produce symmetric waveguides and a directional coupler in PMMA, using fs radiation (85 fs, 800 nm, and 1 kHz) together with a slit beam shaping method in transversal writing [25]. The structures were single-moded at 633 nm and NA measurements resulted in $\Delta n = 4.6 \times 10^{-4}$. The chemical process is thought to rely on chain scission causing volume contraction, but no further explanation is given.

Wochnowski et al. started work on fs laser inscription of grating structures in planar PMMA substrates in 2005 [26]. Volume as well as surface gratings (material removal visible with SEM and SNOM) were found to be totally transparent, which contradicts the observations presented by Baum et al. [19]. The observed diffraction of volume gratings was low (no value stated) but the researchers assumed a positive $\Delta n$ and compared the multiphoton absorption process with the linear absorption process of ns pulses. Since the laser is the same model as the laser used by the authors (140 fs, 775 nm, and 1 kHz), it is likely that the study produced similar structures.

A negative RI change was observed in only one of the cases, reviewed above, and the main difference is the repetition rate used (MHz rather than kHz). Therefore, it is likely that varying repetition rate affects the thermal interaction, thus, influencing the fs laser structuring mechanism. Negative RI changes were also found in other materials, e.g., in crystalline quartz [27] and phosphate glass [28]. However, most inorganic glasses show densification, accompanied by a positive RI change. Laser modification of glass is believed to be based on restructuring of the silica networks [29–31], which is fundamentally different to the modification mechanism of PMMA.

## 12.5 Direct Writing

This section describes the methodology used by the authors to optimize RI modification by writing diffraction gratings into bulk PMMA. Light from the fs laser beam was focused into 3-mm thick PMMA slabs of dimensions $15 \times 15$ mm, using a 80-mm focal length lens mounted on a manual translation stage. The fs laser power was attenuated using neutral density filters to control the incident fluence on the PMMA sample. Gratings were written by moving the slab relative to the

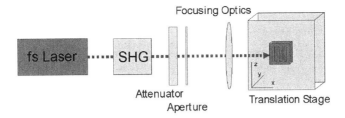

**Fig. 12.1** Direct writing of a grating structure into bulk PMMA mounted on a translation stage using a frequency doubled fs laser

stationary laser beam using an *x*–*y*–*z* translation stage interfaced to a computer as shown in Fig. 12.1. The grating periodicity, number of overwrites, and size were programmed using the stage, and the laser fluence on the sample was controlled by the velocity of the translation stage.

The RI modification, $\Delta n$, was inferred by measuring the diffraction efficiency $\eta$ of a transmitted helium neon laser into the first order [18, 32]. The magnitude of the RI modification was determined from $\eta$ using the equations for Raman–Nath diffraction of sub-surface phase gratings in bulk material [32] and further described in Sect. 12.12. Optimal $\Delta n$ was produced by repeated irradiation with a number of over scans, using a laser fluence below the damage threshold. Grating periodicities ranging from 80 to 20 µm were achieved by manipulating the PMMA sample using the translation stage (direct writing), as described in Sect. 12.5.1. Periodic structures < 0.5 µm were demonstrated using a holographic technique, described in Sect. 12.6.

## 12.5.1 Simple Transmission Gratings (2D)

The authors explored the modification of undoped PMMA using fs radiation at 800 nm, 40 fs pulse duration, and 1 kHz repetition rate, demonstrating significant RI changes ($\Delta n = 5 \times 10^{-4}$) in bulk un-doped, commercial grade PMMA [18]. These studies showed up to 37% total first order diffraction efficiency. Reproducing them with a longer 180 fs pulse duration at 775 nm proved difficult and generated only small RI changes with $\eta < 1\%$, hence a frequency doubled beam was used to evaluate commercial and clinical grade ultra pure PMMA at 387 nm. $\Delta n$ was optimized by laser fluence and number of over scans, and evaluated via diffraction efficiency measurement. First order diffraction efficiencies of up to 40% indicate $\Delta n = 4 \times 10^{-3}$ [19]. $\Delta n$ ranges from $10^{-4}$ to $10^{-3}$, depending on assumed thickness and index profile, of similar magnitude to reported values for other polymers or glasses, and sufficient to form useful photonic structures. The effect of pulse duration and wavelength, related to the bandgap of PMMA for effective $\Delta n$ modification, is discussed further in Sect. 12.9.

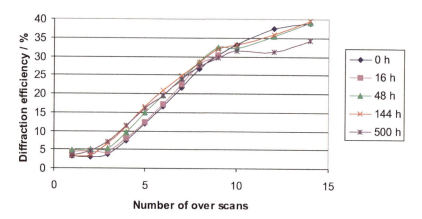

**Fig. 12.2** First order diffraction efficiency of refractive index gratings in clinical grade PMMA Vistacryl CQ (NonUV) at different times after writing ($\Lambda = 30\,\mu m$, $v = 2\,mm\,s^{-1}$, $\Phi = 0.21\,mJ\,cm^{-2}$, varying N over scans

**Fig. 12.3** Phase gratings inside bulk PMMA under white light illumination (*left*: 10 μm period, *right*: 40 μm period)

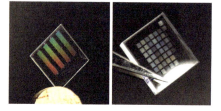

Figure 12.2 demonstrates the efficiency and durability of gratings written line by line ($\Lambda = 30\,\mu m$) in clinical grade PMMA. Permanent modification is shown but $\eta$ decreases at high $\Delta n$, indicating the temporal diffusion or re-polymerization of material degraded by chain scission. Using a 75-mm focal length lens, the modified depth was observed to be ∼3 mm, very much greater than the depth of field (DOF) ∼270 μm, probably due to self-focusing and FL.

Figure 12.3 shows phase gratings inside bulk PMMA (Vistacryl CQ non-UV) under white light illumination (left: 10 μm period, right: 40 μm period). The gratings are clearly visible to the naked eye.

### 12.5.2 Production of Waveguides (1D)

Waveguides up to 3 mm in length were written into clinical grade PMMA (Vistacryl CQ non-UV), by moving the sample parallel to the focused laser beam [19]. Figure 12.4 (left) shows a microscope image in transmitted light. Light guidance was observed within the core of the structure (∼6 μm diameter, corresponding to the laser focal region), indicating an elevated (positive) RI change, confirmed by direct RI measurement using the refracted near field (RNF) technique. Figure 12.4

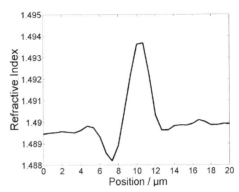

**Fig. 12.4** Waveguide cross-section (beam diameter with $D = 2$ mm aperture, laser fluence $\Phi = 0.35$ J cm$^{-2}$, scan speed $s = 2$ mm s$^{-1}$) and profile of a refractive index structure in clinical grade PMMA (at 633 nm, 23°C)

(right) shows the profile of a structure with a RI increase of $\Delta n_{max} = 4 \times 10^{-3}$ over a 2 μm distance. The apparent index dip on the left hand side of the peak is a measurement artefact, occurring at the transition from lower to higher RI, regardless of the sample orientation. Measurements were performed using a commercial RNF profilometer (from Rinck Elektronik Jena).

The measured maximum RI change confirmed the previous diffraction efficiency measurements from grating structures [19].

## 12.6 Holographic Writing

Refractive index modification of PMMA at sub-micron periodicities $\Lambda < 0.5$ μm was demonstrated using a holographic technique using a phase mask and Schwarzschild objective. The fundamental output from a Ti:sapphire fs laser (Clark-MXR CPA-2010, 1 kHz repetition rate, 180 fs, 775 nm) was attenuated by two diffractive optic attenuators (DOAs) and frequency doubled to 387 nm in BBO. The UV wavelength, rather than 775 nm, was chosen to achieve higher $\Delta n$ combined with a smaller feature size.

The holographic set-up consisted of a custom designed phase mask (100 lines mm$^{-1}$) for 387 nm with 38% diffraction efficiency measured into ±1st order, in conjunction with a fused silica bi-prism of appropriate wedge angle producing two parallel beams for interferometric modification, as shown in Fig. 12.5. A fused silica lens ($f = 75$ mm), UV refractive microscope objective (0.15 and 0.3 NA) or a 0.5 NA (15×) UV reflective Schwarzschild objective focused and overlapped the beams. Careful optical alignment was essential to maintain the temporal coherence of the beams on target by keeping optical path difference $\delta \ll 70$ μm (~coherence length). The symmetric optical set-up simplifies this requirement, avoiding the use

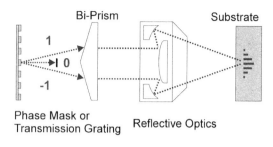

**Fig. 12.5** Holographic writing setup using phase mask, bi-prism, and Schwarzschild objective

**Fig. 12.6** Refractive index change by holographic writing inside bulk PMMA ($\Lambda = 1.4\,\mu$m, $v = 2\,\text{mm s}^{-1}$, $\Lambda_{\text{offset}} = 14\,\mu$m, $\Phi = 0.12\,\text{mJ cm}^{-2}$, NA $= 0.15$)

of a delay line in one arm. The addition of a 0.5× magnification telescope, inserted ahead of the phase grating, increased the divergence, and hence focused spot size and DOF (depth of focus), by a factor of 2.

Figure 12.6 shows a periodic phase grating in PMMA at a period $\Lambda = 1.4\,\mu$m, observed in transmission with an optical microscope. Diffraction efficiency was $\sim 10\%$ in to each first order, measured at 633 nm. As the DOF $= \lambda/2\text{NA}^2$ was $\sim 9\,\mu$m, this 3D volume grating was created by translating the focal spot along the optic axis to a depth of $\sim 100\,\mu$m. Larger areas of a sample were structured by repeated scans, offsetting the holographic fringes by $\Lambda_{\text{offset}}$, a multiple of the grating period $\Lambda$. The diffraction efficiency $\eta$ of sub-surface phase gratings in bulk material was measured to estimate $\Delta n$ using equations for Raman–Nath diffraction [32].

Figure 12.7 shows a periodic phase grating in PMMA at a sub-micron periodicity of $\Lambda = 0.42\,\mu$m in the form of a composite image showing the PMMA sample with inscribed grating on the left hand side and the large angle first order diffraction pattern formed using the writing fs laser wavelength at 387 nm on the right-hand side, at two different SLR camera exposure times (top and bottom). The sub-micron phase grating written in the PMMA sample with the 0.5 NA Schwarzschild objective with measured period $\Lambda = 0.42\,\mu$m. Note the fluorescence from the modified region. We would expect a period $\Lambda = \lambda/2\text{NA} = 0.387\,\mu$m and the difference is due to the physical size of the two beams ($\phi \approx 2.5$ mm) entering the objective whose primary mirror had a 13.5 mm diameter, limiting the effective NA to $\sim 0.46$. The zero order and first order diffraction patterns contain spots due to the 20 $\mu$m offset period used to write the sub-micron grating. The efficiency is low ($< 1\%$ into first order) due to the limited thickness of the grating, but will be optimized in further work.

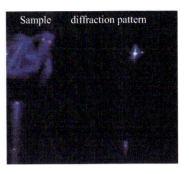

**Fig. 12.7** Composite image showing PMMA sample on LHS and diffraction of 387 nm on RHS

## 12.7 Comparisons of Commercial and Clinical Grade PMMA

Studies were generally performed on commercial grade (Goodfellow) and clinical grade PMMA (Vistacryl CQ Non-UV). The former contains additives that may initiate the photoreaction whilst clinical grade is totally free of UV inhibitors and residual monomer. Comparison of grating diffraction efficiencies in a range of PMMA materials indicated that the intrinsic polymer could be efficiently modified, independent of additives. The following PMMA types were compared, indicating similar susceptibility to RI modification: Vistacryl UV (containing benzotriazole as an UV absorber); clinical grade Vistacryl CQ non-UV (UV inhibitor free, residual monomer $\leq$ 0.4wt%, $\leq$ 0.05% benzoyl peroxide as an initiator) and custom UV-polymerized, ultra pure PMMA (no initiator). Gratings generated in material containing UV absorber and initiator showed weak fluorescence when illuminated with low intensity, 387 nm laser light, indicating the degradation of these additives during the writing process.

The ablation threshold was determined for different types of PMMA (Vistacryl non-UV, Vistacryl without initiator, and Goodfellow commercial grade PMMA) by ablating a series of spots with a constant number of laser pulses but increasing pulse energy. Assuming a Gaussian shaped beam, the threshold fluence can be found in a semi-logarithmic plot of the squared diameter of the craters $D^2$ versus the laser fluence $\Phi$th by extrapolating a fitted line to $D^2 = 0$. The resulting surface damage fluence levels are about 0.1 J cm$^{-2}$ for all three materials ($N = 200$ pulses at a pulse duration of $\tau = 200$ fs). This is in good agreement with previously published values [33, 34].

## 12.8 Pulse Duration, Wavelength, and Bandgap Dependence of Refractive Index Modification

Many studies show the strong influence of laser pulse duration on polymer ablation due to shielding of the substrate by plasma and plume creation (e.g., [35]). However, little is known about the influence of pulse duration on the fs modification of dielectrics. The low intrinsic linear absorption of dielectrics can be increased by non-linear effects such as multiphoton, or excited state absorption at the high peak intensity within ultra-short pulses. The authors investigated RI modification of pure PMMA as a function of pulse duration using fs lasers at 800 and 387 nm wavelength and related it to laser wavelength and the bandgap of PMMA.

### 12.8.1 Pulse Duration Dependence of Refractive Index Modification

Below the ablation threshold, fs laser irradiation causes permanent material changes in RI or absorbance that accumulate for multiple-pulse irradiation. Weakly absorbing material undergoes an incubation reaction that increases absorbance and lowers the damage threshold compared with single-pulse irradiation [36]. Modification of fused silica by repetitive irradiation with sub-ablation threshold fluences (1 kHz repetition rate, 800 nm wavelength) demonstrated a decrease in bulk damage threshold with increasing pulse duration (130–230 fs) and no RI modification for durations greater than 230 fs [37]. This trend opposes that observed above the ablation threshold and appears paradoxical, because short pulses with high peak powers would be expected to cause material damage.

In the sub-ablation threshold regime of PMMA, the authors explored the effect of pulse duration on polymer RI modification [38]. Experiments were performed using a 800 nm 1 kHz Ti:sapphire fs laser with variable pulse duration down to 40 fs. Gratings with periodicity of 40 μm were directly written 100 μm below the surface of bulk clinical grade PMMA (Vistacryl CQ non-UV) using a laser beam (diameter $D = 10$ mm) focused with a 75-mm focal length lens to a spot size of 30 μm and a writing speed of $s = 1.25$ mm s$^{-1}$. Optimal RI modifications were produced by repeated irradiation with a number of overscans $N_s$, using a laser fluence $\Phi$ below the damage threshold. The RI modification $\Delta n$ was inferred from the first order grating diffraction efficiency using a He–Ne laser as described in Sect. 12.5.1.

Figure 12.8 shows the measured diffraction efficiency from the first order at the Bragg angle of a phase grating that has been over scanned up to 32 times with a constant laser fluence $\Phi = 0.14$ J cm$^{-2}$ for different pulse durations at 800 nm wavelength. The shortest pulse duration (45 fs) generated the largest RI modification and thus the largest diffraction efficiency, whereas pulse durations greater than 100 fs showed smaller diffraction efficiency. This demonstrated that pulse duration, and thus peak intensity, was critical in the NIR wavelength range in order to optimize $\Delta n$ whilst maintaining laser fluence below the damage threshold. The number of pulses per spot $N_p$ was estimated from the scanning exposure as $N_p = N_s v d/s$, where $N_s$ is the over scan number, $d$ is laser spot size, $v$ is repetition rate, and $s$ is writing speed. The last point of each data set in Fig. 12.8 represents the number of pulses per spot at which bulk damage was detected (i.e., formation of light scattering, opaque structures, and cracks). The observed decrease of the bulk damage threshold with increasing pulse duration agrees with the pulse duration

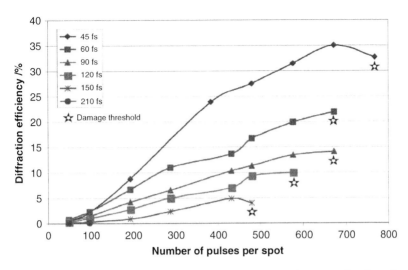

**Fig. 12.8** First-order diffraction efficiency of refractive index gratings as a function of pulse duration and number of pulses per spot $N_p$ for pulse duration between 45 and 210 fs, $\Phi = 0.14\,\text{J cm}^{-2}$, $N_s = 2 - 32$, and wavelength of 800 nm [38]

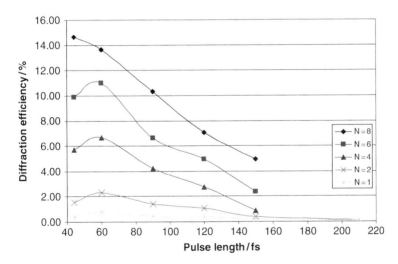

**Fig. 12.9** Rearranged data from Fig 12.8 – diffraction efficiency depending on pulse length

dependency of the sub-ablation threshold modification observed for bulk fused silica [37].

The same data, rearranged as diffraction efficiency versus pulse duration (Fig. 12.9), indicated that for single over scans, the effect of pulse duration is minimal. However, shorter pulses generate more efficient gratings and prevent early onset of surface damage. For repeated irradiation, short pulse duration is essential for optimal $\Delta n$.

Rearranging the data for a given pulse duration (60 fs), so that the photoreaction yield is plotted versus irradiation intensity on a double logarithmic scale (Fig. 12.10), enables discrimination between the order of the multiphoton process [39]. Figure 12.10 shows the logarithm of the diffraction efficiency for sets of over scanned gratings, with a shape following that of an activated process, similar to an Arrhenius-plot that describes photothermal reactions, previously related to fs laser modification and ablation [40]. Repeated exposure by overwriting increases $\Delta n$ due to the accumulated laser fluence, which is expressed as $\Phi_{acc} = N_s \Phi f$, where $f$ is a factor allowing for the integration of a scanning Gaussian beam. Figure 12.10 shows a plot of the diffraction efficiency, versus $\Phi_{acc}$. The gradient in the linear range of the double logarithmic plot indicates the order of the multiphoton process that initiates the photoreaction [39]. For an increasing number of pulses $N_p$, the order of the non-linearity $m$ reduced from three or four, for low $N_p$, to around unity, for high $N_p$ (Table 12.1). Decreased $m$ indicates the accumulation of linearly absorbing photoreaction products (color centers). This indicates that the short pulse radiation creates absorption centres that initiate linear absorption at the laser wavelength, resulting in an incubation reaction. UV-visible transmission measurements showed an increased absorbance of laser modified PMMA near the UV absorption edge, confirming the incubation reaction; attributed to an increase in C=C double bonds due to depolymerization [19].

The authors have shown that multiple pulse irradiation with short pulses efficiently modifies the material, avoiding bulk damage, which occurs when longer pulses are used despite the same energy being delivered. This explains the lower damage threshold observed in multipulse, sub-ablation threshold dielectric bulk

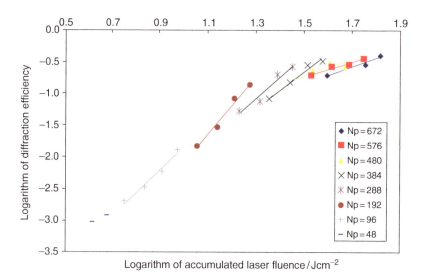

**Fig. 12.10** Yield of photoreaction at constant pulse duration ($\tau = 60$ fs) versus laser fluence, accumulated over a number of pulses $N_p$ at four constant laser fluences ($\Phi = 0.1, 0.12, 0.14$, and $0.16$ J cm$^{-2}$); line gradients indicate the order of the non-linear process (Table 12.2) [38]

Table 12.1 Summary of literature to date on photomodification of PMMA

| Author | Year | Source | $\lambda$/nm | Repetition rate | $\tau_p$ | $\Delta n$ | Mechanism studied /explanation |
|---|---|---|---|---|---|---|---|
| Allison [2] | 1966 | Broadband | – | cw | – | – | Photodegradation via side chain scission |
| Tomlinson [1] | 1970 | UV | 325 | cw | – | $+3 \times 10^{-3}$ | Cross-linking of oxidized monomer |
| Moran [3] | 1973 | He:Cd | 325 | cw | – | $2.8 \times 10^{-3}$ | – |
| Mitsuoka and Torikai [5] | 1990 ff. | He:Cd Xe arc | 260 – 500 | cw cw | – – | – | Photodegradation via side chain scission induced main chain scission |
| Marotz [9] | 1985 | Ar ion | 514 | – | – | $2 \times 10^{-4}$ | Photopolymerization of monomer-doped PMMA |
| Peng and Xiong [7] | 1999 | Ar ion | 248, 280, 325 | cw | – | $3.3 \times 10^{-5}$ – $10^{-4}$ | – |
| Kuper and Stuke [10] | 1989 | Excimer | 193, 248 | 5 Hz | ns–500 fs | $4 \times 10^{-3}$ | Incubation and side chain cleavage |
| Baker and Dyer [12] | ff. | KrF | – | 5 Hz | 16 ns | $4 \times 10^{-3}$ | Incubation and side chain cleavage |
| Scully [16] | 1993 | KrF, Ar ion, 5ωNd:YAG | 216 – | – – | – | $8 \times 10^{-3}$ | – |
| Wochnowski [15] | 2000 2005 | broadband UV excimer | 363 193, 248 | | | – | Combination of degradation and cross-linking |

(continued)

Table 12.1 continued

| Author | Year | Source | $\lambda$/nm | Repetition rate | $\tau_p$ | $\Delta n$ | Mechanism studied /explanation |
|---|---|---|---|---|---|---|---|
| Bityurin [11] | 1999 | fs | 385 | 70 MHz | 200 fs | – | Sub-surface damage |
| Scully [16] | 2002 | Ti:sapphire | 800 | 1 kHz | 40 fs | $5 \times 10^{-4}$ | – |
| Li [21] | ff. | Ti:sapphire | 800 | – | 130 fs | $2 \times 10^{-4}$ | – |
| Zhang | 2002 | Ti:sapphire | 800 | – | 100 fs | – | Microstructures with positive $\Delta n$ tubular waveguides, volume expansion, decreased solubility waveguides |
| Zoubir [22] | 2002 | Ti:sapphire | 800 | 25 MHz | 30 fs | $-2 \times 10^{-3}$ | Developing waveguides; |
| Ohta [23] | 2004 | Ti:sapphire | 800 | 1 kHz | 100 fs | $+1.5 \times 10^{-4}$ | Chain scission and volume contraction |
| Sowa [25] | 2004 | Ti:sapphire | 800 | 1 kHz | 85 fs | $\pm 4.6 \times 10^{-4}$ | Positive $\Delta n$ (directly measurement) related to photodegradation |
| Baum [16] | 2006 2007 | Ti:sapphire | 775/387 | 1 kHz | 180 fs | $+4 \times 10^{-3}$ | |

**Table 12.2** Slope of trend lines of individual data sets and collective data for Fig. 12.10 [38]

| Number of pulses per spot, $N_p$ | Slope of fitted line |
|---|---|
| 48 | – |
| 96 | 3.5 |
| 192 | 4.6 |
| 288 | 3.4 |
| 384 | 2.8 |
| 480 | 1.4 |
| 576 | 0.1 |
| 672 | 1.3 |
| All data points | 2.3 |

material modification in this study and in [37]. Therefore, short pulses with low laser fluence are a means of controlling photomodification via the incubation process, thus controlling avalanche ionization.

## 12.8.2 Effect of Bandgap and Wavelength on Refractive Index Modification

The bandgap of pure PMMA was previously determined by the authors to be 4.58 eV from transmission data [19]. The photomodification of PMMA by the authors using NIR, sub-100 fs, kilohertz repetition rate laser pulses was reported in Sect. 12.5.1 and [18]. Pulses of 40 fs created phase gratings with RI changes of $\Delta n = 5 \times 10^{-4}$ [18], and 85 fs pulses produced waveguides with $\Delta n = 4.6 \times 10^{-4}$ [41]. The diffraction gratings in [18] demonstrated 37% total first-order diffraction efficiency, $\eta$. An attempt to reproduce these results with a longer pulse duration of 180 fs at $\lambda = 775$ nm produced only small RI changes ($\eta = 1\%$ for similar writing parameters). In contrast, a frequency-doubled beam at $\lambda = 387$ nm generated gratings with up to 40% first-order diffraction efficiency, corresponding to $\Delta n = 4 \times 10^{-3}$ [19] as described in Sect. 12.5.1. The shorter wavelength provides sufficient photon energy $h\nu$ to reduce the order of the non-linear absorption process, which is determined by $mh\nu \geq E_g$, depending on the material optical bandgap energy $E_g$.

Three simultaneously absorbed photons at 800 nm ($h\nu = 1.55$ eV) provide sufficient energy to excite electron transitions in pure PMMA, whereas two photons are below the bandgap energy. Short pulses with high peak power enhance the photomodification at 800 nm wavelength. At 387 nm, a single photon provides 3.2 eV, and so the energy of two photons is above the bandgap. Figure 12.11 compares the processes at 800 nm and pulse duration 40 fs and 387 nm with pulse duration 80 fs in a double logarithmic plot of diffraction efficiency versus laser fluence to estimate the order of the non-linear processes. The observed gradients in the linear regions of 2.9 and 2.0 indicate that three-photon absorption at 800 nm

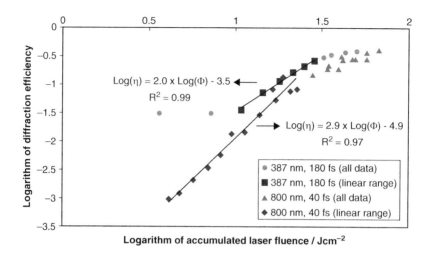

**Fig. 12.11** Determination of the order of the photomodification process in pure PMMA for 800 nm, 40 fs, and 387 nm, 180 fs by plotting photoreaction yield versus integrated fluence; line gradients of 2.9 and 2.0 indicate three- and two-photon absorption at 800 nm and 387 nm, respectively [38]

and two-photon absorption at 387 nm, respectively, are responsible for the RI modification of pure PMMA.

In conclusion, it was demonstrated that optimal fs laser modification relies on an appropriate combination of laser-pulse duration and wavelength in relation to the material bandgap. The results suggest that, for a regime of repeated irradiation with low laser fluence, sub-100 fs laser pulses are preferable, since they enable efficient modification via an incubation effect while avoiding material damage. The authors have shown that at 800 nm, the refractive index is modified more efficiently as the pulse duration decreases below 100 fs, whereas at 387 nm, efficient index modification is accomplished with longer, 180 fs pulses. Results suggest that three and two-photon absorption is responsible for modification of pure PMMA at 800 nm and 387 nm, respectively. Repeated irradiation with short pulses of low laser fluence allows control of the photomodification via incubation, thus reducing bulk damage.

## 12.9 Relating Photochemistry to Writing Conditions

To create useful structures in PMMA, it is necessary to understand the photochemical processes involved in low dose fs laser RI modification to create optimized structures and the longevity of the structures produced.

Refractive index is closely related to optical and thus material density. The RI of a polymer is also determined by its chemical composition, including end groups, additives, and impurities. Furthermore, it is affected by molecular orientation, forces between the polymer chains, and thermal processing of the material during

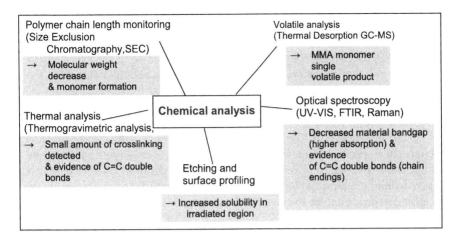

**Fig. 12.12** Schematic of photochemical techniques

manufacture or when irradiated by a laser. Therefore, laser-induced RI modification can be explained by photochemical changes affecting density as well as changes in chemical composition itself.

As discussed in Sect. 12.4, recent fs photomodification studies of PMMA at 800 nm reported contradictory results about the sign of the RI change $\Delta n$ induced. Negative $\Delta n$ were found in the focal volume of an fs oscillator (25-MHz repetition rate) [22], whereas positive $\Delta n$ were observed with 1 kHz regenerative amplifiers [25, 26], indicating that repetition rate influences the modification through thermal effects.

The authors developed a wide range of methodologies (Fig. 12.12) to elucidate the photochemical mechanisms for ultrashort, sub-ablation threshold photomodification effects in PMMA at UV wavelengths that are not linearly absorbed [19]. The writing conditions were 387 nm, 1 kHz repetition rate, and 180 fs pulse duration.

The techniques used included:

- Polymer chain length monitoring using size exclusion chromatography (SEC) to monitor polymer MWD changes, such as molecular weight decrease and monomer formation.
- Volatile analysis (thermal desorption gas chromatography) combined with mass spectroscopy (TD GC-MS) to analyze reaction products formed during laser treatment.
- Thermal analysis (thermogravimetric analysis, TGA) to detect small amount of cross-linking and evidence of C=C double bonds.
- Optical spectroscopy (UV-VIS, FTIR, and Raman) to measure the decreased material bandgap (higher absorption) and evidence of C=C double bonds (chain endings).
- Etching and surface profiling to show increased solubility within the irradiated region.

## 12.9.1 Size Exclusion Chromatography

SEC enabled the MWD for exposed PMMA films to be compared with the original narrow distribution, indicating that the peak broadened and shifted to lower molecular weight values as the sample degraded due to increased laser fluence. This suggested that polymer backbone scission was occurring with subsequent loss of material due to formation of the monomer methyl methacrylate (MMA) and short chain oligomers [19]. This depolymerization process usually creates increased polymer chain specific volume due to increased number of endgroups, causing either volume expansion (which is associated lower optical density and therefore reduced RI) or, within a confined volume, material stresses (which can cause positive or negative $\Delta n$, depending on the material composition).

## 12.9.2 Thermal Desorption Volatile Analysis

Volatile analysis confirmed MMA production below the laser ablation threshold [19], which was previously reported as the main degradation product of PMMA via pyrolytic decomposition [42] caused by UV photodegradation due to cw irradiation and excimer laser ablation [43, 44]. However, MMA formation has not been previously observed below the ablation threshold using ultrashort pulses [19]. The absence of fragments previously associated with polymer side chain cleavage by ablation and sub-ablation threshold degradation by excimer lasers, [44] suggests that direct cleavage of the polymer backbone occurs with fs laser irradiation below the ablation threshold. This process is initiated by end or random chain scission and is propagated by unzipping to form monomers, similar to thermal degradation of PMMA. It is the exact reverse of the polymerization process.

## 12.9.3 Thermogravimetric Analysis

Previous work on TGA of ns laser-deposited PMMA films indicated a slightly raised decomposition temperature compared to the original material, which together with solubility and FTIR experiments indicated that cross-linking took place explaining increased density/RI [45]. TGA was, therefore, used by the authors to explain increased RI at the centre of the fs laser modified PMMA, and identify cross-linked polymers, which exhibit increased RI due to a change in specific volume and densification, since van der Waals' bonds are exchanged for more compact bonds [46]. Proportional and derivative weight losses were compared for laser modified and unmodified PMMA. Weight loss between 175 and 375°C was strongly increased in the laser modified sample, confirming the prediction from SEC that lower molecular weight fragments were created that degraded quicker. Between 375 and 425°C, a slight decrease in decomposition rate of the laser modified material was observed, compared with the blank, indicating cross-linked material, leading to a positive RI change.

### 12.9.4 Optical Spectroscopy

UV-visible absorption measurements of laser-modified clinical- and commercial-grade PMMA by the authors showed increased UV absorption in the transmission spectra, visible to the naked eye as faint but permanent yellowing [19]. Similarly, Küper et al. described the formation of UV-absorbing color centres together with new IR absorption bands due to ns laser-treated PMMA [47]. [C=C] bonds and an ester band formed due to photoinduced cleavage of the [COOCH$_3$] carbonyl side chain, created unsaturated, absorbing backbone bonds, and methylformate as primary decomposition products, giving rise to the absorption bands and color centres.

These observations indicated that the increased UV absorption arose from a large number of end groups formed by partial depolymerization, since half the groups contain an unsaturated carbon bond. When tested with IR spectroscopy, methylformate or other fragments related to the cleavage of the polymer side chain were absent, unlike ablation and sub-threshold degradation with excimer lasers [44]. This suggests direct cleavage of the polymer backbone during fs laser irradiation of PMMA, initiated by end-chain or random chain scission and propagated by unzipping to form monomer, similar to thermal degradation of PMMA. Depolymerization could confirm density decrease of PMMA in the laser focal region causing negative $\Delta n$ [48]; however, chemical changes due to the increased dominance of endgroups at reduced molecular weight and the formation of color centres could also cause positive $\Delta n$. This will be further investigated in the future.

### 12.9.5 Etching of Structures

Refractive index structures were washed in 5% aqueous solution of MIBK, as used in electron-beam lithography, in which PMMA serves as photoresist. For moderate dose exposure, low molecular weight fragments are removed, but for high dose exposure, cross-linking and, therefore, insolubility occurs [49]. The wet chemical etching of fs irradiated polymers was previously used to reveal modification profiles [19, 50]. The laser exposed material is removed, confirming formation of low molecular weight compounds in the laser focal region for kHz repetition rate fs laser modification of PMMA. Partially cross-linked material within the exposed regions may be washed away together with the large portions of easily soluble, low molecular weight components.

## 12.9.6 Summary of Photochemical Analysis

The various photochemical elucidation methods indicate that control of writing conditions may facilitate different photochemical processes, and that analytical methods can be used to identify the photochemical process taking place.

To summarise:

- SEC and thermal desorption volatile analysis identified depolymerization and monomer formation which affect polymer chain specific volume as well as chemical composition and could thus increase or decrease the RI.
- TGA distinguishes lower molecular weight fragments that degrade at lower temperatures from cross-linked material that degrade at higher temperature. Cross-linked material gives increased RI.
- Optical spectroscopy can identify depolymerization products creating absorption bands in UV and IR. Direct cleavage of the polymer backbone to form monomer, similar to thermal degradation of PMMA, was detected.
- Chemical etching of fs irradiated PMMA with MIBK removes low molecular weight fragments in the focal region of fs exposure.

Depolymerization together with a small amount of cross-linking was detected in the fs laser exposed focal region, however, the observed positive RI change that accompanies these chemical and structural changes could not be fully explained to date.

Understanding the photochemical process is essential for optimal writing conditions in the fs, sub-ablation threshold regime. A balance between laser self-focusing and various non-linear defocusing and other optical and thermal effects is required. Such fine laser beam control could enable positive or negative RI modification, e.g., to form depressed cladding structures that confine light within unmodified material that can be specifically doped for other applications, such as lasing or electro-optic modulation. This is a powerful tool because a single laser writing beam could form both passive waveguides/structures and active devices within the same substrate.

## 12.10 Effect of Self-Focusing

The ultrahigh intensity associated with fs pulses leads naturally to non-linear interaction in dielectrics hence inducing an intensity related component to the RI. FL, or self-guiding, results when the Kerr self-focusing (SF) is dynamically balanced by the de-focusing due to the electron plasma initiated by multi-photon absorption. The control of filament length and hence modification depths is sensitive to pulse energy, wavelength, temporal pulse duration, and in particular, effective NA. Much of the research on FL or self-guiding at ultrahigh intensity has been carried out on fused silica [51–54]. For example, Saliminia et al. [54] analyzed the interplay between SF

12 Refractive Index Structures in Polymers

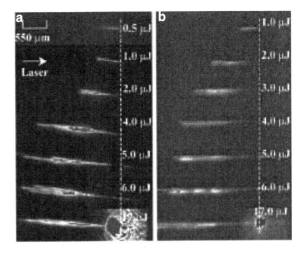

**Fig. 12.13** CCD images of the accumulated laser scattering and plasma fluorescence signals in fused silica at different pulse energies. An objective lens with a power of 1× ($f = 73.5$ mm, NA: 0.03) was used to focus the beam inside the glass. The sample was irradiated by roughly (**a**) 10,000 shots and (**b**) 10 shots. (from [54])

and FL with 810 nm, 45 fs, and 1 kHz laser pulses in fused silica for 0.03<NA<0.85. Sufficiently high pulse energies always resulted in RI modifications ahead of the geometrical focus accompanied by irregular voids due to optical breakdown. At low NA, while carefully controlling pulse energy and pulse number, RI structures with modification depth up to ∼2 mm in length were demonstrated with periodic structuring due to self-focusing, Fig. 12.13. As pulse energy increased, modification regions moved towards the laser source, away from the geometrical focus and as NA increased, modification depths reduced to $< 20\,\mu$m. Peak powers were always well above the critical power for self-focusing $P_c$ approximately few MW in fused silica.

More recently, self-focusing has been studied in PMMA. Uppal et al. [55], for example, created 3D buried structures in PMMA, and studied the effect of NA on the ability to create 15-mm long structures (longitudinal writing) using 150 fs, 800 nm, and 1 kHz laser pulses as well as the effect of pulse energy and writing speed. At low NA∼0.1, with pulse energy $< 2\,\mu$J, self-focusing and self-guiding led to long modification depths and the laser-formed structures were sensitive to polymer degradation and void formation. As the critical power for self-focusing is only ∼23 kW in the case of PMMA ($n_2$∼$2.7 \times 10^{-14}\,\text{cm}^2\,\text{W}^{-1}$) [56], peak powers used were orders or magnitude above this value. Tight focusing using a 0.4 NA objective combined with pulse energies from 0.1 to 1.5 $\mu$J resulted in more symmetric structures, limiting modified depths to 20–250 $\mu$m.

Watanabe [57] recently studied fs filamentary modification of PMMA with 100 fs, 800 nm, and 1 kHz laser pulses and a low NA = 0.1 objective with radiation focused ∼1.5 mm below the sample surface while translating the sample transversely. After exposure, irradiated regions were inspected under an optical microscope. Pulse energies from 0.8 to 1.6 $\mu$J produced RI modified regions from 210 to 420 $\mu$m in length with scattering damage observed above 1.7 $\mu$J in agreement

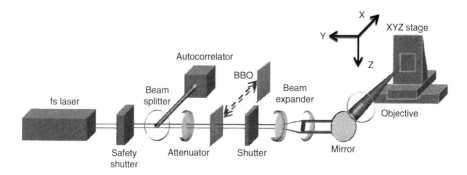

**Fig. 12.14** Schematic of experimental system

with observations of Uppal et al. [55]. The head of the modified regions elongated towards the laser source as pulse energy increased, a characteristic of FL in dielectrics. In addition, the authors simultaneously studied the spectral broadening of the laser pulse during irradiation, showing that the super-continuum broadens with increasing pulse energy.

Filamentation has also been briefly studied in PMMA by the authors in both the NIR (775 nm) and NUV (387 nm) with 160 fs, 1 kHz pulses (Clark-MXR-CPA 2010). Fig. 12.14 shows a schematic of the experimental set-up. The output beam was attenuated, expanded to ∼ 15 mm diameter then focused by a $f = 50$-mm focal length lens ($NA_{eff} \approx 0.15$). For 387 nm work, a BBO doubling crystal was placed in the beam path. The substrates were pure PMMA slabs with size of $30 \times 10 \times 5$ mm$^3$, which were optically polished on all surfaces and mounted on a computer controlled three axis stage (Aerotech). The beam was focused ∼ 2 mm below the surface. The stage was stationary during the laser irradiation and pulse energies in the range of 0.1–3.5 µJ and exposure times in the range of 0.2–3 s were used.

Figure 12.15 (left) shows highly periodic microstructuring with increasing pulse energy in the range $0.1\,\mu J < E_p < 0.6\,\mu J$ due to re-focusing in PMMA at 387 nm when exposed for 0.5 s (500 laser pulses). The modified regions extend towards the laser source with increasing pulse energy and >6 periods with approximately ∼ 100 µm spacing are demonstrated. At 775 nm (right), higher pulse energies are required to modify PMMA and a more chaotic structure results but re-focusing is again evident with damage regions again extending towards the source with increasing pulse energy. The modification threshold at 387 nm is lower since PMMA requires only two-photon absorption to create seed electrons by excitation from the valence band to the conduction band whereas 775 nm modification requires three-photon absorption ($E_b \approx 4.58$ eV).

Filamentation extends the modification in dielectrics well beyond the Rayleigh length but, clearly, may not necessarily guarantee uniform modification throughout

12 Refractive Index Structures in Polymers 339

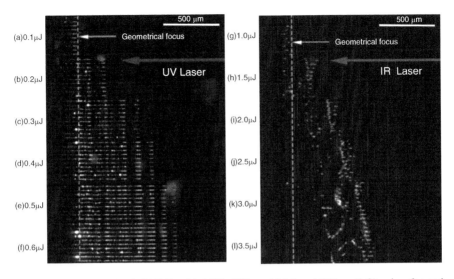

**Fig. 12.15** Filamentation in PMMA with 160 fs, 775 nm (right) and 387 nm (left) pulses focused with NA ∼ 0.15. Note the difference in pulse energies and the highly periodic structuring at 387 nm. Exposure time $t = 0.5$ s (500 pulses). Light scattering shows that there is also dielectric breakdown at higher pulses energies

a material due to the complex non-linear interactions which makes exact balancing of the self-focusing with plasma de-focusing difficult.

## 12.11 Effects of Depth

The effect of depth below the surface on the laser writing parameters is important for fabricating 3D optical components. Laser direct writing techniques are capable of fabricating flexible and complex structures, such as curved waveguides, internal connectors, and out of plane coupling devices. The position of the laser focus location below the surface and the dimensions of structures written is important because self-focusing effects can elongate the modified region, and irradiation on the surface creates ablation rather than RI changes.

Internal fs laser structuring for 3D micromachining is complicated by the air/dielectric interface which introduces depth dependent spherical aberration, with increasing NA, has been studied primarily in fused silica [58, 59]. This results in higher modification thresholds with increasing depth. Hnatovsky et al. [58] showed that the aberrated axial intensity distribution from an uncorrected high NA $= 0.75$ long working distance objective limits the modification depth to $<$ 40 μm from the interface unless the pulse energy is increased, while at lower NA $= 0.2$, spherical aberration was almost negligible, allowing one to write at depths up to ∼10 mm. By using adjustable compensation for spherical aberration

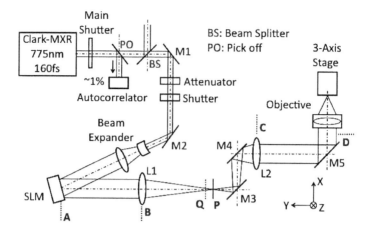

**Fig. 12.16** Energetic 1 mJ, 775 nm, 160 fs pulses are reflected by mirrors M1 and M2, attenuated then expanded to 8 mm diameter before reflection from the SLM liquid crystal device where a CGH is displayed. Diffracted beams are re-imaged to the input aperture (plane D) of the microscope objective with a 4 f optical system where distance AD = 4f and distance AB = BP = PC = CD = f. CGHs are applied to the SLM representing the desired intensity distribution at the focal plane of a long working distance objective (0.15 NA) and carefully focused below optical substrate surface positioned on a three-axis stage

with depth, they demonstrated uniform modifications using 800 nm, 100 fs, and 100 kHz pulses at NA = 0.75 up to ~1 mm from the interface with structure sizes limited axially to <4 μm. Liu et al. [59] showed that the influence of focusing depth was significant when writing waveguides in fused silica with a 0.5 NA objective. The modification threshold increased with depths > 300 μm, in accord with calculations by Hnatovsky [58] and a 3 × 3 waveguide array showed asymmetric coupling due to the effect of spherical aberration with depth.

Based on these observations, it is clear that true 3D structuring in polymers will require dynamic real time aberration compensation with depth during 3D writing, particularly as NA increases. One approach might involve the use of a SLM or other beam shaping techniques (see Chaps. 4 and 5).

## 12.12 Parallel Processing Using Spatial Light Modulator

Single beam processing (laser ablation or $\Delta n$ structuring) with fs laser pulses may be limited by the need to attenuate the laser pulse energy to the μJ level. Recently, we demonstrated highly parallel surface processing with a 1 kHz fs laser source combined with a SLM [60–62] through the application of computer generated holograms (CGHs). This technique lends itself naturally to 3D $\Delta n$ structuring. Figure 12.16 shows a schematic diagram of the experimental set-up used to generate multiple diffracted beams for RI structuring in PMMA.

The output from a Clark-MXR CPA-2010 fs laser system (775 nm wavelength, 160 fs pulse duration, 1 kHz repetition rate) passes through a pick off, autocorrelator

12 Refractive Index Structures in Polymers    341

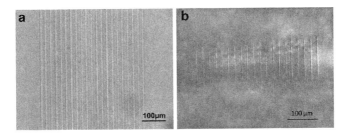

**Fig. 12.17** (a) Parallel modification of PMMA with 21 diffracted beams to create a grating, scan speed 1 mm s$^{-1}$ and 0.6 µJ per pulse per beam, (b) the cross-section of the RI structures showing nearly uniform modifications

and 50/50 ultrafast beam splitter to turning mirror M1 then attenuated and expanded onto a reflective liquid crystal on silicon SLM (Hamamatsu X10468–01) which has 800 × 600 pixels (16 × 12 mm) with diffraction efficiency $\eta > 70\%$. A 4f optical system consisting of two plano-convex lenses L1 and L2 (f1 = f2 = 300 mm) re-images the surface of the SLM to the input aperture of a microscope objective (Nikon, 0.15 NA) allowing the blocking of the remaining energetic zero order reflection near the Fourier plane (Q) of lens L1 using a small target.

From the desired intensity distribution required at the focal plane of the objective, the corresponding CGH is calculated and applied to the SLM and the resulting phase pattern (8 bit grayscale) is observed on a separate monitor representing the phase distribution on the SLM.

Parallel RI modification to create a grating without optical breakdown at 775 nm, 160 fs with 21 nearly uniform beams with a period $\Lambda$ of 22 µm is shown in the optical micrographs of Fig. 12.17. The diffracted spots ($E_p \sim 0.6$ µJ per pulse per beam) were scanned transversely at 1 mm s$^{-1}$ to create a grating and each modified region was scanned once only. With the CGH nearly optimized, the cross-section of modified regions is shown in Fig. 12.17b to have a depth of 123 µm ± 8.1% (1 standard deviation) and this pattern was used to create a thick grating.

By scanning in the transverse ($Y$) direction (21 × 22 µm) then offsetting in the axial ($X$) direction with $\Delta X$ of 100 µm (deepest first), a grating with dimensions of 5 × 5 × 1 mm was created in PMMA in less than 10 min. Figure 12.18 shows a cross-section of this thick volume grating when offsetting to produce a continuous modification. A fast mechanical shutter (Newport, response time $\tau_r \approx 3$ ms), synchronised with the translation stage, avoided unwanted exposure of material. The diffraction efficiency for this thick grating was measured to be $\eta = 68\%$ at the Bragg angle $\theta_B = 0.687°$ in excellent agreement with the value given by $\theta_B = \arcsin(\lambda/2\Lambda) = 0.693°$ where $\lambda = 532$ nm is the probe wavelength. The diffraction efficiency with angle is shown in Fig. 12.19.

The corresponding RI change is $\Delta n \approx 1.6 \times 10^{-4}$ using Kogelnick's coupled wave theory given by [63]:

$$\Delta n = \frac{\lambda}{\pi d} \cos\theta \arcsin\sqrt{\eta}, \tag{12.1}$$

where $d$ ($\sim 1$ mm) is the grating thickness and $\theta$ is the Bragg angle.

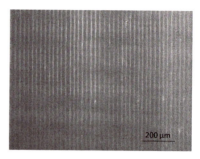

**Fig. 12.18** Cross-section of thick volume grating in PMMA created using 21 parallel beams $\Delta n$ structuring using an SLM at 775 nm in PMMA. $E_p \approx 0.6\,\mu J$ per pulse per beam at transverse scan speed $s = 1\,\text{mm s}^{-1}$ with one over scan and axial offsets of $100\,\mu m$

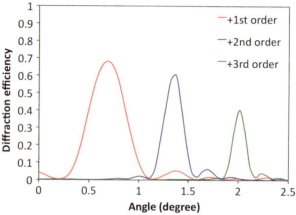

**Fig. 12.19** Diffraction efficiency of $5 \times 5 \times 1$ mm thick, $22\,\mu m$ pitch grating created using parallel processing

The single layer modification depth of $\sim 120\,\mu m \gg$ depth of field DOF $\sim (\lambda/2NA^2) \sim 17.2\,\mu m$ shows that indeed, self-focusing and FL occur since peak power is well above the critical power $P_c \approx 23$ kW in PMMA.

Clearly, there is great scope for speeding up 3D optical circuit microfabrication using SLM technology through the application of dynamic CGH's for parallel processing – and dynamic control of spherical aberration with depth. Recently, two beam dynamic parallel writing of 3D waveguides using an SLM was demonstrated in fused silica by Mauclair et al. [64] (refer to Chap. 4).

## 12.13 Applications of Refractive Index Structures in Polymers

The methods described for writing optimized and permanent RI changes at controlled feature size, RI magnitude, and position facilitate applications in the following research areas:

*Lab-on-a-chip*: 1D and 2D optical structures for which optimal writing conditions are identified can be combined to form simple demonstrator optical circuits in bulk material, for example, Y-splitter, waveguides, and 2D gratings, 3D structures and arrays such as photonic crystals. PMMA based injection moulding methods for manufacturing microfluidic devices facilitate lab-on-a-chip, miniaturized analytical sensors with surface moulded optical devices such as lenses and gratings [65]. Incorporating 3D internal optical circuits enables compact, robust, miniature versions of optical instruments inside a PMMA substrate such as miniature spectrometers.

### 12.13.1 Polymer Optical Fibre Sensors and Devices

Writing periodic RI structures into the core of single mode optical fibre creates filters and reflectors that are sensitive to strain or temperature changes (in fibre Bragg grating – FBG), or enables core modes to interact with cladding measurands (long period grating – LPG). PMMA has high elasticity and thermal expansion coefficients, and is the base material for POFs. It has advantages for strain sensing over glass fibre due to its increased sensitivity by 14% [66], biocompatibility, and good optical transmission in the visible region coincident with high brightness, low cost visible LEDs, and laser diodes now commercially available. In 1999, the first inscription of tunable Bragg gratings in doped, single mode POF by excimer laser radiation, exhibiting $\Delta n \sim 10^{-4}$, was reported [7]. The study of FBG based devices in POF was recorded, such as wavelength tunable filter [67] and fibre interferometry [68].

## 12.14 Summary

This chapter has described RI structuring of PMMA by fs laser irradiation, as a function of wavelength, pulse duration, and associated photochemistry.

The authors have demonstrated RI modification of PMMA using 800 nm and 387 nm, without doping for photosensitivity and shown that undoped, ultra pure, bio-compatible, and clinical grade PMMA can be modified. A permanent, positive RI change of $\Delta n_{max} = 4 \times 10^{-3}$ (comparable to doped material/glass) has been generated. Efficient modification of completely additive free, clinical grade PMMA opens the way for embeddable medical implants and optical devices for use in vivo. A strong dependency on fs pulse duration has been discovered. New insights into the photochemical mechanism of the laser writing process indicate that the RI change is created by a combination of depolymerization, including accumulation of photodegradation products and absorption centre formation, and cross-linking, depending on the exact writing conditions.

Simple integrated structures have been demonstrated (waveguides, bulk gratings, and sub-micron structures). Refractive index modification of PMMA using ultrafast UV optical pulses at a period $\Lambda = 0.42\,\mu m$ indicates that sub-micron (nano)structuring is achievable with a holographic approach, despite the short coherence length associated with fs laser pulses. A custom designed phase grating, bi-prism and Schwarzschild objective achieved sufficient fluence to generate these holographic structures via sub-surface RI modification for fs pulse duration at kHz repetition rates. Such nanostructures formed in 3D, can create photonic crystal structures, further shrinking integrated optical circuits incorporating optical interconnects and highly selective filters.

The effect of self-focusing and FL on $\Delta n$ modification at a range of depths below the surface of bulk PMMA was presented and a SLM for parallel processing demonstrated increased speed and full use of available laser energy. Finally, applications of RI structures in polymers are considered such as microfluidics, lab-on-a-chip, organic optoelectronic devices and gratings in POFs.

**Acknowledgements** The authors acknowledge support from the Engineering and Physical Sciences Research Council, the Unilever-Manchester Advanced Measurement Partnership, Vista Optics and Rinck Elektronik, Jena.

# References

1. W.J. Tomlinson, I.P. Kaminov, E.A. Chandross, R.L. Fork, W.T. Silf-vast, Photoinduced refractive index increase in poly (methyl methacrylate) and its applications. Appl. Phys. Lett. **16**(12), 486–489 (1970)
2. J.P. Alison, Photodegradation of poly (methyl methacrylate). J. Polym. Sci. Part A: Polym. Chem. **4**(5PA1), 1209–1221 (1966)
3. J.M. Moran, I.P. Kaminov, Properties of holographic gratings photoinduced in polymethyl methacrylate. Appl. Opt. **12**(8), 1964–1970 (1973)
4. A. Torikai, M. Ohno, K. Fueki, Photodegradation of poly(methyl methacrylate) by monochromatic light: quantum yield, effect of wavelengths, and light intensity. J. Appl. Polym. Sci. **41**(5–6), 1023–1032 (1990)
5. T. Mitsuoka, A. Torikai, K. Fueki, Wavelength sensitivity of the photodegradation of poly(methyl methacrylate). J. Appl. Polym. Sci. (6), 1027–1032 (1993)
6. A. Torikai, T. Mitsuoka, Electron spin resonance studies on poly(methyl methacrylate) irradiated with monochromatic light. J. Appl. Polym. Sci. **55**(12), 1703–1706 (1995)
7. G.D. Peng, Z. Xiong, P.L. Chu, Photosensitivity and gratings in dye-doped polymer optical fibers. Opt. Fibre Technol. **5**, 242–251, (1999)
8. Z. Xiong, G.D. Peng, B. Wu, P.L. Chu, Highly tunable bragg gratings in single-mode polymer optical fibers. IEEE Photon. Technol. Lett. **11**(3), 352–354 (1999)
9. J. Marotz, Holographic storage in sensitized poly (methyl methacrylate) blocks. Appl. Phys. B-Photophys. Laser Chem. **37**(4), 181–187 (1985)
10. S. Küper, M. Stuke. Ablation of uv-transparent materials with femtosecond uv excimer laser-pulses. In Laser- and Particle-Beam Chemical Processes on Surfaces, volume 129 of Materials Research Society Conference Proceedings, pages 375–384 (1989)
11. N. Bityurin, S. Muraviov, A. Alexandrov, A. Malyshev, UV laser modifications and etching of polymer films (PMMA) below the ablation threshold. Appl. Surf. Sci. **109–110**, 270–274 (1997)
12. A.K. Baker, P.E. Dyer, Refractive-index modification of poly methyl-methacrylate (pmma) thin films by krf-laser irradiation. Appl. Phys. A-Mater. Sci. Process. **57**(6), 543–544 (1993)
13. C. Wochnowski, M.A.S. Eldin, S. Metev, Uv-laser-assisted degradation of poly(methyl methacrylate). Polym. Degrad. Stabil. **89**(2), 252–264, (2005)

14. C. Wochnowski, S. Metev, G. Sepold, UV-laser-assisted modification of the optical properties of polymethylmethacrylate. Appl. Surf. Sci. **154**, 706–711 (2000)
15. A.A. Miller, E.J. Lawton, J.S. Balwit, Effect of chemical structure of vinyl polymers on crosslinking and degradation by ionizing radiation. J. Polym. Sci. **14**(77), 503–504 (1954)
16. P.J. Scully, R. Bartlett, S. Caulder, P. Eldridge, R. Chandy, J. McTavish, V. Alexiou, I. P. Clarke, M. Towrie, A.W. Parker. UV laser photo-induced refractive index changes in poly methyl methacrylate and plastic optical fibres for application as sensors and devices. 14th International Conference on Optical Fiber Sensors, **4185**, 854–857 (2000)
17. P.J. Scully, D. Jones, D.A. Jaroszynski. Writing refractive index gratings in perspex and polymer optical fibre using femtosecond laser irradiation. In Photon 02, Cardiff, 2002. IOP
18. P.J. Scully, D. Jones, D.A. Jaroszynski, Femtosecond laser irradiaton of polymethylmethacrylate for refractive index gratings. J. Optics A Pure Appl. Opt. **5**, S92–S96 (2003)
19. A. Baum, P.J. Scully, M. Basanta, C. L. Thomas, P. Fielden, N. Goddard, W. Perrie, P. Chalker, Photochemistry of refractive index structures in poly(methyl methacrylate) by femtosecond laser irradiation. Opt. Lett. **32**(2), 190–192 (2007)
20. N.M. Bityurin, A.I. Korytin, S.V. Muraviov, A.M. Yurkin, Second harmonic of ti:sapphire femtosecond laser as a possible tool for point-like 3D optical information recording. In Laser Applications in Microelectronic and Optoelectronic Manufacturing IV, volume 3618 of Proceedings of SPIE, p. 122–129 (1999)
21. Y. Li, K. Yamada, T. Ishizuka, W. Watanabe, K. Itoh, Z.X. Zhou, Single femtosecond pulse holography using polymethyl methacrylate. Opt. Exp. **10**(21), 1173–1178 (2002)
22. A. Zoubir, C. Lopez, M. Richardson, K. Richardson, Femtosecond laser fabrication of tubular waveguides in poly(methyl methacrylate). Opt. Lett. **29**(16), 1840–1842 (2004)
23. K. Ohta, M. Kamata, M. Obara, N. Sawanobori, Optical waveguide fabrication in new glasses and pmma with temporally tailored ultrashort laser. In Commercial and Biomedical Applications of Ultrafast Lasers IX, volume 5340 of Proceedings of SPIE, p. 172–178, (2004)
24. C.T. Kauter, B. Koesters, P. Quis, E. Trommsdorff, M. Buck, C.-J. Diem, G. Schreyer, P.R. Szigeti, Herstellung und Eigenschaften von Acrylglaesern *Polymethacrylate* in Kunststoff-Handbuch volume 4, Hanser Munich (1975)
25. S. Sowa, W. Watanabe, T. Tamaki, J. Nishi, K. Itoh. Symmetric waveguides in poly(methyl methacrylate) fabricated by femtosecond laser pulses. Opt. Exp. **14**(1), 291–297 (2006)
26. C. Wochnowski, Y. Cheng, K. Meteva, K. Sugioka, K. Midorikawa, S. Metev, Femtosecond-laser induced formation of grating structures in planar polymer substrates. J. Opt. A-Pure Appl. Opt. **7**(9), 493–501 (2005)
27. F. Korte, S. Adams, A. Egbert, C. Fallnich, A. Ostendorf, Sub-diffraction limited structuring of solid targets with femtosecond laser pulses. Opt. Exp. **7**(2), 41–49 (2000)
28. J.W. Chan, T.R. Huser, S.H. Risbud, J.S. Hayden, D. M. Krol, Waveguide fabrication in phosphate glasses using femtosecond laser pulses. Appl. Phys. Lett. **82**(15), 2371–2373 (2003)
29. M. Douay, W. X. Xie, T. Taunay, P. Bernage, P. Niay, P. Cordier, B. Poumellec, L. Dong, J. F. Bayon, H. Poignant, E. Delevaque, Densification involved in the UV-based photosensitivity of silica glasses and optical fibers. J. Lightwave Technol. **15**(8), 1329–1342 (1997)
30. D.M. Krol, J.W. Chan, T.R. Huser, S.H. Risbud, J.S. Hayden Fs-laser fabrication of photonic structures in glass: The role of glass composition. Proc. SPIE **5662**, 30–39 (2004)
31. W.J. Reichman, D.M. Krol, L. Shah, F. Yoshino, A. Araj, S.M. Eaton, P.R. Herman, A spectroscopic comparison of femtosecond-laser-modified fused silica using kilohertz and megahertz laser systems. J. Appl. Phys. **99**(12), (2006)
32. T.K. Gaylord, M.G. Moharam, Analysis and applications of optical diffraction by gratings. Proc. IEEE **73**, 894–937 (1985)
33. S. Baudach, J. Bonse, J. Krueger, W. Kautek, Ultrashort pulse laser ablation of polycarbonate and polymethylmethacrylate. Appl. Surf. Sci. **154–155**, 555–560 (2000)
34. J. Krueger, S. Martin, H. Maedebach, L. Urech, T. Lippert, A. Wokaun, W. Kautek, Femto- and nanosecond laser treatment of doped polymethylmethacrylate. Appl. Surf. Sci. **247**, 406–411 (2005)

35. J. Ihlemann, F. Beinhorn, H. Schmidt, K. Luther, J. Troe, Plasma and plume effects on UV laser ablation of polymers. Proc. SPIE, **5448**, 572–580 (2004)
36. W. Kautek, J. Krüger, M. Lenzner, S. Sartania, C. Spielmann, F. Krausz, Appl. Phys. Lett. **69**, 3146 (1996)
37. H.C. Guo, H.B. Jiang, Y. Fang, C. Peng, H. Yang, Y. Li, Q.H. Gong, J. Opt. A **6**, 787 (2004)
38. A. Baum, P.J. Scully, W. Perrie, D. Jones, R. Issac, D.A. Jaroszynski, Pulse-duration dependency of femtosecond laser refractive index modification in poly(methyl methacrylate). Opt. Lett. **33**, 651–653 (2008)
39. D.N. Nikogosyan Multi-photon high-excitation-energy approach to fibre grating inscription. Meas. Sci. Technol. **18**, R1–R29 (2007)
40. C. Wochnowski, Y. Hanada, Y. Cheng, S. Metev, F. Vollertsen, K. Sugioka, K. Midorikawa, Femtosecond-laser-assisted wet chemical etching of polymer materials. J. Appl. Polym. Sci. **100**, 1229–1238 (2006)
41. C. Schaffer, A. Brodeur, E. Mazur, Meas. Sci. Technol. **12**, 1784 (2001)
42. J. Liggat, in *Polymer Handbook*, 4th edn., ed. by J. Brandrup, E.H. Immergut, E.A. Grulke, A. Abe, D.R. Bloch, (Wiley, 2005), II/456
43. M.A. Wochnowski, S. Eldin, S. Metev, UV-laser-assisted degradation of poly (methylmethacylate). Polym. Degrad. Stab. **88**, 2975–2978 (2005)
44. G.B. Blanchet, P. Cotts, C.R. Fincher, Incubation: Subthreshold ablation of poly-(methyl methacrylate) and nature of the decomposition pathways. J. Appl. Phys. **88**, 2975–2978 (2000)
45. E. Süske, T. Scharf, H.-U. Krebs, E. Panchenko, T. Junkers, M. Egorov, M. Buback, H. Kijewski, Tuning of cross-linking and mechanical properties of laser-deposited poly(methyl methacrylate) films. J. Appl. Phys. **97**(063501), 1–4 (2005)
46. T.G. Fox, S. Loshaek, Influence of molecular weight and degree of crosslinking on the specific volume and glass temperature of polymers. J. Polym. Sci. **XV**, 371–390 (1955)
47. S. Küper, S. Modaressi, M. Stuke, J. Phys. Chem. **94**, 7514 (1990)
48. A. Zoubir, M. Richardson, L. Canioni, A. Brocas, L. Sarger, Optical properties of infrared femtosecond laser-modified fused silica and application to waveguide fabrication. J. Opt. Soc. Am. B **22**(10), 2138–2143 (2005)
49. I. Zailer, J.E.F. Frost, V. Chabasseur-Molneux, C. J.B. Ford, M. Pepper, Crosslinked PMMA as a high-resolution negative resist for electron beam lithography and applications for physics of low-dimensional structures. Semicond. Sci. Technol. **11**, 1235–1238 (1996)
50. V. Lucarini, J.J. Saarinen, K.E. Peiponen, E.M. Vartiainen, *Kramers-Kronig relations in Optical Materials Research* (Springer, Berlin, 2005)
51. S. Tzortzakis, L. Sudrie, M. Franco, B. Prade, A. Mysyrowicz, A. Courain, L. Berge, Self-guided propagation of ultrafast IR laser pulses in fused silica. Phys. Rev. Lett. **87**(21) (2001)
52. Z. Wu, H. Jiang, L. Luo, H. Guo, H. Yang, Q. Gong, Multiple foci and a long filament observed with focused femtosecond pulse propagation in fused silica. Opt. Lett. **27**(6) (2002)
53. I.M. Burakov, N.M. Bulgakova, R. Stoian, A. Mermillod-Blondin, E. Audouard, A. Rosenfeld, A. Husakou, I.V. Hertel, Spatial distribution of refractive index variations induced in bulk fused silica by single ultrashort and short laser pulses. J. Appl. Phys. **101**, 043506 (2007)
54. A. Saliminia, N.T. Nguyen, S.L. Chin, R. Vallee, The influence of self-focusing and filamentation on refractive index modifications in fused silica using intense femtosecond pulses. Opt. Comm. **241**, 529–583 (2004)
55. N. Uppal, P.S. Shiakolas, M. Rizwan, Three dimensional waveguide fabrication in PMMA using femtosecond laser micromachining system. Micromachining Microfabrication Process Technology XIII. Proc. SPIE **6882**, 68820I (2008)
56. M. Miwa, S. Juodkazis, S. Matsuo, H. Misawa, Femtosecond two-photon stereo-lithography. Appl. Phys. A **73**, 561–566 (2001)
57. W. Watanabe, Femtosecond filamentary modifications in bulk polymer materials. Laser Phys. **19**(2), 342–345 (2009)
58. C. Hnatkovsky, R.S. Taylor, E. Semova, V.R. Bhardwaj, D.M. Raynor, P.B. Corkum, High-resolution study of photoinduced modification in fused silica produced by tightly focused femtosecond laser beam in the presence of aberrations. J. Appl. Phys. **98**, 01357 (2005)

59. D. Liu, Y. Li, R. An, Y. Dou, H. Yang, Q. Gong, Influence of focusing depth on the microfabrication of waveguides inside silica glass by femtosecond laser direct writing. Appl. Phys. A **84**, 257–260 (2006)
60. Z. Kuang, D. Liu, W. Perrie, S. Edwardson, M. Sharp, E. Fearon, G. Dearden, K. Watkins, High throughput diffractive multi-beam femtoseond laser processing using a spatial light modulator. Appl. Surf. Sci. **255**, 2284–2289 (2008)
61. Z. Kuang, D. Liu, W. Perrie, S. Edwardson, M.C. Sharp, E. Fearon, G. Dearden, K.G. Watkins, Fast parallel diffractive multi-beam femtosecond laser surface micro-structuring. Appl. Surf. Sci. **255**(13–14), 6582–6588 (2009)
62. D. Liu, Z. Kuang, W. Perrie, P.J. Scully, A. Baum, S.P. Edwardson, E. Fearon, G. Dearden, K.G. Watkins, High-speed uniform parallel 3D refractive index micro-structuring of poly(methyl methacrylate) for volume phase gratings. Appl. Phys. B: Lasers Optic **101**(4), 817–823 (2010)
63. H. Kogelnik, Coupled wave theory for thick hologram gratings. Bell Syst. Tech. J. **48**, 2909 (1969)
64. C. Mauclair, G. Cheng, N. Huot, E. Audouard, A. Rosenfeld, I.V. Hertel, R. Stoian, Dynamic ultrafast laser spatial tailoring for parallel micro-machining of photonic devices in transparent materials. Optics Exp. **17**(3531) (2009)
65. Y. Chen, L. Zhang, G. Chen, Fabrication, modification, and application of poly(methyl methacrylate) microfluidic chips. Electrophoresis **29**, 1801–1814 (2008)
66. M. Silva-Lopez, Fender, A., MacPherson, W.N., Barton, J.S., Jones, J.D.C., Zhao, D., Dobb, H., Webb, D.J., Zhang, L., Bennion, I., Strain and temperature sensitivity of a single-mode polymer optical fiber, Opt. Lett. **30**, 3129–3131 (2005)
67. K. Kalli, H.L. Dobb, D.J. Webb, K. Carroll, M. Komodromos, C. Themistos, G.D. Peng, Q. Fang, I.W. Boyd, Electrically tunable Bragg gratings in single-mode polymer optical fiber. Opt. Lett. **32**, 214–216 (2007)
68. H. Dobb, K. Carroll, D.J. Webb, K. Kalli, M. Komodromos, C. Themistos, G.D. Peng, A. Argyros, M.C.J. Large, M.A. van Eijkelenborg, Q. Fang, I.W. Boyd, Grating based devices in polymer optical fibre. Opt. Sens. II **6189**, 18901–18901 (2006)

# Part IV
# Microsystems and Applications

# Chapter 13
# Discrete Optics in Femtosecond Laser Written Waveguide Arrays

Alexander Szameit, Felix Dreisow, and Stefan Nolte

**Abstract** The miniaturization of integrated optical devices can achieve geometric dimensions where crosstalk between adjacent waveguides is no longer negligible. In discrete optics, such interactions can be explained using the well-known coupled mode theory [1]. Based on this fundamental theory, systems of coupled waveguides are reduced to systems of discrete, i.e. countable, cells, and the spatial electromagnetic field is separated into the transverse mode profile of each waveguide and a longitudinal dependent amplitude. In the last years, femtosecond laser waveguide writing has been established as a versatile technique to fabricate three-dimensional optical waveguide systems. The following chapter begins with an overview of the basic principles of discrete light propagation. Linear propagation effects are envisaged in the second section. Here, straight and curved lattices are discussed. In the last section, third order nonlinearity is introduced to comprehend nonlinear propagation effects, which provide essential concepts for the development of ultrafast switching and routing devices.

## 13.1 Introduction to Waveguide Arrays

Waveguide arrays are a particular representation of so-called discrete systems, which analysis reaches back to the early twentieth century when plastic deformations of crystals were investigated [2]. Since then, much attention has been paid to these systems in many different fields in physics, so that discreteness itself evolved into an independent discipline. Discrete systems entered optics in 1965, when arrays of evanescently coupled fibers were analyzed theoretically [3]. The first experimental realization of these geometries was accomplished in 1973 by

A. Szameit (✉) · F. Dreisow · S. Nolte
Institute of Applied Physics, Friedrich-Schiller-Universität Jena, Max Wien Platz 1,
07743 Jena, Germany
e-mail: szameit@iap.uni-jena.de; dreisow@iap.uni-jena.de; nolte@iap.uni-jena.de

investigating waveguides on a GaAs substrate [4]. Over the years it became apparent that arrays of evanescently coupled waveguides are a fruitful model to describe general discrete systems and exhibit a variety of exceptional propagation effects which are not found in homogeneous media [5,6].

Due to the anisotropy of the medium caused by the discreteness, refraction and diffraction are anomalous and periodical [7, 8] in strong contrast to homogeneous media. Furthermore, a periodic self recovery of the light distribution at the incoupling facet can be obtained by introducing potentials for the evanescent coupling between the waveguides in planar arrays [9] or the refractive index of the single waveguides. The latter effect is called optical Bloch oscillation which is an analog to a similar effect in solid states physics concerning electron motion in a periodic potential superimposed by a linear potential. Optical Bloch oscillations [10] were observed in one-dimensional waveguide arrays with linear [11, 12] and curved waveguides [13, 14] as well as in two-dimensional lattices [15]. As a further consequence of the analogy to electron motion, tunneling between the different propagation bands of the systems was proposed and experimentally verified in one-dimensional [16] and two-dimensional [15] waveguides lattices. Another possibility of a self-recovery of an initial field distribution is the discrete Talbot effect, which was theoretically and experimentally analyzed in an infinite planar waveguide array only recently [17]. In recent years, there has been a growing interest in the interaction of the propagating light with boundaries and defects. Recently, an analytical solution for finite waveguide arrays was found [18, 19] by introducing secondary sources. Furthermore, the formation of staggered and unstaggered modes at defects in planar waveguide arrays was investigated in detail [20].

The first connection between nonlinear optics and discrete systems was achieved only in 1982, when the most simple geometry was theoretically investigated, the nonlinear coherent coupler [21]. A connection between optical waveguide arrays and nonlinear optics was identified in 1988, when it was found that a local Kerr nonlinearity yields a compensation of the diffractive linear evanescent coupling, leading to the formation of a discrete spatial soliton in a planar waveguide array [22], which was experimentally verified in 1998 [23]. These initial results paved the way to a variety of investigations. The formation and the nonlinear evolution of light in planar arrays were theoretically discussed in detail [24, 25]. Furthermore, new soliton families were predicted such as discrete breathers [26], hybrid discrete solitons [27], dark solitons [28], or Floquet-Bloch solitons [29], just to mention a few, which can be observed only in discrete systems. Subsequently, also the formation of discrete solitons at array boundaries [30] and interfaces [31] was investigated. During the last years, it has been shown that stable nonlinear solutions for two-dimensional waveguide arrays exist, predicting the existence of two-dimensional discrete solitons [32]. Such phenomena are not possible in homogeneous media since in bulk material the solutions are unstable and collapse. The experimental verification of a two-dimensional discrete soliton was achieved in 2003 in an optically induced lattice in a photorefractive material [33]. Various families of two-dimensional discrete solitons were studied [34], showing a variety of new nonlinear solutions like two-dimensional vortex solitons [35]. However, the interest in two-dimensional nonlinear propagation is much more general, since

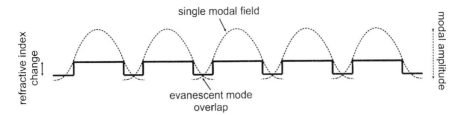

**Fig. 13.1** In a waveguide array given by a periodic refractive index change, the evanescent tails of the modal fields of the individual guides overlap, which causes an energy transfer between adjacent guides

other non-solitary propagation effects like two-dimensional nonlinear Bloch modes [36] were found, demonstrating that the investigation of light evolution in two-dimensional waveguide array is a very promising and fertile field of research.

## 13.2 Fundamental Principles of Discrete Light Propagation

The fundamental feature of waveguide arrays is the periodic modulation of the refractive index $n$ in the transverse directions $x$, $y$. If the index is periodic only in $x$ or $y$, one speaks of a one-dimensional array, otherwise it is called two-dimensional.[1] The idea of the "discrete" light propagation is that the light is localized around particular transverse positions of increased refractive index, where the light is guided and, hence, represents the individual waveguides. Provided that the guides are sufficiently separated, the light fields can be decomposed approximately to the individual fields of each lattice site with exponentially decaying evanescent tails, which slightly overlap between the waveguides, allowing an energy flow also in transverse direction (Fig. 13.1). The entire dynamics of the evolving light field is then given by the conventional longitudinal propagation in the $z$-direction and the transverse energy transfer in the transverse dimensions $x$ and $y$. This results in a symmetry breaking compared to homogeneous materials where all dimensions are equal, allowing for a variety of unexpected phenomena.

The most striking feature of waveguide arrays is the formation of a band structure due to the transverse periodicity [29]. The parabolic dependency between the longitudinal and the transverse wave numbers in homogeneous materials in the paraxial domain is broken into several sub-bands, which are separated by gaps. Light dynamics between the waveguides is only possible for wave numbers residing in the bands, otherwise the transverse energy transport is suppressed and light stays confined in the individual guides.

---

[1]The optical lattices discussed here generally exhibit also an intrinsic longitudinal dimension which is usually omitted in the nomenclature. However, sometimes the notation $(1 + 1)$D and $(2 + 1)$D is used for structures with one or two transverse dimensions, respectively.

The evolution of light in waveguide arrays can be modeled by a coupled mode approach, in which it is assumed that the guides are single-mode, and that the transverse field of an individual guided mode does not change during propagation. The dynamics is then entirely governed by the evolution of the field amplitudes. In the one-dimensional case, the corresponding equations read [1]

$$i\frac{d}{dz}a_m = \varkappa(a_{m-1} + a_{m+1}), \qquad (13.1)$$

where $a_m$ is the amplitude in the $m$th waveguides and

$$\varkappa \sim \omega \int_{-\infty}^{\infty} (\Delta n_{m+1})^2 E_m(x, y) E_{m+1}(x, y) dx dy \qquad (13.2)$$

is the coupling strength between two adjacent guides. The quantity $E_m$ denotes the electric field of the respective guides. In this model, the light evolution in the first band is described, which is given by

$$\beta_z = 2\varkappa \cos(\beta_x d) \qquad (13.3)$$

with the waveguide spacing $d$ and the longitudinal and transverse wave numbers $\beta_z$, $\beta_x$, as shown in Fig. 13.2. In the following, all phenomena occur only in the first band and can therefore be described by the coupled mode approach of (13.1).

In curved lattices the transverse profile $x_0(z)$ causes phase shifts, since light in the outer waveguides with respect to the curvature travels longer paths than in the inner waveguides, which can be explained by an artificial refractive index ramp and mathematically described by an additional term to (13.1) [13].

$$i\frac{da_m}{dz} + \varkappa(a_{m+1} + a_{m-1}) = \frac{2\pi n_0 d}{\lambda}\ddot{x}_0(z)ma_m \qquad (13.4)$$

This phase term is proportional to the second derivative of the bending profile $\ddot{x}_0(z)$ and the waveguide number $m$. Hence, by tuning this term, i.e. the bending profile,

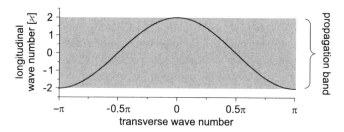

**Fig. 13.2** Band structure diagram of a homogeneous one-dimensional lattice showing the first band. The longitudinal and transverse wave numbers are denoted by $\beta_z$ and $\beta_x$, respectively. The gray area marks the band width of $4\varkappa$

various fundamental phenomena can be investigated and self-imaging devices can be designed [37].

## 13.3 Basic Experimental Techniques

### 13.3.1 Evanescent Coupling

In order to experimentally determine the coupling strength $\varkappa$, one usually considers the simplest case of a coupled waveguide system: the directional coupler, consisting of only two identical parallel waveguides. The corresponding coupled-mode equations (13.1) reduce to

$$\begin{aligned} i\frac{d}{dz}a_1(z) + \varkappa a_2(z) &= 0 \\ i\frac{d}{dz}a_2(z) + \varkappa a_1(z) &= 0. \end{aligned} \tag{13.5}$$

If only guide 1 is excited, the initial conditions are $a_1(0) = 1$ and $a_2(0) = 0$, yielding the solution

$$\begin{aligned} a_1(z) &= \cos(\varkappa z) \\ a_2(z) &= i\sin(\varkappa z). \end{aligned} \tag{13.6}$$

This allows the definition of the coupling length $z = l_c$, corresponding to the first root of $\cos(\varkappa z)$ where the amplitude vanishes completely in the excited guide. The coupling constant can be obtained by comparing the measured nearfield intensity pattern of the individual guides with length $L$ by $\varkappa = k\pi + [L^{-1}\arctan\sqrt{I_B(L)/I_A(L)}]$, where $I_A$ and $I_B$ denote the intensities of each waveguide. Note that due to the periodic character of (13.6), the coupling constant is given with this method only modulo $\pi$.

The coupling constant is dependent on the waveguide spacing and the wavelength. In femtosecond laser written waveguide couplers fabricated with low numerical aperture objectives and without beamshaping, an additional variation with the angular orientation occurs due to the elliptical shape of the waveguide. The dependency of the coupling constant on the waveguide separation is depicted in Fig. 13.3 for different angular orientations and wavelengths, respectively. The function $\varkappa(d)$ can be widely approximated with an exponential decay. Its value and decay trend vary with wavelength and angular orientation and can be tuned by the laser writing parameters. The results in Fig. 13.3 were obtained for writing of 100 kHz repetition rate, 0.3 μJ pulse energy, 150 fs pulse duration, 1250 μm/s, and the 20× focussing objective had a numerical aperture of 0.45 [38]. Henceforth, these parameters are kept fixed except for the writing velocity, which is used to tune the waveguide properties, in particular the refractive index modulation in the range of $\Delta n$ from $2 \times 10^{-4}$ to $1.3 \times 10^{-3}$ [39].

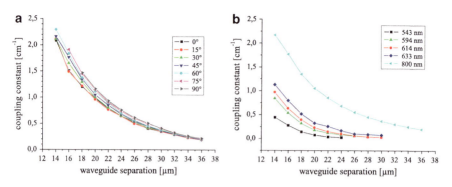

**Fig. 13.3** Measurements of coupling constants dependent on the waveguide separation for (**a**) different angular orientations and (**b**) different wavelengths [38]

## 13.3.2 Waveguide Imaging Microscopy

In Chap. 10, it has been shown that waveguides can be written in active glasses. In such materials, fluorescence light can be used to visualize the light distribution inside waveguide structures. Since the bulk media itself emit this fluorescence, the signal-to-noise ratio is however low. Similar problems occur in etched fluorescent polymer waveguides [16]. A simple technique to access the intensity inside the samples is to use the nonbridging oxygen hole centers (NBOHCs) (see Chap. 3 and [40, 41]), which are formed during the writing process in fused silica. It is therefore advantageous to use a glass with a high content of silanol (SiOH), where many of these color centers are induced [40]. As known from DUV-laser spectroscopy [42], the NBOHC-fluorescence is broad around 650 nm [43]. However, contrary to UV-lasers, ultrashort pulsed lasers can form metastable NBOHCs through nonlinear absorption processes. Beside the absorption maximum in the UV, the NBOHCs provide absorption from 2.0 to 2.4 eV, which can be covered by convenient laser sources (e.g., HeNe laser) in the visible spectral range. Observation of the sample overhead with a microscope yields detailed images of the light distribution inside the waveguides (Fig. 13.4). Another advantage of femtosecond laser induced NBOHCs is that they form only in the irradiated region, thus offering high contrast fluorescence data with low background signal. Additional noise and laser scattering can be removed by spectral filtering. Benefiting from these properties, fluorescence waveguide microscopy is an excellent tool to characterize waveguides and visualize the light propagation inside waveguide arrays. As a significant example, this technique is applied to a simple directional waveguide coupler written with 1500 μm/s. A typical microscope image is shown in Fig. 13.5a for a waveguide spacing of $d = 16\,\mu m$. A measurement of each waveguide's intensity prove that the images can be used for quantitative investigation. The blue line in graph Fig. 13.5b constitutes the waveguide losses of 0.5 dB/cm which are

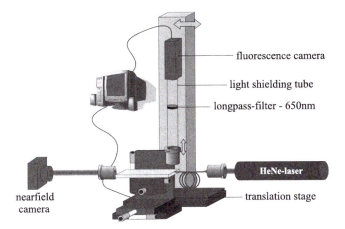

**Fig. 13.4** Nearfield setup for microscope imaging of the waveguide fluorescence. The excitation source is a HeNe laser, and images from the top of the sample and of the nearfield can be recorded by CCD cameras

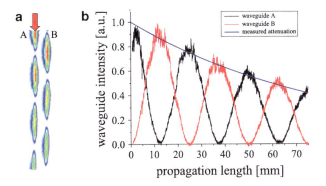

**Fig. 13.5** (a) Fluorescence microscopy image of a directional coupler with $d = 16\,\mu\text{m}$, the arrow marks the excitation of waveguide A (b) measured intensities in each waveguide A and B plotted vs. propagation length and exponentially decaying fit of the total power to determine propagation loss. The excitation wavelength is 633 nm and the waveguide spacing is $16\,\mu\text{m}$

gathered from fitting the coupling curves with the sinusoidal oscillation (13.6) superimposed by an exponential decay.

### 13.3.3 Multi-waveguide Excitation

Conventional methods for exciting waveguides are fiber butt coupling or beam focussing by a microscope objective or lens (see Chap. 7), where light is launched into either one or multiple waveguides by an illumination with a Gaussian intensity

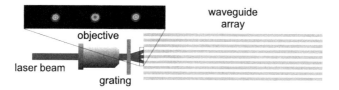

**Fig. 13.6** Schematic experimental setup to excite three waveguides. The DOE is placed between focussing objective and sample, and the inset shows the focal distribution of a tri-pattern. From [46]

distribution. However, in femtosecond laser written waveguide arrays, a broad beam excitation is intricate since the waveguide separation is large with respect to arrays fabricated with other technologies. In addition, peculiar features occur in optical lattices when several distinct locations are excited (see Sect. 13.4.1). Using optical transmission phase gratings, one can focus light to multiple spots (Fig. 13.6). Since only a few percent of the power is converted to higher diffraction orders, the efficiency of such diffractive optical elements (DOEs) can be higher than 95% [41]. When they are placed between the focussing objective and sample, the separation of the foci can be tuned by changing the grating to focus distance to excite distinct waveguides. Beyond two-spot DOEs, elements creating various distributions can be manufactured [44, 45]. An excitation with phase shifts is also easily possible, if the sample is tilted with respect to the optical axis.

## 13.4 Linear Propagation Effects

### 13.4.1 Straight Lattices

The solution of an infinite 1D waveguide array for single waveguide excitation $a_m(0) = \delta_{m0}$ (corresponding to a point source) reads [3]

$$a_m(z) = i^m J_m(2\varkappa z) \tag{13.7}$$

with $J_m(x)$ as the $m$th-order Bessel function of the first kind. As a consequence, at every root $J_m(2\varkappa z) = 0$ the light intensity drops to zero in the respective waveguide. Additionally, the intensity is strongly modulated during propagation and exhibits two strong side lobes (see Fig. 13.7a, where a fluorescence microscopy image is shown). This behavior is in strong contrast to light propagation in conventional continuous materials, where the intensity vanishes nowhere and diffraction is always normal.

In contrast to infinite arrays, boundary effects occur in finite ones, which can be described using a mirror charge formalism [18]. In case of a semi-infinite array (i.e. all waveguides vanish for $m < 0$), the light reflection can be described by [19]:

13 Discrete Optics in Femtosecond Laser Written Waveguide Arrays

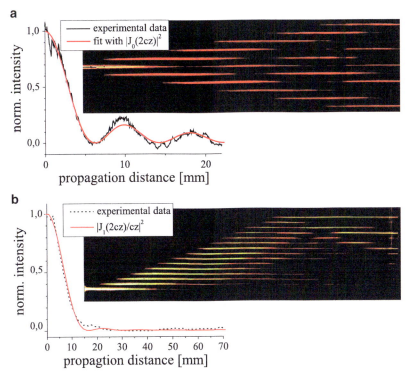

**Fig. 13.7** Straight waveguide arrays excited (**a**) at the center and (**b**) at the boundary of the waveguide. Insets show measured powers of excited waveguides (*black lines*) and are fitted to theoretical predictions (*red curves*). From [47, 48]

$$a_m(z) = i^{m-m'} J_{m-m'}(2\varkappa z) + i^{m+m'} J_{m+m'+2}(2\varkappa z), \tag{13.8}$$

where $m'$ denotes the excited waveguides and $m$ the waveguide under observation. A corresponding fluorescence microscopy image is shown in Fig. 13.7b.

### 13.4.1.1 Second Order Coupling

In most waveguide arrays only coupling to the next neighbors determined by $\varkappa_1$ is regarded. However, for long sample lengths or for closely spaced waveguides, higher order coupling can no longer be neglected. The influence of higher order coupling can be specifically studied using femtosecond laser written 2D zigzag waveguide array structures (Fig. 13.8a). Here, the influence of coupling beyond next neighbors can be tuned at least to the magnitude, where second order exceeds the first order coupling. The influence of second order coupling $\varkappa_2$ has to be taken into account for the design of micro-optical devices, since the relative second order

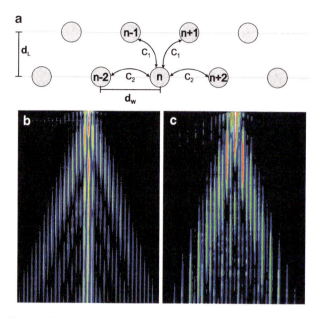

**Fig. 13.8** (a) Sketch of zigzag arrangement. Fluorescence images of waveguide arrays with (b) an in-plane waveguide spacing of $d_W = 24\,\mu m$ and a layer separation $d_L = 12\,\mu m$ yielding 30% second order coupling. (c) Same as (b) but $d_W = 16\,\mu m$, $d_W = 20\,\mu m$ and 280% second order coupling. From [50]

coupling grows exponentially with decreasing waveguide spacing (see (13.2) and Fig. 13.3):

$$\frac{\varkappa_2}{\varkappa_1} = \frac{\exp(-2d/\gamma)}{\exp(-d/\gamma)} = \exp(-d/\gamma), \quad (13.9)$$

where $\gamma$ denotes the strength of the exponential decay. For example, in typical laser inscribed waveguides, $\gamma$ is in the order of a few microns: $1\,\mu m \lesssim \gamma \lesssim 10\,\mu m$. Assuming $\gamma = 5.8\,\mu m$ [48], it turns out that the strength of second order interaction exceeds 10% for a waveguide spacing less than $13\,\mu m$. By means of the zigzag lattice structure [49], one can enlarge this ratio by tuning the angle of the zigzag structure and keeping the projected spacing constant. Interesting scenarios can be studied, e.g., for ratios of $\varkappa_2/\varkappa_1 \approx 0.2$–0.4 the peculiar feature of reduced diffraction is observed Fig. 13.8b.

### 13.4.1.2 Interfaces in Discrete Media

At interfaces of two dissimilar continuous media, the different refractive indices cause reflection and refraction, which are described by Fresnel's formulas and Snell's law. Analogue effects occur at interfaces between two discrete systems, but here two parameters, the coupling constant and different effective refractive indices, cause discrete reflection and refraction [51]. Due to the conservation of momentum, the absolute value of the overall propagation constant $\beta$ must remain constant when light penetrates the interface. This yields after some math the Snell's law for discrete media [51]:

$$\frac{\cos \beta_x^{(2)}}{\cos \beta_x^{(1)}} = \frac{\varkappa_1}{\varkappa_2} - \frac{\delta\beta}{2\varkappa_2 \cos \beta_x^{(1)}}, \qquad (13.10)$$

where $\delta\beta$ is the detuning of the propagation constants and the superscripts denote the two contiguous arrays. The discrete Fresnel formulas for energy reflection and transmission coefficients $R$ and $T$, respectively, are much more complex than their continuous counterparts. For the case of identical propagation constants $\delta\beta = 0$, the Fresnel's law for waveguide arrays simplifies to:

$$R(\delta\beta = 0) = \left| \frac{\varkappa_1 e^{i\beta_x^{(1)}}}{\varkappa_2 e^{i\beta_x^{(2)}}} \right|^2. \qquad (13.11)$$

Assuming energy conservation, the transmission coefficient is given by $T = 1 - R$. By femtosecond laser waveguide array inscription, the coupling constant and the effective index can be simply tuned by modifying the writing velocity [39] and the waveguide separation (see Sect. 13.3.1). Note that the writing velocity modifies both detuning and coupling constant, while the waveguide separation affects only the coupling constant. Two exemplary adjacent arrays without detuning but with different couplings are depicted in Fig. 13.9 to investigate the reflection and refraction behavior at interfaces of discrete media. In Fig. 13.9a, b, light penetrates the interface from an array with weaker coupling to an array with a stronger one. The position of the interface is indicated by the white dotted line, and a slightly increased

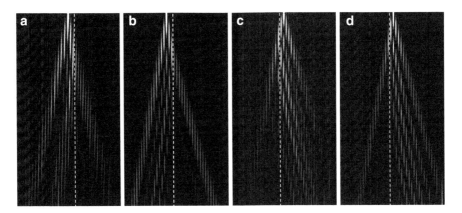

**Fig. 13.9** Dissimilar waveguide arrays with a spacing of 13 μm and 15 μm and homogeneous refractive index modulation of $\Delta n = 2 \times 10^{-4}$. (**a**) Experimental and (**b**) theoretical transmission through an interface to a denser medium ($\varkappa_1 = 1.09\,\text{cm}^{-1}$ and $\varkappa_2 = 1.557\,\text{cm}^{-1}$). (**c, d**) same array as (**a, b**) but for the excitation in the denser fraction [51]

angle of incidence can be found for the transverse wave numbers $\beta_x^{(1)} = \pi/2$ (side lobes). Starting in the denser medium (Fig. 13.9c,d) the angle of incidence is slightly decreased. For both cases the transmission $T$ is diminished with respect to homogeneous lattices, which is clearly depicted by the weaker total power in the nonexcited array.

### 13.4.1.3 Quasi-Incoherent Propagation

The coherence of two point sources and the resulting interference is a fundamental topic in continuous wave optics. The discrete counterpart can be analyzed using a simple DOE, which diffracts light into $\pm 1$st order (see Sect. 13.3.3). Analytically, the intensity $I_m$ in the waveguide $m$ of two point sources at the lattice sites $m'_1$ and $m'_2$ is described by

$$I_m = |a_{m,m'_1}|^2 + |a_{m,m'_2}|^2 + 2\Re\{a_{m,m'_1} a^*_{m,m'_2}\}. \qquad (13.12)$$

The first two terms on the right-hand side contain the intensities of each source, while the latter one is an interference term. Inserting (13.7) into (13.12) the interference term vanishes, if the difference $m_1 - m_2$ is odd [40]. Although excited by coherent light, these two point sources may generate an incoherent propagation pattern. Such "Quasi-Incoherent" propagation occurs for illumination of two neighboring waveguides, and the light pattern is the intensity addition of two sources (Fig. 13.10a). Coherent propagation corresponds to the excitation of two second next neighboring waveguides, resulting in a strong modulation pattern (Fig. 13.10b).

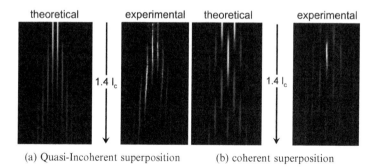

**Fig. 13.10** Theoretical and experimental interference pattern in 50 mm long waveguide arrays with spacings of $d = 20\,\mu$m for two point sources generated by a DOE. The spatial separation of the two foci was adjusted by position of the grating between the sample and microscope objective, so that (**a**) nearest neighbors are excited and (**b**) one waveguide is left for the illumination [40]

## 13.4.1.4 Segmented Arrays

Integrated optical devices can implement various tasks such as self focusing, beam shaping, or self imaging, which is expedient for the miniaturization in telecommunications and data transmission. In particular for self imaging, different approaches have been suggested, i.e., longitudinal zigzag lattices [7], curved arrays [37], and segmented arrays [52]. Zigzag lattices exhibit strong radiation losses due to the abrupt bending, unlike segmented arrays which also work at the array boundaries. The principle idea is to interrupt every second guide at the middle of the sample for a length $d = \lambda/2\Delta n$, which corresponds to an integrated half-wave plate (Fig. 13.11). This half-wave phase shift artificially reverses the evolution in the forward propagation direction by exchanging the slow and fast eigenmodes of the lattice (for details, see [52, 53]) and yields exact image reconstruction at the output plane of the sample. Since an extension to two-dimensional arrays has been also demonstrated (Fig. 13.11d), fiber imaging systems may profit from these image reconstruction mechanisms.

## 13.4.1.5 Two-Dimensional Propagation Effects

Additional peculiarities arise due to the additional dimension and additional degree of freedom in varying the refractive index distribution. Common representatives of two-dimensional waveguide arrays are the square and hexagonal lattice. They have four and six direct neighbors, respectively, and therefore they are distinct from planar lattices. Beyond nearest next neighbors in the square lattice, the second next neighbors are four waveguides at the diagonal lattice positions, which have a distance of $\sqrt{2}\times$ the lattice period. Contrary to 1D structures, the influence of second next neighbors must be accounted for at larger spacings compared to planar

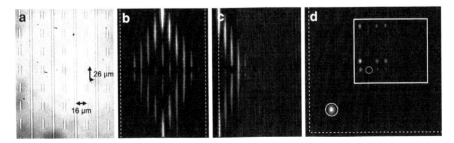

**Fig. 13.11** Segmented waveguide arrays in 50 mm long samples and with a spacing of 26 μm. (**a**) Top view microscopy image of segmented waveguides. The segmentation period is 26 μm and the interrupted length is half as long as the period. Fluorescence measurements of planar waveguide arrays with 100 segmental periods (**b**) excited at the center and (**c**) at the boundary. Self-collimation works in two dimensions as well, which is demonstrated in (**d**). The spacing is decreased to 20 μm. The inset shows the corresponding unsegmented array at a propagation length of 25 mm. From [53]

arrays. Hence, the discrete Schrödinger equation (13.1) has to be extended to its two-dimensional form with eight neighbors:

$$-i\frac{d}{dz}a_{m,n} = \varkappa_1(a_{m-1,n} + a_{m+1,n}) + \varkappa_2(a_{m,n-1} + a_{m,n+1})$$
$$+ \varkappa_3(a_{m-1,n-1} + a_{m+1,n+1}) + \varkappa_4(a_{m-1,n+1} + a_{m+1,n-1}). \quad (13.13)$$

This general equation for 2D lattices can be solved analytically, which is discussed in detail in [19].

Femtosecond laser processing enables the modification of glasses in three dimensions, which exceeds other technologies with homogeneity and ability to form strict boundaries. Therefore, it is ideal to investigate two-dimensional discrete propagation phenomena, as below demonstrated on hexagonal waveguide arrays, when they are excited with a white light source covering the entire visible spectral range. Figure 13.12 shows two arrays with spacings of 16 μm (a) and 20 μm (b), respectively. In the 16 μm spaced array, the red light already touches the boundary, while the green and blue components encounter only the inner waveguides. This behavior is similar in the larger-spaced array, but here due to the overall weaker coupling, the red light does not impact the outer waveguides and the blue remains in the center waveguide.

Furthermore, surface effects and coupling of higher modes are shown in Fig. 13.13, where light is launched into the left outermost waveguide. The light is clearly reflected from the borders. Contrary to expectations, one observes a strong coupling for the blue light, which originates in the multi-mode behavior for green–blue components. In the green and more clearly in the blue component images (Fig. 13.13c, d), the waveguides furthest from the excitation site show in the vertical direction the first higher mode, which is characterized by its zero in the center. Since higher modes are spatially more extended, the coupling will be increased (cf. (13.2)) below the cut-off wavelength for the first mode, which is here around 550 nm.

Another category of 2D waveguide arrays is one where the guides are arranged circularly. This geometry yields an infinite quasi-planar lattice, which ensures

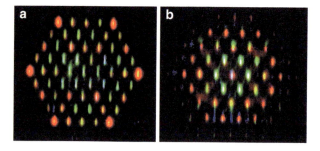

**Fig. 13.12** Hexagonal waveguide arrays with spacings of (**a**) 16 μm and (**b**) 20 μm illuminated with white light and recorded with a color CCD camera. From [54]

13  Discrete Optics in Femtosecond Laser Written Waveguide Arrays         365

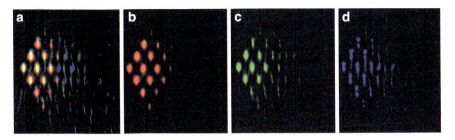

**Fig. 13.13** Hexagonal array with 22 μm spacing. (**a**) Light is launched into the corner waveguide. The decomposition of the RGB image to the (**b**) red, (**c**) green, and (**d**) blue channels depicts the difference in coupling behavior. From [54]

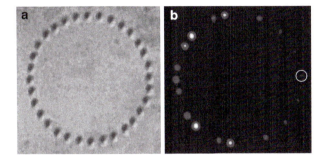

**Fig. 13.14** Quasi infinite planar waveguide array with circular geometry, containing 32 waveguides and with a diameter of 286 μm. (**a**) Microscopy image and (**b**) propagation pattern for an excitation wavelength of 633 nm [38]. White circle indicates coupling location

that no boundary effects occur. Using beam shaping (cf. Chap. 5) to remove the ellipticity of the waveguides, a homogeneous coupling can be achieved (Fig. 13.14).

## 13.4.2  Curved Lattices

The curving of waveguides is desired for the miniaturization of integrated optical components but demands sophisticated fabrication as well. Furthermore, one can benefit from the knowledge of the properties and behavior of bent waveguides [55] for designing curved array structures, which can be applied to artificially expose the system to external potentials. The advantage of femtosecond laser written waveguide arrays is the ability to inscribe along almost arbitrary paths. In the following, we restrict ourselves to negligible bending losses, which require bending radii of at least 30 mm for high refractive index contrast [56] and larger values if the refractive index contrast is lower, e.g., in sinusoidally curved waveguides the maximum curvature radius of 30 mm corresponds to a full transverse shift of 70 μm/cm (Fig. 13.15).

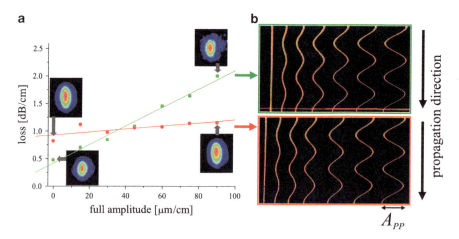

**Fig. 13.15** (a) Curvature losses for waveguides written with 0.22 μJ (*green line*) and 0.28 μJ (*red line*) gained from (b) fluorescence images. From [47]

### 13.4.2.1 Arrays with Periodic Curvature

Dynamic Localization

Well known from solid state physics is dynamic localization, where an AC-field is applied to particles. Contrary to Bloch oscillations (cf. Sect. 13.4.2.2), which occur independently of the field strength, dynamic localization is a resonant effect. The amplitude $F_0$ of the field needs tuning with respect to the external oscillation period $\Lambda$ to match the condition for dynamic localization [57], i.e., of a vanishing effective coupling constant $\varkappa_{\text{eff}} = \varkappa|_{z=\Lambda}$ [58]:

$$\varkappa_{\text{eff}} = \varkappa(\lambda) J_0 \left( \frac{4\pi^2 n_0 d F_0}{\lambda \Lambda} \right), \quad (13.14)$$

with

$$F_0 = \eta \frac{\lambda \Lambda}{4\pi^2 n_0 d}, \quad (13.15)$$

where $\eta$ is any root of the Bessel-function $J_0(\eta) = 0$. The full transverse shift is twice the amplitude, which yields for the first resonance $\eta = 2.405$ weak curvature and therefore low radiation losses in femtosecond laser written waveguide arrays. At $\lambda = 633$ nm, the corresponding propagation pattern including the sinusoidal profile with $\Lambda = 25$ mm and $F_0 = 47$ μm is shown in Fig. 13.16. The light spreads across ~9 waveguides after $\Lambda/2 = 12.5$ mm and is refocused into the initially excited guide at multiples of $\Lambda = 25$ mm. Compared to the self imaging with segmented arrays, one has to note that both exhibit marginal loss, but for a sinusoidally curved structure the overall loss is distributed over the entire sample and for the segmented

13 Discrete Optics in Femtosecond Laser Written Waveguide Arrays

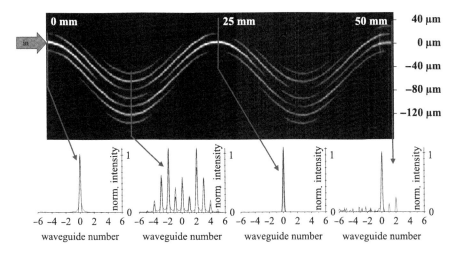

**Fig. 13.16** Dynamic localization: Fluorescence image with two periods of 25 mm and a full amplitude of 95 μm satisfying the dynamic localization condition for 633 nm. The waveguide spacing is 14 μm resulting in a coupling constant $\varkappa = 0.185\,\text{mm}^{-1}$. Additional cross sections are plotted for selected propagation distances (*red lines*) and compared to solutions of the coupled mode equations (*black lines*). From [47]

device only at a confined region loss occurs. Therefore, the segmentation should be preferred for longer transmission lengths.

Polychromatic Self-Imaging

Dynamic localization occurs only at the resonance wavelengths. However, an enlarged spectral operating range would be advantageous for self-imaging devices in telecommunication and data transmission. This can be achieved by using an engineered curvature profile consisting of multiple different harmonic periods, which broadens the spectral width of the self collimation [59]. For a double periodic structure, the first derivative of effective coupling with respect to the wavelength turns to zero, and additionally the second derivative is strongly decreased. Using such structures, self collimation occurs in an extended spectral range, which can span almost one octave [60]. The propagation and spectrally resolved output pattern of arrays, which are adapted to $\lambda = 550\,\text{nm}$ and illuminated by a supercontinuum source covering the full visible range, are depicted in Fig. 13.17. Here, Fig. 13.17 (both center columns) shows single curved profiles, where (13.15) is satisfied using the first and second root of $J_0$, $\eta = 2.405$ and $\eta = 5.520$. Dynamic localization occurs only in a small region around the resonance, while in Fig. 13.17 (right column) the broad output pattern is collimated using a double periodic structure. The self localization is achieved from 400 to 700 nm and can be further extended by multiple structures.

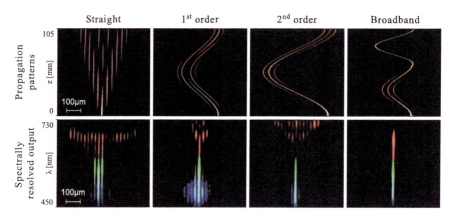

**Fig. 13.17** Polychromatic self imaging in 105 mm long samples with a waveguide separation of 26 μm: Upper row shows propagation images and lower row the spectrally resolved output of (from left to right) straight lattice, 1st order dynamic localization, 2nd order dynamic localization, and broadband self collimation. From [60]

Defect-Free Surface States

Furthermore, in curved lattices it is possible to generate a new type of surface waves, which rely on virtual surface defect caused by the bending in an intrinsically homogeneous array (i.e., all waveguides are identical). One finds that, as long as the condition (13.15) is fulfilled, light is collimated at the array boundary due to resonance of the dynamic localization Fig. 13.18a. Against the expectation, light remains localized also when the period is slightly detuned from the resonance, corresponding to a bound state of the virtual defect [60]. This is illustrated in Fig. 13.18b,c for bending periods of $F/F_0 = 1.05$ and $F/F_0 = 1.10$, respectively. It is important to note that this kind of surface states is of different character than that of the states demonstrated by Tamm [61] and Shockley [62] almost 70 years ago.

### 13.4.2.2 Arrays with Constant Curvature

Bloch Oscillation

The first curved femtosecond laser written waveguide array was demonstrated by Chiodo et al. in 2006 [14]. It has a circular profile and the substrate was an active Er:Yb-doped phosphate glass. The second derivative of the profile yields a constant coefficient on the right-hand side in (13.4) and a linear gradient if the lattice position is included. This linear potential added to the refractive index modulation is analogous to an electric DC-field applied on an electron in an atomic lattice. These quantum particles perform periodic motions, which are known as Bloch oscillations [64]. Like electrons in quantum physics, light waves trapped in

13 Discrete Optics in Femtosecond Laser Written Waveguide Arrays

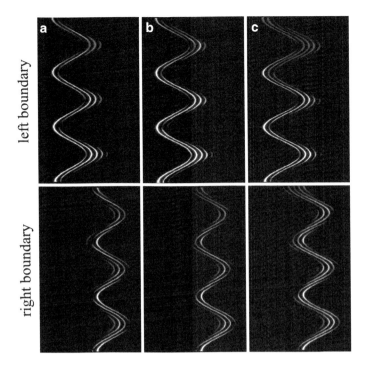

**Fig. 13.18** Surface phenomena in periodically curved arrays with a waveguide spacing of $d = 14\,\mu$m. The surface assists the formation of bound states also in a region around the resonance. The amplitude of the curving profile $F$ was detuned from the resonance amplitude $F_0$ for the left (*upper row*) and right (*lower row*) boundary waveguide. The detuning $F/F_0$ is (**a**) 1.0, (**b**) 1.05, and (**c**) 1.1. From [63]

optical waveguides can oscillate periodically. A waveguide array showing Bloch oscillations with 2.5 periods is depicted in Fig. 13.19b. To enlighten the collimation effect, the propagation in a similar but straight lattice is shown in Fig. 13.19a.

Bloch-Zener Dynamics

Bloch-Zener dynamics appear when the band structure shown in Fig. 13.2 splits into two minibands separated by a small gap $\delta\beta$. At the edge of the Brillouin zone such minibands come close together and tunneling can occur between them. Such a structure can be realized in a binary waveguide array, where for instance every second effective index is increased by tuning the writing velocity. A constant curvature similar to that in Fig. 13.19 enables the Bloch-Zener dynamics. From fluorescence images, the tunneling behavior can be spatially resolved and band occupancies can be measured (cf. Fig. 13.20).

**Fig. 13.19** Bloch oscillations: Measured fluorescence pattern corresponding to the central waveguide excitation of (**a**) straight array with period 10 μm, (**b**) curved array with radius of curvature $R = 7.7$ cm and waveguide separation $a = 10$ μm, (**c**) curved array with $R = 7.7$ cm and $a = 8$ μm. From [14]

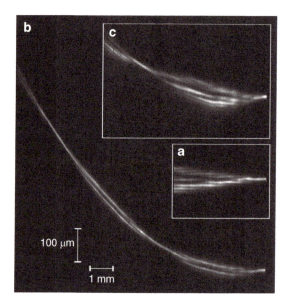

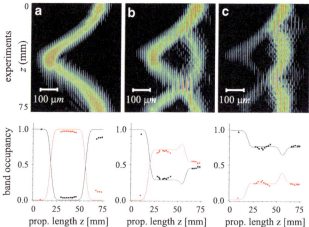

**Fig. 13.20** Bloch-Zener dynamics in a curved binary waveguide array with a waveguide spacing of 11 μm and a curvature radius $R = 1.90$ m corresponding to a Bloch period of 75 mm. *Upper row*: fluorescence images; *Lower row*: measured band occupancy by integrating over the spatially separated beams corresponding to the minibands. (**a**) No band gap, (**b**) small band gap $\delta\beta = 0.44\varkappa$, and (**c**) large band gap $\delta\beta = 0.82\varkappa$. From [65]

### 13.4.3 Quantum-Mechanical Analogies

A fundamental application of waveguide arrays is their utilization as optical models for quantum mechanics or solid state physics. Dynamic localization and Bloch oscillations are examples of such models and have been discussed above already as self-imaging devices. The analogy of the different areas arises due to the same mathematical structure of the Schrödinger equation

$$\left(i\hbar\frac{\partial}{\partial t} + \frac{\hbar^2}{2m}\frac{\partial^2}{\partial x^2} - V\right)\Psi = 0 \qquad (13.16)$$

and paraxial wave equation [66]

$$\left(i\lambda\frac{\partial}{\partial z} + \frac{\lambda^2}{2n_0}\frac{\partial^2}{\partial x^2} + \Delta n\right)E = 0. \qquad (13.17)$$

Comparing (13.16) and (13.17), the correspondence of the wave function $\Psi$ and electric field $E$, reduced Planck's constant $\hbar$ and reduced wavelength $\lambda$ as well as mass $m$ and bulk refractive index $n_0$ is obvious. While a negative quantum mechanical potential $-V$ is necessary to achieve bound states, the optical refractive index modulation $\Delta n$ must be positive to guide waves. But the major difference between both equations is the derivation to time $t$ and space $z$ in the first terms, respectively. Hence, the temporal evolution of the wavefunction is mapped onto space in optics and rapid processes can be modeled enduringly without the need of high time resolutions. Discretizing (13.17) using the coupled mode approach yields (13.1). Additionally, external forces can be included by curving the waveguides, and due to a coordinate transformation the force is mimicked in optics (cf. (13.4)). Such external forces have been already discussed in Sect. 13.4.2 as curved waveguide arrays and can be regarded as analogies of the solid state phenomena Dynamic localization, Bloch oscillations, and Bloch-Zener tunneling. In any case, femtosecond laser waveguide writing is an ideal and fast sample fabrication procedure to prepare quantum analogy models. Further details on quantum–optical analogies can be found in [67].

Landau-Zener Dynamics

The Landau-Zener model describes the dynamics of a crossing two level system. In short, the "speed" of the crossing is responsible for the output occupation of the system. Two regimes have to be considered: diabatic (fast) and adiabatic (slow) transition. Let us first discuss the diabatic case. A system shall be initially in one of its eigenstates. An infinitesimal fast change of the Hamiltonian does not allow the eigenfunction to evolve, and the final state consists of a linear combination of the initial eigenstates. The contrary case is the slow evolution of the Hamiltonian, where the eigenfunctions can follow adiabatically. The final state has new eigenfunctions corresponding to the final Hamiltonian. For Landau-Zener dynamics, this means that a slowly varying potential is applied to our system.

In the optical waveguides, such a potential can be achieved as demonstrated above in curved space. Contrary to a constant field for Bloch oscillations, we need a time-dependent variation, which means that the second derivative on the right-hand side of (13.4) has to be dependent on $z$ and therefore $x_0 \propto z^3$. The corresponding equations read in the curved coordinate system as:

$$i\frac{d}{dz'}\begin{pmatrix} a_1 \\ a_2 \end{pmatrix} = \begin{pmatrix} \eta^2 z' & \varkappa_0 \\ \varkappa_0 & -\eta^2 z' \end{pmatrix}\begin{pmatrix} a_1 \\ a_2 \end{pmatrix}, \quad (13.18)$$

where $\eta$ is a measure for the strength of the curvature, and solutions of this system can be written in terms of parabolic cylinder functions. Waveguide couplers with the desired cubic profile have been fabricated, and Landau-Zener tunneling has been observed. The spatially resolved behavior is illustrated in Fig. 13.21. In the region, where the energies approach each other the system fades into resonance and both states can interact. After the crossing, the resonance is constrained again and the typical oscillatory settling to the final occupation sets in.

Photon Tunneling and the Optical Zeno Effect

Like the quantum Zeno effect, which implies that frequent measurements slow down the decay of a state [69], decay deceleration can be observed in waveguide optics as well. A state represented by a single waveguide is observed multiple times, where the observation is mimicked by coupling to a semi-infinite waveguide array. The multiple observations are created by modulating the coupling between the waveguide and the continuum. This means at points of strong coupling an

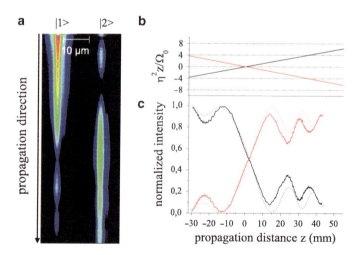

**Fig. 13.21** Landau-Zener dynamics in a cubically curved waveguide coupler with a spacing of 17 μm corresponding to a coupling constant of $\varkappa = 0.063\,\text{mm}^{-1}$: (**a**) Curved space fluorescence measurement, (**b**) applied energy level crossing, and (**c**) measured and calculated spatially resolved occupation of the states $|1\rangle$ and $|2\rangle$. From [68]

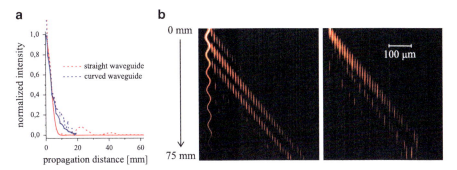

**Fig. 13.22** Photon tunneling experiments showing the optical Zeno effect. (**a**) Measured intensities plotted vs. interaction length, which corresponds to a propagation length normalized to the coupling constant. The *blue line* depicts the modulated state and it clearly turns out that the decay is slower (Zeno effect). (**b**) Recorded fluorescence images showing the decay into the continuum. From [48]

"observation" takes place, while a straight waveguide corresponds to permanent observation. Such a regime is illustrated in Fig. 13.22 for femtosecond laser written waveguides. The modulated waveguide decays faster than the straight one (Fig. 13.22a), with respect to the interaction length as a measure of normalized coupling. Furthermore, the Anti-Zeno effect and fractional decay suppression exist in optics, too. These schemes can be achieved by varying the effective index of the waveguide with respect to the semi-infinite array [48].

## 13.5 Nonlinear Propagation Effects

Intrinsically, for high input powers the light evolution enters another regime, since nonlinear terms in the propagation equations become significant. In this section, we will describe how nonlinearity impacts the evolution of high power ultrashort laser pulses in femtosecond laser written waveguide arrays. Section 13.5.1 deals with the modification of the nonlinear refractive index in the individual waveguides due to the femtosecond-laser radiation. In Sects. 13.5.2 and 13.5.3, we explain the mechanism of discrete spatial soliton formation in laser written structures in one- and two-dimensional structures, respectively.

### 13.5.1 Nonlinear Refractive Index

As described in Chap. 1, during the writing process the molecular structure of the material is altered yielding a change of the linear refractive index $n$. However, it is evident that after the modification of the bulk, also the nonlinear properties

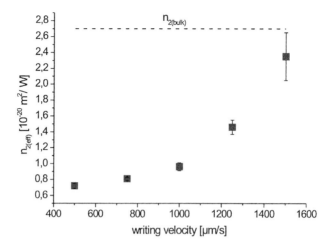

**Fig. 13.23** Dependence of the effective nonlinear refractive index $n_{2(\text{eff})}$ on the writing velocity for fixed pulse duration of 150 fs and pulse energies of 0.3 µJ. From [39]

of refraction are influenced. Following the definition given in [70], the nonlinear refractive index reads as

$$n_2(\omega) = \frac{3}{8} \frac{\Re\{\chi^{(3)}(-\omega; \omega, -\omega, \omega)\}}{n_0(\omega)}, \qquad (13.19)$$

where $\chi^{(3)}(-\omega; \omega, -\omega, \omega)$ is the third-order susceptibility tensor. Hence, the generalized index of refraction can be expressed by

$$n(\omega) = n_0(\omega) + n_2(\omega)|E|^2. \qquad (13.20)$$

It turns out that the modified nonlinear refractive index is a function of the writing velocity, which is shown in Fig. 13.23. While for high writing speeds, the nonlinearity is almost unaffected and reaches the value of the unprocessed bulk material ($n_{2(\text{bulk})} = 2.7 \times 10^{-20}$ m$^2$/W at $\lambda = 800$ nm [71]), with decreasing velocity the nonlinear refractive index drops to $n_{2(\text{eff})} = 0.25 n_{2(\text{bulk})}$ [39]. Therefore, the effective nonlinearity is a parameter which can be precisely tuned by choosing appropriate writing parameters.

### 13.5.2 One-Dimensional Solitons

Including the nonlinearity in the propagation of light waves yields the Kerr lens effect, which focuses the traveling high-intensity beam [72] and enables the formation of optical solitons [73]. However, the formation of discrete solitons

## 13 Discrete Optics in Femtosecond Laser Written Waveguide Arrays

physically corresponds to the nonlinear Kerr lens but differs in the underlying mechanism. The propagation in planar waveguide arrays at high powers can be modeled using the nonlinear extension of (13.1)

$$\left(i\frac{\partial}{\partial z} + \gamma |a_m(z)|^2\right) a_m(z) + \varkappa [a_{m+1}(z) + a_{m-1}(z)] = 0, \quad (13.21)$$

which is the generalized nonlinear version of (13.1). The quantity

$$\gamma = \frac{n_0 \varepsilon_0 \omega}{2 P_0} \iint n_2(\mathbf{r}, \omega) \mathrm{d}x \mathrm{d}y \quad (13.22)$$

determines the strength of the nonlinearity in the individual waveguides. With increasing input power, the nonlinearity of the system becomes significant so that for a valid description of the light propagation the nonlinear term in (13.21), introducing an additional contribution to the propagation constant, can no longer be neglected. Hence, increasing the power moves the propagation constant of the excited waveguide into the semi-infinite gap above the first band, which prevents the light from coupling from the excited waveguide into the adjacent ones (Fig. 13.24).

Since in fused silica waveguides one usually works with an electronic $\chi^{(3)}$ nonlinearity, which exhibits an almost instantaneous response but is rather weak, a high input peak power is required for the excitation of discrete solitons in the fabricated waveguide lattices. Hence, for the demonstration of the nonlinear discrete self-focusing in laser written waveguide arrays, a pulsed Ti:Sapphire CPA laser system (Spitfire, Spectra-Physics) with a pulse duration of about 150 fs, a repetition rate of 1 kHz and pulse energies of up to 3 µJ at 800 nm was used, since such pulses exhibit an extraordinary high peak power allowing for the formation of discrete solitons inside the fs written waveguide arrays in fused silica. The light was coupled into the center waveguide with a 4× microscope objective (NA = 0.10), coupled out by a 10× objective (NA = 0.25) and projected onto a CCD camera. The use of the

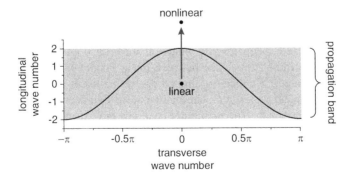

**Fig. 13.24** With increasing input power, the propagation constant of the excited waveguide is shifted into the band gap, forming an isolated state

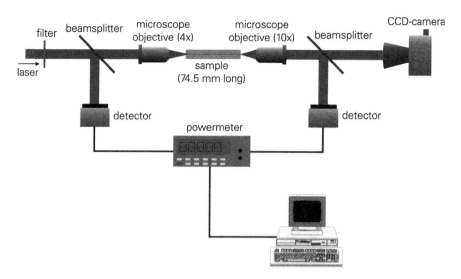

**Fig. 13.25** Setup for the investigation of nonlinear propagation effects in the waveguide array. From [74]

looser focus of the incoupling objective with respect to the writing objective ensured that pulses with a higher peak power could be launched into the array before the damage threshold of the sample was reached. The power of the beam was measured before the incoupling and after the outcoupling objective using a beam splitter, shown in Fig. 13.25. Using this setup, the power coupled into the waveguides was calculated considering Fresnel losses, dumping and the overlap integral of the focal spot of the incoupling microscope objective, and the shape of the propagating mode in every waveguide.

Homogeneous Array

For the demonstration of a one-dimensional discrete soliton in a homogeneous array, a planar 74.5 mm long array consisting of nine waveguides with a separation of 48 μm was fabricated. A microscope image is shown in Fig. 13.26. The experimental task arises from the balance between linear coupling, input peak power, and propagation length. The latter is limited due to the losses of the waveguides, which are approximately 0.4 dB/cm, but accumulate to significant values after some centimeters of propagation. The maximum peak power of the incoupled pulses is limited by the damage threshold of bulk fused silica. Therefore, the waveguides are buried 0.5 mm away from the incoupling facet, since the sample surface has a significantly lower damage threshold than the bulk material. This reduces the applied fluence at the sample surface and allows the incoupling of pulses at a substantially higher peak power into the waveguides.

13 Discrete Optics in Femtosecond Laser Written Waveguide Arrays 377

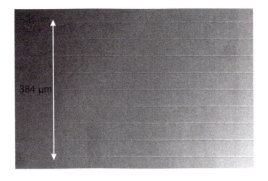

**Fig. 13.26** Microscope image of a planar waveguide array, consisting of nine waveguides with a separation of 48 μm. On the left side the waveguides are buried away from the facet. From [74]

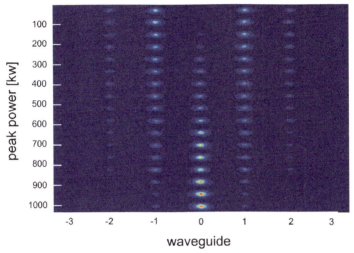

**Fig. 13.27** Measured output pattern as a function of the input peak power. For low peak power, the light diffracts over several waveguides, while for high power a localization in the excited waveguide can be observed. From [74]

Localization is obtained for increasing peak power, which is shown in Fig. 13.27. The output pattern is shown as a function of the peak power. While in the lower peak power range almost all of the guided energy is coupled to the adjacent waveguides due to linear coupling, at a peak power of 1000 kW the output intensity pattern is localized in the center waveguide which has been excited at the entrance [74].

A comparison of the experimental data with a numerical evaluation of (13.21) is shown in Fig. 13.28, which exhibits an excellent agreement in the dependence of the amplitudes in the individual waveguides. It is important to note that for the numerical modeling a reduced nonlinear coefficient of $n_{2(\text{eff})} = 0.25 n_{2(\text{eff})}$ was used, which is consistent with Fig. 13.23 since the array was fabricated with a writing velocity of 500 μm/s [74].

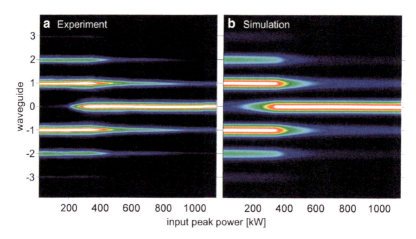

**Fig. 13.28** Comparison of the evolution of the output pattern between the experimental data (*left*) and the numerical analysis (*right*). The amplitude in the $n$th waveguide is shown as a function of the input peak power. From [74]

### Diffraction-Managed Solitons by Curving the Lattice

As shown in Sect. 13.4.2, the beam diffraction in waveguide arrays can be effectively modified using curved lattices, introducing an external potential on the light evolution. The question arises on how a curvature influences the properties of the nonlinear light propagation. We investigated this issue in sinusoidally bent structures, yielding dynamic localization. We fabricated two samples consisting of 13 waveguides with spacings of 34 μm and 40 μm to compare different widths of the light diffusion (Fig. 13.29a, b). The waveguide bending amplitudes are chosen to satisfy the dynamic localization condition (13.15), with $F_0 = 104$ μm and $88$ μm for spacings of 34 μm and 40 μm, respectively, where each sample was L = 105 mm long. The nonlinear dynamics in these structures is depicted in Fig. 13.29.

At low peak powers, in the essentially linear propagation regime, we observe dynamic localization of light in the curved waveguides. When the peak power of the input beam is increased, we observe transitional beam broadening, in excellent agreement with the numerical simulations presented above. The beam experiences significant self-induced broadening (see Fig. 13.29a, b, $P = 1$ MW) because the nonlinearity destroys the dynamic localization condition by changing the refractive index of the waveguide. With propagation, the beam broadens and its intensity is reduced accordingly, such that the effect of the nonlinearity becomes weaker. Since the linear discrete diffraction is fully suppressed in the curved waveguide arrays, the beam broadening stops when the average beam width reaches a certain value.

At higher input peak powers, nonlinear self-trapping of the beam to a single lattice site eventually occurs (see Fig. 13.29a, b, bottom). We found that the peak power required for the formation of the diffraction-managed solitons in curved waveguide arrays is more than two times higher than the critical peak power of

13 Discrete Optics in Femtosecond Laser Written Waveguide Arrays 379

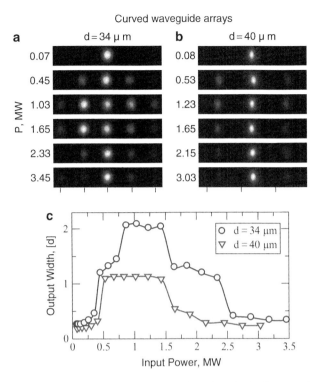

**Fig. 13.29** Comparison of the evolution of the output beam broadening between the experimental (**a**) strong and (**b**) weak coupling. (**c**) The output width is shown as a function of the input peak power. From [75]

lattice solitons in exactly the same but straight waveguide arrays. Therefore, the required peak power for soliton formation can be precisely tuned by changing the linear diffraction properties of the waveguide arrays [75].

Diffraction Management by Longitudinal Detuning

A different concept of influencing the linear diffraction behavior of a propagating beam in a waveguide array is a periodic modulation of the refractive index increase along the longitudinal direction, which is $\pi$-shifted in adjacent waveguides. This also results in the collapse of the entire band structure at a particular resonance frequency and a given amplitude, and therefore a suppressed coupling of the light. This concept holds for waveguides close to the surface and in the bulk of a waveguide array. The according coupled mode equations read

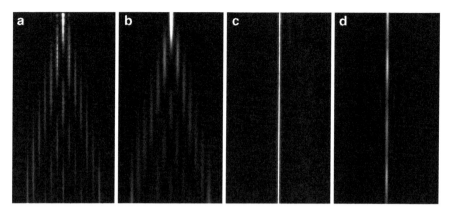

Fig. 13.30 In (a) and (b), the experimental and theoretical propagation patterns inside a homogeneous array are shown. Suppression of the coupling between adjacent guides due to the modulation of the refractive index is shown experimentally in (c) and theoretically in (d). From [76]

$$\left(i\frac{\partial}{\partial z} + (-1)^m \Delta n_{\mathrm{mod}} \sin\{\Omega z\} + \gamma \left|a_m(z)\right|^2\right) a_m(z) + \quad (13.23)$$

$$+ \varkappa \left[a_{m+1}(z) + a_{m-1}(z)\right] = 0, \quad (13.24)$$

where $\Delta n_{\mathrm{mod}}$ is the modulation amplitude and $\Omega$ is the modulation frequency. In our case, we fabricated waveguide arrays with a length of 105 mm and a spacing of 14 μm for the linear experiments and 38 μm for the nonlinear experiments. The index of refraction in the guides was approximately $\Delta n = 4 \times 10^{-4}$. The periodic modulation of the refractive index is achieved by appropriately varying the writing speed of the waveguides. When the modulation amplitude is $\Delta n_{\mathrm{mod}} = 8 \times 10^{-5}$, the corresponding resonance frequency of the modulation in the linear limit and at an excitation wavelength of $\lambda = 633$ nm is $\Omega_{\mathrm{res}} = 0.53$ mm$^{-1}$. In Fig. 13.30, the experimental propagation images (obtained using the fluorescence microscopy method) are shown.

In order to study soliton formation in such structures, we excited the arrays with pulsed radiation at $\lambda = 800$ nm. The modulation amplitude was $\Delta n_{\mathrm{mod}} = 6 \times 10^{-5}$, and the resonance frequency was about $\Omega_{\mathrm{res}} = 0.4$ mm$^{-1}$. The results are summarized in Fig. 13.31. Column (a) shows the beam dynamics in a homogeneous array, while column (b) illustrates linear tunneling inhibition in a modulated array, partial delocalization at an intermediate peak power level, and finally soliton formation at high input peak power. At low peak power, the resonant light propagation results in the complete inhibition of light diffraction. For an increased (intermediate) peak power, the nonlinear influence distorts tunneling inhibition, so that light can couple from the excited into the adjacent guides. Eventually, soliton formation occurs at high input peak power, which is again larger than those in a homogeneous array [76]. Such diffraction-managed solitons may open possibilities for independent control over the strength of diffraction and the nonlinear localization peak power,

# 13 Discrete Optics in Femtosecond Laser Written Waveguide Arrays

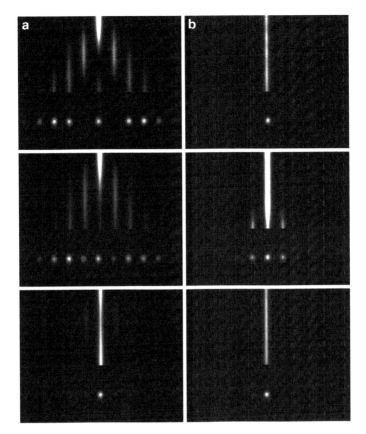

**Fig. 13.31** The beam evolution at low (*first row*), medium (*second row*), and high (*third row*) input power is shown. In (**a**) the array is homogeneous, while in (**b**) the refractive index is modulated. From [76]

being applicable to different physical systems with attractive and repulsive nonlinear interactions.

## *13.5.3 Two-Dimensional Solitons*

The next consequential step in the investigation of spatial nonlinear propagation effects in femtosecond laser written waveguide arrays is the excitation of a two-dimensional discrete soliton. With a pure Kerr nonlinearity, these entities cannot exist in homogeneous materials where they are unstable. In previous experiments, showing 2D discrete solitons in photorefractive crystals [33] and fibers [77], the coupling between the waveguides was either isotropic or statistical. Femtosecond laser written waveguide arrays exhibit the unique possibility of creating a homogeneous anisotropic coupling, giving the opportunity to investigate the formation of two-dimensional discrete solitons in such systems. In this section, we will analyze the formation of two-dimensional discrete solitons in straight and homogeneous arrays in the bulk and close to a surface or interface.

Homogeneous Array

To demonstrate 2D soliton formation in the bulk, a square $5 \times 5$ lattice was used with a length of 74.4 mm and a waveguide separation of 40 μm. The writing velocity was chosen to be 1250 μm/s to obtain a higher effective nonlinear refractive index compared to Sect. 13.5.2. Again, to avoid damage of the device when exciting with high peak power laser pulses, the waveguides are terminated 0.5 mm away from the incoupling facet.

The change in the discrete output pattern with increasing peak power is demonstrated in Fig. 13.32. In Fig. 13.32a, the output pattern in the linear case at $P_{peak} \approx$ 40 kW is shown. At $P_{peak} \approx 700$ kW (see Fig. 13.32b), the linear coupling is reduced and at $P_{peak} \approx 1,000$ kW, depicted in Fig. 13.32c, almost all of the guided energy remains in the excited waveguide [78].

Surface Solitons

The formation of discrete surface solitons is subject to somewhat different physical factors than for discrete solitons without any boundary interaction. Propagation bands and bandgaps are a direct consequence of the periodicity of an infinite waveguide array, yielding the existence of discrete solitons when the propagation constant of the traveling modes are shifted by the nonlinear influence from the bands into the bandgaps. The required periodicity is also approximately satisfied in a finite

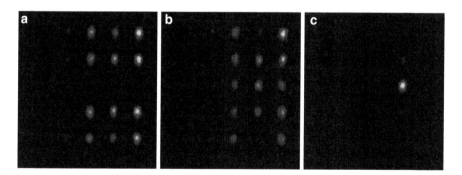

**Fig. 13.32** Measured array output intensity patterns as a function of the input peak power. (**a**) shows the linear output pattern measured at $P_{peak} = 40$ kW, (**b**) the measurement at $P_{peak} = 700$ kW, and in (**c**) peak power is $P_{peak} = 1,000$ kW. From [78]

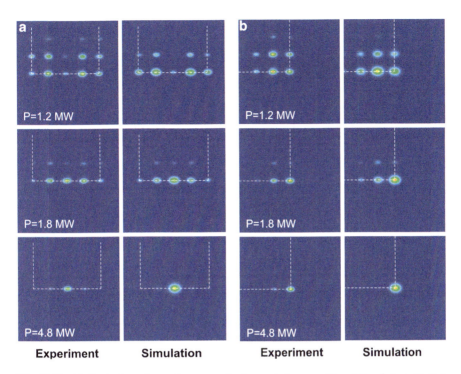

**Fig. 13.33** (a) Excitation of a surface wave in the central waveguide of the first array row. (b) Excitation of a surface wave in the corner waveguide of the array. The *dashed lines* indicate the array surfaces position. From [79]

lattice, when the influence of the boundary on the propagation of the mode can be neglected. The situation completely changes in the presence of a surface, since the periodicity is abruptly interrupted. The former homogeneous band structure is splitting up depending on the distance to the surface, so that the simplified picture of shifting the propagation constant from the band into the bandgap is no longer valid.

For the excitation of discrete surface lattice solitons, a $5 \times 5$ waveguide array with a length of 74.4 mm and a waveguide separation of 40 μm was fabricated at a writing velocity of 1250 μm/s. In Fig. 13.33, the experimental data are shown for surface and edge excitation, and compared to simulations. For an input peak power of $P_{peak} = 1.2$ MW, which corresponds to linear propagation, the output pattern broadens due to discrete diffraction, and almost all of the power in the excited waveguides has been transferred into the adjacent ones. The linear coupling is weakly anisotropic due to the elliptical shape of the waveguides (cf. Sect. 13.3.1). At $P_{peak} = 1.8$ MW, a slightly localized intermediate state is observed. Eventually, at $P_{peak} = 4.8$ MW the light is almost completely trapped in the form of a discrete soliton in the excited waveguide [79].

**Fig. 13.34** Microscope image of a waveguide array exhibiting two different regions with *square* (*left*) and *hexagonal* (*right*) topology. The waveguide excited in the experiments (see Fig. 13.35) is marked with a *red circle*. From [80]

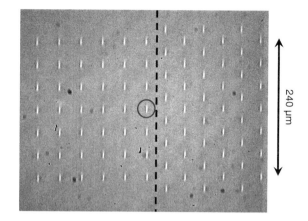

Interface Solitons

A slightly different configuration is the interface between two dissimilar waveguide arrays, where solitons may also form [80]. For the experimental investigation of such 2D interface lattice solitons, we fabricated arrays consisting of hexagonal and square regions with a waveguide separation of 40 μm and a length of 105 mm, where the hexagonal array had a stronger refractive index change than the square one. A microscope image of the array is shown in Fig. 13.34.

Generally the spatial soliton profile will penetrate deeper into the hexagonal array than into the square one. This can be attributed to the fact that the mean refractive index in the hexagonal array is higher due to higher packing density of waveguides, which leads to a slower decay of the soliton tails. Similarly, also for localized linear excitations, the coupling into neighboring guides occurs faster in the hexagonal array. This asymmetry is even more pronounced when the individual waveguides in the hexagonal region exhibit a higher refractive index than those in the square region. A sequence of output intensity distributions for increasing input peak powers is depicted in Fig. 13.35, where a waveguide in the first row of the square array was excited. For low peak powers, the light is almost confined to the square region since it is reflected at the interface (Fig. 13.35a). For increasing power, the light starts to penetrate into the hexagonal region (Fig. 13.35b, c) because nonlinearity increases the refractive index and causes phase matching of both array sections, thereby reducing the reflection at the interface. However, for large input peak powers, one can clearly observe a near-surface localization (Fig. 13.35d, e), so that finally a surface lattice soliton is formed (Fig. 13.35f). As expected, the soliton shape exhibits a characteristic three-point structure due to the penetration in the hexagonal region. The almost perfect symmetry of the observed output pattern in Fig. 13.35 demonstrated the high accuracy of the waveguide writing process. It results in very homogeneous coupling between the waveguides and, additionally, in sharply defined edges and interfaces of the waveguide array. Hence, scattering

13 Discrete Optics in Femtosecond Laser Written Waveguide Arrays                    385

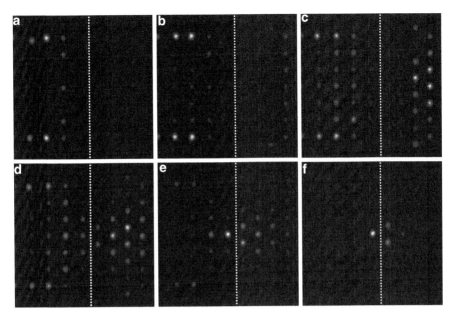

**Fig. 13.35** Sequence of output patterns for light evolution in the vicinity of an interface at (**a**) 62 kW, (**b**) 1.4 MW, (**c**) 2 MW, (**d**) 2.3 MW, (**e**) 2.7 MW, and (**f**) 3.2 MW input peak power. From [80]

or noticeable statistical distortions of the output pattern caused by small waveguide displacements or inhomogeneities are negligible.

## 13.6 Conclusions

In conclusion, we have shown that femtosecond laser written waveguides in integrated optical devices are not only relevant for sophisticated applications but can provide the basis for fundamental research on a general discrete system. Laser written waveguide arrays benefit from homogeneity, the simple fabrication of surface, interfaces, and defects as well as the almost unlimited geometries available in all three dimensions. In addition, the inscription is a fast processing and sample preparation technique. Since color centers form during the writing process in the modified regions, complex light distributions can be analyzed qualitatively and quantitatively by directly monitoring the resulting fluorescence. Thus, a variety of fundamental linear propagation effects which are relevant for discrete optics were constituted, e.g., reflection and refraction at interfaces, the influence of second order coupling, or the discrete interference known as "Quasi-Incoherence". On the basis of versatile methods, different self-imaging devices were fabricated and discussed highlighting their capability for broad spectral functionality. Furthermore, femtosecond laser inscription provides the advantage to fabricate 2D waveguide arrays, which was exhibited on the investigation of polychromatic behavior of hexagonal waveguide lattices. The step to the second dimension provides new peculiar features due to an additional degree of freedom. Beyond the research on fundamental discrete optics and designing integrated optical components, femtosecond laser written waveguide lattices provide a powerful laboratory tool as a system to model quantum mechanics and solid state physics, which has been pointed out by several phenomena, such as Bloch oscillation, dynamic localization, or tunneling dynamics.

When using ultrashort pulses with powers below the threshold for changing the material, femtosecond laser written waveguide arrays provide a sophisticated platform to excite instantaneous nonlinear phenomena. In particular, solitons, which are of fundamental interest, can be excited in various settings. They can be excited in one- and two-dimensional geometries, bulk media, in the vicinity of surfaces, interfaces, and more. The diffraction of light beams can be effectively tuned not only by a curvature of the guides, but also by an interplay of nonlinearity and refractive index landscape of the individual guides. As a further intriguing feature, the nonlinearity itself can be precisely adjusted by the writing parameters.

## References

1. A. Yariv, *Optical Electronics*, 4th edn. (Saunders College Publ., 1991)
2. J. Frenkel, T. Kontorova, J. Phys. (USSR) **1**, 137 (1939)
3. A. Jones, J. Opt. Soc. Am. **55**, 261 (1965)
4. S. Somekh, E. Garmire, A. Yariv, H. Garvin, R. Hunsperger, Appl. Phys. Lett. **22**, 46 (1973)
5. D. Christodoulides, F. Lederer, Y. Silberberg, Nature **424**, 817 (2003)
6. F. Lederer, G. Stegeman, D. Christodoulides, G. Assanto, M. Segev, Y. Silberberg, Physics Reports **463**, 1 (2008)
7. H. Eisenberg, Y. Silberberg, R. Morandotti, J. Aitchison, Phys. Rev. Lett. **85**(9), 1863 (2000)
8. T. Pertsch, T. Zentgraf, U. Peschel, A. Braeuer, F. Lederer, Phys. Rev. Lett. **88**(9), 093901 (2002)
9. R. Gordon, Opt. Lett. **29**(23), 2752 (2004)
10. U. Peschel, T. Pertsch, F. Lederer, Opt. Lett. **23**(21), 1701 (1998)
11. T. Pertsch, P. Dannberg, W. Elflein, A. Braeuer, F. Lederer, Phys. Rev. Lett. **83**(23), 4752 (1999)
12. R. Morandotti, U. Peschel, J. Aitchison, H. Eisenberg, Y. Silberberg, Phys. Rev. Lett. **83**(23), 4756 (1999)
13. G. Lenz, I. Talanina, M. de Sterke, Phys. Rev. Lett. **83**(5), 963 (1999)
14. N. Chiodo, G.D. Valle, R. Osellame, S. Longhi, G. Cerullo, R. Ramponi, P. Laporta, U. Morgner, Opt. Lett. **31**(11), 1651 (2006)
15. H. Trompeter, W. Krolikowski, D. Neshev, A. Desyatnikov, A. Sukhorukov, Y. Kivshar, T. Pertsch, U. Peschel, F. Lederer, Phys. Rev. Lett. **96**(5), 053903 (2006)
16. H. Trompeter, T. Pertsch, F. Lederer, D. Michaelis, U. Streppel, A. Braeuer, U. Peschel, Phys. Rev. Lett. **96**(2), 023901 (2006)
17. R. Iwanow, D. May-Arrioja, D. Christodoulides, G. Stegeman, Y. Min, W. Sohler, Phys. Rev. Lett. **95**(5), 053902 (2005)
18. K. Makris, D. Christodoulides, Phys. Rev. E **73**, 036616 (2006)

19. A. Szameit, T. Pertsch, F. Dreisow, S. Nolte, A. Tuennermann, U. Peschel, F. Lederer, Phys. Rev. A **75**, 053814 (2007)
20. H. Trompeter, U. Peschel, T. Pertsch, F. Lederer, U. Streppel, D. Michaelis, A. Braeuer, Opt. Exp. **11**(25), 3404 (2003)
21. S. Jensen, IEEE J. Quantum Electron. **18**, 1580 (1982)
22. D. Christodoulides, R. Joseph, Opt. Lett. **13**(9), 794 (1988)
23. H. Eisenberg, Y. Silberberg, R. Morandotti, A. Boyd, J. Aitchison, Phys. Rev. Lett. **81**(16), 3383 (1998)
24. H. Eisenberg, R. Morandotti, Y. Silberberg, J. Arnold, G. Pennelli, J. Aitchison, J. Opt. Soc. Am. B **19**(12), 2938 (2002)
25. U. Peschel, R. Morandotti, J. Arnold, J. Aitchison, H. Eisenberg, Y. Silberberg, T. Pertsch, F. Lederer, J. Opt. Soc. Am. B **19**(11), 2637 (2002)
26. S. Flach, C. Willis, Physics Reports **295**, 181 (1988)
27. T. Pertsch, U. Peschel, F. Lederer, Phys. Rev. E **66**, 066604 (2002)
28. Y. Kivshar, W. Krolikowski, O. Chubykalo, Phys. Rev. E **50**(6), 5020 (1994)
29. D. Mandelik, H. Eisenberg, Y. Silberberg, R. Morandotti, J. Aitchison, Phys. Rev. Lett. **90**(5), 053902 (2003)
30. K. Makris, S. Suntsov, D. Christodoulides, G. Stegeman, Opt. Lett. **30**(18), 2466 (2005)
31. K. Makris, J. Hudock, D. Christodoulides, G. Stegeman, O. Manela, M. Segev, Opt. Lett. **31**(18), 2774 (2006)
32. N. Efremidis, J. Hudock, D. Christodoulides, J. Fleischer, O. Cohen, M. Segev, Phys. Rev. Lett. **92**(21), 213906 (2003)
33. J. Fleischer, M. Segev, N. Efremidis, D. Christodoulides, Nature **422**, 147 (2003)
34. N. Efremidis, S. Sears, D. Christodoulides, J. Fleischer, M. Segev, Phys. Rev. E **66**, 046602 (2002)
35. B. Malomed, P. Kevrekidis, Phys. Rev. E **64**, 026601 (2001)
36. D. Traeger, R. Fischer, D. Neshev, A. Sukhorukov, C. Denz, W. Krolikowski, Y. Kivshar, Opt. Exp. **14**(5), 1913 (2006)
37. S. Longhi, Opt. Lett. **30**(16), 2137 (2005)
38. A. Szameit, F. Dreisow, T. Pertsch, S. Nolte, A. Tuennermann, Opt. Exp. **15**(4), 1579 (2007)
39. D. Blomer, A. Szameit, F. Dreisow, T. Schreiber, S. Nolte, A. Tuennermann, Opt. Exp. **14**(6), 2151 (2006)
40. A. Szameit, F. Dreisow, H. Hartung, S. Nolte, A. Tuennermann, F. Lederer, Appl. Phys. Lett. **90**, 241113 (2007)
41. F. Dreisow, A. Szameit, T. Pertsch, S. Nolte, A. Tuennermann, (SPIE, 2007), vol. 6460, p. 64601C. DOI 10.1117/12.711446. URL http://link.aip.org/link/?PSI/6460/64601C/1
42. M. Mizuguchi, L. Skuja, *Hideo Hosono, T. Ogawa, Opt. Lett. **24**(13), 863 (1999). URL http://ol.osa.org/abstract.cfm?URI=ol-24-13-863
43. L. Skuja, J. Non-Cryst. Solids **179**, 51 (1995)
44. H. Herzig, *Micro-optics: Elements, systems and applications* (Crc Pr Inc, 1997)
45. S. Sinzinger, J. Jahns, *Microoptics*, 2nd edn. (Wiley-VCH, 2003)
46. A. Szameit, H. Hartung, F. Dreisow, S. Nolte, A. Tuennermann, Appl. Phys. B. **87**, 17 (2007)
47. F. Dreisow, M. Heinrich, A. Szameit, S. Doering, S. Nolte, A. Tuennermann, S. Fahr, F. Lederer, Opt. Express **16**(5), 3474 (2008)
48. F. Dreisow, A. Szameit, M. Heinrich, T. Pertsch, S. Nolte, A. Tünnermann, S. Longhi, Phys. Rev. Lett. **101**(143602) (2008)
49. N. Efremidis, D. Christodoulides, Phys. Rev. E **65**, 056607 (2002)
50. F. Dreisow, A. Szameit, M. Heinrich, T. Pertsch, S. Nolte, A. Tuennermann, Opt. Lett. **33**(22), 2689 (2008). URL http://ol.osa.org/abstract.cfm?URI=ol-33-22-2689
51. A. Szameit, H. Trompeter, M. Heinrich, F. Dreisow, U. Peschel, T. Pertsch, S. Nolte, F. Lederer, A. Tuennermann, New Journal of Physics **10**(10), 103020 (2008). URL http://stacks.iop.org/1367-2630/10/103020
52. S. Longhi, Opt. Lett. **33**(5), 473 (2008). URL http://ol.osa.org/abstract.cfm?URI=ol-33-5-473

53. A. Szameit, F. Dreisow, M. Heinrich, T. Pertsch, S. Nolte, A. Tuennermann, E. Suran, F. Louradour, A. Barthelemy, S. Longhi, Appl. Phys. Lett. **93**(18), 181109 (2008). DOI 10.1063/1.2999624. URL http://link.aip.org/link/?APL/93/181109/1
54. A. Szameit, D. Bloemer, J. Burghoff, T. Pertsch, S. Nolte, A. Tuennermann, Appl. Phys. B. **82**(4), 507 (2006)
55. W. Karthe, R. Mueller, *Integrierte Optik* (Akademische Verlagsgesellschaft, 1991)
56. L. Tong, R.R. Gattass, I. Maxwell, J.B. Ashcom, E. Mazur, Opt. Comm. **259**(2), 626 (2006). DOI DOI:10.1016/j.optcom.2005.09.040. URL http://www.sciencedirect.com/science/article/B6TVF-4H6XNDS-1/2/d3aff92f775e65065fdb4a17e0413082
57. D. Dunlap, V. Kenkre, Phys. Rev. B **34**, 3625 (1986)
58. S. Longhi, M. Marangoni, M. Lobino, R. Ramponi, P. Laporta, E. Cianci, V. Foglietti, Phys. Rev. Lett. **96**, 243901 (2006)
59. I. Garanovich, A. Sukhorukov, Y. Kivshar, Phys. Rev. E **74**, 066609 (2006)
60. A. Szameit, I. Garanovich, M. Heinrich, A. Sukhorukov, F. Dreisow, T. Pertsch, S. Nolte, A. Tuennermann, Y. Kivshar, Nature Phys. **5**(4), 271 (2009)
61. I. Tamm, Phys. Z. Soviet Union **1**, 733 (1932)
62. W. Shockley, Phys. Rev. **56**(4), 317 (1939). DOI 10.1103/PhysRev.56.317
63. A. Szameit, I. Garanovich, M. Heinrich, A. Sukhorukov, F. Dreisow, T. Pertsch, S. Nolte, A. Tuennermann, Y. Kivshar, Phys. Rev. Lett. **101**(20), 203902 (2008). DOI 10.1103/PhysRevLett.101.203902. URL http://link.aps.org/abstract/PRL/v101/e203902
64. F. Bloch, Z. Phys. **52**, 555 (1928)
65. F. Dreisow, A. Szameit, M. Heinrich, T. Pertsch, S. Nolte, A. Tuennermann, S. Longhi, Phys. Rev. Lett. **102**(7), 076802 (2009). DOI 10.1103/PhysRevLett.102.076802
66. D. Gloge, D. Macruse, J. Opt. Soc. Am. **59**(12), 1629 (1969). URL http://www.opticsinfobase.org/abstract.cfm?URI=josa-59-12-1629
67. S. Longhi, Laser & Photonics Reviews **3**(3), 243 (2009)
68. F. Dreisow, A. Szameit, M. Heinrich, S. Nolte, A. Tuennermann, M. Ornigotti, S. Longhi, Phys. Rev. A **79**(5), 055802 (2009). DOI 10.1103/PhysRevA.79.055802. URL http://link.aps.org/abstract/PRA/v79/e055802
69. P. Facchi, S. Pascazio, Journal of Physics A: Mathematical and Theoretical **41**(49), 493001 (45pp) (2008). URL http://stacks.iop.org/1751-8121/41/493001
70. P. Butcher, D. Cotter, *The Elements of Nonlinear Optics*, 1st edn. (Cambridge University Press, 1990)
71. D. Milam, Appl. Opt. **37**(3), 546 (1998)
72. R. Chiao, E. Garmire, C. Townes, Phys. Rev. Lett. **13**(15), 479 (1964)
73. A. Barthelemy, S. Maneuf, C. Froehly, Opt. Commun. **55**, 201 (1985)
74. A. Szameit, D. Bloemer, J. Burghoff, T. Schreiber, T. Pertsch, S. Nolte, A. Tuennermann, F. Lederer, Opt. Exp. **13**(26), 10552 (2005)
75. A. Szameit, I. Garanovich, M. Heinrich, A. Minovich, F. Dreisow, A. Sukhorukov, T. Pertsch, D. Neshev, S. Nolte, W. Krolikowski, A. Tuennermann, A. Mitchell, Y. Kivshar, Phys. Rev. A **78**(3), 031801(R) (2008)
76. A. Szameit, Y. Kartashov, F. Dreisow, M. Heinrich, T. Pertsch, S. Nolte, A. Tuennermann, V. Vysloukh, F. Lederer, L. Torner, Phys. Rev. Lett. **102**(5), 153901 (2009)
77. T. Pertsch, U. Peschel, S. Nolte, A. Tuennermann, F. Lederer, J. Kobelke, K. Schuster, H. Bartelt, Phys. Rev. Lett. **93**(5), 053901 (2004)
78. A. Szameit, J. Burghoff, T. Pertsch, S. Nolte, A. Tuennermann, F. Lederer, Opt. Exp. **14**(13), 6055 (2006)
79. A. Szameit, Y. Kartashov, F. Dreisow, T. Pertsch, S. Nolte, A. Tuennermann, L. Torner, Phys. Rev. Lett. **98**, 173903 (2007)
80. A. Szameit, Y. Kartashov, F. Dreisow, M. Heinrich, V. Vysloukh, T. Pertsch, S. Nolte, A. Tuennermann, F. Lederer, L. Torner, Opt. Lett. **33**(7), 663 (2008). URL http://ol.osa.org/abstract.cfm?URI=ol-33-7-663

# Chapter 14
# Optofluidic Biochips

Rebeca Martínez Vázquez, Giulio Cerullo, Roberta Ramponi, and Roberto Osellame

**Abstract** In this chapter, we discuss the contributions of femtosecond laser micromachining to the fields of lab-on-chip and optofluidics. We show how femtosecond laser irradiation can be used to fabricate directly buried optical waveguides and, if followed by chemical etching, 3D microfluidic channels. We discuss the integration of waveguides and channels and present examples of integration of femtosecond laser written photonic sensors in microfluidic devices fabricated either by standard technologies or by the same femtosecond laser source. The results pave the way to a new generation of optofluidic devices with unprecedented capabilities.

## 14.1 Introduction

Lab-on-chips (LOCs) are microsystems aiming at the miniaturization onto a single substrate of several functionalities that typically would require an entire biological laboratory [1–3]. LOCs use networks of microfluidic channels to transport, mix, separate, make react and analyze very small volumes (micro- to nanoliters) of biological samples. The main advantages of the LOC approach are high sensitivity, speed of analysis, low sample and reagent consumption, and measurement automation and standardization. Applications of LOCs range from basic science (genomics, proteomics and cellomics) to chemical synthesis and drug development [4], high-throughput medical and biochemical analysis [5], environmental monitoring and detection of chemical and biological threats. Thanks to the miniaturization and integration afforded by LOCs, the life sciences are undergoing a revolution similar

R. Martínez Vázquez · G. Cerullo · R. Ramponi · R. Osellame (✉)
Istituto di Fotonica e Nanotecnologie - Consiglio Nazionale delle Ricerche (IFN-CNR), and Department of Physics - Politecnico di Milano, Piazza Leonardo da Vinci 32, 20133 Milan, Italy
e-mail: rebeca.martinezvazquez@ifn.cnr.it; giulio.cerullo@fisi.polimi.it; roberta.ramponi@fisi.polimi.it; roberto.osellame@ifn.cnr.it

to that triggered by integrated microelectronic systems, which gave birth to the Information Society.

Several substrate materials are used for LOC fabrication, including silicon, glass and polymers. Although polymers have the advantages of a very low cost and of the simplicity of microchannel fabrication by molding or embossing, glass is still the material of choice for many applications [6] due to the following benefits: it is chemically inert, stable in time, hydrophilic, non-porous, optically clear, and it easily supports electro-osmotic flow. In particular, the choice of fused silica as the basic material adds to the previous advantages a very high optical transparency down to the UV range and a very low background fluorescence. In addition, well-established microfabrication processes, based on photolithography and wet/dry etching, are available for this glass [7].

While many different fluidic functions have already been implemented on LOCs, a key unsolved problem is the development of integrated optical functionalities, as for example on-chip optical detection system [8]. In traditional setups used in conjunction with LOCs, optical functionalities are performed using bulk optical equipment, such as mirrors, lenses and microscope objectives, that focus light into a tiny volume in order to manipulate or analyze the sample flowing in the microchannel [9, 10]. Such schemes, however, require accurate mechanical alignment of the optical components to the microfluidic channels, are more sensitive to mechanical vibrations and drifts and allow only a limited number of configurations. The requirement of coupling a miniaturized LOC system with a massive benchtop instrument, such as an optical microscope, frustrates many of the LOC advantages, in particular it strongly limits device portability and prevents on-field or point-of-care applications. Much of the commercial success of the LOC concept will critically depend on the ability to successfully integrate optical detection schemes.

Several efforts have been performed in order to integrate microoptical components in LOCs [11, 12]. Optical waveguides allow one to confine and transport light in the chip, directing it to a small volume of the microfluidic channel and collecting the transmitted/emitted light. However, the integration of optical waveguides or other photonic components with microfluidic channels is not a straightforward process. In fact it requires a local modification of the refractive index of the substrate, which means adding further lithographic steps, thus complicating the fabrication process of the chip. Depending on the substrate of choice, different processes can be used. Approaches reported in the literature include waveguide fabrication by silica on silicon [13–15], ion exchange in soda-lime glasses [16, 17], photolithography in polymers [18, 19] and liquid-core waveguides [20–23]. All these approaches suffer, when applied to LOCs, from several major limitations: (a) they are inherently planar techniques, i.e. they are able to define optical guiding structures only in two dimensions, close to the sample surface; (b) they are multistep methods, requiring masking for waveguide definition followed by the etching/diffusion step; (c) they require clean-room environment, and (d) they may create uneven surfaces which make sealing of the microfluidic channels problematic. Such problems have so far strongly limited the integration of optical waveguides in microfluidic chips.

Recently, there has been considerable interest in the application of femtosecond laser writing to the integration of optical waveguides into LOCs. This technique presents some unique advantages that overcome the previously described limitations: (a) it is a direct maskless fabrication technique, i.e. in a single step one can create optical waveguides or more complicated photonic devices (splitters, interferometers, etc.) by simply moving the sample with respect to the laser focus; (b) it is a three-dimensional (3D) technique, since it allows one to define waveguides at arbitrary depths inside the glass. The first feature allows one to add waveguides, by simple post-processing, inside LOCs that have been fabricated by standard technologies, enabling a straightforward upgrade of chips that have already been optimized for microfluidic functionality. The second feature gives a much greater freedom in the device design, allowing novel and more compact geometries and unprecedented functionalities exploiting the 3D architecture.

In addition to waveguide writing, femtosecond lasers also offer the possibility of direct fabrication of microfluidic channels in fused silica, by the technique known as femtosecond laser irradiation followed by chemical etching (FLICE) [24]. The FLICE technique consists of two steps: (1) irradiation of the sample with focussed femtosecond laser pulses; (2) etching of the laser modified zone by a hydrofluoric acid (HF) solution in water. Femtosecond laser irradiation, by a mechanism discussed in Sect. 14.2, enhances the etching rate by up to two orders of magnitude, enabling the manufacturing of channels with high aspect ratio.

The FLICE technique has to compete with conventional approaches for the fabrication of microfluidic channels, borrowed from semiconductor processing, such as wet chemical etching and deep reactive ion etching [7]. These techniques require a clean room facility and create a 2D open channel network at the surface of the sample, which needs to be sealed by a glass slab to produce the microfluidic channels; moreover, 3D structures are very complex to produce by these approaches, since they would require a multilayer processing. The FLICE approach, although it produces channels that are presently limited in length to a few millimeters, has some distinct advantages: (a) it avoids photolithography and clean rooms; (b) it creates directly buried channels which do not require sealing with a cover glass; (c) it naturally produces channels with a circular cross-section, which are not easily obtained by other techniques; (d) it allows straightforward fabrication of channels with 3D geometries.

The possibility of combining optical waveguide writing with the FLICE technique provides the exciting opportunity of using a single platform, based on femtosecond laser microstructuring, for fabricating microfluidic devices with integrated optical sensing. It should also be noted that the integration, on the same substrate, of optical and microfluidic components has far-reaching scientific and technological implications, that go beyond the specific application of sensing in LOCs. To define this new field of research, the term "optofluidics" has been recently introduced in the scientific literature [25, 26]. Optofluidics exploits the synergy of optics and fluidics for the realization of completely new functionalities.

This chapter provides an overview of the applications of femtosecond laser microstructuring to optofluidics. The capability of direct writing of optical waveguides has been already discussed in the previous chapters and, therefore, we refer to them for a detailed discussion (see for example Chaps. 1, 4, 5, and 7). Here, we introduce the FLICE technique (Sect. 14.2); then we present examples of integration of femtosecond laser written photonic sensors in microfluidic devices fabricated by standard technologies (Sect. 14.3); finally, we present optofluidic devices in which both the optical and the fluidic part are fabricated by femtosecond laser microstructuring (Sect. 14.4).

## 14.2 Femtosecond Laser Microfluidic Channel Fabrication

### 14.2.1 Fundamental Physical Mechanisms

In the following, we review the basic physical mechanisms underlying the FLICE process. When fused silica is irradiated, the modifications induced by the femtosecond laser pulses can be classified into three categories, depending on the laser processing conditions [27]: (a) for a low fluence, a smooth modification is achieved, resulting mainly in a positive refractive index change with a very weak selectivity to etching; (b) for a moderate fluence, sub-wavelength nanogratings are produced, resulting in a high etching selectivity of the irradiated volume with respect to the pristine one (up to two orders of magnitude); (c) for high fluence, a disruptive modification is obtained, with the creation of voids and microexplosions [28]. In particular, regime (a) is the one typically used for waveguide fabrication, while regime (b) is the one employed in the first step of the FLICE technique for microchannel production. Regime (c) can be used for direct laser ablation and it will not be further considered in this chapter.

The increased HF etching rate of fused silica following femtosecond laser irradiation is due to the combination of several mechanisms. A first one is related to the decrease of the Si–O–Si bond angle induced by the hydrostatic pressure or compressive stress created in the irradiated region [29]. This explanation is particularly suited for regime (a), where the formation of waveguides is due to a local increase of the density of the material; in fact, under these conditions, the etching rate has been shown to be proportional to the increase of the refractive index [30]. However, in this regime only a very modest increase in the etching rate is obtained after irradiation. A second mechanism is active in regime (b) and results in a much higher selectivity in etching of the irradiated regions. This can be attributed to the formation of self-ordered nanograting structures, perpendicular to the laser polarization direction [31]. The physical mechanism underlying the formation of nanogratings has been studied in detail in [32, 33] and is based on a transient nanoplasmonic model involving the following steps:

(1) In the focal volume, hot spots for multiphoton ionization occur due to the presence of defects or colour centers (panel a in Fig. 14.1).

**Fig. 14.1** Pictorial illustration of the nanoplasmonic model explaining the formation of plasma nanoplanes aligned perpendicularly to the writing beam polarization [27]. See text for the description of each panel

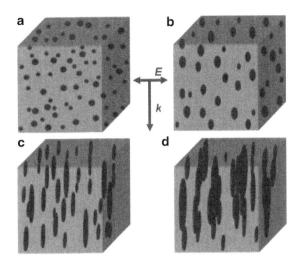

(2) Such hot spots evolve into spherically shaped nanoplasmas over successive laser pulses due to a memory effect [34] equivalent to a reduction of the effective bandgap in the previously ionized region (panel b in Fig. 14.1).

(3) Field enhancement at the boundaries of the nanoplasma droplets results in asymmetric growth of the initially spherical droplets, in a direction perpendicular to the laser polarization, leading to the formation of nanoellipsoids, which eventually grow into nanoplanes (panel c in Fig. 14.1).

(4) The nanoplanes are initially randomly spaced; when the electron plasma density inside them exceeds the critical density, they become metallic and start influencing light propagation in such a way that they assemble in parallel nanoplanes spaced by $\lambda/2n$, where $\lambda$ is the wavelength of the femtosecond writing laser and $n$ is the refractive index of the medium (panel d in Fig. 14.1).

One particular feature of this process, which is very important for applications, is that nanoplanes formed in adjacent focal volumes displaying a lateral offset tend to be coherently linked, resulting in a self-alignment of the planes (that become aligned nanocracks if a suitable fluence is used in the irradiation process). This simple model highlights some peculiar features of the FLICE process:

(1) It is not a single-pulse effect, but it relies on the cumulative action of multiple femtosecond laser pulses impinging on the same spot; it is, therefore, important to select the right combination of intensity and translation speed during the writing process.

(2) The process critically depends on the relative alignment of the writing laser polarization with respect to the translation direction [31]; when the polarization is aligned perpendicularly to the translation direction, then the nanocracks (which are perpendicular to the polarization) are directed along the channel axis, favouring the diffusion of the etchant; if on the other hand the polarization

is parallel to the translation direction, then the nanocracks are perpendicular to the channel axis and etchant diffusion is blocked.

Based on this model, we can now understand the FLICE process as follows: during etching, the nanocracks act as channels for the diffusion of the acid deeper into the fused silica, starting from the end face of the sample. Hence the etching process is a combination of two simultaneous phenomena: the diffusion of the acid along the irradiated region and the etching of the fused silica that gets in contact with such acid. Therefore, the etching process should not be thought as the acid carving its way by progressively removing the irradiated material but as a fast diffusion of acid in the irradiated region that causes an etching of material along the diffusion path. This is the reason why the microchannels produced in regime (b), in which nanogratings are formed, yield much higher aspect ratios than those produced in regime (a), where the nanogratings are absent. Even in regime (b), however, the etching process is self-terminating, due to the exhaustion of the HF acid in the depth of the etched microchannel and to the difficulty in refreshing it.

## 14.2.2 Microchannel Properties

As in waveguide writing, also in the FLICE technique the microfluidic channels are defined inside the glass substrate by translation along (longitudinal geometry) or perpendicularly (transversal geometry) to the laser beam propagation direction. The longitudinal geometry creates channels with circular cross-section but with length limited by the working distance of the focusing objective. The transverse geometry has superior flexibility and allows defining structures of arbitrary length and shape, but has the disadvantage of generating channels with highly asymmetric cross-section. In fact, the size of the laser affected zone (LAZ) perpendicularly to the writing beam axis is given approximately by the beam focal diameter $2w_0$, while along the beam axis it is given by the confocal parameter $b = 2\pi w_0^2/\lambda$ (see also Chap. 5). For typical focussed beam sizes, this results in LAZ dimensions markedly different along the two directions, making the channel cross-section strongly elliptical, which is undesirable for microfluidic applications.

Several solutions to this problem have been proposed and implemented in the context of microfluidic channel fabrication. Shaping of the focal volume has been achieved by the use of a slit, oriented parallel to the laser scanning direction, before the focussing lens [35]. This arrangement creates channels with symmetric cross-section, but has the disadvantage of wasting most of the laser power impinging on the slit. An alternative approach consists in stacking side by side different LAZs, created by subsequent laser scans of the substrate, to obtain a nearly symmetric cross-section of the modified volume [29]. This solution uses more efficiently the laser power but needs multiple scans and longer processing time; in addition, it gives rise to channels with irregular rims that follow the periodicity of the scans. More recently, a spatio-temporal focussing technique has been proposed [36], which is

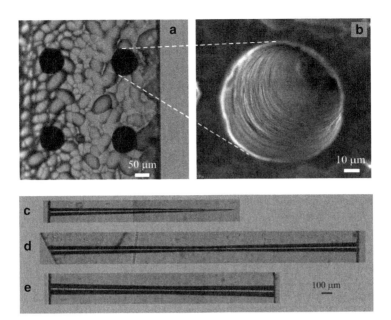

**Fig. 14.2** (a) End-face microscope image of a 2 × 2 matrix of microchannels; (b) SEM image of one channel; (c) top view of a dead-end microchannel; (d) and (e) top views of two-side etched microchannels with different degrees of overlap

based on the spatial dispersion of the spectrum of the writing beam; in this way, the pulse has the minimum duration only in the focus of the objective and this reduces the confocal parameter of the beam, producing spherical voxels.

An alternative approach is shaping the focal volume by introducing a focussing geometry in which the femtosecond writing beam is astigmatically shaped by changing both the spot sizes in the tangential and sagittal planes and the relative positions of the beam waists [37]. This shaping allows one to modify the focal volume in such a way that the cross-section of the LAZ can be made circular and with arbitrary size, making use of all the available laser power [38]. Astigmatic beam shaping has been applied to microchannel fabrication in [39].

Only two materials have been used, up to now, for microchannel fabrication by the FLICE technique: fused silica and Foturan (Schott). The results on Foturan will be presented in Chap. 15; therefore, we will concentrate in the following on the results obtained in fused silica.

Pulses from a regeneratively amplified Ti:sapphire laser (800 nm, 150 fs, 1 kHz) with 3–μJ energy are astigmatically shaped by a cylindrical telescope and focussed below the surface of the fused silica sample by a 50× microscope objective (NA = 0.6, focal length $f = 4$ mm). The laser beam is polarized orthogonal to the translation direction and the sample is moved at a 20 μm/s speed. Following the irradiation, the sample is etched for 3.5 h in a solution of 20% HF in water, immersed in an ultrasonic bath. Figure 14.2 shows examples of cross-sections of the resulting channels, highlighting the reproducibility of the process and its 3D capabilities. For a channel diameter of 80 µm, the estimated average roughness is of the order of 200 nm.

The FLICE technique lends itself to the fabrication of both surface channels [29] and directly buried channels. In the first case, no practical limitations exist on the channel length and shape, as well as on the microfluidic circuit layout. The second case is, however, more appealing, since it avoids the chip sealing step and exploits the 3D capabilities of the FLICE technique. With directly buried channels, however, one is faced with limitations in channel length, shape and aspect ratio. In fact, etching of fused silica occurs through the reaction: $SiO_2 + 4HF \rightarrow SiF_4 + 2H_2O$. As the channel is progressively etched starting from the end face, it is necessary to remove the reaction products and provide fresh acid, which has to diffuse along the channel. With increasing channel length, the amount of fresh acid able to reach its end reduces and the etch rate gradually decreases [40]. This limits the maximum achievable channel length and also results in microchannels with conical shapes, with larger radius at the channel entrance as compared to the buried end (see Fig. 14.2c). The cone apex angle $\alpha$ depends on the ratio between the etching rate of non-irradiated fused silica and diffusion rate of the acid in the irradiated channel. There is usually a trade-off between microchannel aspect ratio and length. Indeed, a low HF concentration [40] decreases the etch rate and results in higher aspect ratio but shorter channel length; on the other hand, high HF concentration [39] results in a lower aspect ratio, due to a faster lateral etching, but delays the self-termination of the etching process and increases the channel length.

Currently, the longest dead-end channels fabricated by FLICE are $\sim 1.8$ mm-long, and have an aspect ratio (length to diameter ratio) of $\sim 20$. It is possible to obtain longer channels by etching from two sides. According to the degree of overlap, different channel shapes can be obtained, such as a longer one with a narrow passage in the middle (Fig. 14.2d) or a shorter one with an almost uniform cross-section (Fig. 14.2e) [41]. The former has a 3 mm length with an entrance diameter of 90 µm and a waist diameter of 50 µm, while the latter has a 2.2 mm length with 110 µm diameter at the ends and 90 µm at the waist. The capability of creating tapered channels is a unique feature of FLICE and can be of use for some applications (see Sect. 14.4.1).

A method to compensate for the conical shape in a single side etching process consists in irradiating a reverse cone with respect to the one normally obtained after etching [42]. The concept behind this fabrication approach is shown in Fig. 14.3 and consists in irradiating a conical spiral, as that represented in Fig. 14.3a, forming a laser-modified conical surface, which is complementary to the cone that is created when etching a straight line. Additionally, a straight line forming the axis of the cone is also irradiated, which helps in removing the inner volume of the cone. The laser polarization is always orthogonal to the cone axis. The expected result is shown in Fig. 14.3b, where the blue cone is that of typical channel etched starting from a straight line, while the red line represents the conical irradiation, which would compensate for this effect, and the dotted line shows the expected cylindrical microchannel. The spiral periodicity $\Lambda$ is chosen such that there is

**Fig. 14.3** (a) Schematic diagram of a conical spiral inscribed into the substrate; (b) representation of a conical microchannel (*blue*), of the compensating conical spiral (*red*) and of the final cylindrical microchannel (*grey*)

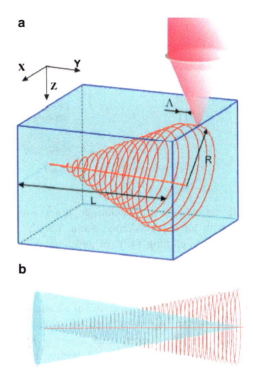

sufficient overlap between two subsequent laser irradiated arcs. It is worth noting that, as mentioned in Sect. 14.2.1, the nanocracks, that are formed in the irradiation of successive overlapping arcs, are naturally self-aligned, thus providing nanopaths for the acid diffusion in the irradiated region. This self-alignment of subsequently written nanocracks, even with a few seconds delay as in the conical spiral irradiation, although surprising, is consistent with the observation in [27], where self-alignment of the nanocracks created by subsequent pulses at a 1 kHz repetition rate is reported. In fact, the 1-ms delay between the two pulses is already sufficient for the material to cool down completely and, therefore, the same mechanism for the nanocracks self-alignment will hold also for much longer delays.

Given the much longer time that it takes to irradiate the spiral structure, microfabrication is performed using a high repetition rate laser. We use the second harmonic (515 nm) of a cavity-dumped Yb:KYW oscillator providing 350-fs laser pulses at repetition rates up to 1 MHz [43]. With 150 nJ pulse energy, 600 kHz repetition rate and focussing with a 0.6 NA objective, it is possible to create straight microchannels with translation speeds up to $1\,\text{cm}\,\text{s}^{-1}$, i.e. nearly three orders of magnitude higher than with the 1 kHz laser. This allows to irradiate a spiral in approximately the same time that it takes to irradiate a straight line with the 1 kHz laser. Figure 14.4 shows an example of this fabrication technique: by irradiating in the glass a conical spiral (Fig. 14.4a) with suitable angle, it is possible to obtain a dead-end etched channel with nearly uniform cross-section (Fig. 14.4b).

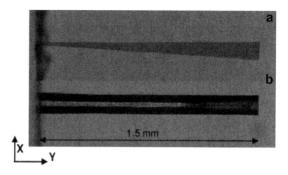

**Fig. 14.4** Microscope images of (**a**) conical spiral inscribed in the glass and (**b**) the etched microchannel

Using the spiralling technique, it was possible to fabricate uniform microchannels with an unprecedented 4-mm length and constant 90-μm cross-section by double side etching [42]. In addition, with its unique control over the channel cross-section, this approach allows also the fabrication of more complex structures, such as microchannels with access holes, interconnecting channels, microchannel adapters and O-grooves.

Recently KOH was proposed as an alternative to HF for highly selective etching of femtosecond laser irradiated microchannels [44]. Experiments were performed using an highly concentrated aqueous KOH solution (10 M, corresponding to 35.8%) at 80°C. The typical saturation behaviour observed for etching of fused silica in HF was not observed and the etching selectivity remained almost constant regardless of the etching period. It was, therefore, possible, using prolonged etching (60 h) in KOH, to fabricate microchannels as long as 1 cm with less than 60-μm diameter, corresponding to an aspect ratio of almost 200. The reasons for this enhanced selectivity in KOH is not yet fully understood, but it is attributed to the structural changes induced in $SiO_2$ by the femtosecond laser irradiation when the high pressures and temperatures produce silicon-rich structures, i.e. an excess of Si–Si bonds [45]. Since KOH is a common etching agent for silicon but minimally attacks silica, one can expect highly selective etching of the laser irradiated volume. Despite the rather long etching times, thus, KOH overcomes some of the problems of HF and may, with further optimization, significantly boost the performance of the FLICE technique.

### 14.2.3 Integration of Optical Waveguides and Microfluidic Channels

Femtosecond laser technology provides the exciting opportunity of using the same system for the fabrication of both microchannels and optical waveguides, and has the capability of easily integrating both structures in the same substrate with a three-dimensional geometry. Here, we review the results presented by Osellame et al. [41]. The experimental setup used for the microfabrication starts with a

**Fig. 14.5** (a) Schematic of the waveguides crossing the microfluidic channel; (b) microscope image of the microfluidic channel with the fluorescence excited by the optical waveguides, which are not visible due to the low refractive index change and scattering losses

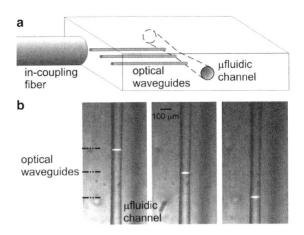

regeneratively amplified Ti:sapphire laser generating 150 fs, 500 μJ pulses at 1 kHz and 800 nm. In order to produce modifications with a circular cross-section, the beam is astigmatically shaped by passing it through a cylindrical telescope [38]. The beam is then focussed by a microscope objective inside the sample. Typically, the focus is located from 100 to 300 μm below the surface of the sample, which is moved perpendicularly to the beam propagation direction by a precision translation stage at a speed of 20 μm s$^{-1}$.

The microchannels are manufactured by laser irradiation with pulse energy of 4 μJ through a 50× objective (0.6 NA) and subsequent etching for 3 h in an ultrasonic bath with a 20% HF solution in water. By using samples of a few millimeters length it is possible to etch the channels from both sides, as discussed in the previous section, achieving pass-through channels. The same femtosecond laser is employed for the optical waveguide writing, but this time, since a lower intensity is required, a milder focussing is used (20× objective, 0.3 NA). High quality waveguides are obtained with nearly circular cross-section and refractive index change $\Delta n = 7 \times 10^{-4}$, resulting in single-mode operation throughout the visible range. Propagation losses at 543 nm, measured using the cut-back technique, were found to be 0.9 dB cm$^{-1}$. This value is very promising as compared to those obtained at similar wavelengths on other kinds of waveguides integrated on LOCs and fabricated with SU-8 polymer (2.5 dB cm$^{-1}$) [14,19] and with SiON technology (1 dB cm$^{-1}$) [13].

To demonstrate the integration capabilities of femtosecond laser fabrication technology, three waveguides spaced by 200 μm were inscribed crossing the central region of a 2.2 mm-long microchannel according to the scheme depicted in Fig. 14.5a.

The device capability of selectively addressing the content of microchannels is demonstrated by filling them with a solution of rhodamine 6G in ethylene glycol. When coupling 543 nm light in one waveguide by means of an optical fiber, yellow fluorescence is visible from the microfluidic channel (Fig. 14.5b). No stray light coming from waveguide scattering reaches the microfluidic channel, demonstrating the high spatial selectivity of the excitation. Coupling each of the three waveguides provides a fluorescent signal at three different points of the microfluidic channel (see the three panels in Fig. 14.5b), paving the way to the possibility of parallel multiple sensing at different positions in the microchannel.

This result demonstrates the integration on a fused silica substrate of microfluidic channels and optical waveguides, both fabricated by femtosecond laser pulses, and has a twofold implication. On one hand, it introduces a powerful method for the direct integration of optical waveguide sensing in LOCs; on the other hand, it indicates the possibility of using a single production tool for the fabrication of both the microfluidic channels and the optical waveguides required for biophotonic sensors.

## 14.3 Integration of Photonic Sensors in LOCs

In this section, we present examples of application of femtosecond laser waveguide writing to the integration of photonic devices in LOCs fabricated by conventional techniques. Here the laser is used as a post-processing tool, to add the optical sensing to a device that has already been optimized for the microfluidic functionality.

### *14.3.1 Cell Sorting*

The first pioneering work on the integration of femtosecond laser written waveguides into a LOC was performed by Applegate et al. [46]. The device is a fluorescence activated cell sorter (FACS) which allows one to trap, detect and sort fluorescently labelled cells inside a microfluidic channel. A first laser beam (the trapping beam) is used to trap cells inside the channel, while a second beam (the sensing beam) is used to excite their fluorescence and then to sort them into the desired channel by releasing, at a suitable time, the trapping beam.

The experimental setup of the FACS device is shown in Fig. 14.6 (left panel). The microfluidic network is fabricated in a polydimethylsiloxane (PDMS) substrate using soft lithography. The main channel is split into four channels (see Fig. 14.6 (right panel)), two of which are used as waste and two for sorting purposes. A line focus, created inside the channel by relay imaging the $1 \times 100\,\mu$m output of a diode laser bar (980 nm, 500 mW), is used to trap the particles in the channel. The PDMS chip lies on a fused silica substrate in which an optical waveguide network is fabricated by femtosecond laser writing. The network is a $1 \times 4$ splitter with four output beams of equal intensity, each with $6\,\mu$m diameter, spaced $30\,\mu$m apart. The waveguides are perpendicular to the channel and, when visible light is coupled into the splitter, four excitation points are available for cell fluorescence (see Fig. 14.6 (right panel)). The four waveguide outputs are aligned with the line focus of the trapping beam.

14 Optofluidic Biochips 401

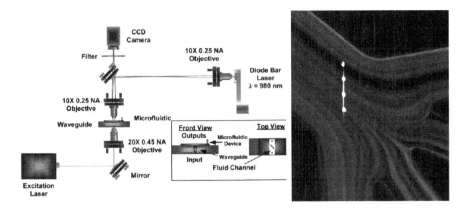

**Fig. 14.6** (*Left panel*) Schematic of integrated FACS system based on integrated optical waveguides combined with laser trapping. The trapping laser is aligned over the waveguide outputs. (*Right panel*) Image of the junction in the microfluidic channel. The two outside channels are waste, while the central two channels are used for fluorescing and non-fluorescing sorted particles. The line focus of the laser diode bar and the four waveguide outputs are also shown

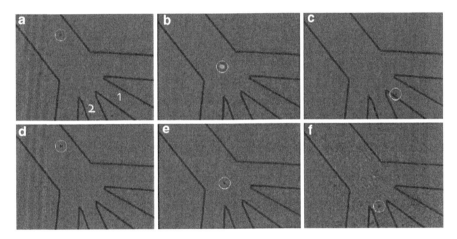

**Fig. 14.7** Images of a non-fluorescing colloid and a fluorescing colloid being sorted into different output streamlines, labelled 1 and 2 in frame *a*. Frames *a–c* show the path of the fluorescing colloid going into the outlet labelled 1. In frame *b* the colloid is fluorescing over the waveguide output aligned with outlet 1. Frames *d–f* show the non-fluorescing colloid path

Figure 14.7 shows an example of device operation, using $10\,\mu m$ diameter Crimson red labelled colloids. In this simple demonstration, shuttering the beam releases fluorescent colloids into outlet 1 while the non-fluorescent colloids are directed into channel 2. In Fig. 14.7 the first particle, which has been labelled, brightly fluoresces over a waveguide output. The emission radiation is detected using a photodiode and the trapping beam is subsequently blocked, causing this colloid to be sent into outlet 1. The second particle does not display fluorescence and thus it is allowed to traverse the entirety of the trap, whereupon it is released into its respective streamline and exits into output 2. This enables sorting of particles that display distinct optical characteristics. This approach can be easily upscaled by exploiting the multipoint excitation capabilities of optical waveguides.

### 14.3.2 Microchip Capillary Electrophoresis

One of the most powerful methods for the analysis of biomolecules is capillary electrophoresis (CE), in which charged molecules are separated in a microchannel as a result of their different mobility under an applied electric field. This technique is normally performed in a glass capillary filled with buffer, resulting in very high sensitivity and sorting accuracy, but at the expense of a bulky equipment and a rather long separation time. CE is particularly suited for on-chip integration [1,2,10], since electrokinetic flow can be used to move and mix liquids, thus avoiding the need to integrate pumps and valves. Microchip CE (MCE) is very promising for clinical applications, since it allows one to perform genetic tests to diagnose a variety of diseases, both exogenous (such as bacterial or viral infections) and endogenous (detection of mutated DNA sequences related to cancer or genetically inherited diseases).

The schematic structure of a MCE chip is shown in Fig. 14.8. It consists of two channels crossing at 90°, the injection channel and the separation channel. Each channel is equipped with access holes at its end, where suitable electrodes allow applying voltage differences to the channels content. Separation of a mixture of different biomolecules occurs according to the following steps:

(1) First the injection channel is filled, using electrokinetic flow, with the analyte molecules (Fig. 14.8a).
(2) Then by a rapid switching of the voltages, a microfluidic "plug", corresponding to the intersection volume between the channels, is injected in the separation channel (Fig. 14.8b); the different analyte species move at different speeds $v$, according to the expression $v = \mu E$, where $E$ is the electric field inside the channel and $\mu$ the so-called electrophoretic mobility, and are identified on the basis of their arrival times at the detection point.

To achieve high resolution MCE, it is necessary to suppress the electro-osmotic flow (EOF) and to fill the channel with a suitable sieving matrix. EOF is typical of glass substrates and occurs due to the formation of a double layer of electrical charges at the microchannel walls [47]. Such layer moves under an applied electric field and contributes to the electro-kinetic mobility. However, the EOF depends strongly on the quality of the microchannel walls, so that small defects in the walls will create "turbulences" in the EOF that will affect the MCE experiment by broadening the injected plugs. High quality MCE thus requires that EOF is suppressed by coating the channel walls with a suitable polymer that shields the surface charges.

**Fig. 14.8** Scheme of a microfluidic channel network suitable for MCE, and different steps of the process: (**a**) filling of injection channel; (**b**) separation of the negatively charged biomolecules

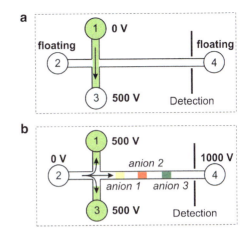

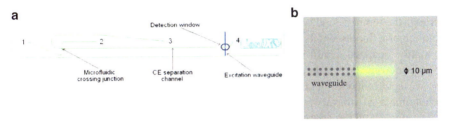

**Fig. 14.9** (**a**) Schematic of the MCE device. The line connecting the four reservoirs represent microfluidic channel fabricated by standard technologies, whereas the vertical thick line represents an optical waveguide inscribed by femtosecond laser and providing the fluorescence excitation source; (**b**) microscope image of the fluorescence excited in the microchannel by an optical waveguide [49]

In addition, the channel must be filled with a sieving medium which is able to induce a size-dependent separation of the different fragments which would otherwise have very similar electrophoretic mobility [48].

Different detection strategies are possible in MCE, but among them the most popular is laser induced fluorescence which, being a background-free technique, allows the measurement of very low analyte concentrations [9]. To perform this measurement first, the analytes must be labelled with a fluorescent molecule, then the CE separation is performed and the fluorescence is excited and collected at the detection point. Femtosecond laser waveguide writing can be used as a post-processing tool to integrate inside MCE chips optical waveguides intersecting the microfluidic channels at various locations, in order to excite their content with high spatial selectivity [49–52].

Figure 14.9 shows a scheme of the MCE chip with integrated optical waveguides. It starts with a commercial fused silica LOC (LioniX BV) manufactured by photolithography followed by wet etching and then sealed by a cover glass slab. The microfluidic circuit consists of two crossing channels following the scheme reported in Fig. 14.8; however, the channels are folded in a complex way in order to reduce the chip footprint. The microchannels are located 500 μm below the chip surface and have a nearly rectangular cross-section with 100 μm width and 50 μm height. The MCE chip is inserted in a commercial cartridge providing reservoir connections and electrical contacts. Optical waveguides were inscribed towards the end of the separation channel (approximately 3 cm from the injection cross) using an amplified Ti:sapphire laser (150 fs, 1 kHz, 4 μJ at 800 nm) and a translation speed of 20 μm s$^{-1}$. Employing astigmatic beam shaping, an elliptical cross-section of the written waveguide was obtained, with a major diameter of $\sim$50 μm in the vertical direction, in order to excite the maximum possible volume of the channel, while the minor diameter in the horizontal direction is $\sim$12 μm in order to retain a high spatial resolution along the flow direction. Alignment of the waveguides with the microchannels is achieved with a positioning accuracy higher than 2 μm in the depth direction. To demonstrate the high spatial selectivity of excitation through the femtosecond laser written waveguides, the channel was first filled with a solution of a fluorescent dye (rhodamine 6G). Figure 14.9b shows that fluorescence is highly localized, confirming the very low leakage out of the waveguide, and it extends across the whole channel width, due to the relatively low divergence of the waveguide mode. Femtosecond laser writing thus enables the integration of a "photofinish" at arbitrary positions in the separation channel. To achieve even higher integration, the fluorescence is collected by a high numerical aperture (NA = 0.5) optical fiber, at 90° with respect to the excitation waveguide, resulting in a compact and portable setup [49, 51].

The MCE chip with integrated optical waveguides was used to separate double-stranded DNA samples, consisting of mixtures of different fragments (oligonucleotides) with varying number of base pairs (bps) [52]. To assess its performances a commercial DNA ladder consisting of molecules with 17 different sizes (in the range of 50–3000 bp) was used. The DNA fragments were labelled with intercalating dyes (Sybr GREEN I) which dynamically link to the double helix (approximately one molecule every 10 bps).

Figure 14.10a shows a typical electropherogram of the DNA ladder, showing that all the peaks are clearly resolved and providing a calibration for this device that correlates each fragment size to its time of arrival.

These results demonstrate the capability of femtosecond laser writing to fabricate high quality optical waveguides in a fused silica MCE chip, which can excite with high spatial selectivity fluorescent molecules flowing in the microfluidic channels. This approach is quite powerful because it allows the integration of photonic functionalities by simple post-processing of commercial LOCs, fabricated with standard techniques. This can allow the development of highly integrated fluorescence excitation/detection schemes which could strongly increase the portability and compactness of the LOCs by overcoming the present limitation where microfluidic systems are coupled to macroscopic bulk optical detection systems.

# 14 Optofluidic Biochips

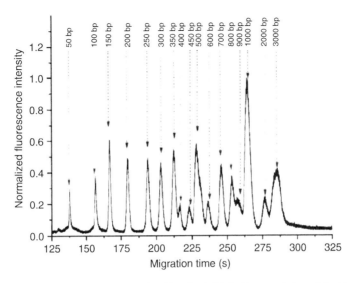

**Fig. 14.10** CE separation in an optofluidic chip of a DNA ladder consisting of double-stranded molecules with 17 different sizes (50–3,000 base-pairs), fluorescence-labelled with an intercalating dye [52]

## 14.3.3 Label-Free Sensing with Mach–Zehnder Interferometers

Fluorescence detection, due to its background-free nature, provides the highest sensitivity, down to the single-molecule limit [53]. However, this approach suffers from laborious labelling processes that may interfere with the function of a biomolecule or living cell, or may prevent on-chip chemical reactions from taking place. On the other hand, the measurement of refractive index variations allows label-free sensing, detecting the analyte in its natural form without any alteration. This detection approach is particularly suited for monitoring the kinetics of transient processes without interfering with the reactions. Various monolithic optical devices for refractive index sensing have been proposed [54]; standard approaches, based on integrated optics, include Mach–Zehnder interferometers (MZIs) [55], Young interferometers [56] and Bragg gratings [57]. These devices are fabricated by various techniques such as plasma vapour deposition, reactive ion etching and liquid-core waveguides [58]. These techniques are, however, only capable of producing 2D optical circuits. In addition, these devices typically exploit interaction between the evanescent field of the guided mode and the fluid in the microchannel, requiring long interaction lengths (millimeters to centimeters) for sensitive detection. On the other hand, several microfluidic applications call for spatially selective label-free detection, e.g. when molecules are separated in a microchannel by capillary electrophoresis or when counting is required as in cell flow cytometry. In addition, spatially selective refractive index detection would enable multipoint sensing, e.g. for monitoring the reaction kinetics in chemical microreactors [59].

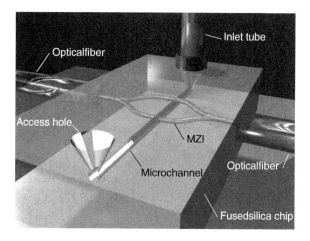

**Fig. 14.11** Schematic of the femtosecond-laser-fabricated microfluidic channel and integrated MZI. The sensing arm crosses orthogonally the channel, while the reference one passes over it [60]

Femtosecond laser writing is employed to fabricate a monolithic device capable of label-free and spatially-resolved optical sensing in a microfluidic chip. It consists of an MZI integrated with a microfluidic channel in a commercial LOC for MCE. The sensing arm of the MZI orthogonally crosses the channel, and the reference arm passes over it (see layout in Fig. 14.11) [60]. The device is capable of refractive index sensing with a spatial resolution of the order of the waveguide mode diameter (~10–15 μm). The innovative 3D layout of the MZI, which is required for spatially resolved sensing, is only made possible by the unique capabilities of femtosecond laser microstructuring. Since this geometry implies a very small interaction length with the analyte, evanescent field sensing would provide a limited sensitivity; therefore, in our configuration, the sensing arm directly intersects the microchannel.

As a preliminary step to integration with the microfluidic channel, the MZIs were fabricated in a plain fused silica substrate. The writing laser is a mode-locked cavity-dumped Yb:KYW laser producing 350-fs pulses at 1030 nm, with energy up to 1 μJ and 1 MHz repetition rate. The writing beam is focussed into the sample by a 50× objective with 0.6 NA, using an average power of 90 mW. The sample is moved at a speed of 100 μm/s by a three-axis air-bearing stage. The obtained waveguides display a refractive index contrast $\Delta n \approx 5 \times 10^{-3}$, corresponding to a single guided mode at the 1.5 μm wavelength used for the experiments, with 15-μm diameter. Unbalanced interferometers were designed, in order to detect fringes in the wavelength dependent transmission when scanning a sufficiently large spectral region with a tunable laser [61].

The MZI was integrated in a commercial LOC (LioniX BV) for MCE, similar to the one discussed in Sect. 14.4.1. The microchannels of this chip have a cross-section with 110 μm width and 50 μm depth and are buried at a depth of 500 μm from the surface. The MZIs are inscribed at the end of the separation channel, in the area evidenced by shading in Fig. 14.12a. The MZIs intersect the channel at a 90° angle in order to achieve spatially selective detection. The high spatial resolution comes at the expense of the sensing length, which is limited by the channel width.

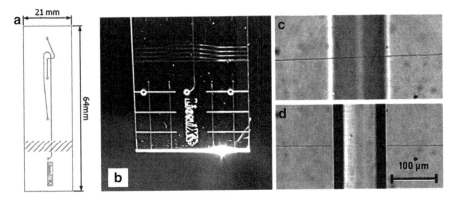

**Fig. 14.12** (a) Schematic of the commercial LOC for MCE (LioniX BV). MZIs are fabricated in a region at the end of the separation channel indicated by the *shaded area*. (b) Picture of the commercial LOC with 4 MZIs inscribed across a microchannel; the grid of microchannels at the bottom of the chip facilitates the sealing of the device and has no fluidic function. Microscope images of (c) the reference arm passing over the microchannel and (d) the sensing arm crossing it

Therefore, to maximize the phase shift induced by the analyte, direct intersection of the sensing arm with the microchannel was chosen. In this way, the sensing region is a cylinder with a height equal to the channel diameter ($\sim$100 μm) and a base with a diameter of $\sim$15 μm, approximately equal to the waveguide mode size. To enable only one arm intersecting the microchannel, the MZI was inscribed in a plane tilted by 7° with respect to the horizontal. The sensing arm of the MZI (which is chosen as the shorter one) orthogonally intersects the microchannel in its center (see Fig. 14.12, panel c), while the reference arm passes 20 μm above the microchannel (panel d). This tilted geometry of the MZI is necessary to achieve spatially selective sensing, since only one arm has to cross the channel. This is a clear example of an application where the 3D capabilities of femtosecond laser microfabrication are needed. In fact, it would have been very difficult to obtain this geometry with standard integrated optics technologies, which are intrinsically planar.

When one of the arms of the MZI crosses the microchannel, the transmission spectrum becomes

$$T = \frac{1}{2}\left[1 + \cos\left(\frac{2\pi}{\lambda}\left(n_0 \Delta s - (n_{\text{channel}} - n_0) L\right)\right)\right], \quad (14.1)$$

where the difference in propagation loss of the two arms has been neglected, for simplicity. Now, the phase difference between the two arms of the MZI is not only due to a length difference (the $n_0 \Delta s$ term) but also to a different refractive index of the content of the microchannel ($n_{\text{channel}}$) with respect to the effective index of the guided mode ($n_0$) in the reference arm (the $(n_{\text{channel}} - n_0)L$ term, where $L$ is the microchannel width). Consequently, the transmission spectrum shows fringes with peaks positioned at

**Fig. 14.13** Measured fringe shift from spectral data for different concentrations of glucose-D in water (*inset*: 0 mM *solid*; 50 mM *dashed*; 100 mM *dotted*); the correspondent refractive index increase is also shown in the right axis

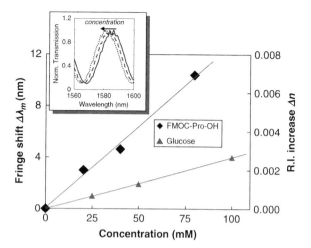

$$\lambda_m = \frac{n_0 \Delta s - (n_{\text{channel}} - n_0) L}{m}, \quad (14.2)$$

where $m$ is the fringe order.

If the content of the microchannel changes its refractive index by a quantity $\Delta n$, the transmission peaks will shift by

$$\Delta \lambda_m = -\frac{\Delta n \cdot L}{m} \quad (14.3)$$

To characterize the sensitivity and the linearity of the interferometer response, the microchannel was filled with different concentrations of aqueous glucose-D solutions, and for each of them the spectral response of the optofluidic device is measured statically (see inset in Fig. 14.13). Figure 14.13 also shows the fringe shift of the transmission spectrum, calculated from the experimental data, as a function of glucose concentration (triangles). The scale on the right represents the corresponding refractive index variation of the solutions with respect to pure water, based on data reported in the literature. This measurements with glucose solutions allow a calibration of the device, correlating the fringe shift with the refractive index change in the microchannel. From these measurements, the device responsivity is estimated to be 1500 nm/RIU.

The standard deviation of the data taken from repeated measurements is $\sigma \approx 4.8 \times 10^{-5}$ RIU; therefore we can extrapolate a limit of detection of $1.5 \times 10^{-4}$ RIU with a signal-to-noise ratio $\sim 3$. Temperature fluctuations of the analyte ($\pm 0.5\,^\circ$C) are the limiting factor for device sensitivity, causing most of the noise in index measurements. In fact, it can be estimated that an order of magnitude improvement in sensitivity could be achieved if the analyte were temperature stabilized.

Having assessed the performance of the device, it was used to detect biochemically relevant molecules such as peptides, used for drug synthesis in the pharmaceutical industry. Figure 14.13 shows the concentration dependent signal of the monopeptide FMOC-Pro-OH in ethanol (diamonds). The previous calibration, based on glucose solutions, quantitatively associates a refractive index increase to the measured fringe shift. Thus, the refractive index dependence on concentration for FMOC-Pro-OH can be determined, and results to be linear with a slope of $8.2 \times 10^{-5}$ RIU/mM.

These results demonstrate a compact and highly integrated interferometric device, directly fabricated on a LOC by post-processing, which allows refractive index sensing of the content of a microfluidic channel. The combination of label-free sensing and high spatial resolution is uniquely enabled by the femtosecond laser writing technique, which allows the fabrication of a 3D interferometer with only one arm crossing the microchannel.

## 14.4 Femtosecond Laser Fabrication of Optofluidic Devices

In the following, we will present several examples of application of femtosecond laser microstructuring to the fabrication of complete optofluidic devices. Here the laser is used for the fabrication of both optical waveguides and microchannels, integrated in unique 3D arrangements.

### 14.4.1 Flow Cytometry

Detection, manipulation and sorting of single cells is required for several research and diagnostics applications in life sciences. Kim et al. [62, 63] have demonstrated an innovative approach to this problem by a glass based optofluidic device entirely fabricated by femtosecond laser microprocessing. The device consists of a three-dimensional flow-through microchannel fabricated inside bulk fused silica glass through the FLICE technique. The Venturi-type microchannels have circular cross-section and channel diameter tapering from $\sim 100\,\mu$m at the entrance/exit ports to $\sim 10\,\mu$m at the flow neck, elegantly exploiting the intrinsic conical shape of the channels (see Sect. 14.3). The circular cross-section of the flow channel neck is designed slightly smaller ($\sim 1$–$2\,\mu$m) than the targeted cell diameter, in order to snugly fit individual cells. Thus, sample cells are self-aligned without resorting to additional flow focusing concepts, such as hydrodynamic [64] or electrokinetic [65] focussing. Note that this device cleverly exploits the unique capability of femtosecond laser processing to fabricate channels with a tapering cross-section (Fig. 14.14c). The microchannel is integrated with perpendicular optical waveguides written both parallel and normal to the top surface of the glass substrate. This approach enables effectively "confocal" cell detection and processing at the microchannel neck. Two different optical detection schemes were adopted: a passive setup measuring the transmission light intensity of a He–Ne laser (Fig. 14.14a) and

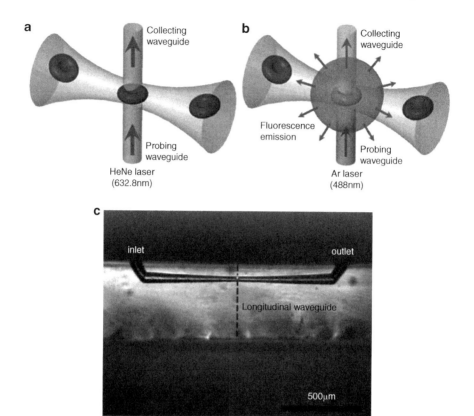

**Fig. 14.14** Schematic of the integrated cell detection using (**a**) light transmission and (**b**) fluorescence. (**c**) Microscope image of the actual device [62]

fluorescence emission detection using fluorescence dyed cells and an Ar laser as excitation source (Fig. 14.14b).

In the first approach, the transmitted light intensity through the collecting waveguide exhibits distinct peaks due to the refractive index difference between cell and base solution. The second approach, due to its background-free nature, provides much higher contrast but requires cell pre-treatment.

The performance of the device is verified by detecting red blood cells (RBCs) for volumetric flow rates up to 0.5 µl/min. Figure 14.15 shows measured particle counting density (number of detected particles per second) in the microchannel as a function of volumetric flow. Each data point depicts an average of counting density measured for 30 s and error bar represents the standard error of the mean. The dashed line represents calculated counting density based on the given particle concentration and volumetric flow rate. The measured counting density seems to follow the trend of the calculated counting density, verifying that the proposed glass

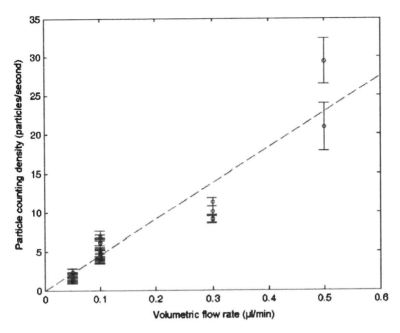

**Fig. 14.15** Measured particle counting density with respect to volumetric flow rate ($\mu$l min$^{-1}$). Each data point is an average of measured values for 30 s and error bar represents the standard error of the mean. *Dashed line* represents calculated particle counting density given particle concentration and volumetric flow rate [62]

based optofluidic device can perfectly work as a cell/particle counter or pre-detector for single cell manipulation on a bio-chip.

Even though the operation range of this device as a cell/particle counter is somewhat lower than commercial cytometers (up to 1000 particles per second), it may be enough throughput for a cell/particle predetector of single cell manipulation process on a chip, using unique geometric and material advantage of the proposed device. The range can also be extended up to the order of cell counting efficiency of commercial cytometer using fine tuning of detection electric circuit and particle speed calibration.

### 14.4.2 Label-Free Sensing with Bragg Gratings

As mentioned in Sect. 14.3.3, label-free refractive index sensing of the content of a microfluidic channel is desirable for many applications. Maselli et al. have recently presented an interesting example of an optofluidic device in fused silica, completely fabricated by femtosecond laser micromachining, which combines microfluidic channels and Bragg grating waveguides (BGWs) for refractive index sensing [66].

Writing the BGWs sufficiently close (less than 2 μm distance) to the very smooth microchannel walls enabled evanescent probing of the waveguide mode into the liquid-filled channel and refractive index sensing through a shift of the narrow (~0.2 nm) BGW resonance.

The microfabrication was performed with the second harmonic (520 nm) [67] of an ultrafast Yb:fiber system, providing 300-fs pulses at 500 kHz repetition rate. The BGWs were created by burst modulating the femtosecond writing laser (see Chap. 9) with an acousto-optic modulator [68], creating a continuous array of refractive index voxels that simultaneously provided low-loss waveguiding (~0.6 dB/cm), efficient mode coupling with standard optical fibers and high-strength Bragg resonance (waveguide reflectance $R \approx 90\%$ and transmittance $T \approx -35$ dB). The sample was translated in a direction parallel to the laser polarization to avoid waveguide birefringence and the Bragg wavelength was controlled by the sample translation speed. The microchannels were fabricated by the FLICE technique, using multiscan irradiation, thus leading to the formation of microchannels with rectangular 11−μm cross-section and positioned 75 μm below the substrate surface. The fabrication parameters were optimized in order to minimize the microchannel roughness, and thus allow the smallest possible microchannel/BGW distance for maximum interaction of the evanescent wave with the fluid. Practically, the minimum achievable channel/BWG distance was 1.5 μm.

Different device geometries were investigated, according to the schemes reported in Fig. 14.16. The first topology (Fig. 14.16a) consists of a straight 25 mm long BGW parallel to a 10 mm long microchannel. The BGW consists of three serially cascaded segments, each one with a distinct Bragg wavelength. The central grating is positioned as close as possible to the channel, while the two side gratings, that have a slightly different Bragg wavelength, serve as references to compensate for temperature and strain variations. The second configuration (Fig. 14.16b) consists of a double S-bend waveguide positioned with the sensing BGW parallel to and slightly shorter than the adjacent microchannel. With this configuration, the sensor grating does not probe the starting and ending walls of the channel that induce back reflections and spectral distortions to the Bragg resonance. Finally, the third design (Fig. 14.16c) is similar to the first, but with two microchannels placed parallel to and on opposite sides of the sensor BGW, with the objective of increasing the amount of evanescent field that penetrates into the microfluidic channels and thus improving the device sensitivity.

The three devices were tested by butt-coupling them to optical fibers: the input one is coupled to a broadband near-IR source (1530–1610 nm), while the output one is connected to an optical spectrum analyzer (OSA). Different analyte fluids filled the channels and induced shifts in the BGW resonance. The device sensitivity strongly depends on the refractive index of the sample to be measured and the highest value of 81 nm/RIU was obtained with the double-channel design and a sample refractive index approaching the value of the fused silica substrate ($n = 1.458$). Given the present minimum $\lambda$ shift of 0.01 nm resolvable by the OSA, the 81 nm/RIU response would offer a minimum resolvable refractive index variation $\Delta n_{min}$ of $\sim 1.2 \times 10^{-4}$. Higher sensitivities can be achieved by: (a) improvement

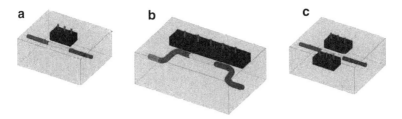

**Fig. 14.16** Schematics of microfluidic sensor geometries: straight BGW single channel (**a**), S-bend BGW single channel (**b**), and straight BGW double channel (**c**) [66]

of the channel wall roughness together with the extension of channels to wrap fully around the BGW; (b) narrowing the Bragg resonance linewidth with shorter length BGWs; (c) engineering spectral defects with $\pi$-phase shifts.

### 14.4.3 Cell Trapping and Stretching

Biophotonic devices based on optical forces are powerful tools for single cell study and manipulation without physical contact [69–71]. The exploitation of optical forces for analysis at single cell level provides significant information, opening new scenarios for the comprehension of basic biological mechanisms and for the early detection of several diseases. In particular, the investigation of the viscoelastic properties of trapped cells and their response to the application of intense optical forces, able to cause a significant deformation of the cytoskeleton, is of great interest. It is widely recognized that alterations of the cytoskeleton deformability are present in many diseases and their measurement can be used as a reliable marker of the cell status [72].

Mechanical properties of single cells can be efficiently tested using a recently developed device based on optical fibers, named optical stretcher (OS) [73]. OS basic idea relies on a double beam laser trap [74] obtained with two counter-propagating fiber beams [75, 76]. Increasing the laser power, the radiation pressure exerted by the two beams over the trapped cell surface leads to an elongation of the cell along the beam axis, providing meaningful information on the cell health. Although the effectiveness of the OS has been widely demonstrated, the typical set-ups, based on assembling optical fibers with glass capillaries or PDMS microchannels, present some criticality mainly due to the fine and stable alignment required between discrete optical and microfluidic components [73, 75].

Differently from standard fabrication technologies, femtosecond laser micromachining combined with chemical etching is able to provide direct writing of both optical waveguides and microfluidic channels [41], ensuring extreme flexibility and accuracy, together with intrinsic three-dimensional capabilities. This innovative

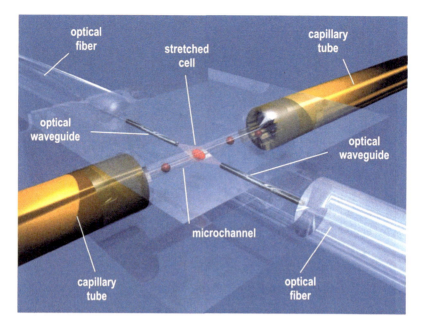

**Fig. 14.17** 3D rendering of the monolithic optical stretcher fabricated by femtosecond laser micromachining. The cells flowing in the microchannel are trapped and stretched in correspondence of the dual beam trap created by the optical waveguides. Connections to capillaries and optical fibers are also shown [77]

technology has thus been used to fabricate a monolithic optofluidic chip that allows performing mechanical analysis on single cells without physical contact and with high reproducibility (Fig. 14.17). The integrated chip is based on a fused silica glass substrate, thus providing high transparency for cell imaging, and represents a significant improvement in terms of stability, robustness and optical damage threshold over existing optical cell stretchers.

The femtosecond laser irradiation set-up for both microchannel formation and waveguide fabrication includes a frequency-doubled cavity-dumped Yb:KYW laser, delivering 230-fs pulses at 1 MHz. The laser beam is focussed through a 50× (0.6 NA) objective into the substrate, which is translated in three-dimensions by computer controlled stages [77].

As previously discussed, femtosecond laser technology provides an unprecedented freedom in designing 3D fluidic components by irradiating the substrate with beam trajectories more complex than just straight lines. Here we exploit this technique for two main purposes: (a) fabrication of large access holes at the end of the microchannel, in order to enable a straightforward plugging of capillary tubes for connection with an external fluidic circuit; (b) shaping of the microchannel in the trapping region, in order to improve the imaging of the cell. Circular shape of the microchannel cross-section acts as a lens when imaging the trapped cells

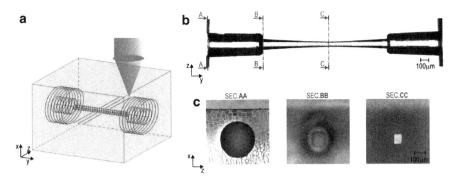

**Fig. 14.18** (a) Sketch of the pre-etching path for channel formation and optical waveguides. (b) Picture of the fabricated microchannel after chemical etching. (c) Pictures of channel cross-section in different regions along the channel axis [78]

by a transmission microscope. This causes a deformation of the cell contour and complicates the determination of small elastic elongations during stretching. In order to avoid this problem, we implement an irradiation path to obtain a square cross-section channel. This is obtained by irradiating two coaxial helixes with rectangular cross-section one inside the other. The microchannel is then terminated by two access holes with circular cross-section that are obtained by irradiating three coaxial helixes with circular cross-section, as shown in Fig. 14.18a. The chemical etching is achieved by immersing the chip in an ultrasonic bath with 20% of HF in water. A 4.5 h etching produces the microchannel as reported in Fig. 14.18b, which is 400-µm buried under the top surface, has a 2-mm length, a central rectangular cross-section of $85 \times 75 \mu m$, and two 800-µm-long access holes. While in the portions of the microchannel closer to the access holes the etching smoothes out the corners of the rectangular cross-section, in the central portion the channel closely follows the irradiation path with a sharp rectangular shape as shown in Fig. 14.18c.

The same fabrication set-up used for pre-etching irradiation is employed, with lower fluence, to directly write optical waveguides into the fused silica substrate. The obtained waveguides have an intensity mode radius of 4.2 µm. Waveguide writing and pre-etching irradiation take place in the same fabrication step and so the accuracy in the alignment between the fluidic and the optical components corresponds to the accuracy of the translation stages, thus better than 100 nm. The optofluidic chip is connected to an external fluidic circuit through two capillary tubes glued into the access holes and it is coupled to the laser source by two bare-end optical fibers. The connected chip is then placed on a transmission microscope plate for imaging.

The integrated OS has been experimentally tested by trapping and stretching RBCs. Once the RBC suspension is flown into the microfluidic channel, optical trapping of single cells can be easily observed with an optical power in the trap of about 20 mW. In addition, by unbalancing the powers emitted by the two fibers,

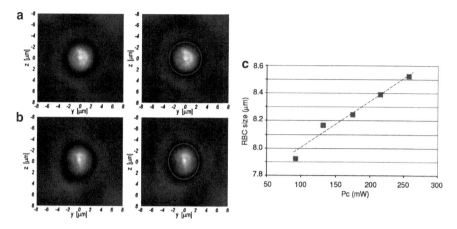

**Fig. 14.19** (**a**) Images of a trapped RBC with contour recognition (*white line*). (**b**) Images of a stretched RBC with visible elongated size along the $z$-axis. (**c**) Plot of the RBC elongated size as a function of the optical power $P_C$ [78]

the trapped cell can be moved with good control along the beam axis. The square cross-section of the channel allows obtaining images of the trapped cells in which the contour is very well-defined and can be automatically recovered. Figure 14.19a shows an image of the trapped cell where the recovered contour is indicated by a red line. By increasing the optical power in the trap up to 300 mW, we observe a progressive stretching of the trapped cell as shown in Fig. 14.19b. Figure 14.19c reports a plot of the cell elongated dimension as a function of the optical power in the trap (Pc).

## 14.5 Outlook and Conclusions

In this chapter we have discussed the contribution of femtosecond laser micromachining to the field of lab-on-chips and optofluidics. We have shown that ultrashort laser pulses, besides their ability to inscribe waveguides and photonic devices (splitters, interferometers and gratings), can also be used, in combination with HF etching, to fabricate directly buried microfluidic channels. The two main assets of femtosecond laser microfabrication are: (a) being a direct-write technique, it can be used as a post-processing tool to add optical (and possibly fluidic) functionalities to a microfluidic device manufactured by conventional methods (see the devices of Sect. 14.3); (b) its unique three-dimensional capabilities allows one to implement novel geometries, enabling the realization of devices with unprecedented capabilities (see the 3D interferometer of Sect. 14.3.3); the capability to produce both optical waveguides and microchannels allows one to design and fabricate a full optofluidic device with extreme control on the components properties and device geometry (see the devices of Sect. 14.4). The field of femtosecond laser processing for optofluidics is still in its infancy and, in order to flourish, it will require a close synergy between microfabrication and biochemistry skills. However, femtosecond laser micromachining has already proven to be an exciting playground to develop new optofluidic devices.

# References

1. R.D. Reyes, D. Iossifidis, P.A. Auroux, A. Manz, Anal. Chem. **74**, 2623–2636 (2002)
2. P.A. Auroux, D.R. Reyes, D. Iossifidis, A. Manz, Anal. Chem. **74**, 2637–2652 (2002)
3. G.M Whitesides, Nature **442**, 368–373 (2006)
4. A.J. deMello, Nature **442**, 394–402 (2006)
5. P. Yager, T. Edwards, E. Fu, K. Helton, K. Nelson, M.R. Tam, B.H. Weigl, Nature **442**, 412–418 (2006)
6. J.F. Dishinger, R.T. Kennedy, Anal. Chem **79**, 947–954 (2007)
7. M. Madou, *"Fundamentals of Microfabrication: The Science of Miniaturization"* (Taylor and Francis Ltd, Boca Raton, 2002)
8. E. Verpoorte, Lab Chip **3**, 42N–52N(2003)
9. J. Enger, M. Goksör, K. Ramser, P. Hagberg, D. Hanstorp, Lab Chip **4**, 196–200 (2004)
10. A.T. Woolley, R.A. Mathies, Proc. Natl. Acad. Sci. U.S.A. **91**, 11348 (1994)
11. K.B. Mogensen, H. Klank, J.P. Kutter, Electrophoresis **25**, 3498–3512 (2004)
12. H.C. Hunt, J.S. Wilkinson, Microfluid Nanofluid **4**, 53–79 (2008)
13. K.B. Mogensen, P. Friis, J. Hübner, N.J. Petersen, A.M. Jorgensen, P. Telleman, J.P. Kutter, Opt. Lett. **26**, 716–718 (2001)
14. J. Hubner, K.B. Mogensen, A.M. Jorgensen, P. Friis, P. Telleman, J.P. Kutter, Rev. Sci. Instrum. **72**, 229–234 (2001)
15. K.B. Mogensen, N.J. Petersen, J. Hübner, J.P. Kutter, Electrophoresis **22**, 3930–3938 (2001)
16. R. Mazurczyk, J. Vieillard, A. Bouchard, B. Hannes, S. Krawczyka, Sens. Act. B **118**, 11–19 (2006)
17. J. Vieillard, R. Mazurczyk, C. Morin, B. Hannes, Y. Chevolot, P.L. Desbene, S. Krawczyk, J. Chromatogr. B **845**, 218–225 (2007)
18. B. Kuswandi, Nuriman J. Huskens, W. Verboom, Anal. Chim. Acta. **601**, 141–155 (2007)
19. K.B. Mogensen, J. El-Ali, A. Wolff, J.P. Kutter, Appl. Opt. **42**, 4072–4079 (2003)
20. C.L. Bliss, J.N. McMullin, C.J. Backhouse, Lab Chip **7**, 1280–1287 (2007)
21. C.L. Bliss, J.N. McMullin, C.J. Backhouse Lab Chip **8**, 143–151 (2008)
22. D. Yin, D.W. Deamer, H. Schmidt, J.P. Barber, A.R. Hawkins, Opt. Lett. **31**, 2136–2138 (2006)
23. D. Yin, E.J. Lunt, M.I. Rudenko, D.W. Deamer, A.R. Hawkins, H. Schmidt, Lab Chip **7**, 1171–1175 (2007)
24. A. Marcinkevicius, S. Juodkazis, M. Watanabe, M. Miwa, S. Matsuo, H. Misawa, J. Nishii, Opt. Lett. **26**, 277–279 (2001)
25. D. Psaltis, S.R. Quake, C. Yang, Nature **442**, 381–386 (2006)
26. C. Monat, P. Domachuk, B.J. Eggleton, Nat. Photonics **1**, 106 (2007)
27. R. Taylor, C. Hnatovsky, E. Simova, Laser Photon. Rev. **2**, 26–46 (2008)
28. E. Glezer, E. Mazur, Ultrafast-laser driven microexplosions in transparent materials, Appl. Phys. Lett. **71**, 882 (1997)
29. Y. Bellouard, A. Said, M. Dugan, Ph. Bado, Opt. Exp. **12**, 2120 (2004)
30. R.S. Taylor, C. Hnatovsky, E. Simova, D.M. Rayner, M. Mehandale, V.R. Bhardwaj, P.B. Corkum, Opt. Exp. **11**, 775–780 (2003)
31. C. Hnatovsky, R. Taylor, E. Simova, V. Bhardwaj, D. Rayner, P. Corkum, Opt. Lett. **30**, 1867 (2005)

32. Y. Shimotsuma, P. Kazansky, J. Qiu, K. Hirao, Phys. Rev. Lett. **91**, 247405 (2003)
33. V.R. Bhardwaj, E. Simova, P.P. Rajeev, C. Hnatovsky, R.S. Taylor, D.M. Rayner, P.B. Corkum, Phys. Rev. Lett. **96**, 057404 (2006)
34. P.P. Rajeev, M. Gerstvolf, E. Simova, C. Hnatovsky, R.S. Taylor, D.M. Rayner, P.B. Corkum, Phys. Rev. Lett. **97**, 253001 (2006)
35. Y. Cheng, K. Sugioka, K. Midorikawa, M. Masuda, K. Toyoda, M. Kawachi, K. Shihoyama, Opt. Lett. **28**, 55 (2003)
36. F. He, H. Xu, Y. Cheng, J. Ni, H. Xiong, Z. Xu, K. Sugioka, K. Midorikawa, Opt. Lett. **35**, 1106 (2010)
37. G. Cerullo, R. Osellame, S. Taccheo, M. Marangoni, R. Ramponi, P. Laporta, S. De Silvestri, Opt. Lett. **27**, 1938 (2002)
38. R. Osellame, S. Taccheo, M. Marangoni, R. Ramponi, P. Laporta, D. Polli, S. De Silvestri, G. Cerullo, J. Opt. Soc. Am. B **20**, 1559 (2003)
39. V. Maselli, R. Osellame, G. Cerullo, R. Ramponi, P. Laporta, L. Magagnin, P.L. Cavallotti, Appl. Phys. Lett. **88**, 191107 (2006)
40. C. Hnatovsky, R.S. Taylor, E. Simova, P.P. Rajeev, D.M. Rayner, V.R. Bhardwaj, P.B. Corkum, Appl. Phys. A Mater. Sci. Process. **84**, 47–61 (2006)
41. R. Osellame, V. Maselli, R. Martinez Vazquez, R. Ramponi, G. Cerullo, Appl. Phys. Lett. **90**, 231118 (2007)
42. K.C. Vishnubhatla, N. Bellini, R. Ramponi, G. Cerullo, R. Osellame, Opt. Exp. **17**, 8685–8695 (2009)
43. A. Killi, U. Morgner, M.J. Lederer, D. Kopf, Opt. Lett. **29**, 1288–1290 (2004)
44. S. Kiyama, S. Matsuo, S. Hashimoto, Y. Morihira, J. Phys. Chem. C **113**, 11560–11566 (2009)
45. M. Boero, A. Oshiyama, P.L. Silvestrelli, K. Murakami, Appl. Phys. Lett. **86**, 201910 (2005)
46. R.W. Applegate Jr., J. Squier, T. Vested, J. Oakey, D.W.M. Marr, P. Bado, M.A. Dugan, A.A. Said, Lab Chip **6**, 422–426 (2006)
47. S. Ghosal, Electrophoresis **25**, 214–228 (2004)
48. J.P. Landers, Anal. Chem. **75**, 2919–2927 (2003)
49. R. Martinez Vazquez, R. Osellame, D. Nolli, C. Dongre, H. van den Vlekkert, R. Ramponi, M. Pollnau, G. Cerullo, Lab Chip **9**, 91–96 (2009)
50. C. Dongre, R. Dekker, H.J.W.M. Hoekstra, M. Pollnau, R. Martinez-Vazquez, R. Osellame, G. Cerullo, R. Ramponi, R. van Weeghel, G.A.J. Besselink, H.H. van den Vlekkert, Opt. Lett. **33**, 2503–2505 (2008)
51. R. Martinez Vazquez, R. Osellame, M. Cretich, M. Chiari, C. Dongre, H.J.W.M. Hoekstra, M. Pollnau, H. van den Vlekkert, R. Ramponi, G. Cerullo, Anal. Bioanal. Chem. **393**, 1209–1216 (2009)
52. C. Dongre, J. van Weerd, G. van Weeghel, R. Martinez Vazquez, R. Osellame, G. Cerullo, M. Cretich, M. Chiari, H.J.W.M. Hoekstra, M. Pollnau, Electrophoresis **31**, 2584–2588 (2010)
53. H.T. Li, L.M. Ying, J.J. Green, S. Balasubramanian, D. Klenerman, Anal. Chem. **75**, 1664–1670 (2003)
54. P. Dumais, C.L. Callender, J.P. Noad, C.J. Ledderhof, IEEE Sensors J. **8**, 457–464 (2008)
55. R.G. Heideman, P.V. Lambeck, Sens. Actuators B Chem. **61**, 100–127 (1999)
56. A. Ymeti, J. Greve, P.V. Lambeck, T. Wink, S.W.F.M.v. Hovell, T.A.M. Beumer, R.R. Wijn, R.G. Heideman, V. Subramaniam, J.S. Kanger, Nano Lett. **7**, 394–397 (2007)
57. A.S. Jugessur, J. Dou, J.S. Aitchison, R.M. De La Rue, M. Gnan, Microelectron. Eng. **86**, 1488–1490 (2009)
58. D. Yin, E.J. Lunt, M.I. Rudenko, D.W. Deamer, A.R. Hawkins, H. Schmidt, Lab Chip **7**, 1171–1175 (2007)
59. S.J. Haswell, R.J. Middleton, B. O'Sullivan, V. Skelton, P. Watts, P. Styring, Chem. Commun. 391–398 (2001)
60. A. Crespi, Yu. Gu, Hugo J.W.M. Hoekstra, C. Dongre, M. Pollnau, R. Ramponi, G. Cerullo, R. Osellame, Lab Chip **10**, 1167–1173 (2010)
61. C. Florea, K. Winick, J. Lightwave Technol. **21**, 246–253 (2003)
62. M. Kim, D.J. Hwang, H. Jeon, K. Hiromatsu, C.P. Grigoropoulos, Lab Chip **9**, 311–318 (2009)

63. D.J. Hwang, M. Kim, K. Hiromatsu, H. Jeon, C.P. Grigoropoulos, Appl. Phys. A **96**, 385–390 (2009)
64. Z. Wang, J. El-Ali, M. Engelund, T. Gotsaed, I.R. Perch-Nielsen, K.B. Mogensen, D. Snakenborg, J.P. Kutter, A. Wolff, Lab Chip **4**, 372–377 (2004)
65. L.M. Fu, R.J. Yang, C.H. Lin, Y.J. Pan, G.B. Lee, Anal. Chim. Acta **507**, 163–169 (2004)
66. V. Maselli, J.R. Grenier, S. Ho, P.R. Herman, Opt. Exp. **17**, 11719 (2009)
67. L. Shah, A.Y. Arai, S. Eaton, P. Herman, Opt. Exp. **13**, 1999–2006 (2005)
68. H. Zhang, S.M. Eaton, P.R. Herman, Opt. Lett. **32**, 2559–2561 (2007)
69. D.G. Grier, Nature **424**, 810–816 (2003)
70. J.E. Molloy, M.J. Padgett, Contemp. Phys. **43**, 241–258 (2002)
71. C. Liberale, P. Minzioni, F. Bragheri, F. De Angelis, E. Di Fabrizio, I. Cristiani, Nat. Photonics **1**, 723–727 (2007)
72. J. Guck, S. Schinkinger, B. Lincoln, F. Wottawah, S. Ebert, M. Romeyke, D. Lenz, H.M. Erickson, R. Ananthakrishnan, D. Mitchell, J. Käs, S. Ulvick, C. Bilby, Biophys. J. **88**, 3689–3698 (2005)
73. J. Guck, R. Ananthakrishnan, H. Mahmood, T.J. Moon, C.C. Cunningham, J. Käs, Biophys. J. **81**, 767–784 (2001)
74. A. Ashkin, Phys. Rev. Lett. **24**, 156–159 (1970)
75. A. Constable, J. Kim, J. Mervis, F. Zarinetchi, M. Prentiss, Opt. Lett. **18**, 1867–1869 (1993)
76. B. Lincoln, S. Schinkinger, K. Travis, F. Wottawah, S. Ebert, F. Sauer, J. Guck, Biomed. Microdevices **9**, 703–710 (2007)
77. N. Bellini, K.C. Vishnubhatla, F. Bragheri, L. Ferrara, P. Minzioni, R. Ramponi, I. Cristiani, R. Osellame, Opt. Exp. **18**, 4679 (2010)
78. F. Bragheri, L. Ferrara, N. Bellini, K.C. Vishnubhatla, P. Minzioni, R. Ramponi, R. Osellame, I. Cristiani, J. Biophoton. **3**, 234–243 (2010)

# Chapter 15
# Microstructuring of Photosensitive Glass

**Koji Sugioka**

**Abstract** Femtosecond laser direct writing followed by thermal treatment and successive wet etching can form three-dimensional (3D) hollow microstructures inside photosensitive glass. The principles and procedures of this process are explained. Next, the fabrication of 3D microfluidic structures and optical microcomponents is reviewed. Finally, the manufacture of functional microchip devices such as a microfluidic dye laser, optofluidics, and a nano-aquarium by integrating the microcomponents in a single glass chip is demonstrated.

## 15.1 Introduction

The ultrashort pulse widths and extremely high peak powers of femtosecond (fs) lasers can induce strong absorption, even in materials such as glass that are transparent to the laser frequency, due to nonlinear multiphoton absorption, resulting in high-quality processing with minimal formation of heat-affected zones (HAZs) described in Chap. 1. If the fs laser beam is focused inside glass with a moderate pulse energy, then the multiphoton absorption can be confined to the region near the focal point, so that internal modifications such as of the refractive index of transparent materials can be performed (cf. Parts II and III). Internal modification followed by selective etching with wet chemicals can form three-dimensional (3D) hollow microstructures embedded inside certain glasses [1–3].

One application of this technique is to the fabrication of biochips such as microreactors, lab-on-a-chip devices, micro-total analysis systems (μ-TAS), and optofluidics discussed in Chap. 14. Palm-size microchips, into which a roomful of laboratory equipment is shrunk and packed, are capable of meeting this demand

---

K. Sugioka (✉)
Laser Technology Laboratory, RIKEN – Advanced Science Institute, Wako,
Saitama 351–0198, Japan
e-mail: ksugioka@riken.jp

[4, 5]. Microfluidic-based systems are increasingly being integrated with various electrical, mechanical, and optical microcomponents on a single chip in order to perform more accurate analyses. Currently, the fabrication of microchips relies on planar microfabrication techniques such as injection molding or on conventional semiconductor processes based on photolithography. Although these techniques are well established and suitable for surface microfabrication, 3D microstructures require multilayer and multistep processes including stacking and bonding of substrates. Furthermore, in order to bond substrates, an adhesive agent is used, which can leak into the microfluidics embedded in the chip, thereby affecting the samples. A fs laser is a promising new tool for the direct fabrication of integrated microchips in glass due to its ability to internally modify transparent materials by multiphoton absorption.

One of the most popular glasses for this application is fused silica (cf. Chaps. 14 and 16), while photosensitive glass has some advantages over fused silica for 3D hollow microstructure fabrication although some ions contained may be a problem for some applications [2, 3]. One advantage of photosensitive glass is its pure photochemical reaction, enabling much smaller photon doses, i.e., lower laser powers and/or higher scanning speeds for laser direct writing [3, 6, 7]. Another advantage of photosensitive glass as compared with fused silica is that smooth surfaces of optical quality can be formed by a post-thermal treatment following the wet etching, which is important for biochip applications.

In this chapter, the principles and microstructuring procedures of fs laser processing of photosensitive glass are explained. Next, state-of-the-art techniques for fabricating 3D microfluidic structures and optical microcomponents are reviewed. Finally, fabrication of functional microchip devices such as a microfluidic dye laser, optofluidics, and a nano-aquarium by integrating the microcomponents in a single glass chip is demonstrated.

## 15.2 Photosensitive Glass

### 15.2.1 *Characteristics*

Photosensitive glass was originally developed by Stooky at Corning for surface microstructuring using ultraviolet (UV) irradiation [8]. There are now over 5,000 varieties of such glasses. One of the most commercially successful is manufactured by Schott glass and sold under the trade name of Foturan [9]. It has a large Young's modulus, a low absorption coefficient in the visible range, and good chemical stability and biocompatibility, which are attractive for fabrication of microsystems such as gas electron multiplier-type (GEM-type) detectors [10], microreactors [11], and miniaturized satellites [12]. It is a lithium aluminosilicate glass doped with trace amounts of silver and cerium. The cerium ($Ce^{3+}$) ion plays an important role as photosensitizer, releasing an electron when exposed to UV light to become $Ce^{4+}$.

## 15 Microstructuring of Photosensitive Glass

Some silver ions are then reduced by these free electrons, creating silver atoms. In a subsequent thermal treatment, the silver atoms first diffuse and agglomerate to form nanoclusters with a size of 8 nm or larger at about 500°C; then the crystalline phase of lithium metasilicate grows at about 600°C in the amorphous glass matrix in the vicinity of the silver clusters which act as nucleating centers. Because this crystalline phase is much more soluble in dilute HF acid than in the glass matrix, it can be preferentially etched away. A UV lamp is typically used for two-dimensional (2D) microfabrication of the photosensitive glass. Although internal microstructuring of the glass is possible using a pulsed UV laser [13–16], multiphoton absorption during fs laser irradiation makes it possible to fabricate complicated 3D microstructures with a high spatial resolution, including in the vertical direction parallel to the laser beam propagation.

Figure 15.1 shows an absorption spectrum of Foturan [17]. The absorption band edge is at around 340 nm. The small peak near 315 nm is ascribed to resonant $Ce^{3+}$ absorption. Therefore, a traditional fs laser with 775–800 nm wavelengths is transparent to this glass, as is two-photon absorption, but three-photon absorptions are likely.

The position of the permanent defect level $E_D$ in Foturan was determined using the usual Tauc relation [18],

$$K(E)E \propto (E - E_D)^2, \qquad (15.1)$$

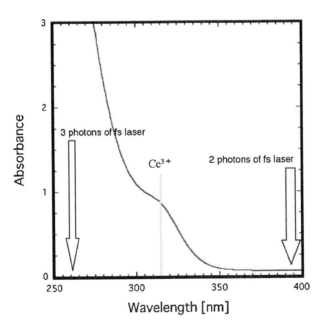

**Fig. 15.1** Absorption spectrum of a typical photosensitive glass (Foturan) between 250 and 400 nm in wavelength

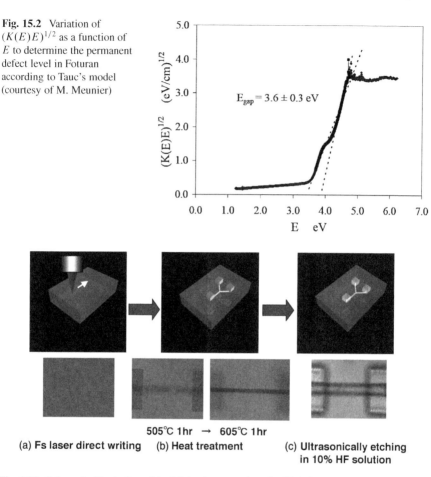

**Fig. 15.2** Variation of $(K(E)E)^{1/2}$ as a function of $E$ to determine the permanent defect level in Foturan according to Tauc's model (courtesy of M. Meunier)

**Fig. 15.3** Schematic illustration of the fabrication procedure for 3D hollow microstructures inside the photosensitive glass, together with optical microscope images of the sample after each step. (**a**) 3D latent images are written into the photosensitive glass by fs-laser direct writing. (**b**) The samples are subjected to a heat treatment to develop the modified regions. (**c**) The samples are soaked in a solution of 10% hydrofluoric (HF) acid diluted with water in an ultrasonic bath to selectively etch the laser-exposed regions and form hollow microstructures inside the glass

where $K(E)$ is the absorption coefficient and $E$ is the photon energy. Figure 15.2 plots the variation of $(K(E)E)^{1/2}$ as a function of $E$, revealing the defect level at $E_D = 3.6 \pm 0.3 \, \text{eV}$ [19].

## 15.2.2 Microstructuring Procedure

Figure 15.3 sketches the fabrication procedure for 3D hollow microstructures inside the photosensitive glass Foturan, including optical microscope images of the sample after each step. First, 3D latent images were written inside the photosensitive glass by fs-laser direct writing in panel (a). Typical laser writing parameters are a few $\mu$J pulse energy and a scanning speed of a few mm s$^{-1}$. At this stage, silver atoms are precipitated but no visible change occurs. After exposure to the fs laser, the samples are subjected to a programmed heat treatment. The temperature is first ramped up to 500°C at 5°C min$^{-1}$ and held at this temperature for 1 h. It is then raised to 605°C at 3°C min$^{-1}$ and held for another hour. Here, the laser-exposed regions become a crystalline phase of lithium metasilicate and show a dark color as seen in panel (b). After the sample is cooled to room temperature, it is soaked in 10% HF acid diluted with water in an ultrasonic bath. The lithium metasilicate is preferentially etched away at a 43× higher etching rate compared to the unmodified regions [20]. Finally, hollow microstructures are formed inside the glass as evident in panel (c). Any internal structure possesses one or more openings at the sample surface and such openings are needed in microfluidic devices for the introduction of samples. Since the crystallites of lithium metasilicate developed by the thermal treatment must be grown to a certain size (a few microns) to form an etchable network, the etching of the crystallites leaves behind a rough surface. To improve the surface smoothness, the glass sample can be baked again. First, the temperature is ramped up to 570°C at 5°C min$^{-1}$, held at this temperature for 5 h, and then reduced to 370°C at 1°C min$^{-1}$. The average surface roughness of 80 nm just after wet etching can be greatly improved to 0.8 nm by this additional thermal treatment [21], although the smoothing mechanism is unclear because the temperature is lower than the melting point of the glass. Perhaps the melting point at the surface is lower than that of the bulk so that a thin layer of liquid forms on the surface during the additional thermal treatment. The surface tension of this liquid layer can then form a smooth surface on the glass.

### 15.2.3 Microstructuring Mechanism

Three-photon absorption of the IR fs laser by the photosensitive glass is likely. Figure 15.4 plots the transmittance of the laser by the glass as a function of the input fluence [22]. Assume that absorption is related to the incident intensity $I$ by

$$dI/dx = -\alpha I^n, \qquad (15.2)$$

where $\alpha$ is the nonlinear absorption coefficient, $n$ is the number of photons absorbed simultaneously, and $x$ is the distance from the incident sample surface. Integrating (15.2) over the sample length, $L$ gives

$$I^{(1-n)} - I_0^{(1-n)} = (n-1)\alpha L, \qquad (15.3)$$

where $I_0$ is the laser intensity at the surface. By fitting the data in Fig. 15.4 between 0.6 and 1 J cm$^{-2}$ to (15.3), the values of $n$ and $\alpha$ are estimated to be 3.09 and

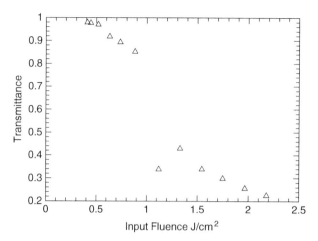

**Fig. 15.4** Transmittance versus input fluence for Foturan irradiated with a fs laser at 800 nm (courtesy of X. Xu)

$6.22 \times 10^{-7}$ cm$^3$W$^{-2}$, respectively. This result confirms three-photon absorption of the IR fs laser by the photosensitive glass, consistent with expectations from the absorption spectrum in Fig. 15.1.

It is also important to know the number of photons necessary for total photoreaction, as quantified in terms of the critical dose $D_c$, defined as the lowest dose necessary to achieve selective etching of the photosensitive glass in the exposed region. The model proposed by Fuqua et al. [15] assumes that to make the photosensitive glass selectively soluble in HF, the density of nuclei needs to reach a critical value $\rho_c$ at which an interconnected network of crystallites forms. The density of nucleation sites $\rho$ is expected to be proportional to the dose. Therefore, $\rho$ is fluence dependent,

$$\rho = KF^m N, \qquad (15.4)$$

where $F$ is the fluence per pulse, $m$ is the power dependence, $N$ is the number of laser pulses, and $K$ is a proportionality constant. By defining dose $D$ to be equal to $\rho/K$, (15.4) can be rewritten as

$$D = F^m N. \qquad (15.5)$$

For some number of pulses, $D_c$ corresponds to a critical fluence $F_c$, so that (15.5) becomes

$$D_c = F_c^m N. \qquad (15.6)$$

For doses exceeding $D_c$, the laser exposed regions are selectively etched, but not otherwise. This critical dose is a constant, depending only on material composition and process parameters.

The critical value $F_c$ is experimentally determined as the threshold fluence for selective etching at a given number of laser pulses. Figure 15.5 is a log–log plot of

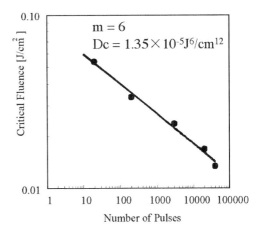

**Fig. 15.5** Dependence of the critical fluence on the number of laser pulses for photosensitive glass microstructuring using an IR fs laser

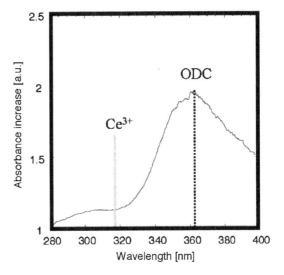

**Fig. 15.6** Absorbance increase of photosensitive glass after exposure to an fs laser, as derived by dividing the absorbance spectrum of the exposed sample by that of the unexposed sample

the critical fluence versus the total number of laser pulses [3, 18]. Fitting the data to (15.5) results in $m = \sim 6$, a six-photon process, and $D_c = 1.35 \times 10^{-5}$ J$^6$ cm$^{-12}$. The six-photon process was also confirmed by Fisette et al. [19,23] but their reported $D_c$ of $1.2 \times 10^{-2}$ J$^6$ cm$^{-12}$ is significantly higher, probably due to the different spot sizes and numerical apertures (NAs) of the focusing lenses used. Summarizing, absorption takes place by three photons while six photons are necessary for the total photoreaction.

Figure 15.6 plots the spectral absorbance increase of the photosensitive glass in the near UV range after exposure to a fs laser, as derived by dividing the absorbance spectrum of the exposed sample by the unexposed one [17]. A significant increase is observed in the vicinity of 360 nm corresponding to absorption by oxygen deficient centers (ODCs) [2, 24], while there is little change around 315 nm which would

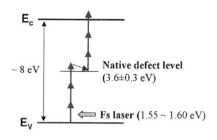

**Fig. 15.7** Processes for free-electron generation in photosensitive glass by fs-laser irradiation

indicate a transformation from $Ce^{3+}$ to $Ce^{4+}$. Therefore, $Ce^{3+}$ hardly contributes free electrons to reduce $Ag^+$ to $Ag^0$ during the fs laser irradiation. Presumably, free electrons are instead generated by interband excitations. That is, when the fs laser irradiates the photosensitive glass, free electrons are generated from valence to conduction band transitions, resulting in bond scission of Si and O in the glass followed by formation of ODCs that pair with non-bridging oxygen hole centers (NBOHCs).

Based on these ideas, the process of generating free electrons in photosensitive glass by Ag atom precipitation during fs laser irradiation is summarized in Fig. 15.7. Free electrons are generated by interband excitations induced by three-photon absorption, in accordance with the absorption edge in Fig. 15.1 and the transmittance measurements plotted in Fig. 15.4. The absorption is dominated by levels associated with impurities and native defects. Specifically, a defect level at $E_D = 3.6 \pm 0.3\,eV$ above the valence band in Foturan was experimentally observed (see Fig. 15.2) and excitation to this defect level requires at least three photons of 775–800 nm wavelength (corresponding to 1.60–1.55 eV). First, electrons are excited from the valence band to this defect level by three-photon absorption. Second, at least three others photons are necessary to excite electrons from the defect level to the conduction band at approximately 8 eV [25] above the valence band and thus 4.4 eV above the defect level. Consequently, the overall process for generating free electrons requires six photons in two steps of three-photon absorption each from the valence band to the conduction band through the defect level [17, 19, 23]. The generated electrons reduce $Ag^+$ to $Ag^0$. The successive processes of thermal treatment and wet etching after Ag precipitation (namely, Ag nanoparticle formation, growth of the lithium metasilicate, and selective removal of the modified regions) proceed by the same mechanisms as it occur in the case of UV light exposure.

## 15.3 Fabrication of Microfluidic Structures

Formation of 3D hollow microstructures in photosensitive glass using an fs laser was first demonstrated by Kondo et al. [2]. Nowadays, fabrication of various microfluidic structures including straight, U-shaped, and Y-shaped channels, vertical microfluidic structures, and microvalves have been demonstrated [3, 19, 20, 22, 26–37].

15 Microstructuring of Photosensitive Glass

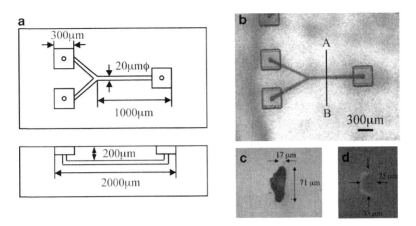

**Fig. 15.8** (**a**) Configuration of a horizontal Y-shaped microchannel structure. Photos of (**b**) a top view of the fabricated structure, and (**c**) a cross-section of the microchannel after it was mechanically cut along line A–B. (**d**) Rounder cross-sectional shape of the microchannel by using a slit with a width of 0.2 mm

Based on the procedure described in Sect. 15.2.2, a planar microfluidic structure with a 3D Y-shaped microchannel in photosensitive glass was fabricated as shown in Fig. 15.8 [3, 20, 26]. Panel (a) is a schematic of the fabricated structure. Three open microreservoirs of 300 × 300 μm size, formed at the surface, were connected by a Y-shaped hollow microchannel embedded 200 μm below the surface. Panels (b) and (c) are optical micrographs of a top view and of a cross-section of the embedded microchannel when mechanically cut along line A–B, respectively. The hollow channel is elliptical with a width and aspect ratio (ratio of height to width) of 17 μm and 4.2, respectively. This elliptical shape results from the mismatch between the focal radius and Rayleigh range at the focus. The large aspect ratio of the microchannel can be controlled by using a narrow slit [38]. By using a slit with a width of 0.2 mm, the aspect ratio of the microchannel was improved to ∼1.3 as shown in panel (d). Changing the width of the slit enables control of the aspect ratio [39, 40] as discussed in Chap. 5. A method to realize a perfectly circular cross-sectional shape is crossed-beam irradiation of fs laser beams, in which the foci of two beams are spatiotemporally overlapped in the glass using two orthogonal objective lenses [41].

Figure 15.9 shows two optical microscope images of a microfluidic device with a moveable microplate in a hollow chamber inside the photosensitive glass [28]. The microplate can be actuated using compressed air. When air is infused from the top left opening, the microplate moves to the right side, as in Fig. 15.9a. In this scheme, the liquid sample injected from the central opening at the top flows into the left channel of the bottom through the chamber as indicated by the dashed arrows. On the other hand, if compressed air is introduced from the top right opening, the microplate moves leftward, as indicated in Fig. 15.9b in which case the liquid flows

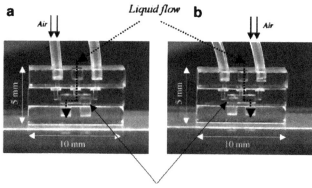

*Movable plate embedded in the photosensitive glass*

**Fig. 15.9** Prototype of a microfluidic device with a freely moveable microplate that can switch the flow direction of a liquid sample. (**a**) By infusing compressed air from the left opening of the top part, the microplate blocks the right side. (**b**) If compressed air is infused from the top right opening, the microplate slides to the left

to the right channel of the bottom part. Consequently, this microplate can act as a switching valve for a stream of liquid. Other micromechanical components such as a microrotator and a microgear can also be fabricated.

## 15.4 Fabrication of Micro-Optic Structures

### 15.4.1 Micro-Optics

The technique for fabricating 3D hollow microstructures inside photosensitive glass using an fs laser can also be applied to create micro-optics, since the post-thermal treatment after the wet etching yields smooth surfaces of optical quality [42–44]. To fabricate a 45° internal micromirror, a hollow microplate was embedded in a glass chip as shown in Fig. 15.10a. It reflects incident light by total internal reflection, since the interior of the hollow microplate is filled with air whose refractive index is smaller than that of the surrounding glass [42]. The optical path is indicated by the arrows in the figure. The structure bends incident light through an angle of 90°. Figure 15.10b shows the beam spot of a HeNe laser reflected by the micromirror onto a receiving screen positioned 10 mm away from the end of the photosensitive glass. The reflectivity of the micromirror was measured to be about 92%, implying that there is almost no loss because the light beam passes through two air–glass interfaces at the glass chip sidewalls with a 4% reflection loss each time.

A microlens was also fabricated in photosensitive glass [44]. Such hollow structures always have openings at either or both ends. One of the internal sidewalls of the hollow structure has a spherical shape, thereby comprising a plano-concave

## 15 Microstructuring of Photosensitive Glass

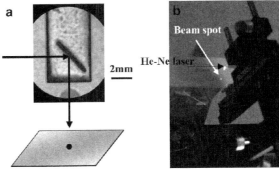

**Fig. 15.10** (a) Top view of a 3D micromirror fabricated inside photosensitive glass with its optical path indicated by the *arrows*. (b) Beam spot reflected by the micromirror onto a receiving screen placed 10 mm away from the end of the glass chip

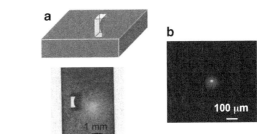

**Fig. 15.11** (a) 3D schematic and optical microscope image of a microlens embedded in photosensitive glass. (b) Image of a HeNe laser beam focused by the fabricated structure

hollow structure which works as a plano-convex lens, as shown in Fig. 15.11a. The figure also shows an optical microscope image of the microlens having a thickness of 2 mm, radius of curvature $R$ of 0.75 mm (design value), and focal length of 1.5 mm calculated from the formula $1/f = (n-1) \cdot (1/R)$ with $n = 1.5$ for the glass. The focal spot size was measured to be approximately 30 μm in diameter for an incident HeNe laser beam of 1.5–1.8 mm in diameter, as shown in Fig. 15.11b. The efficiency of the microlens was measured to exceed 80%. The principal optical losses are the four Fresnel reflections at the air–glass interfaces (two glass chip side walls and two lens surfaces). Subtracting the loss due to the four Fresnel reflections (4% each), the remaining loss is then estimated to be only ∼4%.

### 15.4.2 Optical Waveguides

Optical waveguides can be written in photosensitive glass by internal modification of the refractive index using fs-laser direct writing similar to what has been done in fused silica and other glasses (cf. Parts II and III) [40, 45–48]. After the laser writing, thermal treatment must not be performed, in order to avoid precipitation of silver atoms. Single-mode propagation is achieved by writing the waveguides with a laser pulse energy of 1–2 μJ per pulse and a scanning speed of 200–500 μm s$^{-1}$ at 775 nm wavelength and 150 fs pulse width [48]. Higher pulse energy or slower scanning speed results in the formation of multimode waveguides. The propagation loss at 632.8 nm was estimated to be ~0.5 dB cm$^{-1}$ for all single-mode waveguides regardless of the laser irradiation conditions. That loss is within acceptable limits for optofluidic applications.

### 15.4.3 Integration of Optical Microcomponents

Since the intrinsic properties of the photosensitive glass in the unirradiated regions do not markedly change even after multiple thermal treatments, optical waveguides can be written inside the glass by fs-laser-induced refractive index modifications after fabrication of a microlens and micromirror. Thus, 3D integration of waveguides and micro-optics can be realized in a single glass chip. Figure 15.12a diagrams a micro-optical device in which two waveguides are integrated with a micromirror and a microlens. Waveguide I of 5-mm length is connected to a micromirror at an angle of 45° that connects to another waveguide II of 4-mm length at an angle of 90°. Waveguide II was stopped 2 mm before a plano-convex microlens so as to obtain a focused beam output. A near-field focal spot profile transmitted by the structure in panel (b) is shown in panel (c). The output spot size (dark blue spot) is about 7 μm in diameter. The propagation loss of the optical waveguide and the net loss of the microlens are 0.5 dB cm$^{-1}$ and ~6%, respectively. In addition, the bending loss at the micromirror is measured to be less than 0.3 dB at a wavelength of 632.8 nm [48]. One of the biggest advantages of this structure is that it can bend a light beam in a small area with minimal loss. From a practical point of view, light beams in waveguides often need to be bent when used in optical and microfluidic devices. Usually a curved waveguide is used for that purpose, but then bending losses can become significant. To minimize the bending loss, the curvature of a curved waveguide has to be about 5–6 mm, resulting in an undesirable increase in device size.

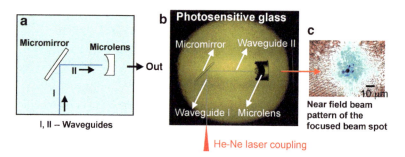

**Fig. 15.12** (a) Schematic of the 3D integration of two waveguides with a micromirror and an optical plano-convex microlens in a single glass chip. (b) Optical microscope image and (c) characterization of the 3D microdevice. The *solid gray lines* in panel (b) mark the locations of the invisible waveguides inside the glass

## 15.5 Fabrication of Microchip Devices

### 15.5.1 Microfluidic Dye Laser

Three-dimensional micro-optic and microfluidic devices can be integrated into a single chip for fabricating hybrid devices, such as microfluidic dye lasers, which are useful as light sources for applications such as fluorescence detection or photo-absorption spectroscopy in lab-on-a-chip devices, μ-TAS, and optofluidics [49, 50]. Figure 15.13a shows a top view of a microfluidic laser that has an optical microcavity composed of four 45° micromirrors vertically buried in the glass, a horizontal microfluidic chamber embedded 400 μm below the glass surface, and a microfluidic channel through the center of the microchamber [51]. Figure 15.13b is a micrograph of a side view of the laser, showing the microchannel with an average diameter of 80 μm and the microchamber with a thickness of 200 μm. All components were made of hollow microstructures and fabricated by one continuous fs-laser direct writing followed by thermal treatment, wet etching, and post-thermal treatment. Figure 15.13c illustrates the optical path in the laser. The optical cavity is composed of a pair of corner mirrors formed by two micromirrors on the left-hand side and two on the right-hand side. Light bounces back and forth in the optical cavity by total internal reflection. Lasing action occurs if the microchamber is filled with a gain medium such as the laser dye rhodamine 6G (Rh6G), and then pumped by the 2nd harmonic of an Nd:YAG laser. A small amount of light leaks

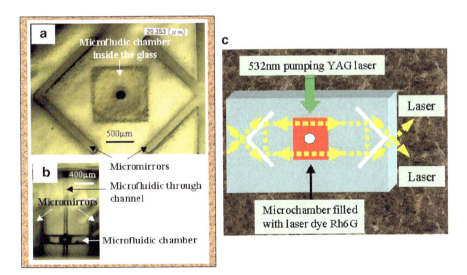

**Fig. 15.13** Optical micrograph of (**a**) a top view and (**b**) a side view of a microfluidic dye laser. (**c**) Illustration of the optical path inside the dye laser

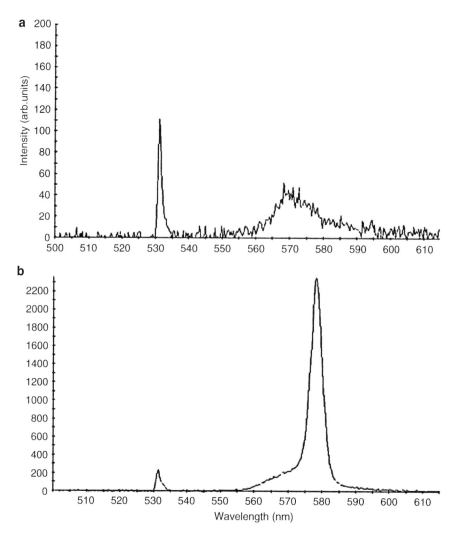

**Fig. 15.14** Emission spectra of the microfluidic dye laser at pump powers of (**a**) 1.66 mJ cm$^{-2}$ and (**b**) 4.49 mJ cm$^{-2}$. The peaks centered at 532 nm are scattered pump light

out of the optical cavity and is emitted tangentially from the internal surfaces of the micromirrors [52].

After filling the microchamber with Rh6G, the dye laser was pumped by the 2nd harmonic of a pulsed Nd:YAG laser. Figure 15.14 plots the resulting emission spectra at two different pump energies. When a low pump fluence of 0.46 mJ cm$^{-2}$ was incident, only spontaneous emission with a broad spectrum was observed (not shown in Fig. 15.14). The threshold for lasing action occurred at a pump fluence of 1.66 mJ cm$^{-2}$ as shown in Fig. 15.14a. Further increase in the pump fluence rapidly increased the output power of the microfluidic laser accompanied by a narrowed bandwidth. A typical emission spectrum with a center wavelength of 578 nm for a large pump energy of 4.49 mJ cm$^{-2}$ is shown in Fig. 15.14c. The output power of the microfluidic laser at this pump energy reached ∼10 mW at a 15-Hz repetition rate.

## 15.5.2 Optofluidics

The integration of microfluidics with optical waveguides and micro-optics can achieve optofluidics possessing high sensitivity. First, microfluidics and micro-optics were fabricated in one continuous process of fs-laser direct writing, thermal treatment, successive wet etching, and post-thermal treatment. Then an optical waveguide was written inside the glass chip by fs-laser-induced refractive index modification. Neither heat treatment nor wet etching was employed afterward.

Figure 15.15a shows 2D and 3D schematic illustrations of an integrated microchip for photonic biosensing, wherein one waveguide with a 6-mm length is connected to a microchamber of $1.0 \times 1.0 \times 1.0$ mm$^3$ volume, and two microlenses of 0.75-mm radius of curvature are arranged for fluorescence and absorption measurements at a distance of 200 μm away. An optical micrograph of the fabricated microchip is shown in Fig. 15.15b.

For fluorescence analysis of a liquid sample, the microfluidic chamber was filled with Rh6G dissolved in ethanol at 0.02 mol L$^{-1}$. A pump beam from the 2nd harmonic of an Nd:YAG laser was sent through the optical waveguide and introduced into the microfluidic chamber. Emission spectra from the Rh6G were

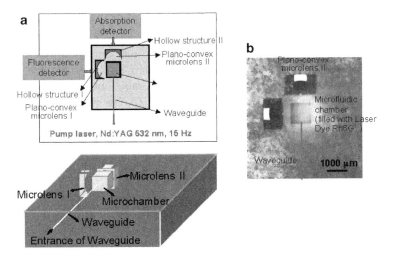

**Fig. 15.15** (**a**) 2D and 3D configurations and (**b**) optical microscope image of two plano-convex lenses and an optical waveguide integrated with a microfluidic chamber in a single glass chip. The *solid gray line* in panel (**b**) marks the location of the invisible waveguide inside the glass

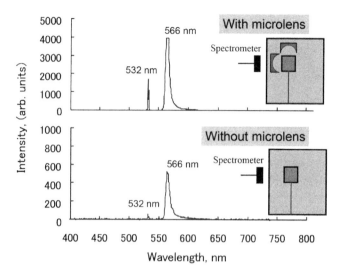

**Fig. 15.16** Emission spectra from Rh6G laser dye pumped by the 2nd harmonic of an Nd:YAG laser at 32 mW. The images in the top and bottom panel were obtained using a microchip with and without a microlens, respectively

collected in the detector through plano-convex microlens I. The head of the spectrometer was placed at the end of the glass chip behind microlens I to collect the fluorescence. The emission spectra were measured at different pump energies. A typical emission spectrum with a central wavelength of 566 nm (where the peak at 532 nm is from the pump laser), corresponding to the emission maximum of Rh6G, is plotted in Fig. 15.16. For comparison, the emission measurement was also performed for a microfluidic chamber integrated with a waveguide but without a microlens. Enhanced emission intensity by a factor of 8 is achieved when a micro-optical plano-convex lens is integrated to collect the fluorescence.

In addition, optical absorption of a liquid sample through plano-convex microlens II was demonstrated using black ink at different concentrations ranging from 0.05% to 1.0% diluted in water in the microfluidic chamber. The absorption spectra showed a strong dependence on the concentration of the ink when white light was guided into the microfluidic chamber through the optical waveguide. The sensitivity was enhanced by a factor of 3 for the microchip integrated with the microlens at concentrations of 0.4% to 0.6%.

### 15.5.3 Nano-Aquarium

Another interesting application of 3D microfluidic structures is the manufacture of microchips designed for dynamic observations of microorganisms, referred to as a nano-aquarium [53]. A large variety of organisms are presently living on the earth, among which micrometer-sized aquatic organisms (microorganisms) have been surviving for over 500 million years, barely changing their shapes. Some of these microorganisms exhibit extremely rapid motions, which are unusual in the macro-world in which we live, and unique 3D movements contradicting gravity. Most such organisms are composed of a single cell and so it is useful to explore their dynamic movements and physiological energy generation mechanisms to better understand the ability and function of single cells in general. However, the observation of microorganisms is challenging for cell biologists [54–57]. A nano-aquarium has several advantages over conventional observation methods using a glass slide with a coverslip or a Petri dish. It can significantly shorten the observation times and it can be used for 3D observations. In addition, the microorganisms can be easily stimulated using mechanical microcomponents integrated into the microchip.

*Euglena gracilis* is a single-celled organism living in fresh water. It has a flagellum emerging from its anterior end, which it whips rapidly to swim. Until now, many biologists have investigated the flagellum movement using a microscope to understand the origins of this functionality. However, only the thrusting movement has been amenable to study [58,59] owing to difficulties in capturing the continuous high-speed movement of the flagellum.

To better enable observation of the flagellum motion, a microchannel buried in photosensitive glass was fabricated whose structure is shown in Fig. 15.17a [53]. The size of one *Euglena gracilis* is about 100-μm long ×40-μm wide and it generally moves forward by whipping its flagellum around its body. Therefore, a microstructure with a 1-mm-long channel and a cross-section of $150 \times 150\,\mu m^2$ embedded 150 μm below the glass surface was constructed. At both ends of the channel, two open reservoirs with a size of $500 \times 500\,\mu m^2$ were connected to enable the introduction of *Euglena gracilis*. A critical factor in the fabrication of this microchip is that the distance between the glass surface and the upper wall of the microchannel needs to be between 130 and 170 μm, as determined by the working distance of the objective lens used for the observation. Furthermore, the etched glass surface must be smooth and the upper wall of the channel must be flat and parallel to the glass surface. For observations, *Euglena gracilis* is introduced into one of the reservoirs using an injection syringe filled with water. The microchannel immediately fills with water and the *Euglena gracilis* swims into the microchannel by itself since there is no longer any water flow in the microchannel. In the microchannel, the *Euglena gracilis* is confined in a limited area so that one can analyze the dynamic observations of the flagellum movement using a microscope from the top of the glass surface, as shown in Fig. 15.17b. The objective lens of the microscope is focused on the center of the microchannel, so that sequential pictures of the flagellum movement can be recorded when the *Euglena gracilis* swims in it. The microchip fabricated for this observation has a microchannel with a rectangular cross-section embedded 150 μm below the glass surface as shown in the inset in Fig. 15.17b. The top internal wall of the microchannel is flat and smooth and is parallel to the glass surface, in order to take clear images. That internal wall was fabricated by multiple scanning of the laser beam with lateral shifts and

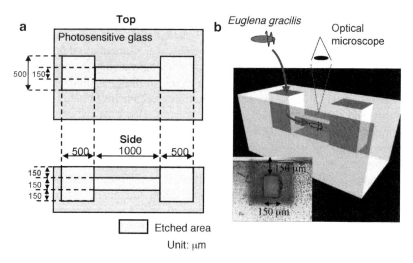

**Fig. 15.17** (a) 2D and (b) 3D schematics of the microchip used for observations of the motion of *Euglena gracilis*. The *inset* in panel (b) is an optical microscope image of the cross-section of an embedded microchannel

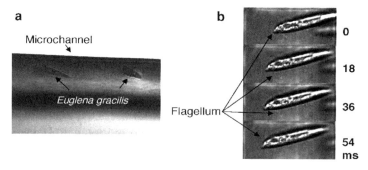

**Fig. 15.18** Microscope images of (a) encapsulated *Euglena gracilis* swimming in the microchannel and (b) sequential pictures of advancing *Euglena gracilis*

additional annealing. It is then straightforward to capture images of *Euglena gracilis* swimming in the microchannel by microscopic observations.

Figure 15.18a shows an optical microscope image of *Euglena gracilis* swimming in the embedded microchannel. Panel (b) shows enlarged sequential pictures of an advancing *Euglena gracilis* obtained from a movie. The *Euglena gracilis* coils its flagellum around its body and rotates very rapidly to swim in a straight line.

The *Euglena gracilis* can be stimulated by irradiating light from any direction to control its motion. It then proved possible to capture a movie of a rotating *Euglena gracilis* with white light shining up from the bottom of the microchip. Observation of the front side of *Euglena gracilis* was demonstrated for the first time using the microchip.

In addition to the observation of *Euglena gracilis*, several kinds of nano-aquariums with different structures and functionality were fabricated for various applications, including understanding the information transmission process in *Pleurosia leavis*, observations of high-speed motion of *Cryptomonas* in flowing water, and the *Phormidium* assemblage to the seedling root for growth promotion of *Komatsuna* [53, 60, 61].

## 15.6 Summary

Photosensitive glass has the useful properties of a large Young's modulus, a low absorption coefficient in the visible, good chemical stability, and biocompatibility for fabrication of various microsystems. Infrared fs-laser direct writing followed by thermal treatment and wet chemical etching can form 3D hollow microstructures embedded in the photosensitive glass. Additional thermal treatment after the wet chemical etching results in smooth etched surfaces of optical quality. A variety of microfluidic structures, micromechanical components such as a microvalve, and micro-optical components such as micromirrors and microlenses were fabricated by this technique. One of the interesting and new applications of microchips with 3D microfluidic structures is the dynamic observation of microorganisms, enabling highly functional observations and overcoming some drawbacks with conventional methods. Integration of microfluidics and micro-optics can be carried out in one continuous procedure without the cumbersome need to align each microcomponent, thereby enabling fabrication of a microfluidic dye laser. An optical waveguide can be further integrated into a glass chip after fabrication of microfluidic and micro-optic components with little difficulty. Such an integrated microchip is of great use for sensitive biochemical analyses and medical inspections based on photonic sensing. In the future, it is anticipated that 3D micromachining inside photosensitive glass using an fs laser will have a significant impact on the manufacture of microchips for biomedical applications.

## References

1. A. Marcinkevicius, S. Juodkazis, M. Watanabe, M. Miwa, S. Matsuo, H. Misawa, Opt. Lett. **26**, 277 (2001)
2. Y. Kondo, J. Qui, J.T. Mitsuyu, K. Hirao, T. Yoko, Jpn. J. Appl. Phys. **38**, L1146 (1999)
3. K. Sugioka, Y. Cheng, K. Midorikawa, Appl. Phys. A **81**, 1 (2005)
4. M.A. Burns, B.N. Johnson, A.N. Brahmasandra, K. Handique, J.R. Webster, M. Krishnan, T.S. Sammarco, P.M. Man, D. Jones, D. Heldsinger, C.H. Mastrangelo, D.T. Burke, Science **282**, 484 (1998)
5. P.S. Dittrich, K. Tachikawa, A. Manz, Anal. Chem. **78**, 3887 (2006)
6. Y. Bellouard, A. Said, M. Dugan, P. Bado, Opt. Exp. **12**, 2120 (2004)
7. Y. Bellouard, A. Said, P. Bado, Opt. Exp. **13**, 6635 (2005)

8. S.D. Stookey, Ind. Eng. Chem. **45**, 115 (1953)
9. http://www.mikroglas.com/foturane.htm.
10. S.K. Ahn, J.G. Kim, V. Perez-Mendez, S. Chang, K.H. Jackson, J.A. Kadyk, W.A. Wenzel, G. Cho, IEEE Trans. Nucl. Sci. **49**, 870 (2002)
11. T.R. Dietrich, A. Freitag, R. Scholz, Chem. Eng. Technol. **28**, 447 (2005)
12. H. Helvajian, P.D. Fuqua, W.W. Hansen, S. Jansen, RIKEN Rev. **32**, 57 (2001)
13. W.W. Hansen, S.W. Janson, H. Helvajian, Proc. SPIE **2991**, 104 (1997)
14. P.D. Fuqua, S.W. Janson, W.W. Hansen, H. Helvajian, Proc. SPIE **3618**, 213 (1999)
15. P.D. Fuqua, D.P. Taylor, H. Helvajian, W.W. Hansen, M.H. Abraham, Mater. Res. Soc. Symp. Proc. **624**, 79 (2000)
16. H. Helvajian, P.D. Fuqua, W.W. Hansen, S. Janson, Proc. SPIE **4088**, 319 (2000)
17. T. Hongo, K. Sugioka, H. Niino, Y. Cheng, M. Masuda, I. Miyamoto, H. Takai, K. Midorikawa, J. Appl. Phys. **97**, 063517 (2005)
18. S. Elliot, *The Physics and Chemistry of Solids* (Wiley, West Sussex, 1998) p.770
19. B. Fisette, M. Meunier, J. Laser Micro/Nanoeng. **1**, 7 (2006)
20. M. Masuda, K. Sugioka, Y. Cheng, N. Aoki, M. Kawachi, K. Shihoyama, K. Toyoda, H. Helvajian, K. Midorikawa, Appl. Phys. A **76**, 857 (2003)
21. Y. Cheng, K. Sugioka, K. Midorikawa, M. Masuda, K. Toyoda, M. Kawachi, K. Shihoyama, Opt. Lett. **28**, 1144 (2003)
22. J. Kim, H. Berberoglu, X. Xu, J. Microlith. Microfab. Microsyst. **3**, 478 (2004)
23. B. Fisette, F. Busque, J.-Y. Degorce, M. Meunier, Appl. Phys. Lett. **88**, 091104 (2006)
24. Y. Kondo, K. Miura, T. Suzuki, H. Inouye, T. Mitsuyu, K. Hirao, J. Non-Cryst. Solids **253**, 143 (1999)
25. W.D. Kingery, H.K. Bowen, D.R. Uhlmann, *Introduction to Ceramics*, 2nd edn. (Wiley, New York, 1976), p. 1032
26. Y. Cheng, K. Sugioka, M. Masuda, K. Toyoda, M. Kawachi, K. Shihoyama K. Midorikawa, RIKEN Rev. **50**, 101 (2003)
27. K. Sugioka, M. Masuda, T. Hongo, Y. Cheng, K. Shihoyama, K. Midorikawa, Appl. Phys. A **78**, 815 (2004)
28. M. Masuda, K. Sugioka, Y. Cheng, T. Hongo, K. Shihoyama, H. Takai, I. Miyamoto, K. Midorikawa, Appl. Phys. A **78**, 1029 (2004)
29. K. Sugioka, Y. Cheng, K. Midorikawa, J. Photopolmer Sci. Technol. **17**, 397 (2004)
30. Y. Cheng, K. Sugioka, K. Midorikawa, Appl. Surf. Sci. **248**, 172 (2005)
31. Y. Cheng, K. Sugioka, K. Midorikawa, Opt. Exp. **13**, 7225 (2005)
32. K. Sugioka, Y. Cheng, K. Midorikawa, F. Takase, H. Takai, Opt. Lett. **31**, 208 (2006)
33. K. Sugioka, Y. Cheng, K. Midorikawa, J. Phys. Conf. Ser. **59**, 533 (2007)
34. K. Sugioka, Y. Hanada, K. Midorikawa, Appl. Surf. Sci. **253**, 6595 (2007)
35. Y.Z. Wu, W. Jia, C.Y. Wang, M.L. Hu, X.C. Ni, L. Chai, Opt. Quant. Electron. **39**, 1223 (2007)
36. Y. Wu, Q.Y. Wang, Lasers Eng. **19**, 39 (2009)
37. J.M. Fernández-Pradas, D. Serranoa, P. Serraa, J.L. Morenza, Appl. Surf. Sci. **255**, 5499 (2009)
38. Y. Cheng, K. Sugioka, K. Midorikawa, M. Masuda, K. Toyoda, M. Kawachi, K. Shihoyama, Opt. Lett. **28**, 55 (2003)
39. M. Ams, G.D. Marshall, D.J. Spence, M.J. Withford, Opt. Exp. **13**, 5676 (2005)
40. K.J. Moh, Y.Y. Tan, X.-C. Yuan, D.K.Y. Low, Z.L. Li, Opt. Exp. **13**, 7288 (2005)
41. K. Sugioka, Y. Cheng, K. Midorikawa, F. Takase, H. Takai, Opt. Lett. **31**, 208 (2006)
42. Y. Cheng, K. Sugioka, K. Midorikawa, M. Masuda, K. Toyoda, M. Kawachi K. Shihoyama, Opt. Lett. **28**, 1144 (2003)
43. Y. Cheng, H.-L. Tsai, K. Sugioka, K. Midorikawa, Appl. Phys. A **85**, 11 (2006)
44. Z. Wang, K. Sugioka, K. Midorikawa, Appl. Phys. A **89**, 951–955 (2007)
45. V.R. Bhardwaj, E. Simova, P.B. Corkum, D.M. Rayner, C. Hnatovsky, R.S. Taylor, B. Schreder, M. Kluge, J. Zimmer, J. Appl. Phys. **97** 0831021 (2005)
46. Z.L. Li, D.K.Y. Low, M.K. Ho, G.C. Lim, K.J. Moh, J. Laser Appl. **18**, 320 (2006)
47. R. An, Y. Li, D. Liu, Y. Dou, F. Qi, H. Yang, Q. Gong, Appl. Phys. A **86**, 343 (2007)
48. Z. Wang, K. Sugioka, Y. Hanada, K. Midorikawa, Appl. Phys. A **88**, 699 (2007)

49. M.A. Burns, B.N. Johnson, A.N. Brahmasandra, K. Handique, J.R. Webster, M. Krishnan, T.S. Sammarco, P.M. Man, D. Jones, D. Heldsinger, C.H. Mastrangelo, D.T. Burke: Science **282**, 484 (1998)
50. P.S. Dittrich, K. Tachikawa, A. Manz, Anal. Chem. **78**, 3887 (2006)
51. Y. Cheng, K. Sugioka, K. Midorikawa, Opt. Lett. **29**, 2007 (2004)
52. Z. Wang, K. Sugioka, K. Midorikawa, Appl. Phys. A **93**, 225 (2008)
53. Y. Hanada, K. Sugioka, H. Kawano, I.S. Ishikawa, A. Miyawaki, K. Midorikawa, Biomed. Microdevices **10**, 403 (2008)
54. K. Okano, E. Hunter, N. Fusetani, J. Exp. Zool. **276**, 138 (1996)
55. K. Yoshimura, C. Shingyoji, K. Takahashi, Cell Motil. Cytoskeleton **36**, 236 (1997)
56. S.L. Fleming, C.L. Rieder, Cell Motil. Cytoskeleton **56**, 141 (2003)
57. D.J. Stephens, V.J. Allan, Science **300**, 82 (2003)
58. K.M. Nichols, R. Rikmenspoel, J. Cell Sci. **23**, 211 (1977)
59. K.M. Nichols, R. Rikmenspoel, J. Cell Sci. **29**, 233 (1978)
60. Y. Hanada, K. Sugioka, H. Kawano, I.S. Ishikawa, A. Miyawaki, K. Midorikawa, Appl. Surf. Sci. **255**, 9893 (2009)
61. Y. Hanada, K. Sugioka, I.S. Ishikawa, H. Kawano, A. Miyawaki, and K. Midorikawa, Lab. Chip. **11**, 2109 (2011)

# Chapter 16
# Microsystems and Sensors

Yves Bellouard, Ali A. Said, Mark Dugan, and Philippe Bado

**Abstract** In this chapter, the use of femtosecond laser to fabricate glass-based microsystems and sensors is examined. Several advanced technology demonstrators are presented, including some very unusual fused silica MEMS. Present technological barriers are discussed.

## 16.1 Introduction

The unconventional laser–matter interaction resulting from the very high-peak powers associated with femtosecond laser pulses provides novel ways to tailor material properties. Applied to dielectrics, such as fused silica, these localized changes of physical properties can be advantageously used to create miniaturized devices that combine optical, fluidics, and mechanical functions.

In this chapter, we discuss the opportunity of using femtosecond laser processes to form microsystems. The first part briefly introduces micro- and nano-systems with an emphasize on limitations of current fabrication practices. The second part reviews the work done to create miniaturized device with femtosecond lasers. The third part presents a generalized concept of integrated microdevices produced using femtosecond lasers.

---

Y. Bellouard (✉)
Mechanical Engineering Department, Eindhoven University of Technology, Den Dolech 2, P.O. Box 513, 5600 MB Eindhoven, The Netherlands
e-mail: y.bellouard@tue.nl

A.A. Said · M. Dugan · P. Bado
Translume Inc., 655, Phoenix Drive, Ann Arbor, MI 48108, USA

R. Osellame et al. (eds.), *Femtosecond Laser Micromachining*,
Topics in Applied Physics 123, DOI 10.1007/978-3-642-23366-1_16,
© Springer-Verlag Berlin Heidelberg 2012

## 16.2 Micro- and Nano-Systems

### 16.2.1 A Brief Overview of Microsystems

Micro- and nano-technologies have emerged as key technologies for modern societies. These technologies are expected to significantly impact our industrialized economies as well as to generally contribute to the advancement of science.

Microsystems perform sophisticated tasks in a miniaturized volume. Shaping or analyzing light signals, mixing, processing or analyzing ultra-small volumes of chemicals, sensing mechanical signals, probing gas, and sequencing bio-molecules are common operations that can be done by these small machines. Rationales for the use of microsystems are numerous. The reduction of consumables (reduced consumption of chemicals in Lab-on-a-Chip for example), a fast response time (critical in airbag sensors), an enhanced portability (RF-MEMS), higher resolution (with Inkjet printer head), higher efficiency (using microchemical reactors), smaller footprint, etc. are typical benefits sought.

Starting from integrated circuits in the 1970s, followed by micromechanical systems in the 1980s–1990s, photonics and fluidics in the 1990s–2000s and recently the addition of organic material and bio-molecules, micro-/nano-systems are becoming complex machines performing sophisticated tasks (Fig. 16.1).

### 16.2.2 Issues on Microsystems Integration and Fabrication

So far, the fabrication of microsystems [1] essentially relies on two technology platforms that are used separately or jointly.

The first technology platform is based on surface micromachining of substrates such as silicon. It relies on "clean-room" processes that, for the most part, were inherited from the microelectronics industry. Devices produced by surface micromachining are fabricated through successive steps of material deposition and

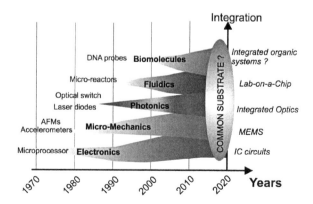

**Fig. 16.1** Roadmap toward microsystem integration

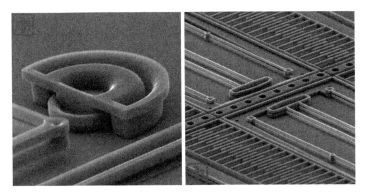

**Fig. 16.2** Illustration of a multilayers MEMS device produced by lithography methods (courtesy of Sandia National Lab)

selective material removal. By carefully selecting the proper combination of materials deposition and selective etching, one can fabricate, layer-by-layer, relatively complex devices. This approach produced elements that are planar or near-planar. Figure 16.2 shows an example of MEMS devices produced with this approach. The multilayer structure observed through sequential steps of etching/deposition is particularly visible on the left picture.

With the increasing number of desired functionalities, the surface micromachining approach faces numerous issues. The growing number of materials and processing steps introduce difficult material compatibility.

The second technology platform is the micro-assembly. Parts produced by surface micromachining or other single-process microfabrication techniques are put together by micro-assembly to form a device. There, the main challenges are to precisely position objects and to assemble them in a reliable manner through bonding or joining processes. With the small size of the components of interest, the assembly process is often quite tedious and sometimes nearly impossible.

In addition to the technical limitations mentioned above, these two microfabrication approaches face economical and societal issues. The spectacular miniaturization trend ongoing for the last thirty years in various fields has not been paralleled by a similar miniaturization of the production means. Ironically, today, the microsystems industry uses very large pieces of sophisticated equipment to fabricate very small parts, thus large capital investments are required to set-up and operate microsystem production facilities. Consequently, only products with potential large markets are considered, and only a few large suppliers can make the necessary financial investments. Small and medium size enterprises are prevented from entering the field although they usually are a strong source of innovative ideas. This issue has a negative impact on our innovation capabilities as well as our abilities to rapidly adapt to new demands. Further, foundries where surface micromachining takes place consume enormous amount of energy, most of it being wasted in operating the machinery and in the control of air temperature, humidity, and purity (which accounts for 90% to 95% of the total energy budget) as required for photolithography. As sustainable growth requirements become more prevalent, these power hungry fabrication techniques will face increasing societal scrutiny.

As will be shown in the next paragraphs, alternative production methods that can bypass some of these issues will be of increasing interest. One of these new approaches is the use of femtosecond lasers to fabricate glass based microsystems and sensors. As we will see, this approach is highly flexible and provides a mean to create highly integrated three-dimensional elements.

## 16.3 Microsystems Fabricated Using Femtosecond Lasers: Review and State of the Art

### 16.3.1 Specificities of Femtosecond Laser–Matter Interaction from the View-Point of Microsystems Design

Femtosecond lasers produce ultra-high peak power pulses, leading to a fundamentally different laser–matter interaction than that associated with conventional long-pulse lasers: non-linear absorption phenomena, such as multiphotons processes, are observed. This opens new and exciting opportunities to tailor the laser–matter interaction, provides excellent spatial resolution, and provides a path to three-dimensional fabrication.

Although the peak power is enormous (in the $GW/mm^2$ or even $TW/mm^2$ ranges), the average power is small. For instance, the devices shown in the illustrations in this chapter were made with no more than 200-mW average power. The femtosecond lasers required for these types of applications are now rather small. The effect of femtosecond laser on dielectrics, although not completely understood, is well documented (see for instance [2,3]) and is addressed in details in the first part of this book. Here, we just summarize some of the main features and in particular the effects on fused silica, a material that is of particular interest for microsystems. Other dielectrics, such as photo-etchable glass [4], have also been considered for microsystems integration. See Chap. 15 for details.

Synthetic amorphous silica (a-$SiO_2$) is a high-quality glass that has outstanding optical properties and is inert to almost all chemicals. Furthermore, although it may sound counter-intuitive, fused silica has excellent elastic properties making it a suitable material for micromechanics as will be further emphasized. $SiO_2$ [5] is one of the most abundant material on earth and is used in a very large number of industrial applications including optics (telecommunications fibres and various optical elements), chemistry (catalyst, catalytic hosts, and absorbents), electronics (insulators and diffusion barrier) and biology (substrates for functionalization) – just to name a few examples. Furthermore, fused silica is biocompatible. In fact, amorphous silica forms the skeletons of most plankton (diatoms).

In fused silica, we observe three different regimes when the material is exposed to femtosecond lasers radiation (Fig. 16.3):

16 Microsystems and Sensors

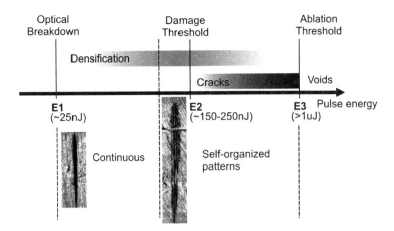

**Fig. 16.3** Effect on fused silica for different pulse energies. One can distinguish three typical regimes. The effective levels of pulse energies depend on the focusing optics (numerical aperture in particular). [The order of magnitude provided for the pulse energies correspond to pulse-length of 100 fs and for an objective with a numerical aperture (NA) of 0.55.]

- Above a first energy threshold $E_1$ (which level depends on various parameters such as focusing optics), the material refractive index and etching susceptibility are locally increased (but no material ablation is observed). Although some aspects of this phenomenon are still debated, this effect is linked to a localized densification of the material [6].
- Above a second energy threshold $E_2$, the formation of self-organized patterns is observed. These self-organized patterns have been reported by various authors [7–10]). Various interpretations for the formation of these patterns have been suggested. Note that cracks may form in these self-organized patterns [27] but are not necessarily present [6].
- Finally, if the energy density is further increased to a given level $E_3$, the formation of voids or the ablation of material is observed.

From the view-point of microsystems fabrication, the first two regimes are potentially the most attractive. In particular, the first regime provides a mean to locally increase the refractive index of the material (which is of particular interest for writing waveguides) [28] as well as a mean to locally increase the etching selectivity [11].

### 16.3.2 Integrated Optics Devices

To fabricate integrated optical devices, the laser is scanned through the glass structure in order to create continuous densified zones. Typically, the laser is stationary and a moving stage is used to displace the specimen.

This method has been used to create a variety of integrated optics devices in fused silica but also in other material stems including lithium niobate. For an in-depth review of these developments, we refer the reader to the Parts II and III of this book.

### 16.3.3 Opto-Fluidics

The observation of an increased etching rate resulting from the laser exposure (first reported by Juodkazis et al. [11]) has triggered numerous developments toward the fabrication of fluidic channels and tunnels [12]. Combined with waveguides, these developments have led to novel concept of small-scale optofluidics devices [13–17] for a variety of applications such as the detection and the screening of algae population [16, 17].

### 16.3.4 Micromechanical Functionality

At about the same time, we suggested to further push down the integration path by inserting micromechanical functionalities into fused silica substrates. To illustrate this concept, a microsensor with an embedded position detection subsystem (waveguide array) was demonstrated. This development opens new avenues where mechanical and optical functions are deeply embedded and created using a single manufacturing process step.

## 16.4 Multifunctional Monolithic System Integration

### 16.4.1 Concept

To reduce the fabrication complexity and to increase the performance and reliability of microsystems, we have focused our research on monolithic integration based on a concept of "system-material." Rather than building up a device by combining and assembling materials, this concept consists in turning a single piece of material into a system through spatially-localized tailoring of its material properties. The material is no longer just an element of a device, but becomes a device on its own. There are many advantages associated with this fabrication approach: it reduces microsystems assembly steps (a common source of significant cost, inaccuracy, and reliability issues) and it opens new design opportunities. This approach was first proposed to fabricate shape memory alloys used in micro-engineering [18]. There, a laser was used to locally introduce active and passive functions in a layer of material that was originally amorphous.

With regards to microsystems, the use of ultrafast lasers to process amorphous fused silica is of particular interest. Femtosecond lasers give a new dimension to the concept of system-material, when applied to dielectrics. Femtosecond laser beam can locally increase the refractive index [28], enhance the etching rate [11], introduce sub-wavelength patterns [7], create voids [20], or change the thermal properties [29, 6] of fused silica. By scanning the laser through the specimen volume, one can distribute, combine, and organize these material modifications to form complex patterns to be used for instance as waveguides or fluidic channels.

With this technique, instead of building up a device by combining layers of materials as common practice, the microdevice structure and function are directly "printed" into a monolithic piece material. Note that due to the non-linear nature of the femtosecond laser–matter interaction, the material modifications can be introduced not only at the material surface but also anywhere in the bulk of the material.

## 16.4.2 Taxonomy of Individual Elements Used in a Monolithic Design

### 16.4.2.1 Waveguides

The laser affected zone (LAZ) shape and size can be determined using either a refractive index map technique or more recently a novel technique based on scanning thermal imaging [29, 6]. Typically, the LAZ has the shape of an ellipsoid stretched along the optical axis. This shape can be correlated to laser beam parameters (beam-size, waist location, and energy). The stretching along the vertical direction depends on the chosen focusing optics. Noteworthy, as femtosecond laser matter interaction involved non-linear processes, the LAZ can be smaller than the laser spot-size itself. These observations are supported by near-field optical profilometry measurements which provide a refractive index map of the region of interest. A refractive map of a single line written in the glass using fs-laser is shown in Fig. 16.4a. To increase the mode-field diameter of the waveguides so that to match a specific wavelength or to get particular waveguiding conditions, one can write multiple laser-written line next to another (Fig. 16.4b) [8].

Figure 16.5 illustrates a rather complex interferometric system written in a fused silica monolith using a femtosecond laser. The interferometer is formed of an input short straight waveguide terminated by a 50–50 splitter. Two arms forming the main section of the interferometers follow this element. The two arms recombined at a coupler that is followed by a short straight output waveguide. All waveguides are single-mode at the design wavelength. Note that, unlike traditional interferometers, the two arms are curved. The minimum curvature radius, approximately 15 mm, is a function of the change in the refractive index that can be introduced with the laser: the higher the change of refractive index, the smaller the radius of curvature.

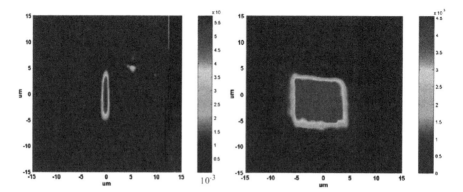

**Fig. 16.4** (**a**, *left*) Refractive index map of a single line written in a fused silica glass. The center of the line that has the highest index of refraction can be correlated with a local increase of glass density which in turn can be related with the local increase of etching rate. (**b**, *right*) Illustration of an enlarged refractive index zone by writing multiple lines adjacent one to another [8]

**Fig. 16.5** An unbalanced waveguided Mach–Zehnder interferometer written in the bulk of a fused silica substrate. (Picture courtesy of Translume.)

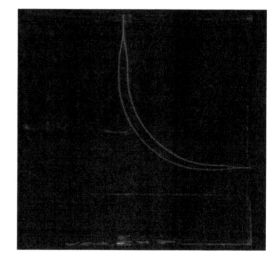

### 16.4.2.2 Channels, Grooves, etc.

To form three-dimensional structures, the following two-step procedure is applied (see Fig. 16.6):

(1) The material is selectively exposed by rasterizing a pattern according to a technique described in [12] and briefly outlined in the next paragraph. The laser used in our experiment is a Ti:sapphire laser (RegA from Coherent) operating at 800 nm. The pulse width is typically 100-fs, and the repetition rate is set at 250 kHz. The average power ranges from 20 to 400 mW, which corresponds to pulse energies ranging from 55 nJ to 1.6 µJ. The linear spot

**Fig. 16.6** Process steps. (**1**) The material is exposed to femtosecond laser irradiation. (**2**) The part is etched with hydrofluoric acid. Exposed regions etch away much faster than unexposed regions

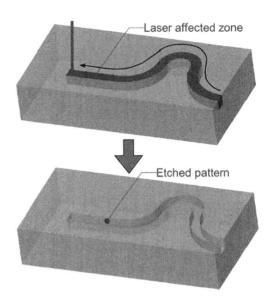

size is approximately 1 μm in diameter at the focus. In our experiments, we used 20× and 50× long-working distance microscope objectives from Mittutoyo. With these objectives, one can work well below the surface of the glass substrate. Typical writing speeds are 0.5–2 mm/s. Affected regions are hit multiple times by the laser (typically 500 to 2,500 times).

(2) After laser-exposure, the part is etched in a low-concentration HF bath. Concentrations between 2.5% and 5% are typically used. Etching time depends on pattern sizes and varies from 1 h to several hours for the deepest structures. Following etching, the part is rinsed in de-ionized water and dried.

Noticeably, the laser polarization has a strong effect on the etching efficiency as first reported by Hnatovsky et al. [21]. The explanation of this effect is still debated. One hypothesis rests on the presence of oriented nanocracks that would promote a rapid penetration of the HF etchant. However, this explanation fails to provide a satisfactory answer as to why index change tracks made at low-energy (where no cracks are observed) also etched at a polarization dependent rate.

Further investigations are needed to fully understand these observations. However, from a manufacturing point-of-view, this effect is easily controllable and has been quantified. In Fig. 16.7, a typical etching time/etching depth curve is shown [22]. These results were achieved by scanning a 100-fs laser beam to form a rectangular pattern in the glass. The pattern was then etched and the depth observed at different time period.

The polarization state not only affects the etching rate but also the edge surface quality of microchannels. This is illustrated further in Fig. 16.8.

As mentioned previously, by spatially arranging laser affected tracks, one can form complex shapes and patterns of nearly arbitrarily dimensions. A collection

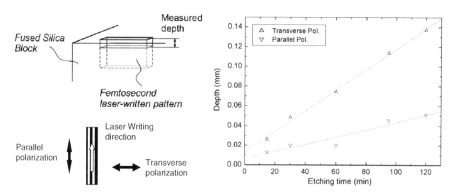

**Fig. 16.7** Etching rate as a function of the polarization (The convention used to define parallel and transverse polarization is shown on the left.)

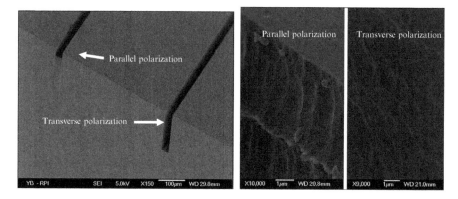

**Fig. 16.8** Effect of polarization: the left photograph shows two trenches etched with the same energy and writing speed parameters but with two different polarizations. Detailed views of the wall morphology are shown on the right

of microchannels with various cross-sections is shown in Fig. 16.9. This ability to describe arbitrary three-dimensional volumes can also be exploited to create features such as micromixing chevrons. This is further illustrated in Fig. 16.10.

### 16.4.2.3 Mechanical Component: Flexures

Flexures are mechanical elements used in micro- and precision-engineering to precisely guide the motion of micro-parts. They consist of slender bodies that deform elastically upon the application of a force. As such they can be considered as frictionless joint between two solid-parts that constraint certain degree-of-freedoms in order to precisely guide the relative motion of two connected parts. The table

16 Microsystems and Sensors 453

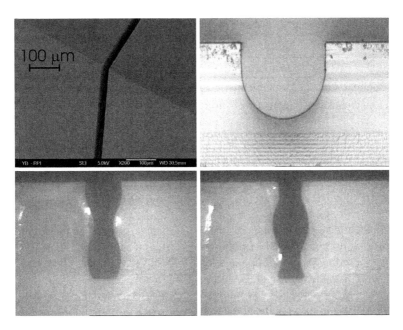

**Fig. 16.9** A collection of microfluidic channels created by femtosecond laser irradiation followed by chemical etching. *Top left*, a deep straight channel. *Top right*, a channel with vertical walls and a rounded floor. Bottom left, a channel with narrowing at mid-height. *Bottom right*, a channel with maximum width at mid-height

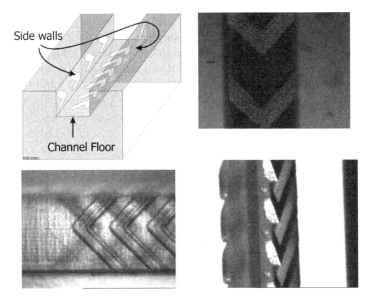

**Fig. 16.10** Microfluidic channel with shaped walls and floor. *Top left*, a schematic representation of the channel with location of various pattern. *Top right*, floor with chevrons. *Bottom left*, sidewalls with chevrons (the back wall is out of focus). *Bottom right*, a side view of chevrons in a sidewall

below summarizes the main differences between a traditional mechanical joint and a flexure.

|  | Mechanical joint | Flexure |
|---|---|---|
| Main characteristics | Assembled | Monolithic |
| Guiding mechanism | Geometrical surfaces | Elasticity of material |
| Pros | Large range of motion possible | No backlash/no play |
| Cons | Backlash/play | Limited range of motion |

One of the early examples of an elastic element to guide a motion is the clock that Christiaan Huygens, a prominent Dutch scientist, imagined in the seventeenth century. This clock revolutionized time keeping, offering a precision never obtained before.

While they were first confined to the niche of precision instruments, there are now much more broadly used with the general trend for miniaturization. They are extensively used in MEMS-design and microrobotics [23]. In fact, the majority of MEMS with movable parts operate with (silicon) flexures.

We have investigated the use of fused silica as a material for miniaturized flexures and more specifically, the mechanical properties of femtosecond laser microfabricated flexures [24, 32]. Using the same approach, we also investigated the strength of silica glass at the micro-scale level. We fabricated test structures, as that shown in Fig. 16.11, in order to evaluate the elastic limits of our fused silica flexures.

The test structures are monolithic structures cut out of a 1 mm-thick fused silica substrate. The structure is made of a slender body itself consisting of a thin part (the notch hinge) and a thicker beam. The test structure also has a frame to protect it during the fabrication process (in particular the etching step). Additional features, such as reference surfaces and a mounting hole for fastening, are also built-in.

These test structures were loaded by pressing a contact pin against the thicker element of the slender beam and translating along the $Y$-axis (as shown in Fig. 16.11). During the test, the contact pin smoothly slides along the beam and

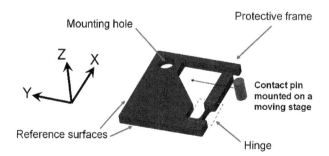

**Fig. 16.11** Schematic of a test structure used to evaluate fused silica flexures. The part is cut out from a fused silica wafer. A contact pin moving along the $Y$-axis is used to load the hinge in bending

16 Microsystems and Sensors

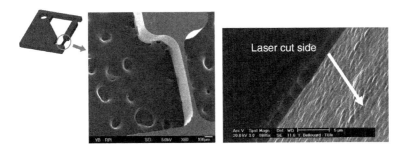

**Fig. 16.12** Micromachined flexure: the RMS roughness is approximately 300 nm at the edges, and 200 nm at mid-height

applies a force that deforms elastically the structure. The pin-beam contact was dry (i.e., no lubricant was used). The pin diameter is sufficiently large compared to the glass roughness to prevent chattering or the application of an unwanted force along the $X$-axis.

During the experiments, a video of the deformed beam was captured and later analyzed using image processing techniques. Stationary features (such as the edge cut at 45° in the middle of the structure) were used as reference points to measure the angular deflection of the beam. We calculated the maximum stress in the beam using known equations for flexures on which a pure moment is applied (the loading case corresponds to a pure bending mode). Figure 16.12 shows a scanning electron microscope image of one of the micromachined hinges (left). As can be appreciated, the surface roughness is in the few hundreds of nanometer (RMS), which is good for a laser-based machining process.

Our tests showed that our micromachined fused silica follows a brittle mechanical behavior. It does not exhibit plasticity (at least not at the scale we are considering). Fused silica is characterized by an asymmetric mechanical load response: while it can be submitted to high compression stress level, it is weak while submitted to tensile stress.

Although glass materials have a high theoretical elastic limits (estimated at 15 GPa or higher), they tend to break at much lower stress levels (typically a few tens of MPa) due to presence of surface flaws that act as crack nucleation sites [25, 31]. When a crack forms, it rapidly propagates through the structure and leads to catastrophic failure. In general, the tensile stress elastic limit dominates the flexure design as it defines the maximum possible excursion for a given hinge thickness, and ultimately governs how small a flexure can be for a given displacement. As an example, 500 MPa is a typical value used as limit when designing flexures made of common steels. Figure 16.13 shows a sequence of photographs of a notch hinge (flexure) being loaded in bending. The maximum bending angle achieved was 62°. The deformation shown in Fig. 16.13 was reversible. This is quite remarkable result for a glass material that demonstrates the unusually high resistance to tensile stress of our micromachined hinges.

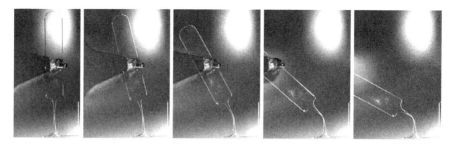

**Fig. 16.13** Flexure bending tests sequence: the flexure is 40 μm wide at the neck. The bending is completely reversible. Note the severe deflection of the microhinge

A preliminary analysis shows that the maximum strength of the flexure is a function of the etching time. We observed that the ultimate tensile strength (UTS) limit linearly increases with the etching time. Similar observations have been made with macroscopic sample in the past [25]. We observed UTS as high as 2.5 GPa for specimens etched for a long period which is quite remarkable for a non-trivial shape [24, 32]. Thus, femtosecond laser combined with etching can be used to fabricate to fused silica devices of unusually high mechanical strength.

### 16.4.3 System Integration: Design Strategies and Interfacing

In the previous section, we described a taxonomy of simple elements that can be made using femtosecond laser processing of fused silica. The combination of these elemental structures opens numerous opportunities in term of system integration. As illustrated in Fig. 16.14, through a combination of these various simple elements, one can create numerous complex devices such as, Lab-on-a-Chip and optomechanical sensors.

Figure 16.15 illustrates the simultaneous integration of a waveguides and a fluidic channel [13] as well as waveguides and channels in a mechanical structure [26]. Figure 16.16 shows a more complex assembly.

In term of design, one has to take into consideration which element, if any, should be etched first. One may also want to isolate some features, such as waveguides, from the etchant. This can be performed by carefully planning the etching pattern and the etchant progression through the microstructure.

### 16.4.4 Illustration: Micro-Displacement Sensors and Micro-Force Sensors

To conclude, we illustrate the concept of monolithic integration with a presentation of a micro-force instrument with an integrated optical micro-displacement sensor.

16 Microsystems and Sensors

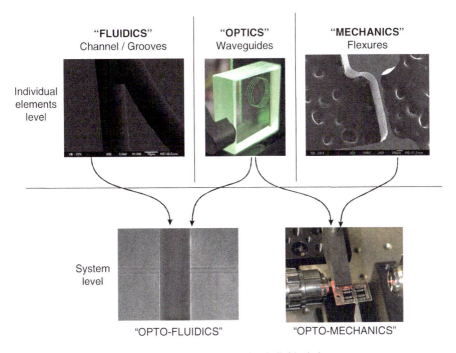

**Fig. 16.14** Monolithic system integration by combining individual elements

**Fig. 16.15** Optical microscope of a microchannel with two waveguides facing each other across the channel. The channel is 30 μm wide and 60 μm deep. The waveguides are 8 μm in diameter and are positioned 50 μm below the glass surface

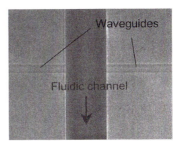

This type of device, when fabricated using conventional practices, is made of several process steps that require careful alignment and present difficult permanent fixturing/bonding challenges.

Our femtosecond laser-written micro-force device is shown in Fig. 16.17. A force applied to the sensor tip induces a linear motion of the mobile platform. The device has two key subsystems: a flexure-based micromechanism that accurately guides the motion of the platform along one axis and a waveguide-based element that senses the corresponding displacement. This displacement-sensing element consists of an array of optical waveguides embedded in the moving platform and two waveguide segments embedded in the stationary frame.

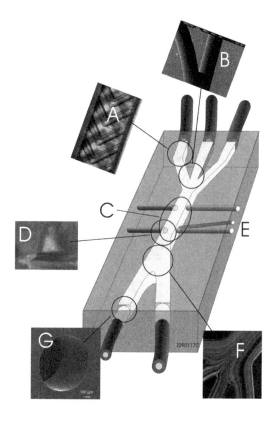

**Fig. 16.16** A schematic representation of a micro-flow cytometer with an overlay of various relevant elements fabricated by femtosecond direct write. This example is a perfect illustration of monolithic integration by combining multiple individual elements

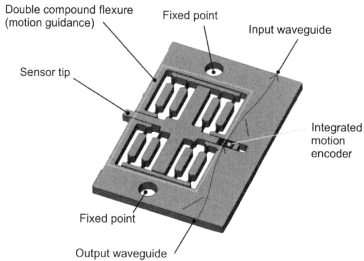

**Fig. 16.17** Drawing of a glass-based force-sensing device. The mobile "sensing" platform has a cross shape and is prolonged by a sensor tip

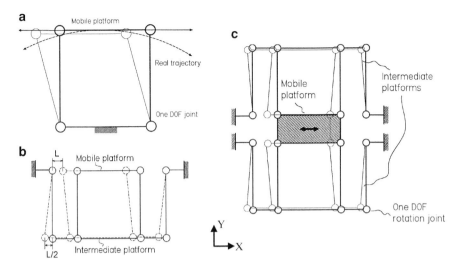

**Fig. 16.18** Sensor kinematics: the *circle* represents ideal mechanical joints with one degree of freedom (rotation in the plane). Figure (**a**) shows a parallelogram four-bars mechanism; (**b**) represents a single compound design, and (**c**) a double-compound design

### 16.4.4.1 Sensor Kinematics

The kinematics model is based on two identical four-bar mechanisms serially connected, as shown in Fig. 16.18c. This kinematics design is well-known in precision engineering [30] and has the property to produce a well-defined linear motion (unlike a parallelogram four-bar mechanism as shown in Fig. 16.18a without adding an internal mobility in the mechanism like a single compound would do Fig. 16.18b. This strategy offers also the additional advantage that it is self-compensated for thermal-expansion.

Figure 16.18 shows idealized rotational joints. In microengineering, due to scale and precision requirements, such a mechanical design is difficult to implement with multiple parts. Rather, a monolithic, flexure-type design is preferred. The principle is to replace traditional (i.e., multipart joints) by elastic hinges that provide the same kinematics. In our case, we use a notch-hinge to emulate the behavior of a rotational joint. The flexure was designed using both analytical and finite element modeling. The design procedure is detailed in [23]. The analytical model predicts that a force of 200 mN is required to reach the full 1-mm excursion.

As mentioned above, it is known that the elastic limit of fused silica depends on the presence or absence of surface flaws. Processes that eliminate these flaws, such as HF etching, can increase the elastic limit by several orders of magnitude. For this work, we used an elastic limit of 300 MPa. Experimentally, we found this value to be conservative.

To refine and optimize the hinge shape, a finite element analysis was conducted as shown in Fig. 16.19. Both static and dynamic analyses were investigated (details

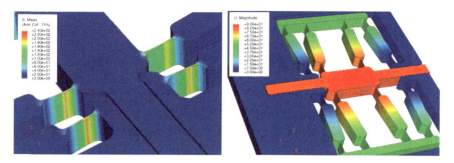

**Fig. 16.19** FEM analysis – stress distribution in four hinges (*left*) and displacement distribution of the entire structure (*right*)

about this analysis can be found in [26]). Good agreement between analytical and FEA model was found. From the finite element analysis, the force to get the full excursion is about 400 mN and the maximum stress is 240 MPa (as opposed to 300 MPa as predicted by the analytical model).

The dynamic analysis indicates that the three first-structural resonance modes are in-plane vibrations). The first natural mode is found at about 405 Hz while the second and third modes are found around 1.2 kHz. Out-of-plane vibrations are activated at much higher frequencies (3.2 to 9.5 kHz).

#### 16.4.4.2 Integrated Linear Encoder

We used the variation of signal intensity induced by lateral misalignment between identical waveguides as the basic principle for the mobile platform displacement. In practice, a waveguide segment is incorporated in the mobile platform so that, at rest it is aligned with two stationary frame waveguides used as transmitting and receiving waveguides for the integrated linear encoder (ILE) signals.

Using a single waveguide segment in the moving platform would limit the sensing range to approximately the width of the mode-field diameter (MFD). To extend the displacement sensing range, the platform contains an array of parallel waveguides. When a waveguide segment of the movable platform is aligned with the input and output stationary waveguides (parts Fig. 16.20a and c), the intensity of the transmitted signal is maximized. Conversely, when the waveguides are misaligned (Fig. 16.20b), the light is no longer guided through the platform (it is only guided in the input segment), which results in a severe loss of transmitted signal. The range of motion sensing can be extended indefinitely with this approach.

In practice, the ILE consists of a fixed 30 μm pitch waveguide array spanning the 1-mm end section of the movable platform. By design, a transmitting and a receiving waveguide in the stationary section of the flexure mount are in direct axial alignment with one of the array waveguides when the stage is unloaded and at rest. The optical signal crosses two identical free space gaps (schematically represented by straight

16 Microsystems and Sensors

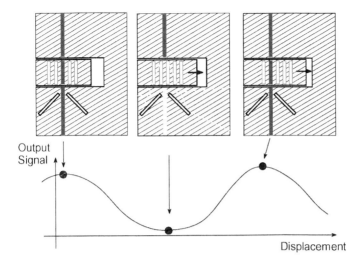

**Fig. 16.20** "Waveguides-based linear encoders principles"

**Fig. 16.21** Magnified view of the actual integrated light encoder. The waveguide array is visible in the lower part of the image and the two input encoders in the upper part

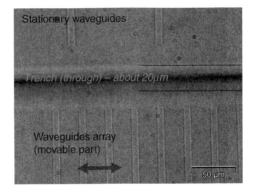

lines in Fig. 16.20) – one located between the output of the transmitting waveguide and the input of the movable array waveguides, and a second located between the output of the array waveguides and the input of the receiving waveguide. These gaps consist of a 30 μm air region sandwiched between two 20 μm glass region where the light is unguided, as shown in Fig. 16.21. The waveguides are 8 μm wide with an index difference (core – cladding) of $\sim 5.25 \times 10^{-3}$. They are highly multimode at the test wavelength of 670 nm.

Light propagation through the structure was simulated using the finite difference method. The fundamental mode at 670 nm is launched into the input waveguide. After the propagation is completed, the output waveguide power is recorded, the array is displaced from the stationary waveguides by 0.6 μm, and the propagation is restarted. This is repeated through a couple of periods of the array. Figure 16.22 shows two of these sequences.

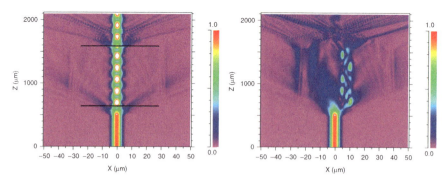

**Fig. 16.22** Wave propagation in the ILE for two configurations as the array is moved from the right to the left. The *horizontal line* indicates the free space gaps

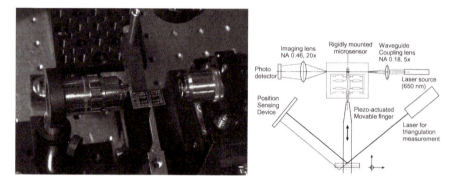

**Fig. 16.23** Experimental setup: (*left*) partial view and (*right*) sketch

Beam expansion of the fundamental mode across the first gap results in a finite overlap with higher order modes at the input of the corresponding array waveguide even when the waveguides are perfectly aligned. From there on, the optical signal propagates in a multimode fashion as can be seen in Fig. 16.22. Despite the multimode nature of the optical signal propagation, the roll-off of a single waveguide transmission with displacement is monotonic and can be scaled with distance. This type of signal response is maintained as long as the array pitch is large enough to prevent coupling between the parallel waveguides.

A sketch and a partial view of the experimental test setup are shown in Fig. 16.23. The device characterization was performed with a free-space optics setup. A mechanical finger is used to apply a displacement on the force sensor tip. The finger is stiff and is attached to a sub-micron accuracy piezo-actuated positioning stage whose stiffness is several orders of magnitude higher than that of the device under test. The finger position is measured using a triangulation measurement system made of a laser beam, a mirror, and a position-sensing device. This measurement system has a 50 nm resolution. As the finger applied a force on

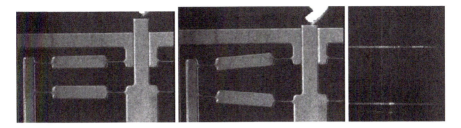

**Fig. 16.24** (*left and center*) Hinges in deformed and non-deformed configurations Sensor prototype: optical microscope view (*right*), close-view of the ILE in a configuration where waveguides are all aligned

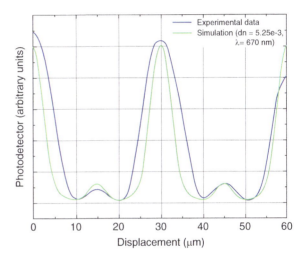

**Fig. 16.25** Experimental results (*blue curve*) compared with simulation results (*green curve*). The figure shows the intensity seen for the last three waveguides (going from *left to right*). The lowest intensity peaks (*on the right*) corresponds to the last waveguide

the sensor tip, the flexure is deformed and the waveguides embedded in the mobile platform moved laterally relative to the stationary waveguide pair located in the frame. A 670 nm laser diode is used as the light source and is injected into the stationary waveguide. At this wavelength the waveguides are multimode. The light intensity transmitted through the sensor is measured by a photodetector. Acquired signals (transmitted intensity and finger position) are further processed. Details of the fabricated microsensor are shown in Fig. 16.24: the two left pictures show an optical microscope view of half of the flexure (in deformed and non-deformed configurations); the right picture is a close view of the ILE.

Waveguides are placed 100 µm below the surface. There is a total of nine waveguides on the movable platform. The experimental results (shown in Figs. 16.24 and 16.25) closely resemble the simulated results. The predicted intermediate peak is clearly visible in the experimental curve. We also notice some peak intensity variation for the largest peaks. This variation is typically within ±15% from waveguide to waveguide. These intensity fluctuations may arise from local changes in wall roughness (typically 300 nm) associated with the two gaps. Inserting an index-matching liquid in the gaps tends to level the peak intensities.

The first resonant frequency was measured on a prototype having hinges with a width of $42 \pm 2\,\mu m$ as measured under an optical microscope. The measurement was done by imposing a sinusoidal mechanical vibration on the piezo-actuator that drives the moving finger main axis. We found the resonant frequency at 209 Hz. For this hinge thickness, the FEM simulation gives a first mode resonant frequency at 206 Hz showing a good agreement with the experimental results. Using the ILE signal, we know the microstage position with a resolution equal or better than 50 nm. (This positioning accuracy is presently limited by our experimental setup and not by the microstage itself)

## 16.5 Summary, Benefits, Future Prospects, and Challenges

Current microsystems technologies face multiple challenges. The increasing functional complexity of these tiny machines poses numerous issues with respect to packaging and reliability. To further increase function integration, new design and manufacturing techniques are needed. A promising approach is based on a concept of system materials where, rather than building a system through an assembling of parts, the material intimate structure is locally modified so that the material can fulfil a specific function not present at first.

In that context, femtosecond lasers can be used to tailor the material properties of fused silica to create not only optical functions (such as waveguides) but also mechanical (flexures) or fluidics (microchannels) elements. These functions can be combined to form complex optofluidics or optomechanical devices. In this chapter, we presented a few illustrative femtosecond laser machined microsystems. These examples demonstrated the viability of the concept of system materials. This novel microfabrication approach provides a means to produce fully integrated devices with advanced functionalities.

There are, however, still numerous challenges to address before this technology is widely accepted. So far, the increase of refractive index remains limited to a fraction of one percent. This imposes numerous design limitations, such as device footprint and ultimately affects the device integration. Machining time is another important limitation, which affects the commercialization of this technology. On-going efforts are targeting these issues.

## References

1. M. Madou, *Fundamentals of Microfabrication: The Science of Miniaturization* (Taylor and Francis Ltd, Boca Raton, 2002) ISBN: 0849308267
2. D. Du, X. Liu, G. Korn, J. Squier, G. Mourou, Appl. Phys. Lett. **64**, 233071 (1994)

3. S.S. Mao, F. Quéré, S. Guizard, X. Mao, R.E. Russo, G. Petite, P. Martin, Appl. Phys. A **79**, 1695–1709 (2004)
4. Y. Cheng, K. Sugioka, K. Midorikawa, M. Masuda, et al., Opt. Lett. **28**, 55–57 (2003)
5. L.W. Hobbs, C.E. Jesurum, V. Pulim, B. Berger, Philosophical Magazine A **78**, 679–711 (1998)
6. Y. Bellouard, E. Barthel, A.A. Said, M. Dugan, P. Bado, Opt. Exp. **16**, 19520–19534 (2008)
7. Y. Shimotsuma, P.G. Kazanksi, Q. Jiarong, K. Hirao, Phys. Rev. Lett. **91**, 247405 (2003)
8. US patent 7,391,947
9. P.G. Kazansky, W. Yang, E. Bricchi, et al., Appl. Phys. Lett. **90**, 151120 (2007)
10. V.R. Bardwaj, E. Simova, P.B. Corkum, et al., J. Appl. Phys. **97**, 83102 (2005)
11. A. Marcinkevičius, S. Juodkazis, M. Watanabe, M. Miwa, S. Matsuo, H. Misawa, J. Nishii, Opt. Lett. **26**, 277–279 (2001)
12. Y. Bellouard, A. Said, M. Dugan, P. Bado, Opt. Exp. **12**, 2120–2129 (2004)
13. A. Said, M. Dugan, P. Bado, Y. Bellouard, A. Scott, J. Mabesa, Photonics West (LASE, 2004)
14. Y. Hanada, K. Sugioka, H. Kawano, I. Ishikawa, A. Miyawaki, K. Midorikawa. "Nano-aquarium for dynamic observation of living cells fabricated by femtosecond laser direct writing of photostructurable glass." Biomedical Microdevices **10**, 403–410 (2008)
15. R.W. Applegate Jr., J. Squier, T. Vestad, et al., Lab. Chip. **6**, 422–426 (2006), DOI: 10.1039/b512576f
16. R. Martinez Vazquez, R. Osellame, et al., Lab. Chip. **9**, 91–96, (2009), DOI: 10.1039/b808360f
17. A. Schaap, Y. Bellouard, T. Rohrlack, Optofluidic lab-on-a-chip for rapid algae population screening. Biomed. Optic. Express **2**, 658–664 (2011)
18. Y. Bellouard, T. Lehnert, J.-E. Bidaux, T. Sidler, R. Clavel, R. Gotthardt, Mater. Sci. Eng. **A273–A275**, 795–798 (1999)
19. Y. Shimotsuma, P.G. Kazanski, Q. Jiarong, K. Hirao, Phys. Rev. Lett. 91, 247405 (2003)
20. E.N. Glezer, M. Milosavljevic, L. Huang, R.J. Finlay, T. Her, J. Paul Callan, E. Mazur, Opt. Lett. **21**, 2023–2025 (1996)
21. C. Hnatovsky, R.S. Taylor, E. Simova, et al., Opt. Lett. **30**, 1867–1869 (2005)
22. Y. Bellouard, A.A. Said, M. Dugan, P. Bado. Proc. SPIE **5989**, 59890V (2005)
23. Y. Bellouard, "Microrobotics: Methods and Applications," Book (CRC/Taylor & Francis Editors, 2009)
24. Y. Bellouard, A.A. Said, M. Dugan, P. Bado, Proc. SPIE **7203**, 72030M (2009)
25. F. Celarie, S. Prades, D. Bonamy, L. Ferrero, E. Bauchoud, C. Guillot, C. Marliere, Phys. Rev. Lett. **90**, 075504 (2003)
26. Y. Bellouard, A. Said, P. Bado, Opt. Exp. **13**, 6635–6644 (2005)
27. R. Taylor, C. Hnatovsky, E. Simova, Applications of femtosecond laser induced self-organized planar nanocracks inside fused silica glass. Laser Photonics Rev. **2**, 26–46 (2008)
28. K.M. Davis, K. Miura, N. Sugimoto, K. Hirao, Opt. Lett. **21**, 1729–1731 (1996)
29. Y. Bellouard, M. Dugan, A. Said, P. Bado, Appl. Phys. Lett. **89**, 161911 (2006); DOI:10.1063/1.2363957
30. R.V. Jones, J. Sci. Instrum. **29**, 345–350 (1952)
31. See 'Treatise on Materials Science and Technology' by Minoru Tomozawa, Robert H. Doremus, Published by Academic Press, 1982, ISBN 0123418224, 9780123418227 as a general reference
32. Y. Bellouard, On the bending strength of fused silica flexures fabricated by ultrafast lasers [Invited]. Opt. Mater. Express **1**, 816–831 (2011)

# Chapter 17
# Ultrashort Laser Welding and Joining

**Wataru Watanabe, Takayuki Tamaki, and Kazuyoshi Itoh**

**Abstract** In this chapter, the most recent results and trends in ultrashort laser welding and joining are reported. The review will cover the possibility of joining transparent slabs made of equal or different glasses, as well as welding between glass and silicon. The role of the laser repetition rate will also be discussed.

## 17.1 Introduction

Development of techniques for welding/joining materials on a micrometer scale is of great importance in a number of applications, including the production of electronic, electromechanical and medical devices, sensors, and microfluidic devices [1, 2]. Laser joining is more suitable for high-flexibility and high-precision manufacturing/integration of small parts compared to widely used joining techniques, such as anodic bonding, fusion bonding, and hydrofluoric acid bonding [3–8]. However, in conventional laser joining, the laser beam penetrates the upper sample and is absorbed at the surface of the lower sample, where it produces localized heating. Alternatively, one of the surfaces can be coated with a light absorbing layer. In this chapter, we review joining techniques for transparent materials by focusing femtosecond laser pulses.

---

W. Watanabe (✉)
Photonics Research Institute, National Institute of Advanced Science and Technology (AIST),
Higashi 1-1-1, Tsukuba, Ibaraki, 305-8565 Japan
e-mail: wataru.watanabe@aist.go.jp

T. Tamaki
Department of Control Engineering, Nara National College of Technology, 22, Yatacho,
Yamatokoriyama, Nara 639-1080, Japan amaki@ctrl.nara-k.ac.jp

K. Itoh
Department of Material and Life Science, Graduate School of Engineering, Osaka University,
2-1, Yamadaoka, Suita, Osaka 565-0871 Japan itoh@mls.eng.osaka-u.ac.jp

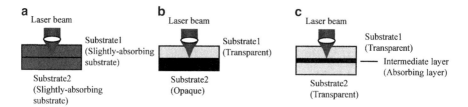

**Fig. 17.1** Schematic diagrams of conventional laser welding/joining

## 17.2 Laser Welding

Laser direct welding using a focused beam can allow localized joining with micrometer precision and has several advantages, such as high flexibility, non-contact welding, and single-step operation [1, 2]. In laser welding of two solids, a delicate balance between heating and cooling is generally maintained within a spatially localized volume overlapping two solids, and a melted material, which is called a liquid pool, is formed and remains stable until solidification. The important feature of laser welding is that it creates the liquid pool and then propagates this liquid pool through the solid interface, eliminating the original seam between the components to be joined. Figure 17.1 shows typical examples of conventional laser welding. Using a $CO_2$ laser (wavelength 10.6 μm), laser welding of two pieces of glass is affected by the liquid pool created by the energy of the laser beam because the glass material has slight one-photon absorption at this wavelength. Due to the absorption, adverse effects such as damage to the material are caused by the melting of the entire section where the $CO_2$ laser beam is irradiated (Fig. 17.1a).

Figure 17.1b shows the schematic diagram of laser welding of a transparent substrate and an opaque substrate. Laser welding with a Nd:YAG laser (wavelength 1064 nm) can be used to weld a glass plate and a silicon plate. One of the items to be joined must be transparent at the wavelength of the laser radiation used, and the other item to be joined must be opaque at that wavelength in order to induce melting and subsequent welding (Fig. 17.1b). The Nd:YAG laser can be used to weld a glass plate and a silicon substrate because the wavelength of Nd:YAG laser has an absorption band only in silicon. Silicon–glass joints have conventionally been performed by the following process: (a) the laser beam is transmitted through transparent glass and (b) is absorbed at the surface of opaque silicon.

Damage such as cracking may be caused by thermal diffusion of the materials outside the focal volume and the difference in coefficients of thermal expansion.

The welding of transparent materials can be performed by using a light-absorbing intermediate layer between the substrates (Fig. 17.1c). In order to weld two transparent substrates such as glass plates, laser welding with a Nd:YAG laser (wavelength 1064 nm) can be used; however, a light-absorbing intermediate layer is needed because transparent materials have a low heat conduction coefficient and absorption coefficient at the wavelengths of the laser.

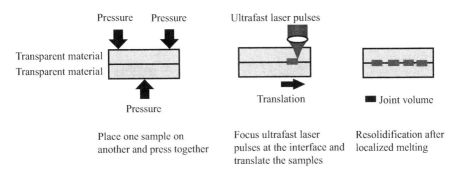

**Fig. 17.2** Schematic diagrams of ultrashort laser microwelding/joining process

## 17.3 Ultrashort Laser Welding of Transparent Materials

In this section, the welding/joining technique by the use of ultrashort laser pulses is described. Two different regimes of the femtosecond laser welding have been identified: the low-repetition rate regime (1–200 kHz) and high-repetition rate regime (> 200 kHz) depending of whether the pulse period is longer or shorter than the time required for heat to diffuse from the focal volume.

### 17.3.1 Ultrashort Laser Welding with Low-Repetition Rate

#### 17.3.1.1 Principle

Tamaki et al. have demonstrated a joining technique for transparent materials using ultrashort laser pulses [9]. In the low-repetition rate regime, material modification is produced by a single pulse. Figure 17.2 shows a process for joining substrates using ultrashort laser pulses from an amplified femtosecond laser [9–11]. Ultrashort laser pulses with low-repetition rate are focused at the interface of two substrates by use of a low-numerical aperture lens. The focal region is elongated along the optical axis due to filamentation [12]. The filamentation is generated by a delicate balance between diffraction caused by plasma formation and self-focusing caused by the Kerr effect. The filamentary propagation of ultrashort laser pulses bridges the two substrates along the laser propagation axis, and the filamentary region is translated two dimensionally (perpendicular to the optical axis).

The intensity in the filament can become high enough to initiate absorption through non-linear field ionization (multiphoton absorption and tunneling ionization) and avalanche ionization. This non-linear absorption results in the creation of an electron-ion plasma that is localized in the focal volume. The melted material, called a liquid pool, is created at the interface, and fills up the original gap between the two materials. By the subsequent resolidification dynamics of the

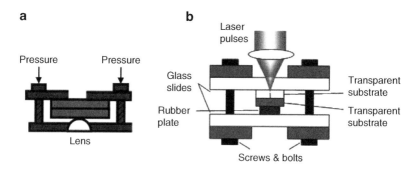

**Fig. 17.3** Schematic of the fixture for holding samples to be joined. (**a**) Fixing of two substrates using sample holder and a pressing lens. (**b**) Fixing of two substrates using rubber plate

liquid pool, the two materials are joined. Therefore, by use of ultrashort laser pulses, the laser microwelding of transparent material can be realized. Because there is minimal heating outside the focal volume, ultrashort laser joining allows space-selective joining without inserting any intermediate layers and can be applied even to transparent materials having insufficient absorption at the wavelength used. Arbitrary joint volumes are produced by two-dimensionally translating the substrate with respect to the focal volume.

In order to perform joining, two substrates are stacked and close contact between them is achieved by pressing the samples together. The gap between the samples in the jointed area was below $\lambda/4$ (where $\lambda$ is the wavelength). When the gap between the samples was above $\lambda/4$, the substrates could not be successfully joined because of ablation occurring at the surfaces of each substrate. For example, by pressing a plano-convex lens on the bottom substrate (Fig. 17.3a) [9], close contact between two substrates is achieved. A substantial amount of bending moment is left inside the samples, sometimes leading to a reduced joint strength or a breakage of the joint. To cope with this problem, the samples are pressed with a rubber plate and a slide glass so that the pressure is applied in a wider area, leaving a reduced bending moment (Fig. 17.3b) [13].

### 17.3.1.2 Welding of Glass–Glass

A silica–glass substrate and a lens are welded together by focusing ultrashort laser pulses from an amplified Ti:sapphire laser system (100 fs, 800 nm, 1 kHz) [9]. The samples were carefully cleaned, stacked one on another, and pressed together by three bolts. Figure 17.4 shows schematic diagram of two stacked substrates.

In order to measure the gap between the glass substrate and the lens, optical interference patterns (Newton's rings) corresponding to the gap were observed by using CCD (charge-coupled device) camera. Subsequently, the gap between the glass substrate and the lens was confirmed to be less than $\lambda/4$ (where $\lambda$ is the wavelength) within a region 400 μm in diameter (Fig. 17.5a). Using an objective

**Fig. 17.4** Schematic diagram of a transparent substrate and a lens

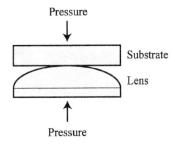

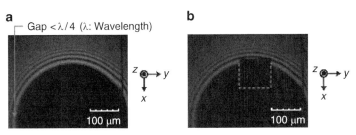

**Fig. 17.5** White-light fringes of the welded samples, silica–glass substrate and lens, before (**a**) and after (**b**) irradiation by ultrashort laser pulses. The incident pulse energy was 1.0 μJ and the translation velocity of the filament was 5 μm/s. The *dashed square* indicates the joint area of 100 × 100 μm$^2$ in the sample. The figure indicates, by irradiation of the laser pulses, a bright area (one-quarter-wavelength gap) that was changed to a black area (zero gap) because one-quarter-wavelength gap was almost filled up

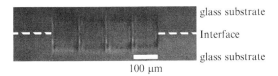

**Fig. 17.6** Microscopic optical image of side view after ultrashort laser joining with pulse energy of 1.0 μJ and a translation velocity of 0.1 mm/s in borosilicate glass

lens with a numerical aperture (NA) of 0.3, the filamentary propagation of ultrashort laser pulses bridges the two samples along the laser propagation axis, and the focal region is translated two dimensionally (perpendicular to the optical axis). After ultrashort laser irradiation, a bright area (one-quarter-wavelength gap) was changed to a black area (zero gap) because the one-quarter-wavelength gap was almost filled up (Fig. 17.5b).

The joining of glass substrates by focusing 1-kHz ultrashort laser pulses at the interface of two samples has been demonstrated [9, 10]. Two silica–glass substrates or two borosilicate–glass substrates were joined using kHz amplified femtosecond lasers. Figure 17.6 shows photomicrograph of a 4 × 4 array of joint volumes (100 × 100 × 30 μm$^3$) after joining the glasses by irradiating ultrashort laser pulses with a pulse energy of 1.0 μJ and a translation velocity of 0.1 mm/s. The joint strength was approximately 15 MPa.

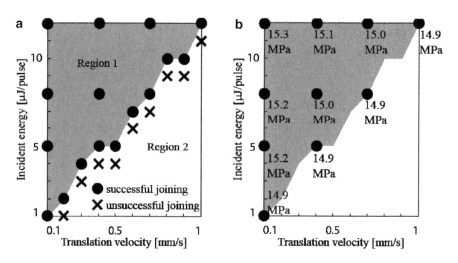

**Fig. 17.7** (a) Dependence of laser-pulse energy and translation velocity on joining of borosilicate and fused silica substrates, whose coefficients of thermal expansion are different. (b) Joint strength after joining of fused silica substrates for various energies and translation velocities

### 17.3.1.3 Welding of Dissimilar Glasses

Watanabe et al. demonstrated the welding of dissimilar glasses. The joining of borosilicate and fused silica glass substrates, whose coefficients of thermal expansion are different, was demonstrated [11]. A borosilicate–glass sample was placed on top of a fused silica sample, as shown in Fig. 17.7a, and the samples were pressed together with an applied force of approximately 40 MPa to achieve intimate contact between them. A femtosecond laser system producing 85-fs, 800-nm, and 1-kHz pulses was used to join the glasses. The laser beam was focused by a 10× microscope objective with a numerical aperture (NA) of 0.3. The sample was two-dimensionally translated with respect to the focal region. A 30 – μm-long filament bridged both glass samples at energy of 1.0 μJ. Refractive-index change was induced in the joint volume.

Figure 17.7a shows thresholds for these parameters as a function of the translation velocity and the pulse energy. In region 1, joining was achieved, whereas in region 2, joining was not possible. The energies required for joining were in proportion to translation velocities. The slope of the line in Fig. 17.7a, which shows the boundary between region 1 and region 2, indicates the threshold for achieving joining.

Figure 17.7b show the dependence on the joint strength by varying the laser-pulse energy and the translation velocity, respectively. At a pulse energy of 12.0 μJ and a translation velocity of 0.1 mm/s, a maximum joint strength of 15.3 MPa was achieved. At the threshold fluence, the joint strength was 14.9 MPa. With increasing energy, the joint strength increased, whereas the joint strength decreased with increasing translation velocity. The joint strength could be thus correlated to the refractive index change induced in the joint volume.

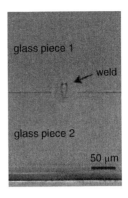

**Fig. 17.8** Ultrashort laser welding of glass. Optical transmission image of a weld region between two pieces of borosilicate glass, using 1-MHz ultrashort lasers. Courtesy of A. Arai and C.B. Schaffer

Ultrafast laser joining can be expanded to join pairs of dissimilar glasses and dissimilar materials, such as fused silica and polymer material, and borosilicate glass and polymer material [11].

## 17.3.2 Ultrashort Laser Welding with High-Repetition Rate

The ultrashort laser microwelding/joining technique can be achieved based on a localized heat accumulation effect in high-repetition rate regime [14]. In this regime, the focused laser pulses collectively act as a point source of heat at the focal volume within the bulk material when the time interval between successive pulses is much shorter than the time scale for diffusion of heat out of the focal volume. The accumulated energy around the focal volume makes it possible to achieve very high temperatures and to melt the material around the focal volume. The liquid pool is created at the interface and fills up the original gap between the two materials. Through the subsequent resolidification dynamics, the liquid pool is resolidified, thus joining the two materials is completed.

Tamaki et al. performed the ultrashort laser microwelding/joining of non-alkali aluminosilicate glass substrates based on a localized heat accumulation effect using an amplified femtosecond Er-fiber laser (1 ps, 1558 nm, 500 kHz) [15].

Bovatsek et al. focused a 1-MHz train of 360-fs duration, 1045-nm wavelength pulses from a laser using a 0.55 NA aspheric lens at the interface between two 170 – μm thick pieces of borosilicate glass [16]. A 1-MHz repetition rate leads to accumulation of thermal energy in the focal region. Optical transmission images reveal a complex pattern of refractive index change within the weld region (Fig. 17.8). High-repetition rate ultrashort lasers allow welding speeds up to 12.5 cm/s.

Horn et al. demonstrated on glass microwelding using focused ultrashort and picosecond laser pulses [17–22]. Two glass plates were joined together by welding with ultrashort laser radiation (Fig. 17.9). The welding zone was investigated dynamically by characterizing the optical phase quantitatively using transient quantitative phase microscopy (QPM) [20–22]. Welding of two plates consisting of commercial borosilicate glass was achieved by focused high-repetition rate

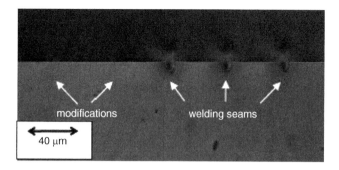

**Fig. 17.9** Cross-section welding seams at the interface when welding of borosilicate glass is achieved by focused high-repetition rate ultrashort laser radiation (350 fs, 1045 nm, 1 MHz). Reprinted with permission from [20]

ultrashort laser radiation (350 fs, 1045 nm, 1 MHz). It was possible to produce crack-free weld seams 20 μm wide and 30 μm high in glass–glass bonding. The refractive index distribution of a weld seam was measured ex situ qualitatively by Nomarsky microscopy and quantitatively by QPM. The time dynamics of optical phase distribution of a weld seam was measured along the direction of the laser radiation during the welding process. A strong decrease in the optical phase during irradiation within the interaction zone was observed and was attributed to free electrons. A hot and molten region close to the interaction zone was detected directly after irradiation by an increased optical phase. The phase decreases due to cooling of the melt within 2 μs.

#### 17.3.2.1 Welding of Glass and Silicon Substrates

To verify that this technique can be extended to semiconductor materials, Tamaki et al. also welded a non-alkali–glass substrate and a silicon substrate. The wavelength of the femtosecond laser system was 1558 nm where silicon and borosilicate glass are transparent.

Glass–silicon welding was achieved by focusing high-repetition rate ultrafast laser radiation (350 fs, 1045 nm, 700 kHz) [21]. At this laser wavelength, silicon is not transparent and glass–silicon welding belongs to the scheme described in Fig. 17.1b. These results are important steps toward the welding of semiconductor materials.

## 17.4 Outlook and Conclusions

Ultrafast laser microwelding/joining technique based on the non-linear absorption of focused ultrashort laser pulses was described. At low-repetition rate regime (e.g., 1 kHz), the energy deposited by each laser pulse diffuses out of the focal volume before the next pulse arrives. At high-repetition rate regime (e.g., 10 MHz),

**Table 17.1** Repetition rate and thermal accumulation in welding of glass

|  | Low-repetition rate (Rep. rate: ~1 kHz) | High-repetition rate (Rep. rate: 200 kHz-1 MHz) |
|---|---|---|
| Mechanisms | Filamentary modifications | Heat accumulation and resolidification |
| Suitable materials | High fictive temperature ex: Fused silica | Low fictive temperature ex: Borosilicate glass |

**Table 17.2** The feasibility of microwelding using ultrafast lasers

| Laser | | | Joined materials | Reference |
|---|---|---|---|---|
| Wavelength | Repetition rate | Pulse duration | | |
| 800 nm | 1 kHz | 85 fs | Similar glass (fused silica) | [9] |
| 800 nm | 1 kHz | 85 fs | Similar glass (fused silica) (borosilicate glass) | [10] |
| 800 nm | 1 kHz | 85 fs | Polymer | [23] |
| 800 nm | 1 kHz | 85 fs | Dissimilar glass (fused silica and borosilicate glass) | [11] |
| 1558 nm | 500 kHz | 950 fs | Similar glass (borosilicate glass), dissimilar materials (borosilicate glass and silicon) | [15] |
| 1045 nm | 1 MHz | 360 fs | Similar glass (borosilicate glass) | [16] |
| 1045 nm | 1 MHz | 350 fs | Similar glass (borosilicate glass) | [20–22] |
| 1045 nm | 1 MHz | 400 fs | Similar glass (borosilicate glass) | [19] |
| | 100 kHz | 325 fs | | |
| 1064 nm | 500 kHz | 10 ps | Similar glass (borosilicate glass) | [18] |
| 1045 nm | 1 MHz | 350 fs | Dissimilar materials (borosilicate glass and silicon) | [21] |
| 800 nm | 1 kHz | 85 fs | Glass and metal (fused silica) | [13] |

energy accumulates in the focal volume making it possible to achieve very high temperatures around the focal volume. Table 17.1 summarizes the repetition rate and thermal accumulation in welding of glass. Table 17.2 shows the reported ultrafast laser welding/joining of various materials. Especially, in [18], Miyamoto et al.

have reported local melting process of glass by picosecond laser pulses and its application to microfusion welding. The understanding of melting process and/or mechanism [24] is very important to optimize ultrashort pulse laser processing and microwelding/joining, so more research is necessary.

The ultrashort laser welding/joining technique can directly join transparent substrates without an intermediate layer due to the non-linear absorption around the focal volume of the laser pulses. This technique can be applied to join dissimilar materials, whose coefficients of thermal expansion are different. This ultrashort laser joining technique is a versatile tool for joining dissimilar materials and opens up possibilities in the production of electronic, electromechanical, and medical devices. In particular, laser welding dissimilar materials can find applications in the assembly of sensors, microsystem components, microfluidic devices, and sealing of microelectromechanical systems (MEMS) and organic light-emitting diodes.

The laser microwelding technique between transparent materials and metal will be one of the indispensable manufacturing techniques. The ultrafast laser welding technique between glass and metal [13] is a candidate process for microsystem products, which posed new challenges to the assembling and packaging process, the downsizing, and hermetic sealing.

**Acknowledgements** This work was supported in part by the Industrial Technology Research Grant Program in 2008 from New Energy and Industrial Technology Development Organization (NEDO) of Japan.

# References

1. M. Wild, A. Gillner, R. Poprawe, "Locally selective bonding of silicon and glass with laser," Sensor. Actuat. A-Phys. **93**, 63–69 (2001)
2. W.Y. Tan, F.E.H. Tay, "Localized laser assisted eutectic bonding of quartz and silicon by Nd:YAG pulsed-laser," Sensor. Actuat. A-Phys. **120**, 550–561 (2005)
3. P.H. Carr, "Reflection of gigacycle-per-second ultrasonic waves from an optical-contact bond," J. Acoust. Soc. Am. **37**, 927–928 (1965)
4. H.I. Smith, "Optical-contact bonding," J. Acoust. Soc. Am. **37**, 928–929 (1965)
5. T. Rogers, J. Kowal, "Selection of glass, anodic bonding conditions and material compatibility for silicon-glass capacitive sensors," Sensor. Actuat. A-Phys. **46–47,** 113–120 (1995)
6. H. Nakanishi, T. Nishimoto, R. Nakamura, A. Yotsumoto, T. Yoshida, S. Shoji, "Studies on SiO2-SiO2 bonding with hydrofluoric acid. Room temperature and low stress bonding technique for MEMS," Sensor. Actuat. A-Phys. **79**, 237–244 (2000)
7. T.M.H. Lee, D.H.Y. Lee, C.Y.N. Liaw, A.I.K. Lao, I.-M. Hsin, "Detailed characterization of anodic bonding process between glass and thin-film coated silicon substrates," Sensor. Actuat. A-Phys. **86,** 103–107 (2000)
8. P.W. Barth, "Silicon fusion bonding for fabrication of sensors, actuators and microstructures," Sens. Actuators A-Phys. **21–23,** 919–926 (1990)
9. T. Tamaki, W. Watanabe, J. Nishii, K. Itoh, "Welding of transparent materials using femtosecond laser pulses," Jpn. J. Appl. Phys. **44**, L687–L689 (2005)
10. W. Watanabe, S. Onda, T. Tamaki, K. Itoh, "Direct joining of silica glass substrates by 1 kHz femtosecond laser pulses," Appl. Phys. B **87**, 85–89 (2007)

11. W. Watanabe, S. Onda, T. Tamaki, K. Itoh, J. Nishii, "Space-selective laser joining of dissimilar transparent materials using femtosecond laser pulses," Appl. Phys. Lett.**89,** 021106 (2006)
12. K. Yamada, T. Toma, W. Watanabe, J. Nishii, K. Itoh, "In situ observation of photoinduced refractive-index changes in filaments formed in glasses by femtosecond laser pulses," Opt. Lett.**26,** 19–21 (2001)
13. Y. Ozeki, T. Inoue, T. Tamaki, H. Yamaguchi, S. Onda, W. Watanabe, T. Sano, S. Nishiuchi, A. Hirose, K. Itoh, "Direct welding between copper and glass substrates with femtosecond laser pulses," Appl. Phys. Express **1,** 082601 (2008)
14. C.B. Schaffer, J.F. Garcia, E. Mazur: "Bulk heating of transparent materials using a high-repetition-rate femtosecond laser," Appl. Phys. A **76,** 351–354 (2003)
15. T. Tamaki, W. Watanabe, K. Itoh, "Laser micro-welding of transparent materials by a localized heat accumulation effect using a femtosecond fiber laser at 1558 nm," Opt. Exp. **14,** 10460–10468 (2006)
16. J. Bovatsek, A. Arai, C.B. Schaffer, "Three-dimensional micromaching inside transparent materials using femtosecond laser pulses: new applications," presented at CLEO/QELS and PhAST 2006, California, USA, 21–26 May, 2006
17. A. Horn, I. Mingareev, I. Miyamoto, "Ultra-fast diagnostics of laser-induced melting of matter," JLMN-J. Laser Micro/Nanoeng. **1,** 264–268 (2006)
18. I. Miyamoto, A. Horn, J. Gottmann, "Local melting of glass material and its application to direct fusion welding by ps-laser pulses," JLMN-J. Laser Micro/Nanoeng. **2,** 7–14 (2007)
19. I. Miyamoto, A. Horn, J. Gottmann, "High-precision, high-throughput fusion welding of glass using femto-second laser pulses," JLMN-J. Laser Micro/Nanoeng. **2,**57–63 (2007)
20. A. Horn, I. Mingareev, A. Werth, "Investigations on melting and welding of glass by ultra-short laser radiation," JLMN-J. Laser Micro/Nanoeng. **3,**114–118 (2008)
21. A. Horn, I. Mingareev, A. Werth, M. Kachel, U. Brenk, "Investigations on ultrafast welding of glass–glass and glass-silicon," Appl. Phys. A **93,**171–75 (2008)
22. A. Horn, I. Mingareev, J. Gottmann, A. Werth, U. Brenk, "Dynamical detection of optical phase changes during micro-welding of glass with ultra-short laser radiation," Meas. Sci. Technol. **19,** 015302 (2008)
23. T. Tamaki, T. Inoue, W. Watanabe, Y. Ozeki, K. Itoh, "Laser micro-welding of dissimilar materials using femtosecond laser pulses," The 8th International Symposium on Laser Precision Microfabrication, Vienna, Austria
24. D. Lee, E. Kannatey-Asibu Jr., "Numerical analysis on the feasibility of laser microwelding of metals by femtosecond laser pulses using ABAQUS," J. Manuf. Sci. Eng. **130**, 061014 (2008)

# Index

Aberrations, 78–84
Absorption, 271
Absorption cross-section, 271
Active optics, 114
Adaptive loops, 75
All-silica core fibre, 208, 221
Amorphization, 142
Amplified spontaneous emission, 273
Anisotropic bubble formation, 137
Anisotropy of laser machining, 129
Annealing, 209–211, 216, 217
Astigmatic beam shaping, 108, 395
Avalanche, 5
Avalanche ionization (AI), 20, 26
Avalanche photoionization, 6

Band structure, 353, 354
β Barium Borate, 301
Beam engineering, 67–88
    spatial beam shaping, 73
    temporal pulse shaping, 73
Beam shaping techniques, 35
Bend loss, 157, 173, 181
Bending angle, 455
Bending phase, 184, 185
BGW sensor, 253
Bi-doped silica glass, 281
Biochemical analysis, 389
Birefringence, 212–214
Bismuth, 269
Bloch oscillations, 368, 370, 371
Bloch-Zener dynamics, 369–371
Bonding or joining processes, 445
Borosilicate, 158, 164, 165, 175, 180
Bragg condition, 200, 204, 231, 248
Bragg gratings, 54, 411

Bragg grating waveguides (BGWs), 229
Bubble formation, 139
Bulk, 32
Burst writing, 238

Capillary electrophoresis, 402
Carbon nanotubes, 288
Cavity-dumped Yb:glass laser, 281
C-band, 283
Cell sorting, 400
Cell trapping, 413
Chalcogenide optical fibre, 222
Change in refractive index, 60
Chirped BGWs, 250
Chirped fibre grating, 205, 209
Chromatic aberration, 11
Cladding mode, 199–201
Cleavage of the polymer backbone, 334, 336
Collateral thermal damage, 132, 134
Commercial and clinical grade PMMA, 325
Compensate, 396
Compression stress, 455
Computer generated holograms (CGHs), 340
Confocal microscope, 49, 51
Confocal parameter, 394
Conical shapes, 396
Core mode, 200
Coupled mode approach, 354, 355, 371
Coupled mode equations, 355, 379
Coupling coefficient, 177, 181, 183–185
Coupling constant, 355, 356, 361
Coupling loss, 97, 160, 161, 164, 190, 191, 272
Critical electron density, 25, 27
Critical power, 12
Cumulative thermal effect, 131
Curved waveguides/lattices, 352, 354, 365, 378

R. Osellame et al. (eds.), *Femtosecond Laser Micromachining*,
Topics in Applied Physics 123, DOI 10.1007/978-3-642-23366-1,
© Springer-Verlag Berlin Heidelberg 2012

DBR fibre laser, 205, 216
Deformable mirror, 115
Depolymerization, 336
Depth of the plasma layer, 26
DFB laser, 286
Diffraction gratings, 319, 320
Diffraction-managed solitons, 378, 380
Direct maskless fabrication, 391
Direct write technology, 228
Direct writing, 320, 339, 421
Direction-dependent, 147
Directional coupler, 175, 176, 179–184, 186, 320
Directional dependence, 130, 131, 134, 135, 137, 142, 145, 148
Directional waveguide coupler, 355–357
Discrete
  light propagation, 353
  reflection and refraction, 361
  spatial soliton, 352, 374, 376, 381–383
  systems, 351
Dispersion, 12, 28, 252
Doped crystals, 302
Doped silica core fibre, 208, 216
Drude model, 25
Drug development, 389
Dynamic localization, 366, 368, 370, 371, 378

Effective refractive indices, 242
Effect of bandgap and wavelength, 331
Effect of polarization, 452
Elastic limits, 455
Electrooptic modulator, 307
Electrooptical coefficient, 305
Emission cross-section, 271
Enhancement, 275
Environmental monitoring, 389
Er:Yb-doped phosphate glass, 281
Erbium, 268
Etching process, 394
Evanescent coupling, 99, 355
Evanescently coupled type II waveguides, 310
External gain, 273
Extrinsic color centers, 44

Femtosecond (fs) laser, 421–423, 425–430, 439
Fibre Bragg grating, 53, 54, 200, 205
Fibre laser, 216–218
Fibre sensor, 218–220
Fictive temperature, 202, 209–211
Filament, 163, 164, 175, 187, 188

Filamentation, 322, 336, 338, 342, 469
Finite difference method, 461
FLICE process, 393
FLICE technique, 391, 396
Flow cytometry, 409
Fluorescence spectroscopy, 44
Form birefringence, 128, 130, 134, 135, 139
Foturan, 422–424, 426, 428
Free electron plasma, 7
Fringe shift, 408
Fs-laser direct writing, 424, 425, 431, 433, 435
Fused silica, ($SiO_2$), 390, 446

Grating, 156, 193

Heat accumulation, 164–168, 172, 473
HF etching rate, 392
High-repetition rate, 473
High-throughput, 389
Hollow microstructures, 421, 424, 425, 433
Holographic writing, 323
Hydrofluoric acid (HF), 391, 396

Images, 52, 59
Index matching fluid, 274
Influence of the pulse duration, 298
Insertion loss (IL), 156, 158, 159, 161, 166–168, 173, 174, 179, 181, 190, 272
Integrated linear encoder, 460
Internal gain, 271
Internal modification, 421, 431
Intrinsic color centers, 44
Inverse Helmholtz technique, 104

Keldysh parameter, 5
Kerr lens self-focusing, 13
Kinematics model, 459
KOH solution, 398
Kramers–Kronig mechanism, 8

Lab-on-chips (LOCs), 342, 389, 421
Landau-Zener dynamics, 371, 372
Laser affected zone (LAZ), 394, 449
Laser processing, 80
  bulk modifications, 76–80
  refractive index engineering, 76–80
Liquid pool, 468
Lithium fluoride, 302

Index 481

Lithium niobate, 127, 139, 140, 145, 149, 296
Lithium tantalate, 302
Long period grating, 199–200, 204
Longitudinal and transverse writing geometries, 106
Longitudinal geometry, 394
Longitudinal writing geometry, 107
Low-repetition rate, 469

Mach–Zehnder interferometer (MZI), 184, 186, 306, 405
MEMS devices, 445
Micro-assembly, 445
Microchannel, 394, 396, 429, 433, 437, 438
Micro-displacement sensors, 456
Micro-explosion, 138
Microfluidic device, 425, 429, 430, 432, 433
Microfluidic dye laser, 421, 422, 433, 434, 439
Microfluidic structures, 421, 422, 428, 429, 436, 439
Microfluidics, 422, 435, 439
Micro-force sensors, 456
Microlens, 430–432, 435, 436, 439
Micromirror, 430–433, 439
Micro-optics, 430, 432, 435, 439
Micro-reflectivity technique, 101
Microrobotics, 454
Microscopy, 75
Microsystems, 444
Micro-total analysis systems, 421
Microwelding/joining, 474
Mie scattering, 216
Miniaturized flexures, 454
Mode-field diameter (MFD), 95, 160, 167–169, 172, 180, 190, 460
Mode-locking, 288
Modulation duty cycle, 240
Monolithic system integration, 448
Multi-photon absorption (MPA), 4, 20, 26
Multiphoton ionization, 392
Multiphoton process, 201, 207, 213, 214
Multiple beam filamentation (MBF), 37
Multiple BGW, 248
Multiple pulse interaction, 10
Multiscan technique, 120
Multiscan writing approach, 305
Multi-waveguide excitation, 357

Nano-aquarium, 421, 422, 436, 439
Nanograting, 9, 132, 134, 137, 139, 161, 163
Nanoplasmonic model, 392, 393
NBOHC defect, 45, 46, 54

Near-field optical profilometry, 449
Negative, 60
Neodymium, 267
Net gain, 273
Newton fringes, 24
Noise figure, 273
Nonbridging oxygen hole centers (NBOHCs), 45, 55, 356
Non-linear absorption, 446
Nonlinear refractive index, 373
Nonlinear response, 304
Nonreciprocal writing, 139

One-dimensional, 354
One-dimensional waveguide arrays, 352
Optical bloch oscillation, 352
Optical breakdown, 6
Optical functions, 67–86
Optical microcomponent, 421, 422, 432
Optical waveguide, 93, 94, 390, 398, 431, 432, 435
Optimal pulse, 79
Optofluidics, 421, 422, 433, 435
Oriented nanocracks, 451
OSA, 275
Overlap factor, 236
Oxyfluoride silicate glass, 279

Parallel processing, 84–86
Passive optical network (PON), 155, 183, 184
Phase mask, 200, 201, 206–209, 222
Phosphate, 163, 167
Phosphate glass, 277
Photoattenuation, 214–216
Photobleaching, 55
Photochemistry, 332
Photodegradation, 319
Photonic biosensing, 435
Photonic crystal fibre, 209, 220
Photonic sensors, 400
Photosensitive glass, 421–432, 437, 439
Photosensitivity, 209
Photothermal effect, 146, 148
Piezo-actuated positioning stage, 462
Planar lightwave circuit (PLC), 156, 157, 162, 163, 173, 185, 186, 191, 192
Plasma defocusing, 13
Plasma emission images, 37
POHC, 46
Point-by-point writing, 201–206, 222
Polarised fibre laser, 216–218
Polarization, 452

Polarization vector, 12
Polymer backbone scission, 334
Polymer optical fibre (POF), 222, 317, 343
Polymer, poly(methyl methacrylate) (PMMA), 315, 316, 318, 319, 332, 337
Ponderomotive force, 133
Population densities, 271
Positive, 60
Programable pulse shaping, 73
Propagation, 70
    equation, 270
    loss, 98, 157, 159–161, 163, 164, 169, 172, 174, 180, 192, 234, 252, 272, 298
    loss coefficient, 271
Pulse duration dependence, 326
Pulse front tilt, 134–137, 149
Pulse tailoring, 73–74
Pump–probe techniques, 22, 23, 28

Quantitative phase microscopy (QPM), 102, 473
Quantum-mechanical analogies, 370
α-quartz, 295
Quasi-incoherent propagation, 362
Quill writing, 128, 135, 136, 145

Raman spectroscopy, 47
Rare-earth, 266
Rare-earth doped phosphate glasses, 60
Refracted near field (RNF), 100, 161, 162, 166, 175
Refractive index change, 56, 69–73, 319, 320, 336, 339, 341
    transformation mechanisms, 69–73
Refractive index modification, 432
Refractive index modulation, 236
Refractive index profile, 93–95, 97, 100–102, 104, 105, 120
Refractive index sensing, 406
Resolidification, 469
Reverse cone, 396
Ring cavity, 288
Ring structures, 47

Sampled grating, 205, 206
Sapphire fibre, 198, 212, 219
Scanning electron microscope image, 455
Second order coupling, 359
Segmented arrays, 363
Self-focusing (SF), 13, 36–38, 163, 322, 336, 337, 339, 342

Self-ordered nanograting structures, 392
Self-trapped exciton defects, 52
Sensing, 411
Shadowgraphy, 33
Shift, 408
Silicon, 300
Single-longitudinal-mode, 283
Single-mode fiber (SMF), 158–160, 162, 172–174, 180, 181, 189, 190, 192
Single-pulse Writing, 231
Single-step, 229
Slit, 162, 163, 168, 172, 177, 190
Slit beam shaping technique, 111
Soliton, 380
Spatial light modulator (SLM), 116
Spatially resolved sensing, 406
Spatial soliton, 384
Spatiotemporal focussing, 118
Spherical aberration, 11, 172, 339
Straight waveguides/lattices, 358, 379, 382
Strain-optic effect, 254
Stress fields, 296
Stress-induced birefringence, 142
Stretching, 413
Surface, 27
    micromachining, 444
    roughness, 455
    states, 368

μ-TAS, 433
Tellurite glass, 281
Tensile stress, 455
Thermal accumulation, 146
Thermal deterioration, 299
Thermal diffusion, 162, 164–167, 172, 193
Thermal expansion, 254
Thermal lensing, 276
Thermal stability, 207, 209–212, 218, 244
Three-dimensional (3D) hollow microstructures, 421, 424, 428, 430, 439
Three-dimensional (3D) integration, 432
Three-dimensional (3D) microfluidic structures, 421, 422, 436, 439
Three-dimensional (3D) microstructures, 422, 423
Three-dimensional (3D) technique, 391
Tilt of the pulse front, 145
Tilted front of the ultrashort laser pulse, 129
Transient plasma dynamics, 29, 35
Transverse geometry, 394
Transverse writing geometry, 107
Tunneling, 5

Two-dimensional, 352
Two-dimensional waveguide array, 353, 363
Type I grating, 202, 204, 207–214, 216–219
Type I modifications, 299
Type II grating, 202, 205, 209, 210, 212, 213, 215, 216, 218, 219
Type II modifications, 299

Ultimate tensile strength (UTS), 456
Ultrafast Imaging, 27, 32
Ultra-high peak power, 446
Ultrashort laser pulses, 67–88

Void formation, 10
Voxels, 229

Waveguides, 319, 320, 322
    arrays, 351, 353, 354, 358, 378, 382
    asymmetry, 97, 100, 120
    cross-section, 94–97, 100, 105, 108, 110–112, 114, 118, 120
    imaging microscopy, 356
    lasers, 308
Wavelength division multiplexing (WDM), 157, 182, 184, 185, 275
Wavelength tuning, 247
Welding/joining, 467, 469
White light, 52
    emission, 137, 139
    image, 54
    microscopy images, 58
    transmission images, 54, 61
Working distance, 13
Writing, 229

Y-junction, 173
Ytterbium, 268, 269

Zeno effect, 372, 373